NanoScience and Technology

The series NanoScience and Technology is focused on the fascinating nano-world, mesoscopic physics, analysis with atomic resolution, nano and quantum-effect devices, nanomechanics and atomic-scale processes. All the basic aspects and technology-oriented developments in this emerging discipline are covered by comprehensive and timely books. The series constitutes a survey of the relevant special topics, which are presented by leading experts in the field. These books will appeal to researchers, engineers, and advanced students.

More information about this series at http://www.springer.com/series/3705

Vinod Kumar Khanna

Integrated Nanoelectronics

Nanoscale CMOS, Post-CMOS and Allied Nanotechnologies

Springer

Vinod Kumar Khanna
MEMS and Microsensors Group
CSIR-Central Electronics Engineering Research Institute
Pilani, Rajasthan
India

ISSN 1434-4904 ISSN 2197-7127 (electronic)
NanoScience and Technology
ISBN 978-81-322-3623-8 ISBN 978-81-322-3625-2 (eBook)
DOI 10.1007/978-81-322-3625-2

Library of Congress Control Number: 2016947018

Printed on acid-free paper

This Springer imprint is published by Springer Nature
The registered company is Springer (India) Pvt. Ltd.

To my late father Shri Amarnath Khanna
for imparting me education and giving
wise counsel,
To my beloved mother Smt. Pushpa
Khanna, for her blessings to live happily,
to keep well and excel.
To my daughter Aloka whose endeavors
are good and wonderful,
Making the family flamboyant and cheerful,
To my wife Amita for backing me up
audaciously with grin
and standing by me through thick and thin
Rain or sunshine,
Tempestuous weather or fine!

Preface

Almost since their inception, integrated circuits have been synonymous with NMOS and CMOS technologies; the bipolar ICs reigned for some period. The primeval and perennial CMOS has borne the brunt and onslaught of brutal miniaturization, and has been stubbornly holding on to maintain its supremacy. It has also received ample support from its nanotechnological cousins like nanophotonics, nanomechanics, nanobiotechnology and spintronics. Receiving this support, it has blossomed into beautiful flowers and fruits of “CMOS applications”, which have cast a tremendous influence on human life at large. The CMOS IC tree has also burgeoned into relatively new areas, spreading into and embracing all spheres of our lives. Indeed, so profound and overwhelming has been the CMOS proliferation!

Over the past few decades, the researchers have also been constantly bothered by the question: What next after CMOS? So scientists have diverted their attention towards exploratory research in quest of potential CMOS alternatives. A noteworthy consequence of this progress has been that nanoelectronics no longer remains a solo silicon semiconductor technology. One must broaden thinking to include other materials such as carbon one-dimensional materials like carbon nanotubes. There are two-dimensional materials, e.g., graphene and dichalcogenides. Several polymers and biological materials like DNA are important too. Nanoelectronics is no longer the forte of physicists and electrical engineers. Biophysicists, chemists, and biologists equally partake in this field. It appears that physics, engineering, chemistry, biology, and related rivulets have all merged together into nanoelectronics to provide the solutions to unabated shrinkage of devices and circuits.

Transistor switches are not the only possible means of implementing logic operations. One must envisage other models like quantum dot cellular automata or magnetic cellular automata. Wired circuits may not be always required. Wireless field-coupled modus operandi must be foreseen. Room temperature operation should not be compulsorily followed. Superconductive nanoelectronics has its own benefits. Innovative circuit designs and system architectures must be visualized for increased energy efficiency in information processing.

In view of these developments, nanoelectronics of today is an interdisciplinary science and technology poised at the intersection and crossroads of several tributaries of knowledge. This book seeks to present an interdisciplinary perspective of nanoelectronics in contrast to hitherto followed view of looking at it merely from physics and electrical engineering angle. In this respect, it differs from existing books, which concentrate more on the electrical engineering aspects, and less on cross-disciplinary view. In line with the above thinking, the readers will find five parts of this book: (i) A part laying down the groundwork for understanding the subject matter. (ii) A part treating about MOSFET and CMOS technologies, the scaling issues, problems faced in short-channel devices and their remedies, the structural innovations in the form of partially- and fully-depleted SOI MOSFETs, FINFETs and multigate architectures. (iii) A part on sister nanotechnogies of CMOS, viz., those from optical, magnetic, mechanical and biological families. (iv) A part on post-CMOS futuristic technologies like resonant tunneling diodes, tunnel FETs, single-electron transistors, one-dimensional material platforms, e.g., silicon nanowires and carbon nanotubes, two-dimensional materials, like graphene and transition metal dichalcogenides, quantum dot cellular automata, magnetic cellular quantum automata, rapid single flux quantum devices, and finally molecular electronics. (v) A concluding part on nanofabrication techniques and diagnostic instrumentation.

The book is targeted to an assorted audience, which includes but is not limited to graduate students of electrical and electronics engineering, physics, chemistry, nanotechnology, semiconductor fabrication technology, and related courses. Professional engineers and scientists engaged in this frontier field will also be immensely benefitted.

Pilani, India Vinod Kumar Khanna

Acknowledgments

It is a pleasure to express my unfathomable gratitude and earnest appreciation to all those who have explicitly or implicitly helped me in this project. *Prima facie*. I am indebted to Almighty God. I thank God for granting me health, wellbeing, vigor, energy, and patience.

> Neither any words nor any actions are adequate to express thankfulness,
> All words fall short and all efforts seem too less!

Without God's grace, it is impossible to do any work, big or small.

I gratefully acknowledge the authors of all books, journal articles and web pages, which have been consulted in this compilation. Many of these esteemed names appear in the references list at the end of each chapter. If any name is missed, it may please be forgiven as an inadvertent omission.

I wish to thank the Director, CSIR-CEERI, Pilani, as well as scientists and colleagues at CSIR-CEERI for encouragement, guidance and support.

I am thankful to my acquisitions editor and project co-coordinator at Springer for offering me this opportunity, and for their kind cooperation and ardent efforts rendering possible timely completion of the project.

The acknowledgments section will remain lacking and unfinished if I do not record my gratefulness to my family. They provided me the valuable time and environment during holidays and evenings for writing this book.

Thanks to everyone from the bottom of my heart!

Pilani, India Vinod Kumar Khanna

About This Book

Integrated Nanoelectronics presents an interdisciplinary perspective of nanoelectronics, unifying the aspects of contemporary CMOS and ancillary nanotechnologies like nanophotonics, NEMS, biotechnology, and spintronics with futuristic "beyond CMOS" technologies, which are likely to take the baton forward after CMOS has exhausted its full potentiality of miniaturization gimmicks. From the present scenario of research trends, it is obvious that fundamental innovations in device and circuit designs and architectures together with introduction of novel paradigms in information representation, storage, transmission, processing, and computation are envisioned to prompt vital breakthroughs. Nanoelectronics of next decade and further is more likely to depend on inputs from diverse fields including physics, chemistry, and biology, and their cross-fertilization. The multidisciplinary growth of nanoelectronics is therefore stressed all throughout the text.

Commencing from the preliminary background material laying down the foundation for easy understanding, the progress in mainstream CMOS nanoelectronics is sketched out. Several supportive nanotechnologies are carrying the burden shoulder-to-shoulder with CMOS. The role played by these nanotechnologies in sensors, actuators, and other fields is enumerated. The steady march of CMOS and its companion technologies over a long triumphant period is getting bogged down by physical and practical limitations. In preparation for the future, several new avenues are explored in the form of resonant tunneling diodes, tunnel FETs, single-electron transistors, quantum dot cellular automata, magnetic quantum cellular automata, rapid single flux quantum logic, and molecular electronics. Readers will also find chapters on advanced nanofabrication and nanocharacterization facilities, which constitute the heart and soul of a nanoelectronics laboratory.

The book presents a lucid description of vast variety of topics reinforced by an elegant mathematical treatment. Students, engineers, and scientists alike will be immensely benefited from the comprehensive coverage of state-of-the-art nanotechnologies and their future trends in this explosively growing field.

Contents

About the Author

Vinod Kumar Khanna born on November 25, 1952 at Lucknow, Uttar Pradesh, India, is an Emeritus Scientist at CSIR-Central Electronics Engineering Research Institute, Pilani, Rajasthan, India, and Emeritus Professor, AcSIR (Academy of Scientific and Innovative Research), India. He is former Chief Scientist and Head, MEMS and Microsensors Group, CSIR-CEERI, Pilani, and Professor, AcSIR. During his service tenure of more than 34 years at CSIR-CEERI, starting from April 1980, he worked on various research and development projects on power semiconductors devices (high-current and high-voltage rectifier, high-voltage TV deflection transistor, power Darlington transistor, fast switching thyristor, power DMOSFET and IGBT), PIN diode neutron dosimeter and PMOSFET gamma ray dosimeter, ion-sensitive field-effect transistor (ISFET), microheater-embedded gas sensor, capacitive MEMS ultrasonic transducer (CMUT), and other MEMS devices. His research interests were micro- and nanosensors, and power semiconductor devices. From 1977 to 1979, he was Research Assistant in the Physics Department, Lucknow University.

Dr. Khanna's deputations abroad include Technische Universität Darmstadt, Germany, 1999; Kurt-Schwabe-Institut für Mess- und Sensortechnik e.V., Meinsberg, Germany, 2008; and Institute of Chemical Physics, Novosibirsk, Russia, 2009. He also participated and presented research papers at IEEE-IAS Annual Meeting, Denver, Colorado, USA, 1986.

Dr. Khanna received his M.Sc. degree in Physics with specialization in Electronics from the University of Lucknow in 1975, and Ph.D. degree in Physics from Kurukshetra University, Kurukshetra, Haryana, in 1988 for his work on thin-film aluminum oxide humidity sensor. A fellow of the Institution of Electronics and Telecommunication Engineers (IETE), India, he is a life member of Indian Physics Association (IPA), Semiconductor Society (SSI), India and Indo-French Technical Association.

Dr. Khanna has published nine books, six chapters in edited books, and 181 research papers in national/international journals and conference proceedings; he holds two US and two Indian patents.

Abbreviations, Acronyms, Chemical Symbols and Mathematical Notation

$(H_2C_2S_2C)_2$	Tetrathiafulvalene
$(NC)_2CC_6H_4C(CN)_2$	Tetracyanoquinodimethane
$(NH_4)_2MoS_4$	Ammonium tetrathiomolybdate
$(NH_4)_2WS_4$	Ammonium Tetrathiotungstate
°C	Degree centigrade
1-D, 2-D, 3-D	One-, two-, and three-dimensional
$1T\text{-}MoS_2$	Tetragonal molybdenum sulfide
$2H\text{-}MoS_2$	Hexagonal molydenum sulfide
A	Ampere
AC	Alternating current
AchE	Acetylcholinesterase
AD FINFET	Asymmetric drain–source doped FINFET
ADSE FINFET	Asymmetric drain-spacer-extended FINFET
AES	Auger electron microscopy
AFM	Atomic force microscope
Ag	Argentum (Silver)
Al_2O_3	Aluminum oxide
AlAs	Aluminum arsenide
ALD	Atomic layer deposition
AlF_3	Aluminum fluoride
AlGaAs	Aluminum gallium arsenide
AlGaN	Aluminum gallium nitride
AlN	Aluminum nitride
AMR	Anisotropic magnetoresistance
APS	Aminopropylsilatrane
APTES	(3-Aminopropyl) triethoxysilane
Ar	Argon
ASG	Asymmetric gate-workfunction FINFET
ATP	Adenosine Triphosphate
Au	Aurum (Gold)

AuNP	Gold nanoparticle
BBF	Blown bubble film (method)
BCP	Block copolymer
BESOI	Bond and Etch-Back SOI
$BiFeO_3$	Bismuth ferrite
BJT	Bipolar junction transistor
BN	Boron nitride
BOX	Buried oxide
BRET	Bioluminescence Resonance Energy Transfer
C	Carbon, Coulomb
$C_{10}H_8S_2$	Dithienylethene
$C_{12}H_{26}S$	Dodecanethiol
$C_{18}H_{35}NH_2$	Oleylamine
C_2H_2	Acetylene
C_2H_4	Ethylene
C_2H_6O	Ethanol
$C_2H_6O_2$	Ethylene glycol
$C_4H_5N_3O$	Cytosine
$C_5H_{10}O_4$	Deoxyribose
$C_5H_{10}O_5$	Ribose
$C_5H_5N_5$	Adenine
$C_5H_5N_5O$	Guanine
$C_5H_6N_2O_2$	Thymine
C_6H_6	Benzene
CAD	Computer-aided design
CaDPA	Calcium dipicolinate
CCD	Charge control device
CdSe	Cadmium selenide
CdTe	Cadmium telluride
CE	Chemical enhancement
CH_2	Methylene group
$–CH_2CH_3$	Ethyl group
$–CH_3$	Methyl group
CH_3SH	Methyl mercaptan, Methanethiol
CH_4	Methane
CHE	Channel hot electron
–CHO	Carbonyl group
CIP	Current-in-plane
CLIO	Cross-linked iron oxide
cm	Centimeter
CMC	Ceramic matrix composite
CMCS	*O*-carboxymethyl chitosan
CMOS	Complementary metal–oxide–semiconductor
CMP	Chemical mechanical polishing
–CN	Cyanide group

CNT	Carbon nanotube
CO	Carbon monoxide
Co	Cobalt
$CoFe_2O_4$	Cobalt ferrite
Con A	Concanavalin A
COOH	Carboxyl group
CPP	Current-perpendicular-to-plane
Cr	Chromium
CRET	Chemiluminescence resonance energy transfer
Cu	Cuprum (Copper)
CVD	Chemical vapor deposition
DAHC	Drain avalanche hot carrier
dB	Decibel
DBQW	Double barrier quantum well structure
DC	Direct current
DIBL	Drain-induced barrier lowering
DLS	Dynamic light scattering
DMR	Diagnostic magnetic resonance
DNA	Deoxyribonucleic acid
DRAM	Dynamic random-access memory
DSA	Directed self-assembly
DTT	Dithiothreitol
DUV	Deep ultra violet
E-beam	Electron beam
EBL	Electron beam lithography
EC	European Commission
E-DNA	Ethylated DNA
EDTA	Ethylenediaminetetraacetic acid ($C_{10}H_{16}N_2O_8$)
EDX	Energy dispersive X-ray analysis
EME	Electromagnetic enhancement
EOT	Equivalent oxide thickness
EPL	Electron projection lithography
ESCA	Electron spectroscopy for chemical analysis
EUV	Extreme UV lithography
eV	Electron volt
F	Farad
FD	Fully depleted
Fe	Ferrum (Iron)
FESEM	Field-emission scanning electron microscope
FET	Field-effect transistor
FIB-SEM	Focused ion beam scanning electron microscope
FINFET	Fin field-effect transistor
fM	Femtomolar
FRET	Förster (fluorescence) resonance energy transfer
FT-IR	Fourier transform infrared spectroscopy

FUSI	Fully silicided (polysilicon gates)
g	Gram
GaAs	Gallium arsenide
GaN	Gallium nitride
GaP	Gallium phosphide
Gbps	Gigabit per second
GBps	Gigabyte per second
GCE	Glassy carbon electrode
Ge	Germanium
GHz	Gigahertz
G-line	436 nm
GMR	Giant magnetoresistance
GNP	Gold nanoparticle
GNR	Graphene nanoribbon
GOD	Glucose oxidase
H	Hydrogen
H_2O_2	Hydrogen peroxide
h-BN	Hexagonal boron nitride
HBT	Heterojunction bipolar transistor
HDD	Hard disk drive
He	Helium
$Hf[NEt_2]_4$	Tetrakis(diethylamido)hafnium
HfO_2	Hafnium oxide
HOMO	Highest occupied molecular orbital
HOPG	Highly oriented pyrolytic graphite
HRP	Horseradish peroxidase
Hz	Hertz
I	Iodine, Intrinsic
IBM	International Business Machines Corporation
IC	Integrated circuit
IG FINFET	Isolated gate FINFET
IgG	Immunoglobulin G
I-line	365 nm
InGaAs	Indium gallium arsenide
InP	Indium phosphide
IR	Infrared
Ir	Iridium
ISO	International Standards Organization
J	Joule
JJ	Josephson junction
JTL	Josephson transmission line
K	Kelvin
kΩ	Kilo ohm
keV	Kilo electron volt
LA	Laser ablation

La_2O_3	Lanthanum oxide
LaB_6	Lanthanum hexaboride
LB	Langmuir–Blodgett
LbL	Layer-by-layer
LDD	Lightly doped drain
LDV	Laser Doppler vibrometer
LPP	Laser-produced plasma
LSPR	Localized surface plasmon resonance
LUMO	Lowest unoccupied molecular orbital
M	An atom of transition metal in the formula MX_2
m	Meter, mass
mA	Milliampere
MBE	Molecular beam epitaxy
MBs	Molecular beacons
MCBJ	Mechanically controlled break junction
MEMS	Microelectroechanical systems
meV	Millielectron volt
Mg	Magnesium
mg	Milligram
MG/HK	Metal gate/High κ (gate stack)
MgF_2	Magnesium fluoride
MgO	Magnesium oxide
MIPG	Metal inserted polysilicon gate
mm	Millimeter
mM	Millimolar
Mn	Manganese
Mo	Molybdenum
MOCVD	Metal-organic chemical vapor deposition
MODFET	Modulation-doped field-effect transistor
MOM	Methoxymethyl (group)
MoO_3	Molybdenum trioxide
MoS_2	Molybdenum disulfide
MOSFET	Metal–oxide–semiconductor field-effect transistor
MPa	Megapascal
MQCA	Magnetic quantum cellular automata
MR	Magnetoresistance
MRAM	Magnetic random-access memory
MRI	Magnetic resonance imaging
MRS	Magnetic resonance switch
MTJ	Magnetic tunnel junction
MV	Megavolt
mV	Millivolt
MWCNT	Multiwalled carbon nanotube
N	Nitrogen, Newton
nA	Nanoampere

NA	Numerical aperture
Nb	Neobium
NbS_2	Neobium sulfide
$NbSe_2$	Neobium selenide
NDR	Negative differential resistance
NEMS	Nanoelectromechanical systems
$-NH_2$	Amine group
Ni	Nickel
NIL	Nanoimprint Lithography
nm	Nanometer
NMOS	N-channel metal–oxide–semiconductor (transistor)
$-NO_2$	Nitro group
NO_2	Nitrogen dioxide
NSOM	Near-field scanning optical microscopy
nTP	Nanotransfer printing
NW	Nanowire
O	Oxygen
–OH	Hydroxyl group
OPH	Organophosphorous hydrolase
Pa	Pascal
PALS	Phase analysis light scattering
Pd	Palladium
PD	Partially depleted (MOSFET)
PDMS	Polydimethylsiloxane
PEALD	Plasma-enhanced atomic layer deposition
PECVD	Plasma-enhanced chemical vapor deposition
PLA	Programmable logic array
pm	Picometer
PMDAH	Pyromellitic dianhydride
PMDA–ODA	Poly(pyromellitic dianhydride-co-4,4′-oxydianiline)
PMMA	Poly(methyl methacrylate)
PMOS	P-channel metal–oxide–semiconductor (transistor)
PREVAIL	Projection reduction exposure with variable-axis immersion lenses
ps	Picosecond
PS-*b*-PB	Polystyrene-block-polybutadiene
Pt	Platinum
PTL	Pass-transistor logic
PVD	Physical vapor deposition
PZT	Lead zirconate titanate ($Pb[Zr_xTi_{1-x}]O_3$), $0 \leq x \leq 1$
Q	Quality factor
Qbit	Quantum bit
QD	Quantum dot
QDCA	Quantum dot cellular automata
RAM	Random-access memory

RAMA	Reconfigurable array of magnetic automata
RBM	Radial breathing mode
RC	Resistance–capacitance
RCA	Radio Corporation of America
RDF	Random dopant fluctuation
RF	Radio frequency
RFSQ	Rapid single flux quantum
rGO	Reduced graphene oxide
RMG	Replacement metal gate
RNA	Ribonucleic acid
R-S	Reset-set (flip-flop)
RTD	Resonant tunneling diode
Ru	Ruthenium
S	Sulfur, Siemen
Salicide	Self-aligned silicide
SAM	Self-assembled monolayer
Sc	Scandium
Sc_2O_3	Scandium oxide
SCALPEL	Scattering with angular limitation in projection electron-beam lithography
Se	Selenium
SELBOX	Selective back oxide
SEM	Scanning electron microscope
SERS	Surface-enhanced Raman spectroscopy
SET	Single-electron transistor
SG FINFET	Shorted gate FINFET
SG-workfunction FINFET	Symmetric gate-workfunction FINFET
-SH	Thiol group
SHE	Substrate hot electron
Si	Silicon
$Si(OC_2H_5)_4$	Tetraethoxysilane
$Si(OCH_3)_4$	Tetramethoxysilane
Si_3N_4	Silicon nitride
SiC	Silicon carbide
SiGe	Silicon–Germanium
SIMOX	Separation by implanted oxygen
SiNW	Silicon nanowire
SiO_2	Silicon dioxide
Si–OH	Silanol group
SOI	Silicon-on-insulator
SOI-MOSFET	Silicon-on-insulator metal–oxide–semiconductor field-effect transistor
SPL	Scanning probe lithography
SPM	Scanning probe microscopy
SPR	Surface plasmon resonance

SQUID	Superconducting quantum interference device
SRAM	Static random-access memory
$SrCO_3$	Strontium carbonate
SrS	Strontium sulfide
$SrTiO_3$	Strontium titanate
SS	Subthreshold swing
ssDNA	Single-stranded DNA
SSRW	Super steep retrograde well
STI	Shallow trench isolation
STM	Scanning tunneling microscope
STT-RAM	Spin transfer torque random-access memory
SWCNT	Single-walled carbon nanotube
Ta	Tantalum
Ta_2O_5	Tantalum pentoxide
TaC	Tantalum carbide
TAE	Tris acetate
TaS_2	Tantalum sulfide
TCNQ	Tetracyanoquinodimethane
TD	Tunnel diode
Te	Tellurium
TEM	Transmission electron microscope
TEOS	Tetraethoxysilane
TFET	Tunnel field-effect transistor
Ti	Titanium
TiN	Titanium nitride
TiO_2	Titanium oxide
$TiSe_2$	Titanium selenide
TJ	Tunnel junction
TMA	Trimethylaluminum
TMD	Transition metal dichalcogenide
TMOS	Tetramethoxysilane
TPa	Terapascal
TTF	Tetrathiafulvalene
UV	Ultraviolet
UV-NIL	Ultraviolet nanoimprint lithography
V	Volt, Vanadium
VD	Vapor deposition
VLS	Vapor–Liquid–Solid
VPE	Vapor phase epitaxy
VSe_2	Vanadium selenide
W	Tungsten, Watt
WS_2	Tungsten disulfide
WTe_2	Tungsten telluride
X	A chalcogen atom in the formula MX_2
XPS	X-ray photoelectron microscopy

XRD	X-ray powder diffraction
ZnO	Zinc oxide
ZnS	Zinc sulfide
ZrO_2	Zirconium oxide
μCP	Microcontact printing

Roman Alphabet

A	Area
a	Radius, Lattice constant of a crystal
A, B	Inputs to QDCA cell
$\vec{a}, \vec{b}$	Unit vectors
a_{exciton}	Exciton Bohr radius
a_{H}	Bohr radius
B	Magnetic field
c	Velocity of light
C	Capacitance
C_{a}, C_{b}	Capacitances of tunnel junctions TJ_{a}, TJ_{b}
C_{g}	Gate capacitance
C_{gate}	Gate capacitance
C_{L}	Load capacitor
C_{ox}	Capacitance of gate oxide per unit area
D	Dispersion, Degree of inhibition
d	Diameter, Grain size, Distance
D_{t}	Translation diffusion coefficient of a particle
E	Electric field, Young's modulus
E_{F}	Fermi level
E_{f}	Final energy
E_{g}	Energy bandgap
E_{i}	Initial energy
$E_{n,l}^{\text{Conf}}(a)$	Confinement energy of a nanocrystal of radius a
E_0	Resonant energy level
E_{Ry}^{*}	Rydberg energy of the exciton
E_{se}	Stored energy
E_{Thermal}	Thermal energy
f	Frequency
f_{B}	Frequency of the Bragg cell
$f(ka)$	Henry's function
g	Electron g-factor
h	Planck's constant
$\hbar$	Reduced Planck's constant = $h/2\pi = 1.05 \times 10^{-34}$ J s
I	Current, Intensity of light

I_C	Critical current
I_{DS}	Drain–source current
I_{DSat}	Drain–source saturation current
I_{off}	Off-state current
I_{on}	On-state current
I_P	Persistent current in a superconductor
I_{Subth}	Subthreshold leakage current
I_{valley}	Valley current
J	Mechanical equivalent of heat
k	Wave vector of the crystal
k_1, k_2	Constants
k_B	Boltzmann's constant
k_y	Strengthening coefficient
L	Inductance, Length
l	Secondary quantum number
L_t, L_b	Lengths of top and bottom parallel sides of the trapezium
L_g	Gate length
m	Mass, an integer
m^*	Effective mass of charge carriers (electrons or holes)
m_e^*	Effective mass of the electron
m_h^*	Effective mass of the hole
m_0	Rest mass of the electron = 9.11×10^{-31} kg
m_s	Spin quantum number
$\langle N \rangle$	Average value of N
n	Principal quantum number, an integer, refractive index
N_A	Acceptor doping concentration
N_c	Effective density of states in the conduction band
N_D	Donor doping concentration
n, m	Indices describing atomic structure of a CNT
N_ρ	Transistor density
N_S	Number of atoms located on the surface S
N_V	Number of atoms in volume V
N_v, N_V	Effective density of states in the valence band
P	Power, Polarization
P_{peak}	Peak power dissipated per transistor
P_ρ	Power density
Q	Charge, Heat generated
q	Electronic charge = 1.6×10^{-19} C
Q_a^i, Q_b^i	Charges associated with tunnel junctions TJ_a and TJ_b
q_e	Electronic charge
Q_g^i	Charge on the gate terminal
R	Radius, Resistance

r	Radius
$\vec{r}$	Chiral vector
R_a, R_b	Resistances of tunnel junctions TJ_a, TJ_b
r_{diff}	Differential resistance
r_h	Hydrodynamic radius of a particle
r_j	Radius of curvature of source/drain diffusion
$R(r)$	Bessel functions
R_{source}	Parasitic resistance of source
R_T	Tunnel resistance
S	Surface area
$\langle SN_C \rangle$	Surface coordination number
T	Temperature on Kelvin scale, Periodic time
t	Thickness, Time
T_M	Melting point of bulk material
T_m	Melting point of a collection of particles
t_{ox}	Thickness of oxide
t_{tr}	Transit time
u_k	A periodic function
V	Volume, Volt, Voltage
V_a^i and V_b^i	Voltages across tunnel junctions TJ_a and TJ_b
V_{DD}	Drain voltage
V_{DS}	Drain–source voltage
V_g	Gate voltage
V_g^i	Gate voltage in initial condition
V_{GS}	Gate–source voltage
V_i	Input signal
$\langle VN_C \rangle$	Volume or bulk coordination number
V_{out}	Output voltage
V_{PT}	Punch-through voltage
V_S	Supply voltage
v_{satn}	Saturation velocity of electron
v_{satp}	Saturation velocity of hole
V_T	Thermal voltage
V_{Th}	Threshold voltage
W	Width
W_C	Charging energy
W_e	Energy stored in the electric field of a capacitor
x	Volume fraction
x_d	Width of depletion region
x_{dD}, x_{dS}	Depletion-layer widths surrounding the source and drain junctions
x_j	Junction depth
$Y(\theta, \phi)$	Spherical harmonics

Greek Alphabet

α	Fine structure constant = 7.3×10^{-3}, one-half the angular aperture of the lens
β	Stewart–McCumber parameter of a SQUID
β_n, β_p	User-defined unitless parameters in Caughey-Thomas equation
γ	Interfacial/surface tension
δ	Depth of focus
ΔE	Uncertainty of energy, Zeeman energy
ΔE_{se}	Change in stored energy
Δf_D	Doppler frequency shift
Δf_r	Change in resonance frequency
Δg	Change in surface stress
ΔH	Latent heat of fusion
Δm	Extra mass
ΔQ	Change in charge
Δt	Uncertainty of time, Time interval
ΔV_{Th}	Change in threshold voltage
ΔW	Change in energy
ε_0	Absolute permittivity
ε_r	Relative permittivity
ε_{Si}	Relative permittivity of silicon
ζ	Intrinsic gate delay of a transistor, Zeta potential
η	Dynamic viscosity of a liquid
θ	Chiral angle, Angle of scattering
κ	Dielectric constant
λ	Scaling factor, Decay constant, Wavelength
μ	Mobility of charge carriers
μ_{ac}	Mobility limited by surface acoustic phonons
μ_B	Bohr magneton
μ_b	Carrier nobility in the bulk
μ_E	Electrophoretic mobility of a particle
μ_n	Mobility of electrons
μ_{n0}, μ_{p0}	Low-field electron and hole mobilities
μ_{sr}	Mobility limited by surface scattering
ν	Poisson's ratio, Frequency
ν_m	Change in frequency
ρ	Resistivity
ρ_S	Density of states
σ_0	Constant
σ_y	Yield stress
τ	Time constant
ϕ	Magnetic flux
$\phi_{env}\ (x)$	Envelope function

Φ_0	Quantum of magnetic flux
ϕ_0	Built-in potential
$\chi_{n,l}$	Roots of the Bessel function
$\psi_{\text{Bloch, bulk}}(x)$	Bloch wave function in bulk crystal
$\psi_{\text{Bloch, nano}}(x)$	Bloch wave function in nanocrystal

Chapter 1
Getting Started to Explore "Integrated Nanoelectronics"

Abstract A prelude to the subject is provided by delving into the present scenario and research trends in nanoelectronics and relating them to the organization of contents of the book. The chapter will provide a snapshot into the diversity of topical coverage and their mutual interrelationship, serving as a launching pad to begin exploration of this vast field.

1.1 What "Integrated Nanoelectronics" Is About?

The impact of nanotechnology on electronics needs hardly to be emphasized. Nanotechnology makes semiconductor chips smaller and light in weight. It makes them cheaper. It reduces power consumption. It makes data communication faster. Integrated nanoelectronics or nanotechnology of integrated circuits is revolutionizing our lifestyles and improving the overall quality of human life everyday [1–3].

For long, the MOSFET device used in the CMOS configuration has been the workhorse of integrated circuit industry. The CMOS structure has been constantly miniaturized and is fast reaching the physical limits. The objective of this book is to survey the MOSFET geometries and architectures, along with necessary process modifications that have emerged with constant downscaling of the primitive designs. A family of nanotechnologies is also supporting CMOS and can be incorporated with CMOS to enhance capabilities. These include the optical, mechanical, and biosensors, together with magnetic memories. These CMOS-supportive nanotechnologies constitute another major area of interest that will be discussed.

As the physical barriers are insurmountable for CMOS, this lone star in the sky of nanoelectronics may no longer be able to bear the burden of nanoelectronics. Many new stars are appearing on the horizon. Although these new stars are not equally bright but they may gain in intensity as time passes by. These upcoming technologies fall under the "Beyond CMOS" category. These potential stars, which are likely to supersede CMOS to fulfill the needs of nanoelectronics, form another interesting area.

V.K. Khanna, *Integrated Nanoelectronics*,
NanoScience and Technology, DOI 10.1007/978-81-322-3625-2_1

The above ideas are crammed into the following poem:

Integrated Electronics Marches On Gloriously....

From 1947-the invention of bipolar transistor,
and 1958-the first integrated circuit,
Electronic integrated circuits or Integrated electronics,
Built on monolithic silicon blocks,
With interconnected active and passive components
Have overcome many roadblocks.
Leapfrogging many hurdles and rocks,
Advancing from macro to microelectronics,
Down to nanoelectronics,
And reaching integrated nanoelectronics.
By applying nanotechnolgy to electronics,
Based on CMOS technologies,
And supporting nanotechnologies,
Photonics, NEMS and biotechnologies.
Also beyond CMOS,
There are carbon nanotubes and nanowires.
2-D materials and quantum dot cellular automata,
Electronic spin and single electronics,
RFSQ and molecular electronics.
Devices and circuits are fast shrinking,
Progressing from miniaturization to super-miniaturization,
New structures and architectures are evolving.
Self-assembly may hold the key to nanofabrication.
Future options are achievable and realistic,
Making us very hopeful and optimistic.
Let us strive with all our might,
The future is very bright!

1.2 Subdivision of the Book

The book is divided into five parts. Part I introduces the elementary topics. Part II is devoted to CMOS nanoelectronics. Part III addresses nanotechnological allies of CMOS. Part IV deals with "beyond CMOS" nanoelectronics. Part V treats nanofabrication and nanocharacterization aspects.

1.3 Organization of the Book

1.3.1 Part I: Preliminaries

Part I contains two chapters. Chapter 2 introduces the subject matter of this book. Nanoelectronics and associated nanotechnologies are defined and their mutual interrelationship is described. Chapter 3 explains the basic terminology of nanomaterials. The unique properties of these materials responsible for their special behavior are highlighted.

1.3.2 Part II: CMOS Nanoelectronics

Part II spanning over Chaps. 4–7 is concerned with the classical MOSFET and its downscaled versions. In Chap. 4, the readers are familiarized with the MOSFET and CMOS structures, the related self-aligned fabrication processes and CMOS logic gates. Then the influence of constant field and constant voltage scaling on MOSFET performance is outlined. Knowledge of basic MOSFET electronics is assumed.

Chapter 5 describes the unfavorable phenomena which come into play when the channel length of a MOSFET becomes comparable in dimension with the space charge region widths of source and drain junctions. These MOSFETs are referred to as short-channel devices. Many unwanted effects such as drain-induced barrier lowering, velocity saturation, carrier mobility degradation, etc., begin to seriously impoverish the device characteristics in these low-dimensional MOSFETs. Nonetheless, these MOSFETs have been able to meet the downscaling challenges by making suitable structural and doping profile changes and using novel materials such as high-κ dielectrics.

The succeeding Chap. 6 takes a look at a different class of MOSFETs which are fabricated on silicon-on-insulator (SOI) wafers containing a thick bottom silicon layer called the handle wafer over which there is a buried oxide layer with a silicon layer at the top in which the device is built. These SOI-MOSFETs have been realized in partially depleted and fully depleted forms. Pros and cons of the two forms are brought out.

In Chap. 7, popular MOSFET varieties, namely the FINFET and tri-gate FET are dealt with. These MOSFET varieties mark a changeover from the ancient two-dimensional MOSFET to the present three-dimensional edifices. They were felt essential because the gate terminal was losing its effectiveness in the planar MOSFET at scaling limits. So, the main intent of these three-dimensional MOSFETs is to make the gate more effective in controlling the channel.

1.3.3 Part III: CMOS-Supportive Nanotechnologies

Part III comprises Chaps. 8–11. Chapter 8 is concerned with nanophotonics. Interaction of light with metallic nanoparticles gives rise to surface plasmon resonance phenomena which form the basis of several useful sensors. Nanophotonics also offers the opportunities of building logic circuits. Laser capabilities are significantly improved by inserting quantum dots in the active region. Photonic nanocrystals offer interesting applications as optical filters. The next Chap. 9 presents a brief treatment of nanoelectromechanical systems, which are a step further to microelectromechanical systems in miniaturization. The operation of several nanosensors for pressure, displacement, etc., is discussed along with NEMS actuators like nanogripper, nanotweezer, and so forth. The focal theme of Chap. 10 is nanobiosensors which represent the confluence of nanoelectronics with biotechnology. Research endeavors in this interdisciplinary field have led to many sensors for biological analytes such as glucose, DNA hybridization, and toxins like organophosphorous pesticides. Chapter 11 on spintronics provides a succinct overview of magnetic memories based on the giant magnetoresistance effect as well as those utilizing magnetic tunnel junctions. These memories have made ingress into the present-day computers, vastly augmenting the storage capacity.

1.3.4 Part IV: Beyond CMOS Nanoelectronics

Part IV makes a departure from the mainstream CMOS and treats the alternative options that have been proposed over the years as successors of CMOS. None of these options have gained popularity up to a level of even a small fraction of that enjoyed by CMOS today. In Chap. 12, attention is directed towards resonant tunneling diodes, which are superfast devices based on quantum-mechanical tunneling through thin insulating films for terahertz frequencies utilizing the negative differential resistance. Another tunneling-based device is also treated, viz., the tunnel field-effect transistor. Tunnel FETs are steep subthreshold slope switches permitting supply voltage scaling for extremely low power computing. Their working principle is band-to-band tunneling mechanism which can be controlled through gate voltage. Chapter 13 is a preparatory chapter for understanding the single electron transistor to be described in Chap. 14. Concept of tunnel junction is developed and its equivalent circuit model is discussed. Coulomb blockade or Kondo effect is a charge quantization effect related to the electrostatic energy that must be added or subtracted from a conducting island for electron transference to it or removal of electron from it. A quantum dot circuit operation is analyzed as a prelude to single electron transistor action. Difference in operational behavior between circuits made with big and small quantum dots is pointed out in terms of energy band diagrams. Chapter 14 is devoted to theoretical formulation of single electron transistor. The mathematical framework is followed by energy band concepts. Two types of logic circuits possible using SET devices are briefly indicated. Voltage-based NOT, AND, and OR circuits are explained. Chapter 15 describes the use of semiconductor nanowire, particularly, silicon nanowire as a building block element of nanoelectronic circuits. Vapor–liquid–solid (VLS) technique of silicon nanowire growth is presented. Simultaneous growth of a large number of nanowires at exactly defined locations allows fabrication of circuits in a controlled fashion. Another important structural unit of nanoelectronic circuits is the carbon nanotube, which is the topic of Chap. 16. Single and multiwall CNTs are distinguished. Arm chair, zigzag, and chiral CNT classes are demarcated. Special electrical, mechanical, and optical properties of CNTs are discussed. Amongst the various available methods of CNT deposition such as arc discharge, laser ablation, etc., chemical vapor deposition stands out prominently as a technique for nanoelectronics fabrication. A remarkable feature of CNTs is that they allow device fabrication by a doping-free process through appropriate choice of materials for contact electrodes. Pass-transistor logic seems to be more favored for CNT nanoelectronics. Both semiconductor nanowires and carbon nanotubes are one-dimensional nanostructures. From these one-dimensional nanomaterials, we move to two-dimensional materials, notably graphene and transition metal dichalcogenides. Chapter 17 concentrates on graphene. Many methods have been devised to prepare graphene such as by cleaving multilayer graphite using adhesive tape or a diamond wedge, by sonicating graphite, exfoliation of graphene by electrochemical synthesis, by heating silicon carbide at high temperatures, etc. Graphene is an allotropic modification of carbon. It is a planar sheet of carbon atoms, which is one atomic layer thick. The atomic arrangement in graphene is shaped like a hexagonal honeycomb lattice. Graphene is associated with many special attributes, a few of which are it being the thinnest compound, the lightest but strongest material, and the best thermal and electrical conductor. Due to its incredible properties, graphene has been hailed as a supermaterial. It offers manifold opportunities to electronic engineers to manufacture flexible circuits. Its shortcoming is the absence of an energy bandgap which must be opened up by structural modification destroying some of its superb qualities. Undoubtedly, graphene is touted to bring breakthrough results reshaping nanoelectronics but we have to wait and watch. Similar to graphene is the family of two-dimensional dichalcogenide materials such as molybdenum disulphide and tungsten disulphide but unlike graphene they show an energy gap allowing fabrication of transistors with high on/off current ratio and thickness of one or few atomic layers. These materials are derived by exfoliation, and prepared by chemical vapor deposition or molecular beam epitaxy. Chapter 18 is devoted to nanoelectronics using TMD materials. The ensuing two chapters take the reader into the world of transistorless logic paradigms. Chapter 19 deals with quantum dot cellular automata which consist of cells of four quantum dots in which electrons reside in either of the two pairs of diagonally opposite quantum dots, corresponding to the binary states of the system. The spacing between quantum dots is typically 20 nm while intercell separation is 60 nm. Electrons can tunnel between quantum dots. Due to their mutual electrostatic repulsion, the assignment of electron positions on the two diagonals constitutes the fundamental principle of this logic. QDCA is a field-assisted computing technology. It does not use any wires as in transistor circuits. The two basic logic gates are the majority gate and the NOT gate. The QDCA scheme promises higher chip density together with fast speed but presently faces fabrication difficulties. Chapter 20 discusses an analogous logic using nanomagnets. Information is conveyed and processed with the help of these nanomagnets. Similar to the quantum dot cellular automata is magnetic cellular quantum automata (MCQA). These automata retain information after switching off electrical power. Moreover, they are more radiation-resistant. Chapter 21 is concerned with cryogenic rapid flux quantum logic. In this logic, overdamped Josephson junctions are used as switching elements in place of transistors. Transmission of picosecond duration voltage pulses correlated with the transference of a single quantum of magnetic flux pertains to logic high state whereas their non-transmission represents logic low state. These voltage pulses are transmitted at 100 GHz and above, promising high speed. Chapter 22 deals with molecular electronics which uses either single molecules or small ensembles of molecules to perform digital logic operations. The field is still in infancy. Reproducible fabrication of molecular structures remains a big challenge.

1.3.5 Part V: Nanomanufacturing

Part V contains three chapters which are concerned with fabrication of nanoelectronic devices/circuits and evaluation of the materials and processes. Nanofabrication follows two principal approaches. One way is to start with big blocks and reduce their sizes to get smaller functional elements. This is the top-down approach described in Chap. 23. Recapitulating optical lithography and recounting its limitation according to the wavelength of light, we move on to extreme UV lithography, followed by electron beam lithography using electron waves, and then to electron beam projection lithography, which is a method to avoid the throughput disadvantage of electron beam lithography. Soft lithography, nanoimprint, and block copolymer lithographies are also described.

Another nanofabrication strategy is to start from smaller units and these units grow bigger to yield the required device, just as a seed grows into a big plant. This is the bottom-up approach discussed in Chap. 24. These methods include the sol-gel technique, and various forms of physical and chemical vapor deposition. Atomic layer deposition is a widely used bottom-up method. Other common methods are molecular self-assembly and DNA nanoengineering.

Whatever be the course followed, top-down or bottom-up, the semiconductor manufacturer needs accurate instruments for in-process measurements and monitoring. These tools are described in Chap. 25. First and foremost come the different types of scanning probe microscopies, e.g., scanning tunneling and atomic force microscopies. Besides electronic microscopic techniques such as those employing scanning and transmission modes are finding widespread applications. Then there are many X-ray-based techniques like energy dispersive X-ray analysis X-ray powder diffraction, X-ray photoelectron spectroscopy, etc. Amongst spectroscopic techniques, FT-IR, UV-visible, and Raman spectroscopies are the main instruments of any characterization laboratory. Facilities for determining the size and distribution of nanoparticles help in controlling the properties of nanodispersions. Laser Doppler vibrometry is an optical method for measuring the displacements and velocities of vibrating objects.

1.4 Discussion and Conclusions

The aims and scope of the book were briefly presented and the subject matter to be covered was summarized chapter-wise to enable the inquisitive reader to forge ahead smoothly.

Review Exercises

1.1 Give three examples illustrating the impact of nanotechnology on the performance of integrated circuits.
1.2 Name a semiconductor device which has been the main "burden bearer" of IC industry.
1.3 Name three nanotechnological allies of CMOS technology which are used to enhance its capabilities.
1.4 Explain what is meant by a short-channel MOSFET? What phenomena plague the performance of such MOSFETs? What basic idea is applied to deal with these phenomena?
1.5 What does the acronym "SOI" stand for? Describe the arrangement of layers in an SOI wafer.
1.6 What is the dimensionality shift in MOSFET geometry from planar to FINFET and multigate devices? Why was it necessary?
1.7 Which nanophotonic phenomenon concerned with metals is widely used in nanosensors? Is it possible to perform logic operations optically?
1.8 NEMS complement nanoelectronics through sensors and actuators. Give some examples.
1.9 Nanobiotechnology complements nanoelectronics via nanobiosensors. Give examples.
1.10 What types of spintronic memories are used in nanoelectronics? Cite two examples.
1.11 Which tunneling-based devices are useful for nanoelectronics?
1.12 What is Coulomb blockade effect? Is it possible to make voltage-based logic circuits using single electron transistor?
1.13 Is it possible to grow silicon nanowires of required diameters at defined positions on a wafer?

1.14 Can carbon nanotube transistors be fabricated without doping?
1.15 Why is graphene said to be a supermaterial? What is its principal drawback?
1.16 In what respect is a two-dimensional dichalcogenide superior to graphene?
1.17 Can we altogether avoid the use of transistors for computation? What computational paradigms do not use transistors?
1.18 What are the switching devices used in rapid single flux quantum logic? How are logic high and logic low states represented in this system?
1.19 What are the rudimentary functional units used in molecular electronics?
1.20 What fabrication approach is followed in normal semiconductor chip manufacturing? Bottom-up or top-down. Can optical lithographic technique be used for patterning layers at nanoscale?
1.21 Can you name two bottom-up fabrication techniques?
1.22 What types of scanning probe microscopes do you know?
1.23 Name two kinds of X-ray-based measurement techniques.
1.24 Name three types of spectroscopic methods.
1.25 What is the instrument used to characterize vibrating objects?

References

1. Chau R, Doyle B, Datta S et al (2007) Integrated nanoelectronics for the future. Nat Mater 6:810–812. doi:10.1038/nmat2014
2. Bogue R (2010) The fabrication and assembly of nanoelectronic devices. Assembly Autom 30(3):206–212
3. Collaert N (ed) (2013) CMOS nanoelectronics: innovative devices, architectures and applications. Pan Stanford Publishing Pte. Ltd., 438 pp

Part I
Preliminaries

Chapter 2
Nanoelectronics and Synergistic Nanodisciplines

Abstract The fields of nanoscience and nanotechnology are introduced. The key terminology is defined. The reader is familiarized with the five fundamental inter-related nanodisciplines: nanoelectronics, nanomagnetics, nanophotonics, nanomechanics, and nanobiotechnology. Salient features of these disciplines are described. Three sub-domains of nanoelectronics known as more Moore, more-than-Moore, and beyond CMOS are explained. The association of different nanoscience disciplines with nanoelectronics is brought out. A synthetic treatment of these disciplines is stressed and the key idea of the book is elaborated.

2.1 Meaning of "Nano" and "Nanometer"

The word "nano" originates from the Greek "nanos" or Latin "nanus", meaning "dwarf". "Nano" is a prefix implying "very small". It is placed before many words to form compound words, e.g., nanoplankton, nanomole, nanosecond, nanoliter, nanogram, nanowatt, nanometer, etc.

Quantitatively, "nano" means a factor of one-billionth: 1 nanometer (nm) = 10^{-9} m. A few popular examples involving comparisons of magnitudes are given below from the web to enable the curious reader to visualize how large is a nanometer [1]: (i) Thickness of human hair = 5×10^4 to 10×10^4 nm; (ii) Thickness of a sheet of newspaper = 10^5 nm; (iii) If 1 nm represents the size of a marble, then 1 m indicates the size of earth; (iv) If a sphere of diameter 1 nm is a soccer ball, then the soccer ball is comparable to the earth in size; (v) If 10^7 marks are made in a cm length, then each mark indicates a nm; (vi) In 1 s, the finger nail grows by 1 nm; (vii) 1 nm = length of 7 oxygen atoms or 3–4 water molecules placed along a line (diameter of oxygen atom = 0.155 nm, and diameter of water molecule = 0.275 nm); (ix) Diameter of a red blood cell = 7000 nm; (x) Positioning a nanometer-size structure on a 1 nm long line is equivalent to positioning a peppercorn in a distance as long as between Tokyo and Beijing.

V.K. Khanna, *Integrated Nanoelectronics*,
NanoScience and Technology, DOI 10.1007/978-81-322-3625-2_2

2.2 Nanoscience

Nanoscience is the scientific study of phenomena concerning extremely small objects, materials, structures, or devices having sizes on the nanometre scale, with at least one dimension in the 1–100 nm range. It is essentially a multidisciplinary science encompassing physical and life sciences or their combinations, and dealing with the exploration of phenomena occurring in low-dimensional structures in the range of nanoscale. It also entails the manipulation of objects of these dimensions. Nanoscience extends the capabilities of existing science into the nanoscale clearly indicating what bulk phenomena happen in the same way as in the macroworld and what phenomena are distinctly different in the range of small sizes. It enunciates and elaborates the laws obeyed by the phenomena when one is dealing with these ultra-miniaturized structures. To reiterate, the nanoworld behaves in a strikingly dissimilar manner to the macroworld, and the exclusive laws of the nanoworld come under the purview of nanoscience. Although, as said above, the prefix "nano" implies 10^{-9} units, in the perspective of nanoscience, the "units" are restricted solely to nanodimensions, and not applicable to any other unit of measurement, e.g., for time, energy or power.

2.3 Nanotechnology

Nanotechnology is engaged in the maneuvering of structures or objects in the 1–100 nm size range in at least one dimension. Its aim is to develop products for potential practical applications. In nanoscience, the physical, chemical, and biological understanding of the behavior and characteristics of objects of small dimensions was acquired. The acquired knowledge is gainfully applied by nanotechnology to useful applications. The expertise thus developed appends to nanotechnology. Thus by controlling the shape and size of objects at nanoscale, nanotechnology designs, fabricates, and characterizes new materials, structures, devices, and complete systems. It continuously improves upon these prototypes to evolve better products useful for mankind.

2.4 Plurality of Nanosciences and Nanotechnologies

Nanoscience and nanotechnology, although referred in singular number, are actually a blending of several sciences and technologies. Thus, it is more correct to call them nanosciences and nanotechnologies. So the terms "nanoscience" and "nanotechnology" must be considered to imply a set of sciences and a combination of enabling technologies, instead of a single science and a single technology. They have evolved by applying nanoconcepts to a diversity of sciences and technologies.

2.5 Nanomaterials

To be designated as a nanomaterial, a material must have at least one external dimension in the size range from 1–100 nm. Nanomaterials can be natural, namely, those that exist in the natural world, and artificial or engineered, viz., those that are created by human activities. Nanomaterials may be thin film coatings on surfaces with thickness <100 nm, or cylindrical structures of diameter <100 nm called nanowires and nanotubes or small objects with all three dimensions <100 nm in size, e.g., quantum dots [2]. More rigorous definitions will be introduced in Chap. 3.

Nanomaterials are finding myriads of applications in consumer products. Among the products which have benefitted maybe mentioned toothpastes, batteries, paints, and clothing. The new products seek to improve the quality of human life. The intent of using nanomaterials is to make items of everyday use cleaner, less expensive, and lighter but stronger. There is an intensive ongoing quest for highly efficient, high precision, or more aesthetic materials. Target-driven pharmaceuticals are being sought. Superior medical diagnostic tools are being manufactured. Superfast computers have been introduced. Cleaner methods of energy production are being improvised.

2.6 Uniqueness and Specialty of Nanomaterials

2.6.1 Quantum Size Effect

When compared with bulk materials in coarse forms but having similar chemical composition, nanomaterials are found to exhibit additional or different properties. Bulk copper is a soft, malleable, and ductile metal. Contrarily, copper particles of <50 nm diameter are ultra hard in nature. Pure bulk gold is yellow in color. But a 20-nm gold particle has a wine red color. The color of bulk silver is metallic gray. But a nanosized particle looks yellowish gray. In bulk form, platinum has a silver-white color and palladium is white. But platinum and palladium particles are black at nanolevel sizes. Moving towards the nanoscale, size-dependent properties are observed. These effects will be discussed later. Examples are quantum confinement effect in semiconductor particles, surface plasmon resonance in some metal particles, and superparamagnetism in magnetic particles.

The reason is not far to seek. The bulk properties of a material represent the averaged effects of the quantum forces for the large number of particles constituting the material. As one moves towards successively lower scales, the averaging process fails to represent the actual behavior of the material. Then individual atoms or molecules become important. What is actually observed is the result of properties of a few atoms or molecules. The properties of a small number of isolated atoms or molecules are expected to be different from those of conglomerates of a large number of these entities.

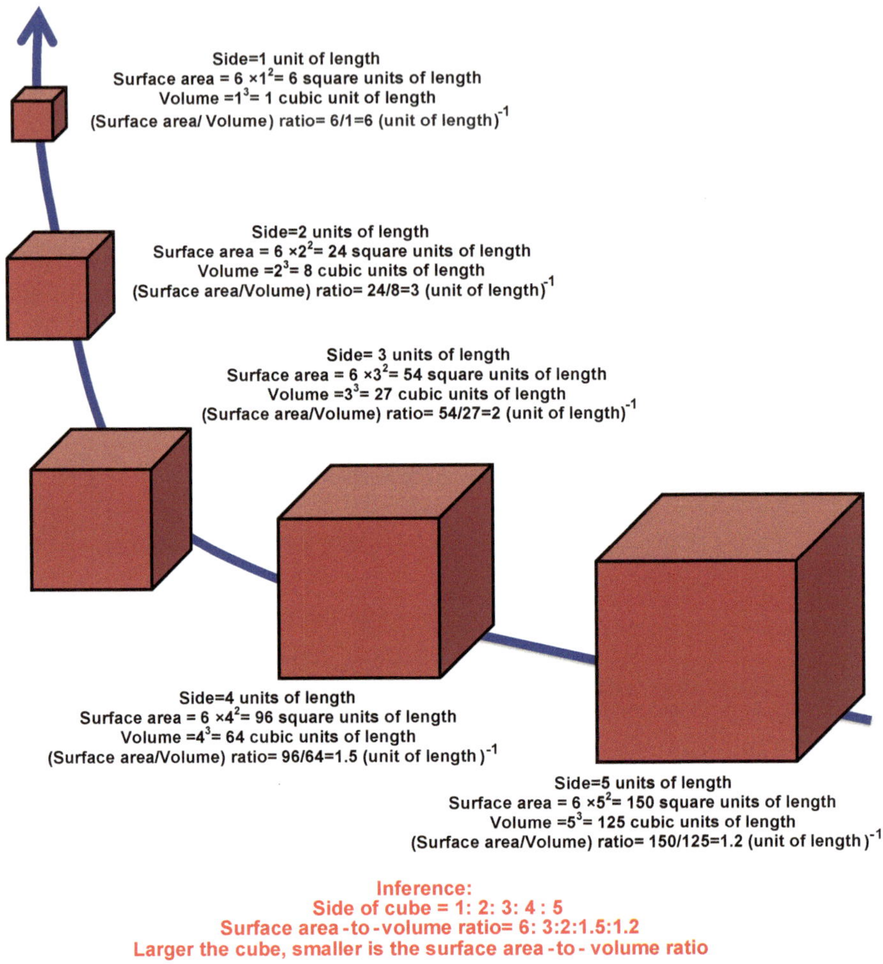

Fig. 2.1 Illustrating the impact of size of a body on the numerical value of surface area-to-volume ratio for the body

2.6.2 *Surface-Area-to-Volume Ratio*

Compared to the bulk form of matter, in the nanoworld, the same mass of material that has a comparatively larger surface area. To give an example (Fig. 2.1), let us consider a given mass of a material that has the shape of disk of diameter $(d_1) = 1$ cm and thickness $(t_1) = 1$ mm. Then its surface area $A_1 = \pi(d_1^2/4) = 3.14\,(1^2/4) = 0.785$ cm^2. Its volume $V_1 = \pi(d_1^2/4)\,t = 3.14\,(1^2/4) \times 0.1 = 0.0785$ cm^3. If now this disk is beaten to form a disk of thickness 10 nm $= 10 \times 10^{-9} \times 100$ cm $= 1 \times 10^{-6}$ cm, then for the same volume, $0.0785 = \pi(d_2^2/4)\,t$ where d_2 is the diameter of the thin disk formed by beating.

The value of d_2 is found to be = $\sqrt{\{(0.0785 \times 4)/(3.14 \times 10 \times 10^{-9} \times 100)\}}$ = 316.23 cm. The surface area of this thin disk is obtained as $A_2 = \pi(d_2^2/4)$ = 3.14 $(316.23^2/4)$ = 78501.11. Thus for the same volume 0.0785 cm^3, the surface area increases by the factor = 78501.11/0.785 = 100001.41 times.

A consequence of the relatively larger surface area is that the properties of a material in nanoscale are enormously different from those in its bulk form, e.g., an inert material in bulk state may exhibit pronounced catalytic properties when reduced to nanodimensions.

2.7 Nanoelectronics

Nanoelectronics = Nano + Electronics. Electronics is concerned with: (i) controlling the flow of electrons through vacuum, inert gas ambient or a semiconductor in solid state to build devices such as diodes and transistors, and with (ii) the design and assembly of circuits using the components fabricated in step (i) for performing assigned functions in information processing, computing, communication, and power conditioning. Nanoelectronics is devoted to the application of nanoscience and nanotechnology to electronics. Briefly stated, it represents the use of nanoconcepts and methods in electronics. It entails the design, fabrication, and applications of electronic devices, circuits, and systems whose building block components are of nanoscale size (Fig. 2.2). It aims to increase the capabilities of customary electronic devices by reducing their size, weight, and power consumption. It aspires to shrink the size of transistors in integrated circuits and increase the density of memory chips with a projected density far ahead of today's systems. The purpose is to stuff the functionalities of present-day equipment such as computers into the palm of a hand. Apart from the above, display screens on electronic devices are being improved to make them comparatively thinner and lighter than they are now. Nanoelectronics is strongly pushing the miniaturization of devices to the extent of hitting their basic limitations from physics viewpoint.

As commonly known, Moore's law predicts that the number of transistors/in.^{2} in an integrated circuit doubles every year. We shall discuss more this law in later chapters. Three distinct sub-domains of nanoelectronics are identified. These are referred to as [3]: More Moore, More-than-Moore, and Beyond CMOS. The phrase "More Moore" covers the capabilities achieved as packing density of devices continues to increase in compliance with Moore's law. By "More than Moore" is meant the incorporation of additional functionalities such as sensors, RF, and power conditioning circuits, which do not scale in accordance with Moore's law. Under "Beyond CMOS" subheading fall the newer devices like single-electron transistors and molecular electronic devices, which are likely to take over from CMOS for achieving higher levels of integration than possible with CMOS.

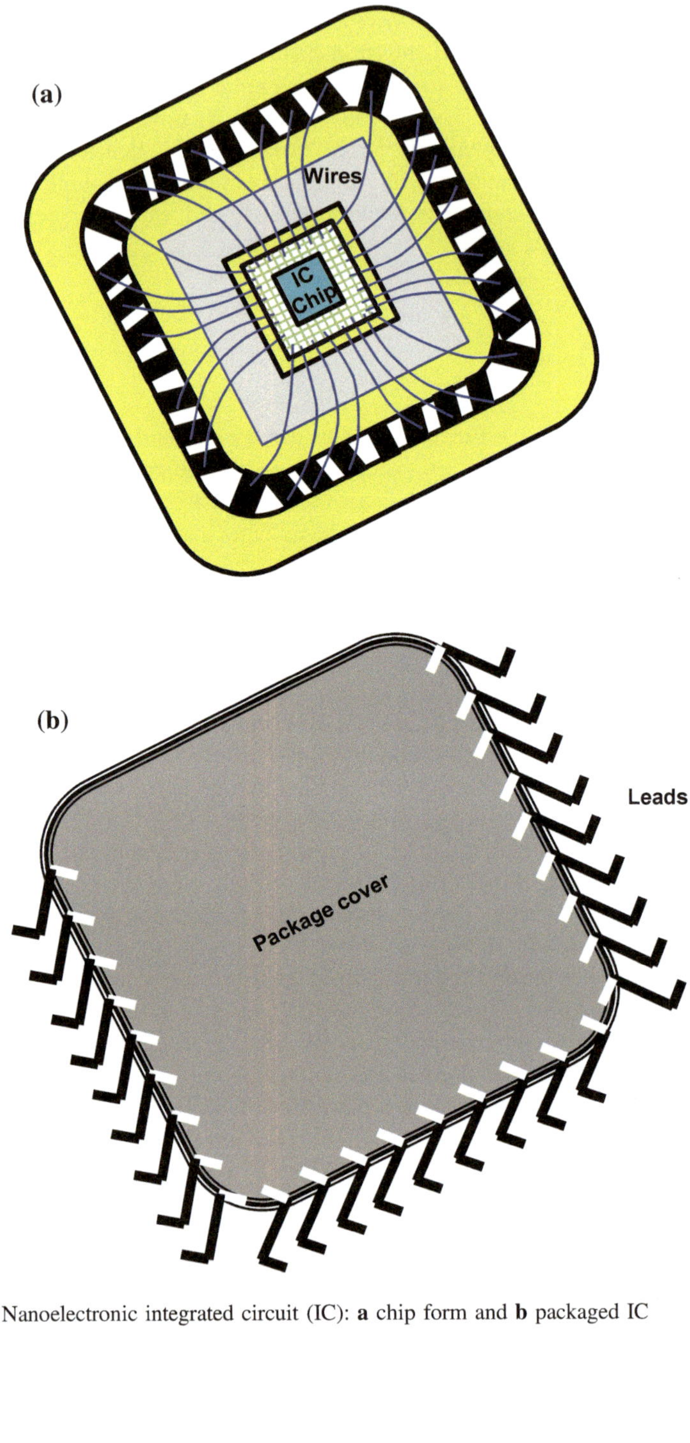

Fig. 2.2 Nanoelectronic integrated circuit (IC): **a** chip form and **b** packaged IC

2.7.1 More Moore Sub-domain

Building circuits with nanoscale components results in component counts reaching giga-scale magnitudes. Sophisticated CMOS technologies will be further improved to minimize cost per unit function. This will affect 70% of the market comprising digital logic circuits, memory, and microprocessor chips. The route leading to giga-scale complexity is the "More Moore" road.

2.7.2 More-than-Moore Sub-domain

Micro- and nanoelectronic devices of non-digital type are included. Notable devices consist of sensors and actuators for mechanical, thermal, chemical, and biological signals along with signal conditioning circuits on CMOS substrate. Other examples are micro- and nanofluidic devices and biosensors, radio frequency devices and circuits, power switching devices and circuits, light-emitting diodes and driving circuits, ultrasonic transducers and other imaging devices with associated circuitry, energy harvesters and ancillaries. Thus "more than Moore" sub-domain is essentially a technological synthesis between purely electronic devices and circuits with mechanical, and biochemical devices along with analog/RF circuits.

2.7.3 Beyond CMOS Sub-domain

This sub-domain comprises electronics based on new state variables such as spin, molecular state, photons, etc. Examples are spintronics, molecular electronics, etc. Thus, "beyond CMOS" sub-domain will bring fundamentally new nanoscale devices into nanoelectronics.

2.7.4 Convergence of Nanosciences

Nanoelectronics can be considered as the core subject, embroidered with the nanotechnology-related ingredients contained in more Moore, more-than-Moore, and beyond-CMOS sub-domains. A closer introspection reveals that sister branches of nanomagnetics, nanophotonics, nanomechanics, and nanobiotechnology should be fused with nanoelectronics in the form of branches augmenting its capabilities (Fig. 2.3). By doing so, useful information relevant to all the above domains is unified with nanoelectronics except for the RF and power conditioning portions. Therefore, all the nanoingredients of the aforesaid sub-domains will be covered under the umbrella of nanoelectronics and complementing nanosciences. In the

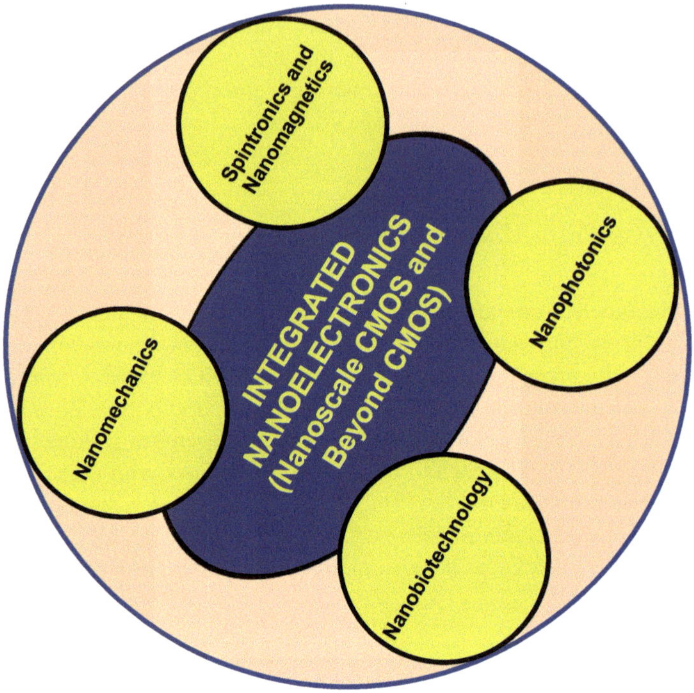

Fig. 2.3 Nanoelectronics and allied nanosciences

forthcoming subsections, we shall look at the sister branches referred to above, the so-called complementary nanosciences.

The goal is to provide a cohesive panoramic overview of an interdisciplinary field, which results from the merger of allied nanoscience disciplines, carefully appreciating their shared and unshared features. At the grass roots, all these disciplines assist each other to construct a holistic nanoscience. Instead of looking at them separately, a blended perspective is needed to understand the correct picture. An inter-diffusing perspective of the scenario will hasten progress and help in designing applications, which hitherto could not be contemplated. By cooperative interaction of the participating nanosciences, results of greater impacts will be achieved than by their individual use. This is an effort which is said to be "synergistic".

2.8 Spintronics and Nanomagnetics

Spintronics deals with the utilization of erstwhile-ignored property of an electron, namely its spin characteristic, for information processing. It is subdivided into metallic, semiconductor, and insulator spintronics. Both spin and charge properties of electron are gainfully used in spintronics.

Regarding nanomagnetics, we recall that magnetism is the study of physical phenomena related to movement of electric charges, which produces forces of attraction or repulsion between objects, especially between iron and certain materials. Nanomagnetics = Nano + magnetics. Nanomagnetics is the scientific study pertaining to nanomagnetism, which is the branch of magnetism dealing with low-dimensional systems that have at least one dimension in the nanoscopic range. These systems exhibit different behaviors from those in the bulk, with regard to magnetic ordering, magnetic domains, magnetization reversal, etc. The differences originate from various factors such as: (i) broken translation symmetry in the nanometric regime; (ii) from the higher percentage of atoms on the surface; (iii) the comparable sizes of nanoscopic objects to some fundamental or characteristic lengths of the constituent materials. Nanomagnetism finds practical applications, encompassing fields from geology to magnetic recording, from ferrofluids (colloidal liquids that become strongly magnetized in a magnetic field) used in loudspeakers to small particles used in medicine for targeted drug delivery to specific organs and tissues [4].

2.9 Nanophotonics or Nano-optics

Nanophotonics is the combination of photonics with nanotechnology. The term "photonics" has originated from the Greek word "photo", which means light. Photonics is the study of light whose fundamental constituent particle is the discrete packet or quantum of energy known as the photon. Photonics performs operations on light which are similar to those carried out by electrons in electronics, i.e., the role of photons in photonics is identical to that of electrons in electronics. Many tasks accomplished by photonics have their corresponding analogs in electronics, e.g., information processing, its transmission to remote locations and reception from these locations, etc.

A comprehensive field is optics. It is the branch of physics and engineering dealing with the study of the behavior and properties of light. The scope of optics includes the propagation of light, its deflection at interfaces, and interactions with different forms of matter. Optics studies light in a classical formalism as comprising rays traveling in straight lines (geometrical optics). It discusses about reflection and refraction of light. The wave theory of light is applied to explain diffraction of light. Light is treated as an electromagnetic wave. Photonics deals with quanta of light which are not considered in optics.

Nanophotonics is an offshoot, which has sprouted from a combination of photonics, optics, optical engineering, electrical engineering, and nanotechnology. It is a generic technology dealing with the study of the behavior of light on the nanometer scale, and of the interaction of light with nanoscale structures. At this scale, their structural, physical, and optical features are drastically modified relative to bulk counterparts. Nanophotonics has been extensively explored for unveiling and exploiting light-matter interactions. Of particular interest are those occurring at a scale below the diffraction limit of light, representing the boundary of conventional photonics [5]. Nanophotonics can provide high-speed and large-bandwidth, optoelectronic components of ultra-small size. Hence, it promises to revolutionize data storage, computation, telecommunications, and sensing fields.

2.10 Nanomechanics

Nanomechanics = Nano + mechanics. Mechanics is a branch of physics and mathematics. It deals with the motion or displacements of material objects, and the analysis of forces responsible for the same.

Nanomechanics is the study of the mechanical, i.e., elastic, thermal, and kinetic properties of nanostructures and nanomaterials. It involves classical mechanics, solid-state physics, statistical mechanics, materials science, and quantum chemistry. Its subbranches are: nanotribology, nanoelectromechanical systems (NEMS), and nanofluidics.

Nanotribology is the investigation of interfacial processes that take place on molecular and atomic scales. Some of the processes studied are adhesion, friction, viscous drag, scratching, wear, etc. A vital process is nanoindentation, which performs mechanical characterization of the surface by making a small notch or recess in nanometer range. Also examined is thin film lubrication at two interacting surfaces in relative motion. It may be noted that "tribology" is derived from two Greek words: "tribo" and "logy", "tribo" meaning rubbing and "logy" knowledge. Therefore, nanotribology is the study of friction, wear, contact mechanics, and lubrication at atomic length and time scales [6]. It finds applications in magnetic storage devices such as computer hard disk drives (HDD). Another process in which it is utilized is chemical mechanical polishing (CMP) for planarization in semiconductor fabrication.

Nanoelectromechanical systems (NEMS) are scaled-down versions of microelectromechanical systems (MEMS) [7, 8]. This downscaling is done to sub-micrometer dimensions. NEMS are regarded as the logical miniaturization step succeeding MEMS. These systems are made from electromechanical devices having critical dimensions from 100 to a few nm. NEMS-based devices can have fundamental frequencies $\sim$100 GHz. Their mechanical quality factors are $\sim 10^3$ to 10^5. They can have active mass in the range of 10^{-15} g. They exhibit force sensitivity $\sim 10^{-18}$ N and mass sensitivity $\leq 10^{-18}$ N. Heat capacities $<10^{-24}$ cal are obtainable. Typical power consumption is $\sim 10 \times 10^{-18}$ W. Integration level in these systems approaches 10^{12} elements cm^{-2}. These systems have immense applications as force sensors, chemical sensors, biological sensors, and ultra-high frequency resonators [Ke].

Nanofluidics is the examination and application of fluid flow in and nearby objects of nanoscale sizes [9]. The transport of fluid in and around objects with at least one characteristic dimension below 100 nm facilitates the occurrence of unusual, distinctive phenomena. Such phenomena are not observed at macrofluidic or microfluidic size scales [10].

Nanofluidic structures are used in circumstances where the specimens must be handled in tremendously small quantities. Nanofluidics is used in clinical diagnostics in lab-on-a-chip devices. It is also applied to nano-optics. Tunable microlens array is thereby made.

2.11 Nanobiotechnology

Nanobiotechnology = Nano + biotechnology. Biotechnology is the exploitation of living organisms, processes, and systems in industrial plants to manufacture products for upgrading the quality of human life. Nanobiotechnology applies nanotechnology to biological sciences to unify the design of materials and devices with the inimitable specificity afforded by biomolecules. In this way, new biomaterials are designed. Sensors working on the changes in conformation of biomolecules are fabricated. More effective particles for drug delivery are produced.

The difference between nanobiotechnology and bionanotechnology must be emphasized. Nanobiotechnology exploits the advancements in nanotechnology for improving biotechnology. Bionanotechnology utilizes the advantages of natural or biomimetic systems. It designs and produces novel nanoscale structures. An intense bilateral exchange of expertise across the precincts of these fields is taking place. It is focused around new materials and tools, predominantly from the physical sciences. It employs new phenomena, largely from the biological sciences. In these negotiations, the physical sciences put forward tools for synthesis and fabrication of devices. These tools are used for characterization of cells, subcellular components, and materials used in cellular and molecular biology. In turn, the biological sciences display the elegant group of prevailing functional nanostructures [11].

2.12 Discussion and Conclusions

Preliminary terms of nanoscience and nanotechnology were defined in this chapter. The vast scope of the subject of the book was introduced. The book seeks to develop the central theme of nanoelectronics aided by complementary nanosciences of nanomagnetics, nanophotonics, nanomechanics, and nanobiotechnology. The roles of different nanosciences were described in terms of the sub-domains defined according to Moore's law. With this notion of burgeoning nanoelectronics tree and its branches, the reader will be taken to further journey in this escalating field.

Review Exercises

2.1 Explain the meaning of the word "nano". What is a nanometer? Give two examples, which will help in imagining how large is a nanometer?

2.2 Define nanoscience and nanotechnogy. In these definitions, can the prefix "nano" be assigned to any physical parameter? If not, what is the restriction?

2.3 Define nanomaterial. Give two examples of nanomaterials and cite two applications of these materials.

2.4 Why are properties of nanomaterials different from those of bulk matter? Give two examples in which a difference of properties is observed as one goes to nano dimensions.

2.5 Correct the statement: As one moves to the nanoscale, copper particles become soft, gold particles look green, and silver particles appear blue in color.

2.6 A cube of side 1 cm is beaten to form a sheet of thickness 1 nm. By what factor will its surface area increase?

2.7 What is the difference between electronics and nanoelectronics? What are the advantages gained in moving to the nanoscale in electronics?

2.8 Explain the concepts of the sub-domains: (i) More Moore, (ii) More-than-Moore and (iii) Beyond CMOS.

2.9 What property of an electron is utilized in spintronics besides its electrical charge?

2.10 What is nanomagnetics? What are its applications?

2.11 How does photonics differ from optics? Define nanophotonics. What are its applications?

2.12 What is nanomechanics? Name its sub-branches.

2.13 What kinds of processes are studied in nanotribology? Where is nanotribology used?

2.14 Explain the following terms and give their applications: (i) nanoelectromechanical systems and (ii) nanofluidics.

2.15 What is nanobiotechnology? How does nanobiotechnology differ from nanotechnology?

References

1. Nano Werk Ten things you should know about nanotechnology: definition and usage of the term, Copyright ©2015 Nanowerk. http://www.nanowerk.com/nanotechnology/ten_things_you_should_know_2.php, Accessed 22 Aug 2015
2. Auffan M, Rose J, Bottero J-Y et al (2009) Towards a definition of inorganic nanoparticles from an environmental, health and safety perspective. Nat Nanotechnol 4:634–641
3. Jammy R (2010) More moore or more than moore? In: Sematech symposium, Taiwan, September 7, Copyright ©2010 SEMATECH, Inc., pp 1–39. https://www.inf.pucrs.br/~moraes/prototip/artigos/moreMoore.pdf. Accessed 22 Aug 2015
4. Guimarães AP (2009) Principles of Nanomagnetism, Chap. 1: The basis of nanomagnetism. Springer, pp 1–20
5. Naruse M, Tate N, Aono M et al (2013) Information physics fundamentals of nanophotonics. Rep Prog Phys 76(5):056401. doi:10.1088/0034-4885/76/5/056401
6. Bhushan B, Israelachvili JN, Landman U (1994) Nanotribology: friction, wear and lubrication at the atomic scale. Nature 374:607–616
7. Roukes ML (2000) Nanoelectromechanical systems. Technical digest of the 2000 solid-state sensor and actuator workshop, Hilton Head Isl., SC, 6/4-8/2000, pp 1–10
8. Ke C, Espinosa HD (2005) Nanoelectromechanical systems and modeling. In: Rieth M, Schommers W (eds) Handbook of theoretical and computational nanotechnology, vol 1, pp 1–38
9. Eijkel JCT, van den Berg A (2005) Nanofluidics: what is it and what can we expect from it? Microfluid Nanofluid 1:249–267
10. Schoch RB, Han J, Renaud P (2008) Transport phenomena in nanofluidics. Rev Mod Phys, 839–883
11. Whitesides GM (2003) The 'right' size in nanobiotechnology. Nat Biotechnol 21:1161–1165

Chapter 3
Nanomaterials and Their Properties

Abstract The melange of definitions of nanomaterials is discussed. Terminology laid down by the International Organization for Standardization (ISO) and European Commission (EC) concerning nanomaterials is described. Ultrafine grained materials with grain size in nanoscale range show unusually higher mechanical strength than coarse-grained materials. Two vital characterizing parameters representing the degree of dominance of surface effects in materials are dispersion and coordination number. Due to predominance of surface effects, nanoparticles are efficient catalytic agents. Melting points of these particles are lower than those of the bulk material, and phase transitions are hazily defined. The onset of quantum size effect in nanomaterials depends on the dimension of the nanomaterial compared to exciton Bohr radius. Due to quantum confinement, the bandgap of a semiconductor nanocrystal is wider than that of the bulk semiconductor. Dependence of bandgap on nanocrystal size leads to emission of light of different wavelengths from these quantum dots. In metals, interaction of light with surface plasmons leads to resonance oscillations at particular frequencies, thereby producing different color effects. Notable magnetic properties of nanomaterials include the display of superparamagnetic behavior, the exhibition of magnetism in materials that are generally believed to be nonmagnetic, and the giant magnetoresistance effect.

3.1 Bewilderment from a Multitude of Nanomaterial Definitions

A clear-cut definition of "nanomaterial" cannot be introduced straightforward. A multiplicity of definitions can be found in the literature. Each definition is valid within the limited parlance of a particular sector or organization. Hence, each definition forestalls the power and strength of any other definition. An internationally harmonized scenario cannot be seen.

Many modern consumer end products utilizing nanomaterials have originated from existing products by incorporating nanomaterials into solid, viscous, or liquid matrices. So, a variety of prevailing definitions leads to the circumstances that the

V.K. Khanna, *Integrated Nanoelectronics*,
NanoScience and Technology, DOI 10.1007/978-81-322-3625-2_3

same substance is agreed to be a nanomaterial under the legislation of a governing body but excluded by another body. Such conflicting situations have led to the great puzzlement of consumers, industries, and law enforcing agencies. Integrity of the market has to be assured. Bogus claims must be thwarted. It is therefore essential that a universal definition should be framed, which is acceptable not only nationally but in global market.

3.2 ISO (International Organization for Standardization) Definitions

3.2.1 *Nanomaterial*

It is a generic term [1]. It includes nano-object and nanostructured material. It applies to a material either having any external dimension in the nanoscale or having its internal structure/surface structure in that scale. A nano-object can be nanostructured.

3.2.2 *Nanoscale*

This is the scale stretching over the size range from circa 1 to 100 nm. In this size range, the material typically but not wholly, displays such properties, which are not clearly evident and deduced from the properties exhibited by it in a larger size. The size limits are considered approximate for appearance of such properties. Further, the lower limit in this definition (~ 1 nm) is purposefully laid out. It is meant to keep away from designation of single and small groups of atoms as nano-objects or elements of nanostructures. These possibilities exist in the absence of a lower limit.

3.2.3 *Nano-object*

A generic term applicable to all discrete nanoscale objects, it indicates a material possessing one, two or three external dimensions in the nanoscale.

An established general definition for particles is related to nano-objects.

3.2.3.1 Particle

A particle is a minute portion of matter. This portion must have defined physical boundaries. An interface is also considered to be a physical boundary. A particle can move as a complete unit in itself. This generalized definition of particle is valid for nano-objects.

For particles, which are clustered together in agglomerates and as aggregates, the definitions are:

3.2.3.2 Agglomerate

The agglomerate is a collection of particles, which are loosely bound to each other. It may also be a blending of particles and aggregates (defined below). The mixture must conform to the condition that the external surface area of the agglomerate = the sum of the surface areas of the separate components.

3.2.3.3 Aggregate

An aggregate signifies a particle containing firmly attached or merged particles. The interparticle binding takes place in such a manner that the following condition is satisfied: external surface area of the aggregate is significantly < the sum of calculated surface areas of the individual components.

Agglomerates and aggregates are treated as "secondary" particles. This is done to tell them apart from the original individual particles, which are labeled as "primary" particles.

The vocabulary dealing with a few types of nano-objects is familiarized. It includes six distinct shapes together with a supplementary specific case of the quantum dot. Nanomaterials of different dimensions are illustrated in Fig. 3.1.

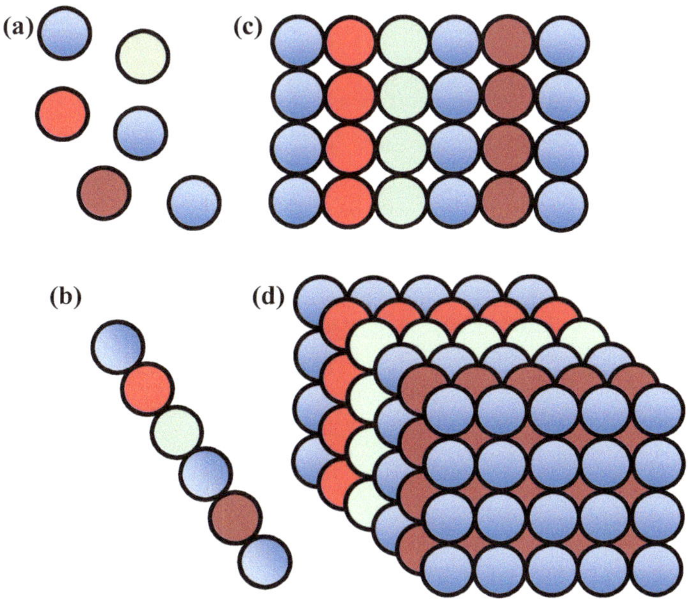

Fig. 3.1 Nanomaterials of different dimensionalities: **a** 0D, **b** 1D, **c** 2D, and **d** 3D

3.2.3.4 Nanoparticle

A nanoparticle is a nano-object. This nano-object has all three external dimensions in the nanoscale. In other words, a nanoparticle is a discrete nano-object. All three Cartesian dimensions of this nano-object are <100 nm.

3.2.3.5 Nanoplate

A nanoplate is a type of nano-object. It has one external dimension in the nanoscale. Its other two external dimensions are much larger. It is correct to refer to a nanoplate as a two-dimensional nano-object.

3.2.3.6 Nanofibre

A nanofibre is a kind of nano-object. It has two similar external dimensions in the nanoscale. Its third dimension is much larger. Nanofibre may be treated as a one-dimensional nano-object.

3.2.3.7 Nanotube

A nanotube is essentially a hollow nanofibre.

3.2.3.8 Nanorod

A nanorod is basically a solid nanofibre.

3.2.3.9 Nanowire

A nanowire is a nanofibre. It may be either electrically conducting or semiconducting.

The three nano entities, namely, nanotube, nanorod, and nanowire are all one-dimensional nano-objects.

3.2.3.10 Quantum Dot

It is a crystalline nanoparticle of a semiconductor material, i.e., a semiconductor nanocrystal. Often, the definition is relaxed to include nanocrystals of conductors such as aluminum. Its size-dependent properties arise from quantum confinement effects on the electronic states. To explain quantum confinement, let us note that:

When the size of the particle becomes comparable to the wavelength of the electron, the random motion of the electron is confined to discrete energy levels instead of energy bands. This effect of confinement or restriction of electron motion is called quantum confinement. The result of the discreteness of energy levels is that the forbidden energy gap is widened. Hence the energy bandgap increases.

Briefly, nanoparticles are materials that are nanoscale in all three dimensions. Thus quantum dots are nanoparticles. So are the colloidal solutions containing finely dispersed particles with linear dimensions <100 nm. Nanowires and nanotubes are materials that are nanoscale in two dimensions. But they are extended in the third dimension. Nanolayers, such as a thin films or surface coatings are materials that have one dimension in the nanoscale. But they are extended in the other two dimensions. Some of the geometrical features seen on computer chips belong to this class.

3.3 EC (European Commission) Definitions

3.3.1 Nanomaterial

There are two subclauses in the article containing the principal recommendation of the commission [2, 3]: (i) "Nanomaterial" means a naturally occurring material, or a parenthetically formed material or a material manufactured by some process. It contains particles, which are either present in an unbound state or as an aggregate or in the form of an agglomerate. One or more external dimensions of these constituent particles must be in the size range 1–100 nm for 50% or greater proportion of their total number with regard to distribution of sizes. (ii) In dealing with situations where safety, environment, health, or competition is of primary interest, the 50% threshold is superseded by a number x satisfying the inequality $1 \geq x \geq 50$.

It is further clarified that by exemption from subclause (i), fullerenes, graphene flakes, and single-walled carbon nanotubes, which have one or more external dimensions <1 nm, i.e., lower than the 1 nm limit, are to be considered as nanomaterials. For explanation of subclause (i), following definitions are laid down:

3.3.2 Particle

It is a minute chunk of matter whose boundaries are precisely and unerringly specified.

3.3.3 Agglomerate

It is a gathering of particles, which are either weakly bound or aggregates. The resulting surface area of these particles equals the sum of the surface areas of the independent components.

3.3.4 Aggregate

It comprises particles that are strongly bound or merged together.

Every time it is scientifically possible and whenever entreated in legislature, obedience of subclause (i) is evaluated from specific area per unit volume. This parameter should be >60 m^2 cm^{-3}. Nonetheless, if a material is judged to be a nanomaterial from subclause (i), it is a nanomaterial. This is acceptable even if its surface area per unit volume is <60 m^2 cm^{-3}.

The definition verbalized by EC is more specific and less ambiguous than that prescribed by ISO. This is because it is formulated for legal purposes. Several methods can be applied for measuring the sizes of nanoparticles: (i) Ensemble methods: These methods measure large numbers of particles simultaneously. Dynamic light scattering, small-angle X-ray scattering, and X-ray diffraction are three such methods. (ii) Counting methods: One such method is particle tracking analysis. Imaging techniques, e.g., electron and atomic force microscopies also fall in this class. (iii) Fractionation methods: These include centrifugal liquid sedimentation, field-flow fractionation, size exclusion chromatography, etc.

In applying these methods, three difficulties are faced: First, measurements on the constituent particles inside aggregates are complicated. Second, the experimentally measured signals cannot be straightforwardly correlated with number size distributions. Third, it is not easy to detect and count particles of <10 nm size. None of the present methods can ascertain the fulfillment of the criteria for being a nanomaterial when all the kinds of potential nanomaterials are taken into account.

Thus far, the focus in this chapter has been towards laying down clear-cut definitions on nanomaterials. From the next section onwards, let us divert our attention to seek the underlying causes for the divergence in behavior of nanomaterials from matter in bulk state.

3.4 Mechanical Strength of Nanomaterials

Bulk materials have a grainy structure. Randomly oriented grains are interconnected by grain boundaries. Dislocations are line defects in the material where the atoms are abnormally placed in the crystal. They are either of the two types: edge and screw dislocation. When a stress is applied to the material, the dislocations move resulting in plastic deformation. Therefore, in order to produce a strong material, the motion of dislocations must be opposed. Hindrance to dislocation movement is offered by grain boundaries. They try to block the motion of dislocations. This implies that a material containing more grain boundaries will be more capable of impeding the dislocations. Such a material is one containing more grains. To produce more grains in the same material, one must reduce the grain size. Hence, smaller is the grain size, more is the resistance to dislocation movement and as a result, stronger is the material. For bulk materials, dependence of the yield stress σ_y on the grain diameter d is expressed by the Hall-Petch equation

$$\sigma_{y=}\sigma_0 + k_y/\sqrt{d} \tag{3.1}$$

where σ_0 and k_y are constants for the material. The constant k_y is called the strengthening coefficient. Theoretically, one would expect that the yield stress will become infinite as the grain size approaches zero. But this does not happen. The yield stress attains a maximum value at a grain size of $\sim$10 nm. Grains of size <10 nm are prone to another mechanism known as grain boundary sliding.

3.5 Characterizing Parameters for the Influence of Surface Effects on Material Properties

In materials science, two parameters are of paramount importance to get an idea about the extent or degree of surface effects in controlling the properties of the material. These parameters are dispersion and coordination number. Dispersion of a material is the fraction of atoms exposed to the surface. In other words, it is the ratio of surface to bulk atoms expressed as the number N_S of atoms located on the surface divided by the total number N_V of atoms present in the given volume of the material. Hence, dispersion D is given by

$$D = N_S/N_V \tag{3.2}$$

D is written in terms of the surface area S and the volume V of the material as

$$D = S/V \tag{3.3}$$

In case a spherical particle of radius r is under consideration, the dispersion D_{sphere} is obtained as

$$D_{\text{sphere}} = \frac{4\pi R^2}{(4/3)\pi R^3} = \frac{3}{R} \tag{3.4}$$

As the radius of the particle decreases, the dispersion increases. Hence, the surface effects begin dominating.

Coordination number $\langle N_C \rangle$ of an atom in a molecule, ion, or crystal is the total number of its neighboring atoms. It is obtained by counting the bonds of an atom with neighboring atoms. Surface coordination number $\langle SN_C \rangle$ refers to the number of atoms associated with an atom located at the surface. Bulk coordination number $\langle VN_C \rangle$ signifies the number of atoms adjoining an atom situated in the interior of the material. Needless to say that surface coordination number is smaller than the bulk coordination number because an atom at the surface is surrounded by atoms on three sides only: left, right, and bottom whereas an atom inside the material is surrounded by atoms on all four sides: left, right, top, and bottom, by other atoms.

On decreasing the size of particles, their dispersion, i.e., the fraction of surface atoms increases. As these surface atoms have smaller coordination number than bulk atoms, surface phenomena dominate over bulk phenomena in these small size particles.

3.6 Catalytic Effects of Nanomaterials

From the coordination number viewpoint, considering a cubic crystal, the atoms at the corners of the cube have the lowest coordination number. Therefore these atoms have unsatisfied bonds. The atoms with unsaturated bonds show the greatest tendency to combine with adsorbate molecules. After the corner atoms come the atoms at the edges of the cube. Their coordination number is higher than that of the corner atoms whereby they are less active in adsorption activities as compared to the corner atoms. At the last place are the atoms on the plane faces of the cube, which are least active due to their high coordination number than previous two classes of atoms. These phenomena indicate that smaller the coordination number, more active is the material in surface exchanges and hence catalytic activity. The small coordination number of nanoparticles considerably improves their catalytic behavior. Gold nanoparticles of size ~2–3 nm are good catalytic agents, losing their noble behavior which has prompted its widespread use as a non-tarnishable metal. Work function of a single platinum atom is 9 eV while that of bulk platinum is 5.3 eV. Acceptance or donation of charges depends on these values. Thus by changing the size of a congregation of particles, their chemical properties are favorably tuned.

3.7 Thermal Properties of Nanomaterials

3.7.1 *Melting Point Depression*

Due to the lower coordination numbers, the strength of stabilization of atoms on nanoparticle surfaces is drastically reduced. Hence, a lower temperature suffices to liberate the molecules from their bonds resulting in fall in melting points. A 2.5 nm

Au nanoparticle has a melting point of 930 K $\ll$ the melting point of bulk gold (1336 K). The melting point T_m of a collection of particles, each having a radius r, is related to the melting point T_M of the bulk material having the latent heat of fusion ΔH, by Gibbs–Thomson equation [4]

$$(T_m - T_M)/T_M = -2V\gamma/(r\Delta H) \quad (3.5)$$

where V denotes the volume occupied by 1 mole of the liquid, i.e., its molar volume, and γ stands for the tension existing at the boundary between the solid and liquid, viz., the interfacial tension. Besides lowering of melting point, another conspicuous effect is that the phase transition in a collection of a smaller number of particles loses sharpness. It is rather ill defined. Some portion of the collection is solid while the remaining portion is liquefied.

3.7.2 *Negative Thermal Capacity*

An interesting phenomenon that occurs with such small groups of atoms is negative thermal capacity. Negative thermal capacity arises from the fact that temperature is an indicator of kinetic energy whereas thermal capacity is related to total energy. When a portion of kinetic energy is converted into potential energy, a decrease in temperature is observed. This may be understood with reference to S8 ring and S8 linear chain structures. On application of heat to S8 ring, vibrational and rotational motions increase in amplitude and temperature rises. But when heat is localized to break a bond to convert a hot S8 ring to the cold S8 linear chain, the temperature falls.

3.8 Exciton Bohr Radius: A Characteristic Length for Quantum Confinement

The occurrence of the quantum size effect begins as soon as the size of the nanostructure becomes smaller than a characteristic length called the exciton Bohr radius. The exciton is a quasi-particle defined as a localized, electrically neutral bound state of an electron–hole pair. This pair is held together through attraction by Coulomb electrostatic force. Bohr radius a_H is the average radius of an electron orbiting around the nucleus of a hydrogen atom in its lowest energy level. It is based on the Bohr model of the atom, and is given by

$$a_H = \hbar/(m_0 c\alpha) \quad (3.6)$$

where $\hbar$ is reduced Planck's constant = 1.05×10^{-34} J s, m_0 is the rest mass of the electron = 9.11×10^{-31} kg, c is the velocity of light = 3×10^8 m/s and α is fine structure constant = 7.3 × 10^{-3}, giving a_H = 0.053 nm. The exciton Bohr radius $a_{exciton}$ is the distance between the electron and the hole in an exciton. It is expressed as a function of a_H as

$$a_{\text{exciton}} = a_H \varepsilon_r m_0 / \mu \tag{3.7}$$

where ε_r is the relative permittivity of the material and

$$\mu = m_e^* m_h^* / \left(m_e^* + m_h^*\right) \tag{3.8}$$

In this equation, m_e* is the effective mass of the electron and m_h* is the effective mass of the hole. For common semiconductors,

$$50\,\text{nm} \leq a_{\text{exciton}} \leq 2\,\text{nm} \tag{3.9}$$

This means that quantum size effect is likely to take place in these materials in the distance range ~2–50 nm. From the viewpoint of dimensions, there are three kinds of confined structures, namely, (i) a quantum well in which the size of the material is reduced in one direction only so that the exciton can move freely in the other two directions; (ii) a quantum wire in which the material size is decreased in two directions so that free exciton movement is possible in only one direction; and (iii) a quantum dot in which the material is squeezed in all the three directions whereby there is no direction of free movement and the exciton's motion is restricted in all the directions. None of these restrictions applies in an infinite or bulk material.

3.9 Electronic and Optical Properties of Nanomaterials

3.9.1 *Bandgap Broadening of a Spherical Semiconductor Nanocrystal: The Quantum Dot*

In a bulk semiconductor crystal, the motion of electrons in a periodic crystal lattice is described by the Bloch wave function $\psi_{\text{Bloch, bulk}}(x)$. The Bloch wave function at a position x is given by

$$\psi_{\text{Bloch, bulk}}(x) = \exp(ikx) u_k(x) \tag{3.10}$$

where k is the wave vector of the crystal. The function u_k in the Bloch wave function is a periodic function. This function has the same periodicity as the crystal. Hence, we can write

$$u_k(x) = u_k(x+a) \tag{3.11}$$

The symbol a represents the lattice constant. Physically, the Bloch wave function $\psi_{\text{Bloch, bulk}}(x)$ is a plane wave exp (ikx). This plane wave is modulated by a periodic function $u_k(x)$.

In a semiconductor nanocrystal or quantum dot, the Bloch wave function is corrected for the spatial confinement of the charge carriers and the exciton. This correction is done by multiplying the Bloch wave function $\psi_{\text{Bloch, bulk}}(x)$ by the envelope function $\phi_{\text{env}}(x)$ to account for the confinement effects. Hence, the Bloch wave function in the nanocrystal $\psi_{\text{Bloch, nano}}(x)$ becomes

$$\psi_{\text{Bloch, nano}}(x) = \psi_{\text{Bloch, bulk}}(x) \times \phi_{\text{env}}(x) \tag{3.12}$$

The envelope is a solution of the Schrodinger equation for the problem of particle in a three-dimensional box. If the confinement is identical in all the directions, the nanocrystal is a spherical box representing a quantum dot. The solutions are

$$\phi_{\text{env}}(\theta, \phi, r) = Y_l^m(\theta, \phi)R(r) \tag{3.13}$$

where $Y(\theta, \phi)$ are spherical harmonics and $R(r)$ are Bessel functions. The solutions exhibit a striking similarity to the wave functions of the hydrogen atom. But in a hydrogen atom, the electron is constrained by the attractive force due to the positively charged nucleus containing a single proton. In a spherical nanocrystal, the electron is confined within the spherical potential well of diameter D. Substitution of Eq. (3.13) into the Schrodinger equation yields the solutions for the discrete energy levels of the electron restricted within the spherical potential well as

$$E_{n,l}^{\text{Conf}}(D) = 2\hbar^2\chi_{n,l}^2/(m^*D^2) \tag{3.14}$$

where m^* is effective mass of charge carriers (electrons or holes) and $\chi_{n,l}$ are the roots of the Bessel function, which are absolute values increasing with principal quantum number $n = 1, 2, 3, \ldots$ and secondary quantum number $l = 0, 1, 2, 3, \ldots$. The first energy level has the quantum numbers $n = 1$, $l = 0$. The second energy level has the quantum numbers $n = 1$, $l = 1$. Quantum numbers of the third energy level are $n = 1$, $l = 2$. Note that the restriction $l \leq (n - 1)$ for a hydrogen atom is inapplicable to the nanocrystal because of the difference in potential functions for the two cases.

The total bandgap of the nanocrystal = the fundamental bandgap of bulk semiconductor + the confinement energy of electrons + the confinement energy of holes, which is expressed as

$$E_{\mathrm{g}}^{\mathrm{total}} = E_{\mathrm{g}}^{\mathrm{bulk}} + 2\hbar^2\chi_{n,l}^2/\left(m_{\mathrm{e}}^*D^2\right) + 2\hbar^2\chi_{n,l}^2/\left(m_{\mathrm{h}}^*D^2\right) \tag{3.15}$$

Since the electron and hole are treated independently, it is implicitly assumed that the Coulomb interaction between them is inadequate for the formation of a bound exciton. This assumption is valid only when the radius a of the nanocrystal is $\ll$ exciton Bohr radius a_{exciton}. This is called the strong confinement regime. In this regime, the Coulomb energy is $\ll$ the confinement energy of the carriers. The kinetic energy of the charge carriers is $\gg$ Coulomb electrostatic energy. Hence, they are uncorrelated and independent. However, when $a > a_{\mathrm{exciton}}$, the confinement energy originates from the quantization of motion of the center of mass of the exciton. Instead of the electron in spherical potential, one has to consider an exciton in a spherical potential well. The discrete energy levels for the exciton are given by an equation similar to that for the electron. The solitary differentiating aspect of this circumstance is that the effective mass of charge carriers (electrons or holes) is replaced by the effective mass of the exciton. The confinement energy for a nanocrystal of radius a is given by [5]

$$E_{n,l}^{\mathrm{Conf}}(a) = \left\{\hbar^2\pi^2/\left(2a^2\right)\right\}\left\{\left(1/m_{\mathrm{e}}^*\right) + \left(1/m_{\mathrm{h}}^*\right)\right\} = \hbar^2\pi^2/\left(2\mu a^2\right) \tag{3.16}$$

Then the total bandgap of the nanocrystal reduces to the form

$$E_{\mathrm{g}}^{\mathrm{total}} = E_{\mathrm{g}}^{\mathrm{bulk}} + \hbar^2\pi^2/\left(2\mu a^2\right) \tag{3.17}$$

According to this equation, if the radius a of the nanocrystal decreases, i.e., for a smaller size nanocrystal, the second term increases and hence the bandgap of the nanocrystal is larger than the bandgap of the bulk semiconductor. Absorption spectra studies have shown that the energy of the exciton in the nanocrystal is shifted towards the blue color with respect to the bulk value resulting in the so-called "blue shift". This is because the blue color corresponds to higher frequency and hence larger energy.

Theoretical calculation based on effective mass approximation was presented by Brus who gave the equation [6]

$$E_{\mathrm{g}}^{\mathrm{total}} = E_{\mathrm{g}}^{\mathrm{bulk}} + \hbar^2\pi^2/\left(2\mu a^2\right) - 1.786q^2/\left(\varepsilon_0\varepsilon_{\mathrm{r}}a^2\right) \tag{3.18}$$

where the second negative term arises from the Coulomb attraction between the electron and the hole having charges $\pm q$. The numerical factor derives from the computations of overlap integrals of the wavefunction. Its value differs marginally for different materials due to variation of dielectric constant ε_{r}.

Another effect, which has been overlooked so far, is the electron–hole spatial correlation effect. Kayanuma treated this effect [7]. Its inclusion resulted in the equation

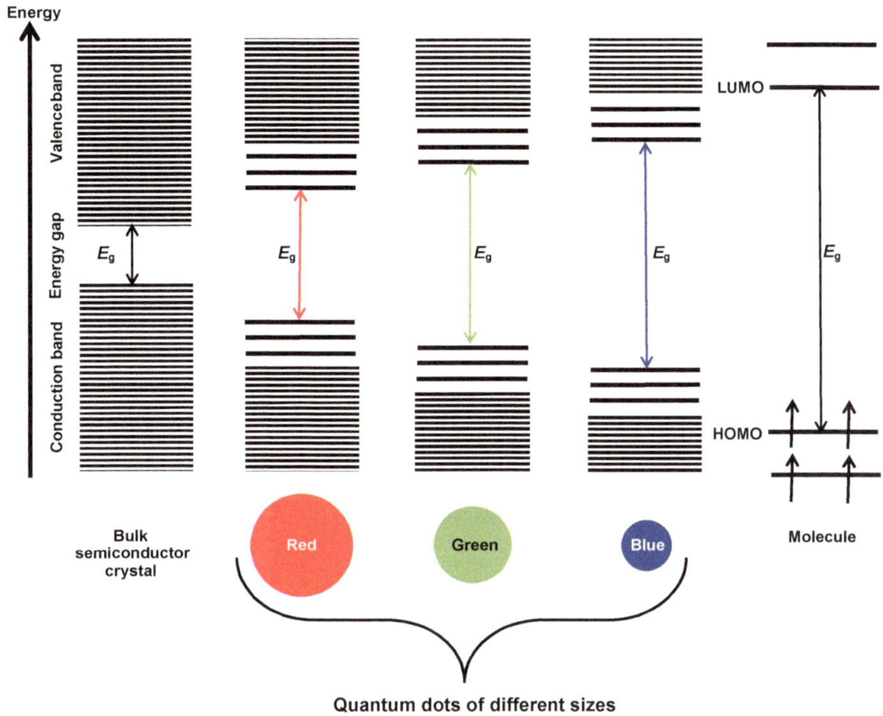

Fig. 3.2 Energy level splitting and bandgap widening with decreasing size of semiconductor crystal due to quantum confinement

$$E_g^{\text{total}} = E_g^{\text{bulk}} + \hbar^2\pi^2/(2\mu a^2) - 1.786q^2/(\varepsilon_0\varepsilon_r a^2) - 0.248E_{Ry}^* \qquad (3.19)$$

where E_{Ry}^* is the exciton Rydberg energy. The last subtractive term is consequential only for semiconductors having a small dielectric constant.

Figure 3.2 shows the effect of decrease in size of a semiconductor crystal on its energy band diagram.

3.9.2 Interaction of Light with Metallic Nanoparticles

In contrast to a semiconductor nanocrystal where light is absorbed/emitted according to the bandgap of the material, in a metallic nanoparticle a different phenomenon takes place. When the oscillating electric and magnetic fields in the electromagnetic wave representing the light beam pass near the free electrons of a metallic nanoparticle, they cause an oscillatory motion of the electronic charge resulting in the phenomenon called surface plasmon resonance in which the free electrons of the metallic nanoparticle collectively oscillate in synchronization with the frequency of the incident light. During surface plasmon resonance with 30 nm Au nanoparticles, the blue-green light of wavelength 450 nm is absorbed whereas red light ~700 nm is reflected so that the Au nanoparticles look reddish in color [8]. On increasing the size of Au nanoparticles beyond 30 nm, the red light is absorbed and blue light is reflected. Hence, the Au nanoparticles acquire a pale blue or purple color appearance. On decreasing the Au nanoparticle size to 2–5 nm, a yellow color is perceived. 40 nm Ag nanoparticles are blue while 100 nm Ag nanoparticles are yellow. Thus by varying the size of Au or other nanoparticles, their optical properties can be tailored in the desired manner. Colorimetric sensors based on Au nanoparticles exploit such color changes to assess the quality of food.

3.10 Magnetic Properties of Nanomaterials

3.10.1 Superparamagnetic Nanoparticles

Paramagnetic materials such as Mg, Mo, Ta, etc., are materials which experience a small force in an external magnetic field due to alignment of unpaired spins producing a feeble attraction. They lose their magnetic behavior after withdrawal of the field due to thermal agitation. In contrast, superparamagnetic iron oxide nanoparticles <50 nm in size contain a single ferromagnetic or ferromagnetic domain. These single domain magnets align their magnetic moments in the presence of the external magnetic field creating a strong attractive interaction. This interaction is much stronger than observed in a paramagnetic material. The magnetic susceptibility of nanoparticles is $\gg$ that of paramagnets; hence they are said to be superparamagnetic. On removing the magnetic field, the magnetization of nanoparticles exhibits random flipping under the influence of ambient temperature. Consequently, a net zero magnetic moment and zero residual magnetism are observed. The average time between flips is called Neel relaxation time. Generally, any ferromagnetic or ferrimagnetic material transfers to the paramagnetic state above a temperature known as Neel temperature but supermagnetic nanoparticles do so below the Neel temperature of the material. Thus superparamagnetic nanoparticles bridge ferromagnetism with paramagnetism. They share with ferromagnetism the property of reaching a high magnetization level in a low-intensity magnetic field but do not share with it the property of retention of magnetism after field removal. They share with paramagnetism the property of lose of magnetism after the external field is withdrawn but do not share with it the small magnetization generated in a paramagnetic material in the presence of the field.

3.10.2 Magnetism in Gold Nanoparticles

Bulk gold is a diamagnetic material. But 2 nm size gold nanoparticles coated with dodecanethiol ($C_{12}H_{26}S$) show ferromagnetic behavior. This magnetism is size-dependent. Its strength increases as the nanoparticles of larger diameters are used in the range 0.7–3 nm. The peak value occurs at 3 nm diameter. After 3 nm limit, the magnetism becomes weaker [9], showing properties like bulk gold. Spin-orbit coupling between the thiol compound bound to the surface with the surface atoms of gold nanoparticle is responsible for this magnetism. Magnetic moments produced by the spin polarization of surface atoms of Au nanoparticles are not quenched fully. In opposition, such quenching is achieved in bulk lattice. Thus stabilization of gold nanoparticles with thiols provides a route towards producing ferromagnetic properties.

3.10.3 Giant Magnetoresistance (GMR) Effect

It is a change in electrical resistance between two ferromagnetic materials separated by a thin layer of a nonmagnetic material having thickness in the nanoscale. In the absence of an external magnetic field, the magnetizations of the two magnetic layers are oriented in opposite directions. But on applying a magnetic field, they are aligned in the same direction. When these magnetizations are anti-directional, the resistance is maximum but when they become unidirectional, the resistance decreases to the minimum value. The effect is noticed only when the intervening nonmagnetic layer between the two magnetic layers has thickness in nanometer regime so that coupling between the magnetic layers can take place. The GMR is a quantum-mechanical effect.

3.11 Discussion and Conclusions

Definitions of nanomaterials were clarified. Associated terminology was discussed. Reasons for departure of nanomaterial properties from bulk behavior were described. Mechanical, chemical, thermal, optical, electronic, and magnetic properties of nanomaterials were treated. As nanomaterials differ in properties from the bulk materials, they may be considered as a separate class of materials with novel applications.

Review Exercises

3.1 Explain the problems arising from the multiplicity of definitions of a nanomaterial.
3.2 What is the range over which the nanoscale extends? What are the reasons for specifying 1 nm as the lower limit of this scale? What will happen if the lower limit is not defined?
3.3 What is a nano-object? Define a particle.
3.4 Distinguish between agglomerate and aggregate.
3.5 Define a nanoparticle. What is a nanoplate? Differentiate between nanotube and nanofibre.
3.6 What are nanorods and nanofibres?
3.7 What is a quantum dot? How does the size of a quantum affect its properties?
3.8 How does the EC definition of a nanomaterial differ from its ISO definition? Explain the subclauses (i) and (ii) in EC definition.
3.9 Name the three materials, which are exempted from subclause (i) in EC definition of nanomaterial.
3.10 What is the guideline regarding specific area per unit volume for classifying a material as a nanomaterial? Explain in what respects the EC definition is more unequivocal than ISO definition?
3.11 Name three methods of measuring the size of nanoparticles.
3.12 Why is a nanomaterial mechanically stronger than bulk material? Write Hall-Petch equation and explain the symbols used.
3.13 Define dispersion of a material. What is meant by coordination number of an atom in a material? How do these parameters help in explaining the surface effects in materials?
3.14 Differentiate the catalytic activity of nanoparticles from that of materials in bulk form.
3.15 Why does a nanomaterial show a lower melting point than a bulk material? Why is phase transition in the nanomaterials not as sharp as for bulk state?
3.16 Write the Gibbs–Thomson equation. What does this equation describe?
3.17 Explain the reason for negative heat capacity of a nanomaterial with an example.
3.18 What is an exciton? Define: (i) Bohr radius and (ii) Exciton Bohr radius. How are they related? What is the typical range of values of the exciton Bohr radius in common semiconductor materials?
3.19 Define: (i) quantum well, (ii) quantum wire, and (iii) quantum dot. How is the exciton motion restricted in these nanostructures?
3.20 Write the equation for the Bloch wave function in a bulk semiconductor and explain the symbols used. How is this equation modified for a nanocrystal?
3.21 How are the solutions of the Schrodinger wave equation for a nanocrystal similar to those for the hydrogen atom? In what ways do they differ? What is the impact of this difference on the restriction of values for the secondary quantum number for the quantum dot?

3.22 What is strong confinement regime for the charge carriers? What happens in the weak confinement regime? Write the equations for the confinement energy of a nanocrystal in both types of confinement. Explain the symbols used.

3.23 Write the Brus equation for the energy bandgap of a nanocrystal and explain the symbols used in this equation. What correction was introduced by Kayanuma in the Brus equation?

3.24 How does light interact with metallic nanoparticles? Why do gold nanoparticles of different sizes have different colors?

3.25 What is superparamgnetism phenomenon observed in nanoparticles? What are the common features between superparamagnetism and paramagnetism, and between superparamagnetism and ferromagnetism?

3.26 Is gold a magnetic material? Explain the origin of magnetic properties in gold nanoparticles coated with thiols.

3.27 What is giant magnetoresistance effect? Why is this effect called a nanoscale phenomenon?

References

1. Lövestam G, Rauscher H, Roebben G et al (2010) JRC reference reports: considerations on a definition of nanomaterial for regulatory purposes. Publications Office of the European Union, Luxembourg, pp 1–36
2. Lidén G (2011) Commentary: the European commission tries to define nanomaterials. Ann Occup Hygiene 55(1):1–5
3. Linsinger TPJ, Roebben G, Gilliland D et al (2012) JRC conference report: requirements on measurements for the implementation of the European commission definition of the term 'nanomaterial'. Publications Office of the European Union, Luxembourg. ©European Union 2012, pp 1–52
4. Roduner E (2006) Size matters: why nanomaterials are different. Chem Soc Rev 35:583–592
5. Koole R, Groeneveld E, Vanmaekelbergh D, et al (2014) Chapter 2 Size effects on semiconductor nanoparticles. In: de Mello Donega C (ed) Nanoparticles: workhorses of nanoscience. Springer, pp 13–51
6. Brus LE (1984) Electron–electron and electron–hole interactions in small semiconductor crystallites: the size dependence of the lowest excited electronic state. J Chem Phys 80(9):4403
7. Kayanuma Y (1988) Quantum-size effects of interacting electrons and holes in semiconductor microcrystals with spherical shape. Phys Rev B 38(14):9797–9805
8. Gold Nanoparticles: Properties and Applications. http://www.sigmaaldrich.com/materials-science/nanomaterials/gold-nanoparticles.html, Accessed 4 Feb 2016
9. Mujica V, Marquez M, Ratner MA (2007) Size dependence of ferromagnetism in gold nanoparticles: mean field results. Phys Rev B 76:224409-1–224409-6

Part II
CMOS Nanoelectronics

Chapter 4
Downscaling Classical MOSFET

Abstract The classical MOSFET serving as the main vehicle carrying integrated circuit technology forward with the help of its opposite polarity NMOS and PMOS devices combined into the well-known CMOS configuration has been constantly downscaled. Riding on the classical MOSFET workhorse, integrated circuits have steadily marched a long way towards the nanoscale. Constant field and constant voltage scaling schemes have been applied. The downscaling succeeded to a large extent in meeting the predictions of the Moore's law before succumbing to physical limitations. Various problems encountered in moving towards smaller geometry devices are outlined and restrictions on downscaling supply and threshold voltages are laid down. The extent of solutions possible with classical MOSFET structure is indicated. Through such technological innovations, the classical MOSFET progressed unless it was realized that revolutionary process and structural improvements were necessary. The chapter surveys the scaling issues and looks at the solutions to the problems in the perspective of classical MOSFET device.

4.1 Moore's Law

It is a succinct statement of an observation made in 1965 by the American entrepreneur Dr. Gordon E. Moore, a cofounder of the microprocessor company Intel. According to this observation, the number of transistors per square inch incorporated in an integrated circuit chip had doubled approximately after every x number of months since the invention of the integrated circuit. Based on this observation, it was forecasted that the trend will perpetuate for the imaginable future [1]. Some people assert that it takes 18 months while others say that the period is 24 months for doubling of the number of transistors. So, $x = 18$ or 24 months while in the original version, x was given as 12 months. Some persons understand that the law applies to the doubling of processing power instead of the number of transistors.

V.K. Khanna, *Integrated Nanoelectronics*,
NanoScience and Technology, DOI 10.1007/978-81-322-3625-2_4

However, the exponential growth suggested by Moore's law is doubtful to continue forever. Many experts believe that Moore's law will hold for another two decades. Some studies have shown that economic barriers and physical limitations are likely to stall the progress by 2017.

4.2 The Classical, Planar, Single-Gate Bulk MOSFETs

4.2.1 The MOS Device and its Electrical Characteristics

The starting point is the four-terminal classical MOSFET device (Fig. 4.1). Its four terminals are the source, drain, gate, and substrate terminals [2]. Electrical characteristics of an N-channel MOS transistor are displayed in Fig. 4.2.

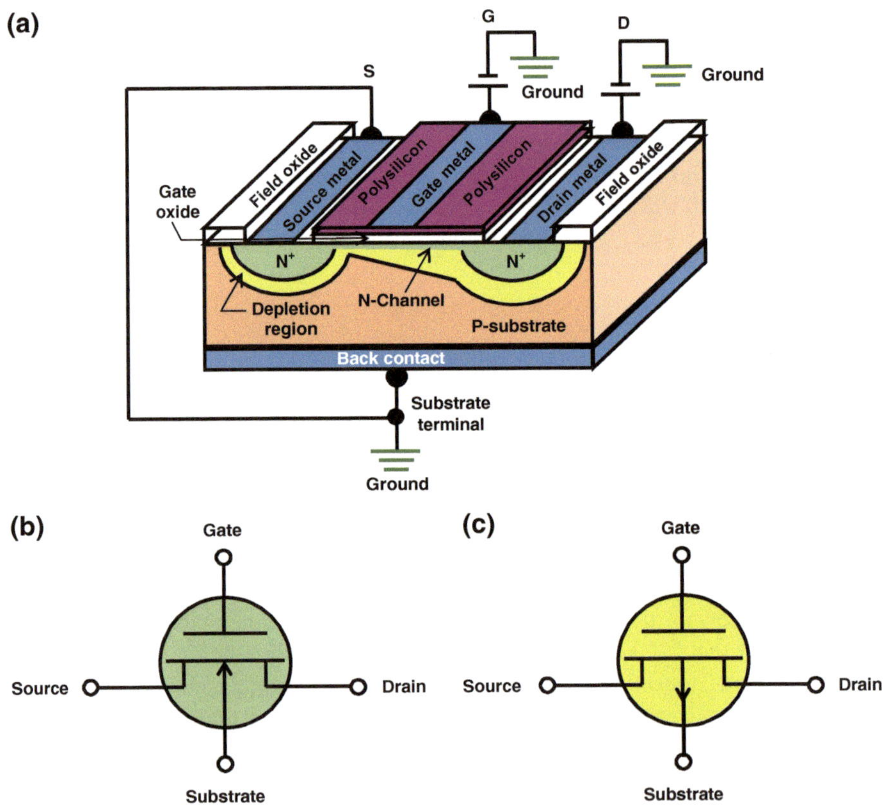

Fig. 4.1 Classical N-channel MOSFET: **a** Cross-section and **b** its circuit diagram symbol. **c** In a P-channel MOSFET, the direction of *arrow* is reversed

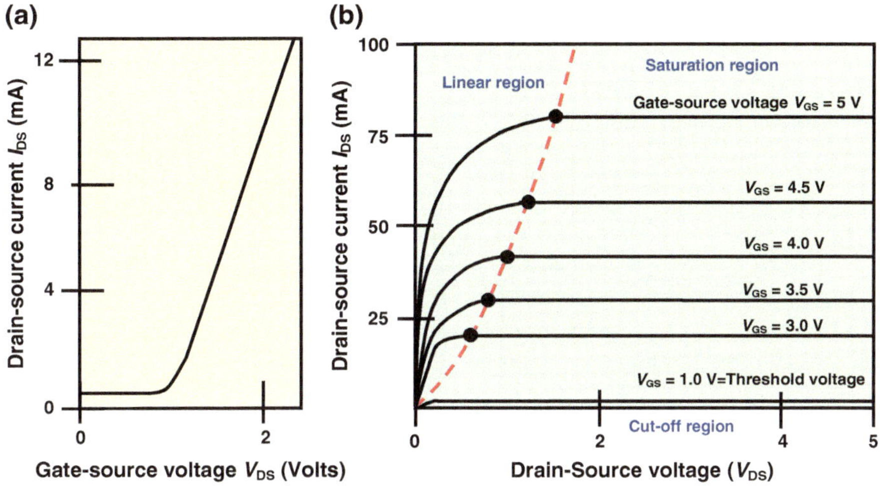

Fig. 4.2 N-channel MOSFET characteristics: **a** Transfer and **b** Output characteristics

4.2.2 Self-aligned Polysilicon Gate MOS Process

Prior to the introduction of self-aligned process, vacuum-evaporated aluminum was used as the gate of the MOSFET in place of polycrystalline silicon. The main idea of this process is that the thin gate oxide underneath polysilicon, a refractory gate contact, acts as a diffusion mask during source and drain junction formation. Hence, a separate photolithographic step is not required for forming these junctions. The benefit of using polysilicon is that it can withstand the required post-implantation high-temperature annealing treatment. A metal like aluminum will be seriously degraded in such an environment. But why is self-alignment necessary? It is required because the source and drain diffusions and the channel region must be connected together to ensure MOSFET operation. If the gate length in the mask is taken exactly = channel length, during photolithography, a slight misalignment may lead to the creation of a gap between source diffusion, channel, and drain diffusion leading to a broken channel. To avoid such a gap, the gate length in the mask is kept slightly larger than the channel length. Although alignment errors are overcome, the parasitic capacitances due to overlapping of gate/source and gate/drain regions irresistibly creep in. These capacitances make the device slow in operation, impacting its performance at high frequencies. Self-alignment reduces the overlap capacitances between gate/source and gate/drain, resulting in faster devices. The aforementioned technology using polycrystalline silicon as the gate contact is called silicon gate technology. The main steps of this process are illustrated in Fig. 4.3.

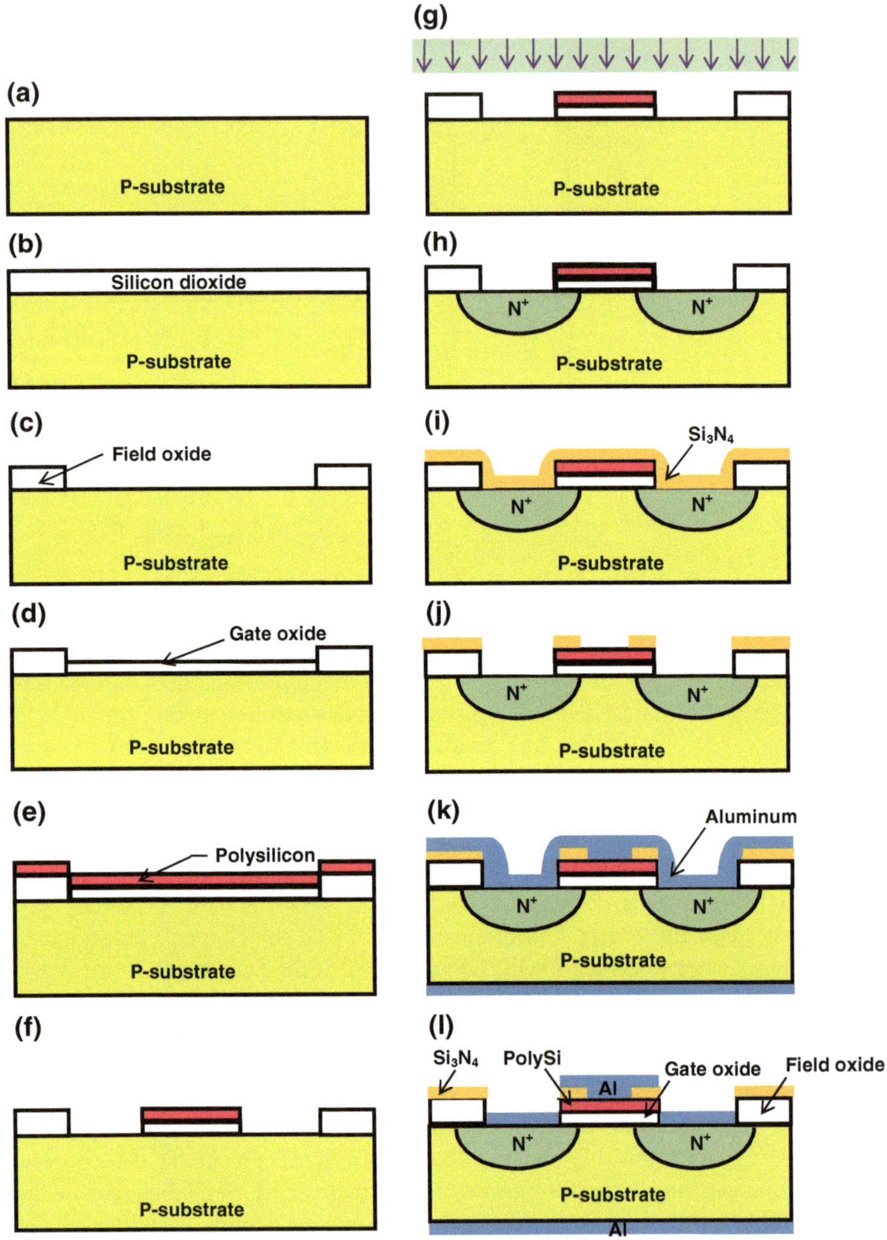

Fig. 4.3 Simple self-aligned polysilicon gate process for N-channel MOSFET fabrication in which polysilicon gate acts as a mask for source/drain formation. Lightly doped source/drain structure formation is excluded. **a** Starting silicon wafer. **b** Field oxidation. **c** Oxide etching. **d** Gate oxidation. **e** Polysilicon deposition. **f** Polysilicon and oxide etching. **g** Source/Drain implant. **h** High temperature annealing. **i** Silicon nitride deposition. **j** Nitride etching. **k** Metal deposition. **l** Metal etching

4.2.3 Self-aligned Silicide (Salicide) Process

This is a process for formation of contact electrodes on source, drain, and gate terminals of a MOSFET without photolithography (Fig. 4.4). Silicide is the name given to an alloy of a metal with silicon. "Salicide" is the shortened or contracted term for "self-aligned silicide" process. "Salicidation" refers to the process of forming silicide by reaction between metal and silicon. The salicide process is referred to by this appellation because it eliminates the requirement of photolithographic alignment in contact formation. A transition metal such as tungsten or titanium is thermally evaporated so as to cover the complete processed wafer including the conducting and insulating regions. Then, it is subjected to annealing at high temperature. During annealing, the metal film reacts with underlying silicon at the contact regions forming tungsten disilicide WSi_2 or titanium disilicide $TiSi_2$. But it does not undergo any reaction whatsoever with dielectrics like silicon dioxide or silicon nitride. After the annealing is completed, the unreacted titanium left over the dielectric covered areas is removed by etching. Thus a conducting metal silicide film is selectively deposited at the contact regions of the device without recourse to any lithographic alignment.

Apart from tungsten or titanium, other metals of interest for salicide process are Pt, Co, Ni. The salicidation has to be carefully watched and monitored, e.g., Cobalt forms three silicides CoSi, $CoSi_2$, and Co_2Si. Amongst these, only $CoSi_2$ has a low resistance so that formation of this phase must be ensured by controlling the salicidation process.

Another silicide process used in MOS industry is polycide process in which the silicide is defined over the polysilicon gate electrode.

4.3 Complementary Metal-Oxide-Semiconductor (CMOS) Technology

4.3.1 CMOS Structure and Advantages

The pervasive circuit component in integrated circuits is the CMOS structure in which both N-channel and P-channel MOSFETs are fabricated on the same wafer. Static power dissipation of CMOS configuration is extremely low because one of the NMOS and PMOS transistors connected in series is always off. Power is only dissipated when the transistors switch between on- and off-states. Further, circuit complexity is reduced. Noise immunity is high. Also, it provides a high density of logic function.

4.3.2 CMOS NOT Gate

Referring to Fig. 4.5, when input = logic 0, i.e., gate voltage is negative with respect to substrate, Q_1 is on and Q_2 is off. Output terminal is shorted to V_{DD} and a

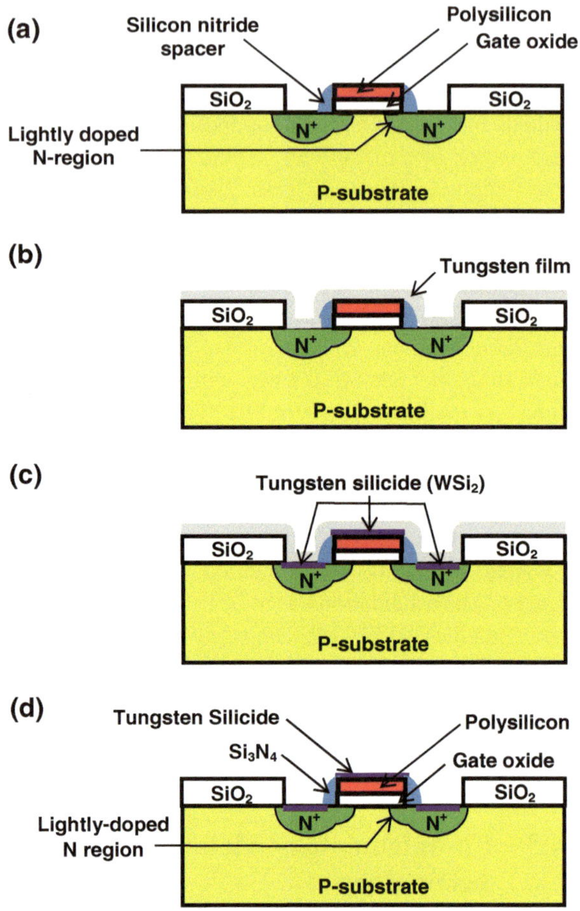

Fig. 4.4 Self-aligned silicide (Salicide) process for simultaneously siliciding polysilicon gate and silicon source/drain regions. **a** Sidewall spacer formation. **b** Tungsten deposition. **c** Metal annealing to form silicide. **d** Removal of unreacted metal

high resistance connection with ground. So output = V_{DD} = logic 1. Conversely, when, input = logic 1, i.e., gate voltage is positive with respect to substrate, Q_1 is off and Q_2 is on. Output terminal is shorted to ground and has a high resistance connection with V_{DD}. So, output = Ground voltage = logic 0. In both cases, the output is converse of input which is the behavior of an inverter or NOT gate.

4.3.3 CMOS NAND Gate

The circuit contains two inverter circuits (Fig. 4.6). One inverter circuit consists of the transistor pair (Q_1, Q_3). This circuit is controlled by input X. The other inverter circuit comprises the transistors (Q_2, Q_4). This circuit is controlled by input Y. The source and drain terminals of transistors Q_1, Q_2 are connected in parallel. The

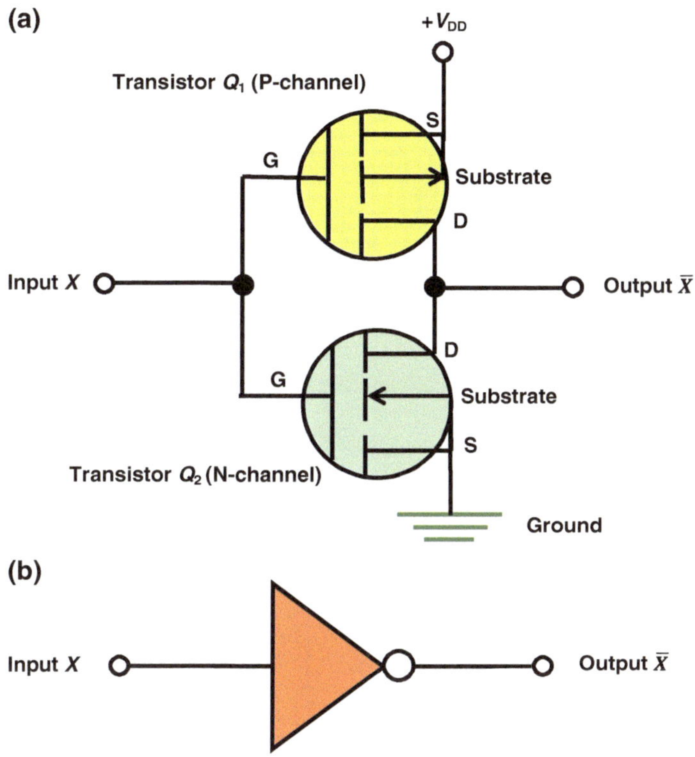

Fig. 4.5 CMOS NOT gate or inverter: **a** Circuit and **b** Circuit diagram symbol

source and drain terminals of transistors Q_3, Q_4 are series-connected. When inputs $X = 0$, $Y = 0$, i.e., gate voltages are negative with respect to substrate, transistors Q_1, Q_2 are on; while Q_3, Q_4 are off. Output = V_{DD} = logic 1. When input $X = 0$ but $Y = 1$, Q_1 is on, Q_2 is off, Q_3 is off and Q_4 is on. Therefore, output = V_{DD} = logic 1. When input $X = 1$ but $Y = 0$, Q_1 is off, Q_2 is on. Q_3 is on, Q_4 is off. Then output = V_{DD} = logic 1. When inputs $X = 1$, $Y = 1$, Q_1 is off, Q_2 is off, Q_3 is on, Q_4 is on. So, output = Ground voltage = logic 0.

4.3.4 *CMOS NOR Gate*

The reader may analyze and understand the operation of this circuit (Fig. 4.7) on similar lines to NAND gate functioning.

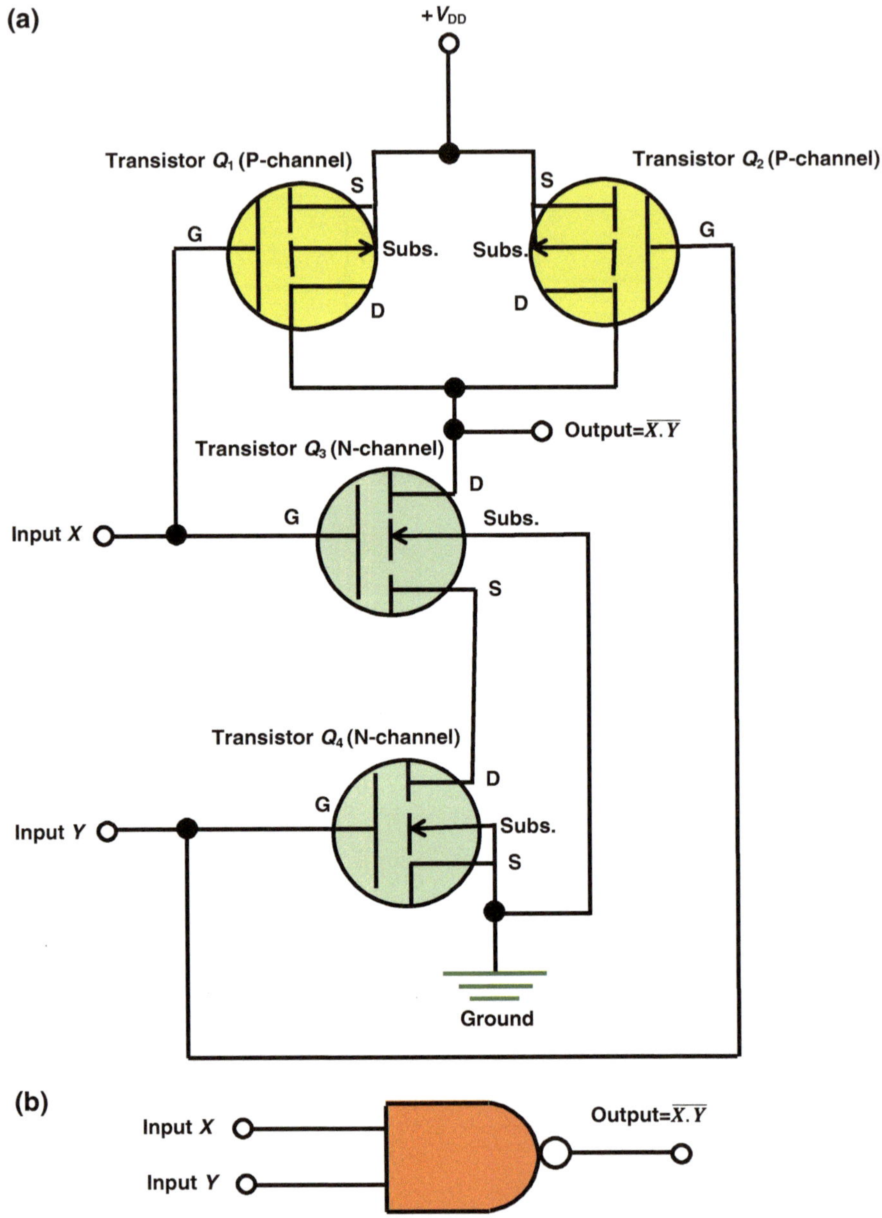

Fig. 4.6 CMOS NAND gate: **a** circuit and **b** circuit diagram symbol

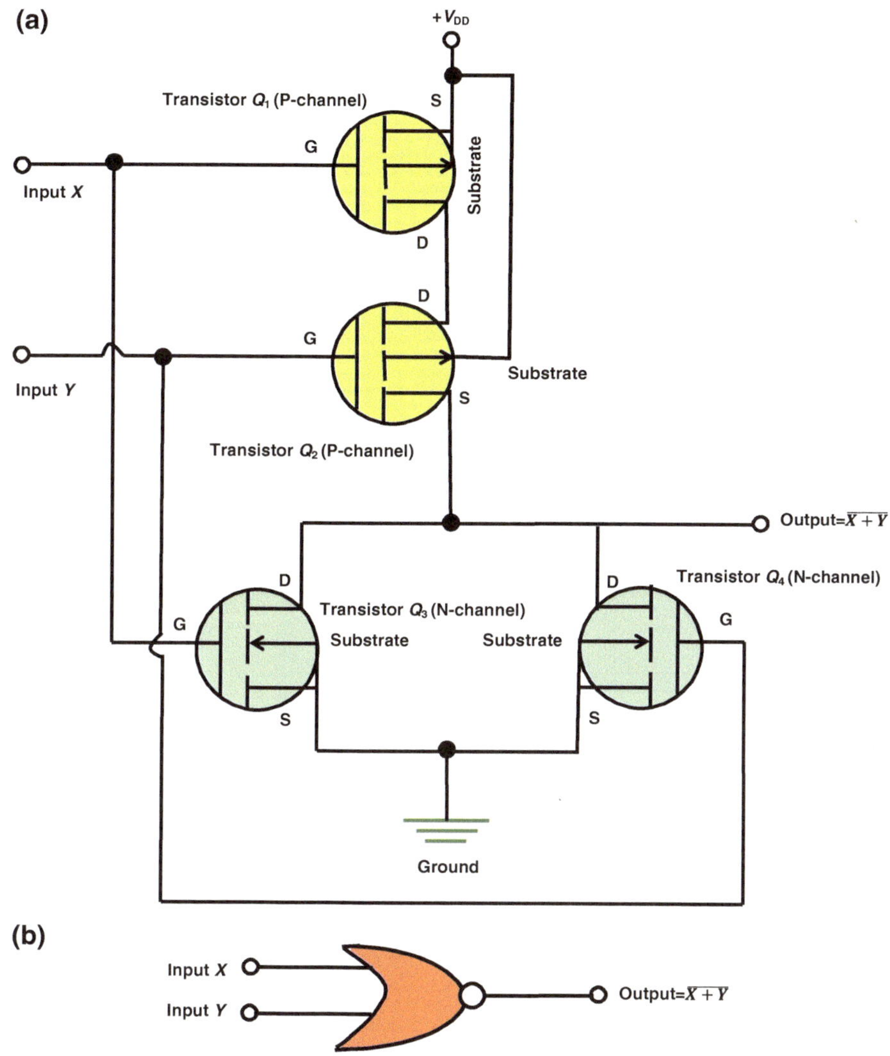

Fig. 4.7 CMOS NOR gate: **a** circuit and **b** circuit diagram symbol

4.3.5 CMOS Process

One of the two devices in CMOSFET, either NMOSFET or PMOSFET must be fabricated in a local substrate created during process, called the well. Generally, the NMOSFET is fabricated on the P-type starting wafer and the PMOSFET is made in an N-well. The process is known as an N-well CMOS process. Figures 4.8 and 4.9 depict the key steps of this process. A few steps are not shown for simplification. In a P-type silicon wafer, the N-well is produced by ion implantation. The impurities

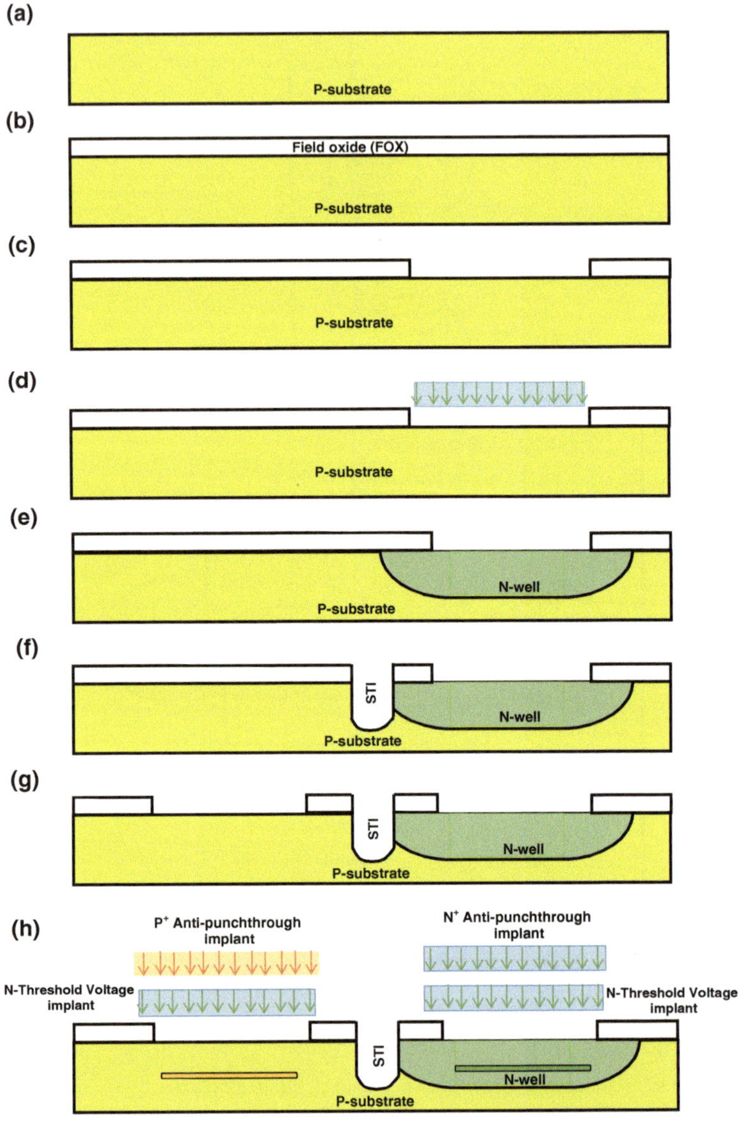

Fig. 4.8 Main fabrication steps in a CMOS process. **a** Starting silicon wafer. **b** Field oxidation. **c** Oxide etching. **d** N-well implant. **d** N-well implant. **e** High-temperature annealing. **f** Shallow trench isolation (STI). **g** Oxide etching. **h** Threshold voltage and anti-punchthrough implants. **i** Gate oxidation. **j** Polysilicon deposition. **k** Polysilicon and oxide etching. **l** Shallow N- and P- implants for lightly doped source/drain. **m** Nitridation for sidewall spacers. **n** Nitride etching. **o** N^+ implant. **p** High-temperature annealing. **q** Nitridation. **r** Nitride etching. **s** P^+ implant. **t** High-temperature annealing. **u** Nitridation. **v** Nitride etching. **w** Tungsten metallization for siliciding (salicide and polycide). **x** Tungsten etching

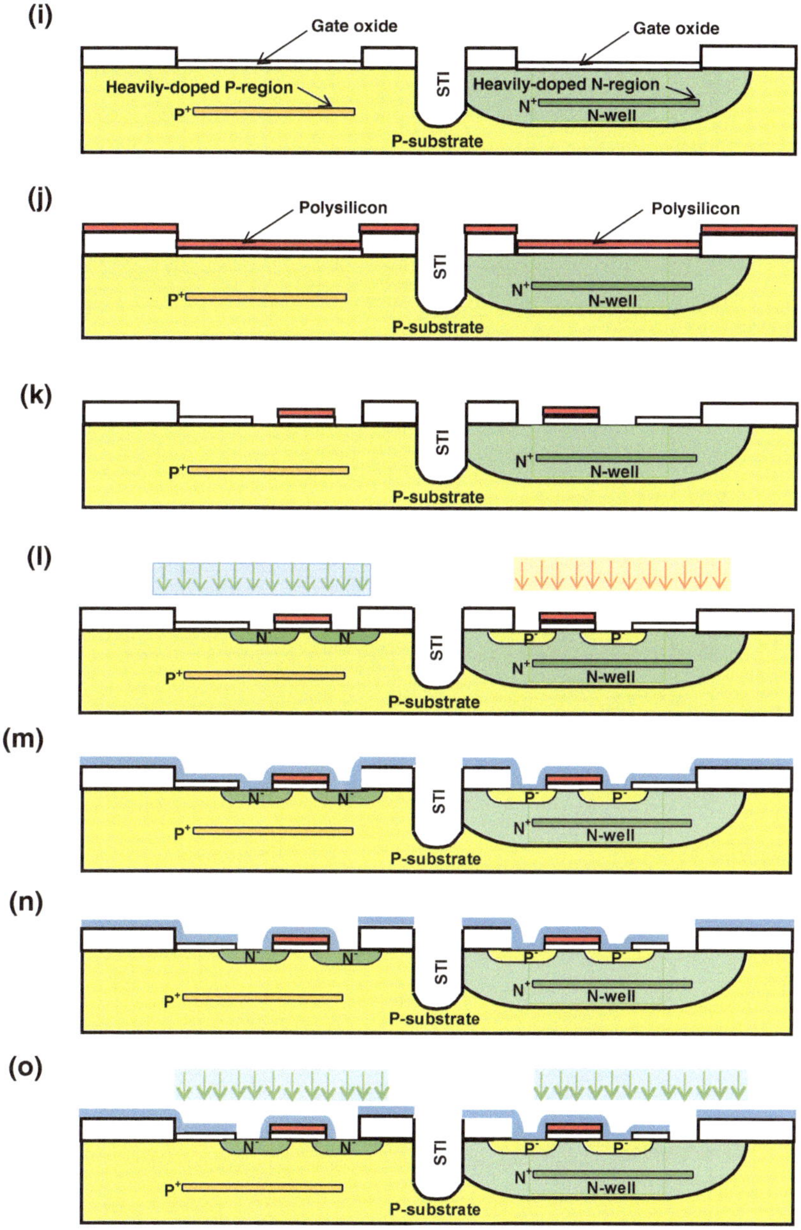

Fig. 4.8 (continued)

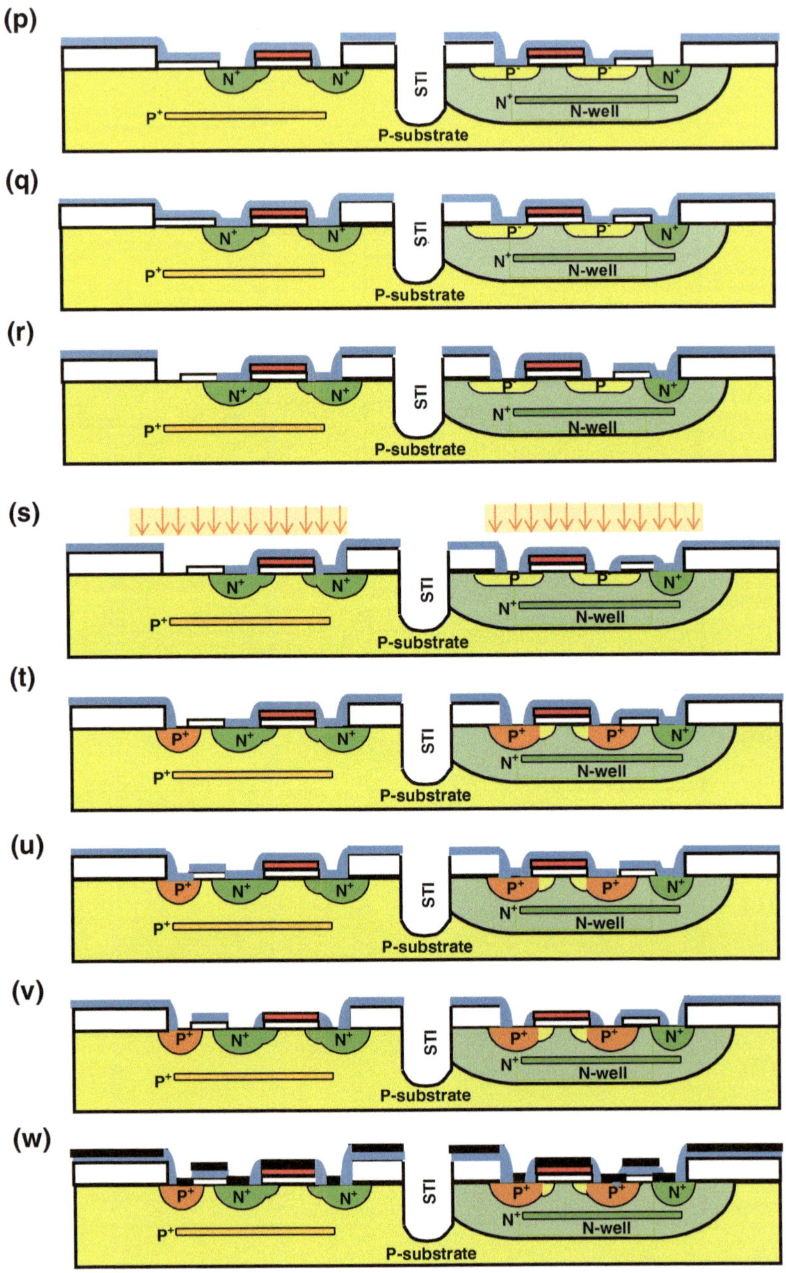

Fig. 4.8 (continued)

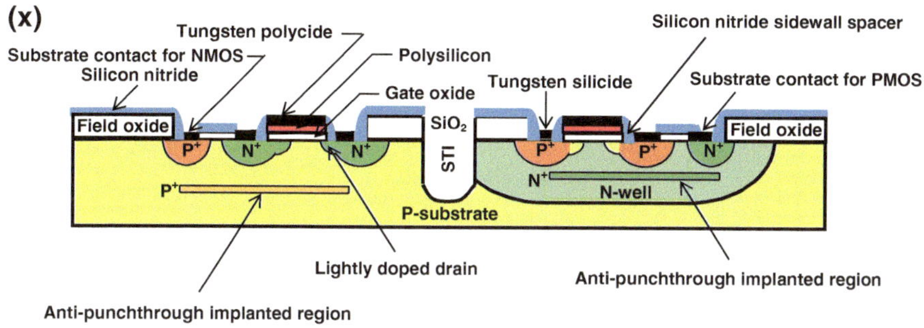

Fig. 4.8 (continued)

are driven deep inside by thermal diffusion. After the well has been formed, shallow trench isolation (STI) process is carried out to separate the different regions/transistors on the chip. Details of this process are given in Sect. 4.3.6.

After completion of STI process, two implantations, namely threshold voltage shift and anti-punchthrough implants are performed. The threshold voltage shift implant is essential because the naturally achieved threshold voltage for NMOSFET is typically 0 V whereas for PMOSFET, it is −1.2 V. Therefore, the NMOSFET must be made more difficult to invert while inversion of PMOSFET must be made easier. The anti-punchthrough implant is necessary to produce a region of high doping concentration under the channel so that possibility of punchthrough breakdown between depletion regions of source and drain is avoided.

Subsequent to these implantations, the important steps are thin gate oxide and polysilicon gate electrode deposition. By ion implantation, lightly doped source/drain regions are produced. Section 5.10.4 presents the steps involved in the process for producing lightly doped source/drain.

To prevent the portions of source/drain adjoining the channel from acquiring high carrier concentrations during the ensuing source/drain diffusion, sidewall spacers of silicon nitride are formed next to the gate oxide-polysilicon structures. Then highly doped source/drain regions are formed.

For metallization, salicide process is used. In Sect. 4.2.3, the steps in this process were explained.

4.3.6 Shallow Trench Isolation (STI) Process

Trenches are etched at the preselected places in the chip. These are the places where transistors or other devices are to be separated (Fig. 4.10). Subsequent to under-etching of the pad oxide, a silicon dioxide layer is grown over the trench

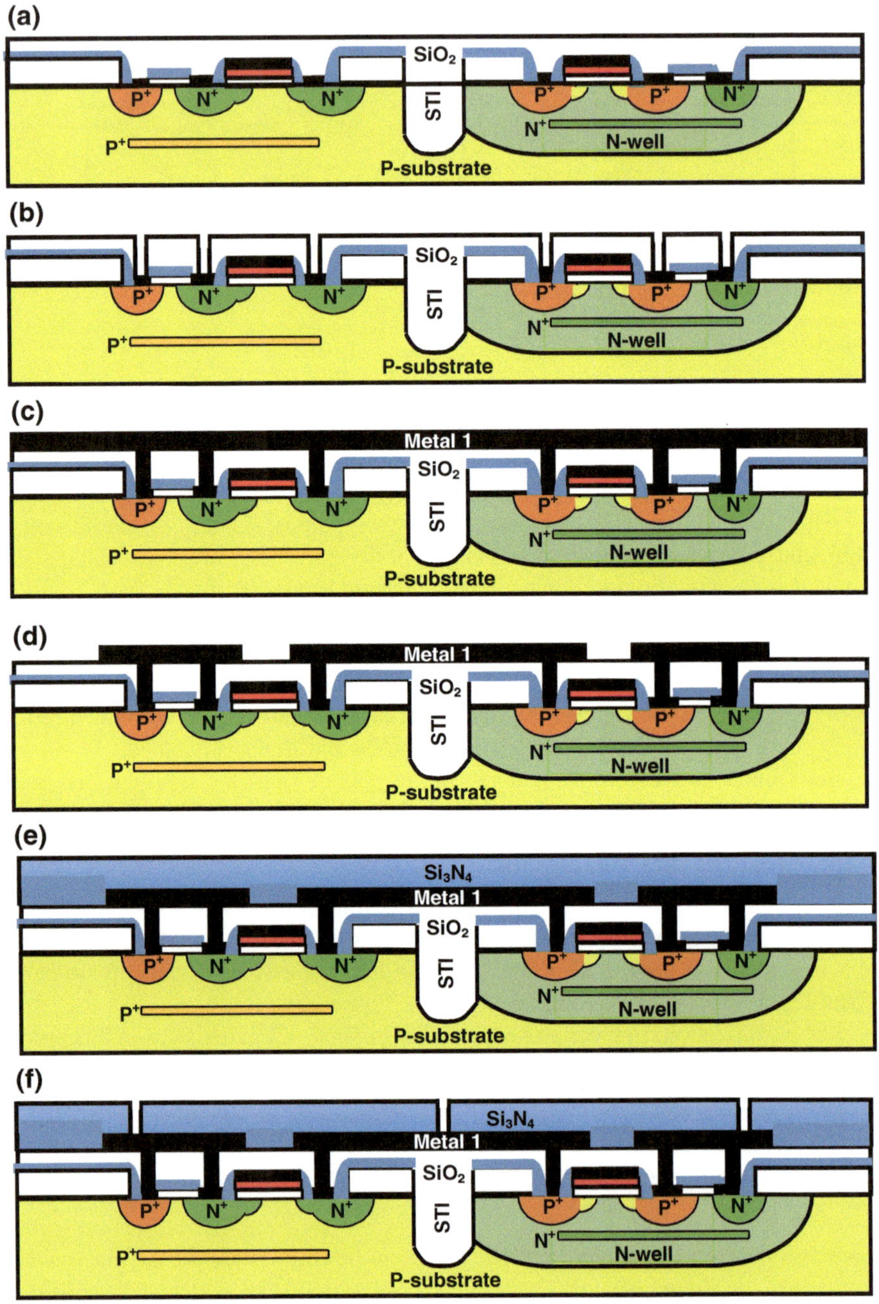

Fig. 4.9 Metal interconnection process for CMOS circuit. **a** Covering of wafer by thick oxide. **b** Contact window etching. **c** Metal 1 deposition over full wafer. **d** Photolithography for selective metal 1 etching. **e** Covering full wafer with silicon nitride. **f** Photolithography and etching using via mask. **g** Covering the full wafer with metal 2. **h** Photolithography and selective etching of metal 2. **i** Passivation of entire chip with glass. **j** Opening contact windows for bonding pads

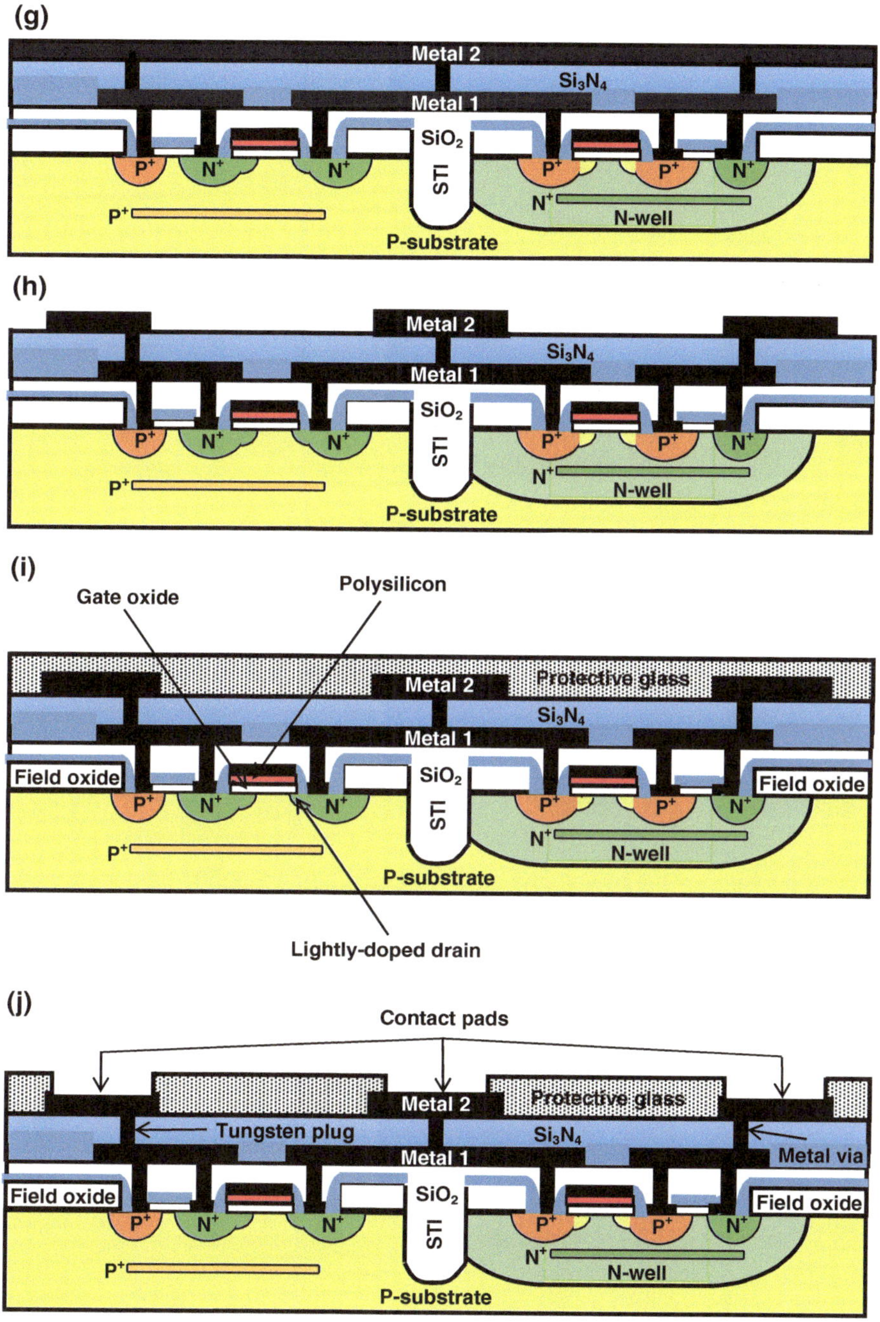

Fig. 4.9 (continued)

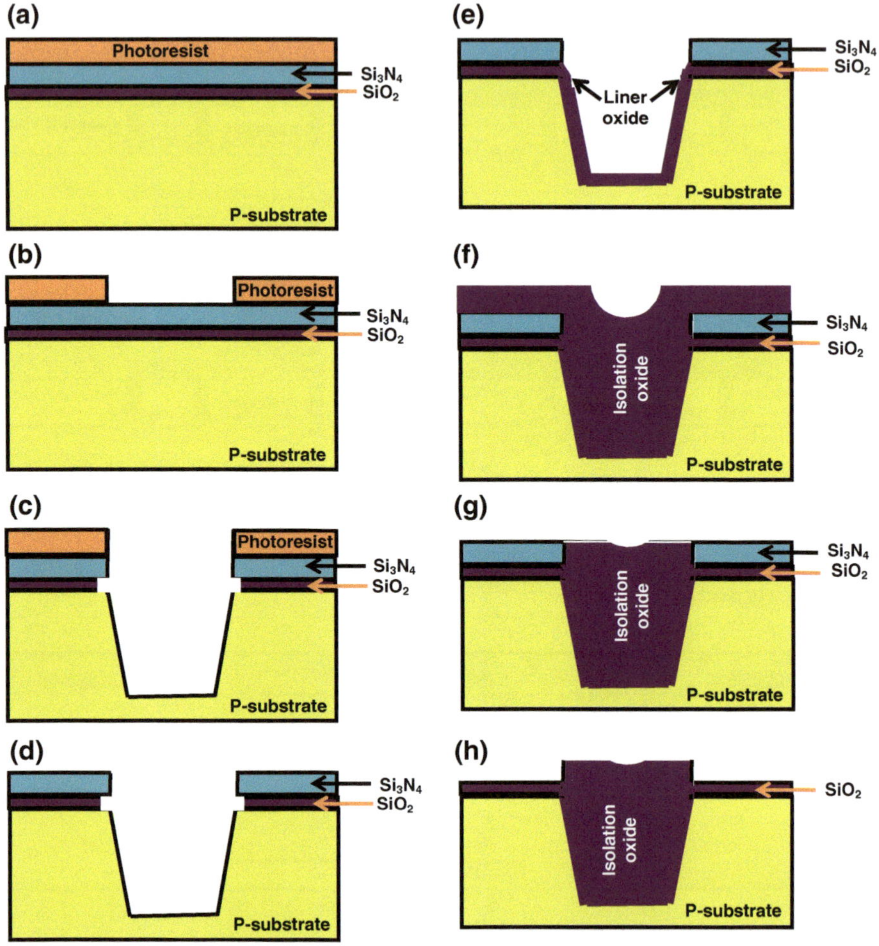

Fig. 4.10 Shallow trench isolation process. **a** ($Si+SiO_2+Si_3N_4$) coated with photoresist. **b** Photolithography and photoresist development. **c** Si_3N_4, SiO_2, and Si etching. **d** Photoresist removal. **e** Thermal oxidation for liner oxide growth. **f** Oxide filling by chemical vapor deposition. **g** Planarization by CMP. **h** Nitride stripping

surface. This layer is known as the liner oxide. Following liner oxide growth, the trenches are plugged with silicon dioxide. As this oxide protrudes out from the wafer surface, the surface of the wafer is planarized by chemical mechanical polishing. By planarization process, the excess silicon dioxide is removed. Finally, the nitride mask is also gotten rid off. The method is also called box isolation technique.

4.4 Scaling Trends of Classical MOSFETs

Scaling is the process of miniaturizing devices. This miniaturization is done in such a way that their electrical characteristics are either maintained constant or show improvement. Thus it seeks to achieve the same or better performance of devices with smaller size [3]. Shrinkage of device dimensions is aimed at increasing the transistor density and operating frequency. It also reduces power dissipation and gate delays. Proper scaling of MOSFET does not only mean a reduction of the gate length and width. It also calls for a cutback of all other dimensions including the oxide thickness and the depletion layer widths. Altering the depletion layer widths affects the substrate doping density as well. The first complete scaling scheme was introduced by Dennard et al. in 1974 [4]. The method is called constant electric field scaling. However, this scheme was soon followed by different scaling schemes. This happened due to certain problems encountered in this scheme. One such scheme is constant voltage scaling. Let us first examine constant field scaling.

4.4.1 Constant Field Scaling

4.4.1.1 The Principle of Constant Field Scaling

In this method, both the horizontal and vertical dimensions of the device are scaled down by the factor $1/\lambda$, where λ is the dimensionless scaling factor. To preserve the electric field within the device constant, the voltages must also be scaled down as $1/\lambda$. So, the electric filed = the ratio between voltage and distance, will remain unchanged. Then the initial field will be recreated in the device. Now, suppose the device dimensions (W, L, t_{ox}) and voltages (V_{DD}, V_{Th}) are scaled down by λ. At the same time, let the doping concentrations be scaled up by λ (>1). Then all electric fields in the scaled transistor will remain at their previous values. Hence, this approach is known as constant field scaling.

4.4.1.2 Effects of Constant Field Scaling

Applying the principles of constant field scaling, let us find how this scaling impacts the MOSFET drive current I_{DSsat}, intrinsic gate delay ζ, transistor density N_{ρ}, peak power P_{peak} dissipated per transistor and power density P_{ρ}. With this in mind, we write down the relevant MOSFET equations. The scaled-down parameters will be denoted by starred symbols and the unscaled parameters by unprimed symbols. Carrier mobility changes will be ignored.

(i) *MOSFET linear current* A MOSFET is operating in the linear, ohmic or triode region at values of the gate-source voltage V_{GS} > threshold voltage V_{Th} and drain-source voltage $V_{DS} < (V_{GS} - V_{Th})$. For a long-channel

NMOSFET confined to linear mode, the drain-source current I_{DS} is related to the drain-source voltage V_{DS} as

$$I_{DS} = (\mu_n C_{ox}/2)(W/L)\{2(V_{GS} - V_{Th})V_{DS} - V_{DS}^2\} \tag{4.1}$$

In this equation, μ_n is not the bulk mobility of electrons but the electron mobility in the channel region of the MOSFET. The symbol C_{ox} stands for the capacitance of gate oxide taking area of the capacitor as unity while *W, L* specify the width and length of the channel respectively.
The scaled-down MOSFET gate oxide capacitance per unit area (C_{ox}) is

$$C_{ox}^* = \left(\frac{\epsilon_0 \epsilon_{ox}}{t_{ox}/\lambda}\right) = \lambda C_{ox} \tag{4.2}$$

where ε_0, ε_{ox} are the permittivity of free space and relative permittivity of silicon dioxide on the gate, respectively; and t_{ox} is the thickness of this oxide layer.
From the C_{ox}^* equation, the scaled-down MOSFET linear current is

$$I_{DS}^* \propto C_{ox}^*(W^*/L^*)(V_{GS}^* - V_{Th}^*)(V_{DS}) = (\lambda C_{ox})\left(\frac{W/\lambda}{L/\lambda}\right)\left(\frac{V_{DS} - V_{Th}}{\lambda}\right)\left(\frac{V_{DS}}{\lambda}\right) \propto I_{DS}/\lambda \tag{4.3}$$

(ii) *MOSFET saturation current* The MOSFET is operating in the saturation region when the gate-source voltage V_{GS} > threshold voltage V_{Th} and drain-source-voltage V_{DS} is $\geq (V_{GS} - V_{Th})$. The saturation current I_{DSsat} is

$$I_{DSsat} = (\mu_n C_{ox}/2)(W/L)(V_{GS} - V_{Th})^2 \tag{4.4}$$

Using the C_{ox}^* equation, the scaled-down MOSFET drive current is

$$I_{DSsat}^* \propto C_{ox}^*(W^*/L^*)(V_{GS}^* - V_{Th}^*)^2 = (\lambda C_{ox})\left(\frac{W/\lambda}{L/\lambda}\right)\left(\frac{V_{DS} - V_{Th}}{\lambda}\right)^2 \propto I_{DSsat}/\lambda \tag{4.5}$$

(iii) *Intrinsic gate delay* Intrinsic gate delay is the delay internal to the gate, and for which the gate is solely responsible. It is the delay produced by the gate in producing a signal at the output after supplying the signal at the input without any external load applied to the output, i.e., without any output loading.
Since the scaled-down MOSFET gate capacitance is

$$C_{gate}^* = L^* W^* C_{ox}^* = (L/\lambda)(W/\lambda)\left(\frac{\epsilon_0 \epsilon_{ox}}{t_{ox}/\lambda}\right) = C_{gate}/\lambda \tag{4.6}$$

the intrinsic gate delay is

$$\zeta^* = \frac{C^*_{\text{gate}} V^*_{\text{DD}}}{I^*_{\text{DSsat}}} = \frac{(C_{\text{gate}}/\lambda)(V_{\text{DD}}/\lambda)}{(I_{\text{DSsat}}/\lambda)} = \left(\frac{C_{\text{gate}} V_{\text{DD}}}{I_{\text{DSsat}}}\right)/\lambda \tag{4.7}$$

where V_{DD} is the positive drain supply.

(iv) *Transistor packing density* Bearing in mind that the area occupied by one transistor on the chip $\propto L^* W^*$, it is evident that the number of transistors accommodated per unit area is

$$N_{\rho} \propto \frac{1}{L^* W^*} \propto \frac{1}{(L/\lambda)(W/\lambda)} = \frac{\lambda^2}{LW} \tag{4.8}$$

(v) *Peak power dissipated per transistor* Each transistor consumes a peak power, whose value is given by

$$P^*_{\text{peak}} = I^*_{\text{DSsat}} V^*_{\text{DD}} = (I_{\text{DSsat}}/\lambda)(V_{\text{DD}}/\lambda) = P_{\text{peak}}/\lambda^2 \tag{4.9}$$

(vi) *Power density* It is estimated by dividing the peak power by the area as under

$$N_{\rho} = \frac{P^*_{\text{peak}}}{L^* W^*} = \frac{P_{\text{peak}}/\lambda^2}{(L/\lambda)(W/\lambda)} = \frac{P_{\text{peak}}}{LW} \tag{4.10}$$

(vii) *Substrate doping density* Depths up to which depletion regions of junctions between source-substrate and drain-substrate extend can be expressed in terms of the voltage they are subjected to

$$x_{\text{d}} = \sqrt{\left(\frac{2\varepsilon_0 \varepsilon_{\text{Si}}}{q}\right)\left(\frac{N_{\text{A}} + N_{\text{D}}}{N_{\text{A}} N_{\text{D}}}\right)|\phi_0 - V|} \tag{4.11}$$

where q is the electronic charge; N_A, N_D are the acceptor and donor doping concentrations, respectively; ϕ_0 is the built-in potential of the above junctions and V is the applied reverse voltage, ranging from 0 to $-V_{\text{DD}}$. Assuming $N_A \gg N_D$, we have

$$x_d = \sqrt{\left(\frac{2\varepsilon_0 \varepsilon_{\text{Si}}}{q}\right)\left(\frac{N_D}{N_A N_D}\right)|\phi_0 - V|} \propto \sqrt{\frac{1}{N_A}|V|} \tag{4.12}$$

so that

$$x^*_d = \frac{1}{\lambda} x_d \propto \left(\frac{1}{\lambda}\right)\sqrt{\frac{1}{N_A}|V|} = \sqrt{\left(\frac{1}{\lambda^2}\right)\frac{1}{N_A}|V|} = \sqrt{\frac{1}{\lambda N_A}\left|\frac{V}{\lambda}\right|} \tag{4.13}$$

Hence,

$$N_A^* \rightarrow \lambda N_A \text{ and } N_D^* \rightarrow \lambda N_D \tag{4.14}$$

Consequent upon this type of scaling, the currents will be low. Hence, with the voltage already decreased, the total power per transistor ($P = I \times V$) will also be reduced. However, the power density will remain constant. This is because the number of transistors per unit area will increase. If the chip size remains the same, the total chip power will remain constant. This is usually the situation. The circuit speed will be increased by the factor λ. The concentrations of P- and N-regions are raised by λ. On the whole, by this arrangement, the nonideal effects will be mitigated to some extent.

Based on the above principle, the device dimensions have been considerably trimmed down. The supply and threshold voltages have also been decreased. These reductions have enabled higher circuit density and better performance. Despite the scaling, the device reliability has been maintained. Constant power dissipation per unit area has also been achieved. On the whole, constant field scaling yields the largest reduction in the power-delay product of a single transistor.

4.4.1.3 Drawbacks of Constant Field Scaling

The main drawback of this scaling scheme is that very often the extent of scaling desired exceeds practical limitations, e.g., it may not be possible in many situations to scale all the parameters in the required proportions due to experimental difficulties. As an example, the substrate doping has an upper limit of 10^{18} cm^{-3}. This limit is determined by the solid solubility of the dopant. Solid solubility of a dopant is the maximum concentration of the impurity that can be introduced at a given temperature. When this temperature is reached, the impurity precipitates out. Then it forms a separate phase. The peak value of solid solubility in the range of diffusion temperatures used in semiconductor processing (800–1300 °C) plays the decisive role in fixing the limit of doping concentration. The melting point of silicon is 1414 °C. So, beyond the solid solubility at a temperature at a safe distance from the melting point of silicon, further increase of doping is impossible.

4.4.1.4 Problems Faced in Constant Field Scaling

Problems arise as the supply voltage (V_{DD}) approaches near 1 V. At this point, the scaling diverges from the ideal constant field scaling. The reason is the difficulty of further lowering the threshold voltage (V_{Th}). This fundamental problem appears from the non-scalable characteristic of the thermal voltage ($V_{\mathrm{T}} = k_B T/q$). The subthreshold swing (*SS*), defined in section below, is somewhat fixed at the constant temperature [5]. This, in turn, causes the subthreshold leakage current to increase exponentially as V_{Th} reduces. Therefore, a lowest possible value of V_{Th} is determined by the application constraints. These restrictions are related to power consumption and circuit functionality. In addition, it is known that V_{Th} variation increases in nanoscale MOSFETs. So, it is necessary to make sufficient margin for V_{Th} variation. Furthermore, especially for high-performance logic technology, it is required to keep a certain level of $V_{DD} - V_{Th}$. This level determines the drive current. As a result, further downscaling of the supply voltage becomes difficult as one decreases the minimum feature size. This diminution cannot be compromised with easily. Thus, threshold voltage V_{Th} scaling poses particularly challenging problems. These problems are concerned with safeguarding the functionality of the circuits and the noise margins relative to each another. These are also scaled down by the same factor as the supply voltage V_{DD}. Devices of 0.25 μm channel length or longer have roughly 0.7 V threshold voltage. Those of 0.1 μm channel length have 0.33–0.40 V threshold voltage.

4.4.2 *Constant Voltage Scaling*

4.4.2.1 Need of Constant Voltage Scaling

As per the technology roadmaps, the industry has concurred years ahead on the supply voltage values. This concurrence has provided manufacturers sufficient time beforehand to design and assemble power supplies. Imagine that a unique power supply is required to be designed and made for each particular application or channel length. Then too much time will be wasted. Also, money will be unnecessarily expended. So, it is unfeasible and uneconomical for the required performance improvement. This means that standard power supply voltages have to be considered when designing a device. Constant voltage scaling is therefore a more sensible application of the more ideal method of constant electric field scaling. Here, V_{DD} is kept constant. However, all the dimensions, including those vertical to the surface are scaled down. Also, concentration densities are scaled up.

4.4.2.2 Effects of Constant Voltage Scaling

(i) *MOSFET linear current*

$$\begin{aligned} I_{DS}^{*} &\propto C_{ox}^{*}(W^{*}/L^{*})(V_{GS}^{*} - V_{Th}^{*})(V_{DS}) \\ &= (\lambda C_{ox})\left(\frac{W/\lambda}{L/\lambda}\right)\left(\frac{V_{DS} - V_{Th}}{1}\right)\left(\frac{V_{DS}}{1}\right) \propto \lambda I_{DS} \end{aligned} \tag{4.15}$$

(ii) *MOSFET saturation current*

$$I^*_{\mathrm{DSsat}} \propto C^*_{\mathrm{ox}}(W^*/L^*)\left(V^*_{\mathrm{GS}} - V^*_{\mathrm{Th}}\right)^2 = (\lambda C_{\mathrm{ox}})\left(\frac{W/\lambda}{L/\lambda}\right)\left(\frac{V_{\mathrm{DS}} - V_{\mathrm{Th}}}{1}\right)^2 \propto \lambda I_{\mathrm{DSsat}} \tag{4.16}$$

(iii) *Intrinsic gate delay*

$$\zeta^* = \frac{C^*_{\mathrm{gate}} V^*_{\mathrm{DD}}}{I^*_{\mathrm{DSsat}}} = \frac{\left(C_{\mathrm{gate}}/\lambda\right)\left(V_{\mathrm{DD}}\right)}{\lambda I_{\mathrm{DSsat}}} = \left(\frac{C_{\mathrm{gate}} V_{\mathrm{DD}}}{I_{\mathrm{DSsat}}}\right)/\lambda^2 \tag{4.17}$$

(iv) *Transistor density*

$$N_\rho \propto \frac{1}{L^* W^*} \propto \frac{1}{(L/\lambda)(W/\lambda)} = \frac{\lambda^2}{LW} \tag{4.18}$$

(v) *Peak power dissipated per transistor*

$$P^*_{\mathrm{peak}} = I^*_{\mathrm{DSsat}} V^*_{\mathrm{DD}} = (\lambda I_{\mathrm{DSsat}})(V_{\mathrm{DD}}) = \lambda P_{\mathrm{peak}} \tag{4.19}$$

(vi) *Power density*

$$N_\rho = \frac{P^*_{\mathrm{peak}}}{L^* W^*} = \frac{\lambda P_{\mathrm{peak}}}{(L/\lambda)(W/\lambda)} = \frac{\lambda^3 P_{\mathrm{peak}}}{LW} \tag{4.20}$$

(vii) *Substrate doping density*

$$x^*_d = \frac{1}{\lambda} x_d \propto \left(\frac{1}{\lambda}\right)\sqrt{\frac{1}{N_A}|V|} = \sqrt{\left(\frac{1}{\lambda^2}\right)\frac{1}{N_A}|V|} \tag{4.21}$$

Therefore,

$$N^*_A \to \lambda^2 N_A \quad \text{and} \quad N^*_D \to \lambda^2 N_D \tag{4.22}$$

The drain current I_{DSsat} increases by the factor λ. Intrinsic gate delay decreases as λ^2. So, higher clock speeds can be obtained. Power density increases λ^3 times. The doping concentrations are aggressively increased by λ^2.

4.4.2.3 Advantages of Constant Voltage Scaling

Constant voltage scaling is not blemished by the low threshold voltage problem. Moreover, voltage compatibility with previous circuit technologies is assured. Therefore, this scaling method is preferred over competing approaches. But constant voltage scaling is riddled with problems, as we shall see below. So, scaling does not really work at small dimensions. Sometimes, it appears that constant field scaling was better, as has been done by departing from 5.0 to 3.3, 2.5, 1.5 V.... Let us relook at the issue.

4.4.2.4 Disadvantages of Constant Voltage Scaling

The scaling principle is purely geometrical. Only lateral dimensions are reduced. So, longitudinal electric field increases. Chances of electrical break down increase. Indeed, the main disadvantage of this scaling scheme is that the supply voltage itself is not scalable. Hence, higher fields are created in the device with the reduction of minimum feature length. The increase in current I_{DS} by λ and hence current density by λ^3, causes electromigration in metallization and interconnects. Similarly, the power per transistor increases by λ. So, the power density per unit area increases by λ^3. This means that, the chip power increases by λ^3 keeping the chip area same. The enormous increase in power makes constant voltage scaling highly impractical, mainly because of localized heating and associated heat dissipation problems. Further, the device doping has to be increased more aggressively by a factor of λ^2 than in the constant field scaling. This is a mandatory requirement to preclude channel punchthrough. The punchthrough occurs when the depletion region around the source touches that around the drain. By increasing the doping concentration by a multiplier λ, the depletion region thickness is divided by λ. This is the same ratio as the channel length. However, the doping can be increased only up to a maximum value. This value depends on the solid solubility limit of the dopant in silicon. Again, this makes the constant voltage scaling easier said than done. Phenomena like velocity saturation and mobility degradation of carriers can no longer be ignored. Hot carrier effects may cause havoc. Increased leakage currents with lower breakdown voltages, and other reliability problems begin to creep in. Some of these effects will be described below. Moreover, this method consumes more power. Therefore, better cooling methods must be used than for constant electric field scaling. A comparison is made between constant field and constant voltage schemes in Table 4.1.

Table 4.1 Constant field and constant voltage scaling

S. No.	Physical parameter	Symbol	Constant field scaling	Constant voltage scaling
1.	Gate length	L	λ^{-1}	λ^{-1}
2.	Gate width	W	λ^{-1}	λ^{-1}
3.	Electric field	E	1	λ
4.	Junction depth	x_j	λ^{-1}	λ^{-1}
5.	Gate oxide thickness	t_{ox}	λ^{-1}	λ^{-1}
6.	Substrate doping density	N_A or N_D	λ	λ^2
7.	Gate oxide capacitance	C_{ox}	λ	λ
8.	Gate capacitance	C_{gate}	λ^{-1}	λ^{-1}
9.	Threshold voltage	V_{Th}	λ^{-1}	1
10.	Drain-source voltage	V_{DS}	λ^{-1}	1
11.	Drain-source current	I_{DS}	λ^{-1}	λ
12.	Power	P	λ^{-2}	λ
13.	Transit time	t_{tr}	λ^{-1}	λ^{-2}
14.	Transit frequency	f_T	λ	λ^2
15.	Power dissipation	P	λ^{-2}	λ
16.	Power-delay	$P\Delta t$	λ^{-3}	λ^{-1}

4.5 Scaling Limits for Supply and Threshold Voltages in Classical MOSFETs

4.5.1 Subthreshold Leakage Current

At $|V_{GS}| < |V_{Th}|$, an N-channel MOSFET is in the off-state. The off-state current I_{OFF} is the drain-source current I_{DS} measured at gate-source voltage $V_{GS} = 0$ V and drain-source bias $V_{DS} = V_{DD}$, the DC voltage applied to the drain terminal. In fact, the current I_{OFF} is never zero. An undesirable leakage current always flows between the drain and the source terminals. This MOSFET current observed at $|V_{GS}| < |V_{Th}|$ is called the subthreshold leakage current I_{Subth}. Subthreshold current of a MOSFET is the current that flows between the drain and source terminals of the device when the gate-source voltage is lower than the threshold voltage of the MOSFET. It is the weak inversion conduction current. It is dominated by the diffusion current flowing between the drain and source when $|V_{GS}| < |V_{Th}|$. Its value determines the standby leakage power dissipation of the MOSFET. It is important to keep I_{Subth} very small. Then only, it is possible to minimize the static power. By static power is meant the power that a circuit consumes even when it is in the standby mode. If the I_{OFF} is 10 nA per transistor in a mobile smartphone containing 10^8 transistors, the smartphone chip will consume $10 \times 10^{-9} \times 10^8 = 1$ A, draining out the battery very fast.

4.5.2 Subthreshold Slope and V_{DD}, V_{Th} Interrelationship

Subthreshold current is plotted on a semilogarithmic graph between I_{DS} and V_{GS} (Fig. 4.11). When the voltage V_{GS} is below the threshold voltage V_{Th}, the drain-source current $I_{DS} = I_{Subth}$ is an exponential function of V_{GS}. The current I_{Subth} is caused by the diffusion of electrons from the channel in the P-substrate across the forward-biased substrate-source P–N junction. Therefore, it resembles the current of diode under forward bias operation. It increases exponentially with both increasing V_{GS} and decreasing V_{Th}

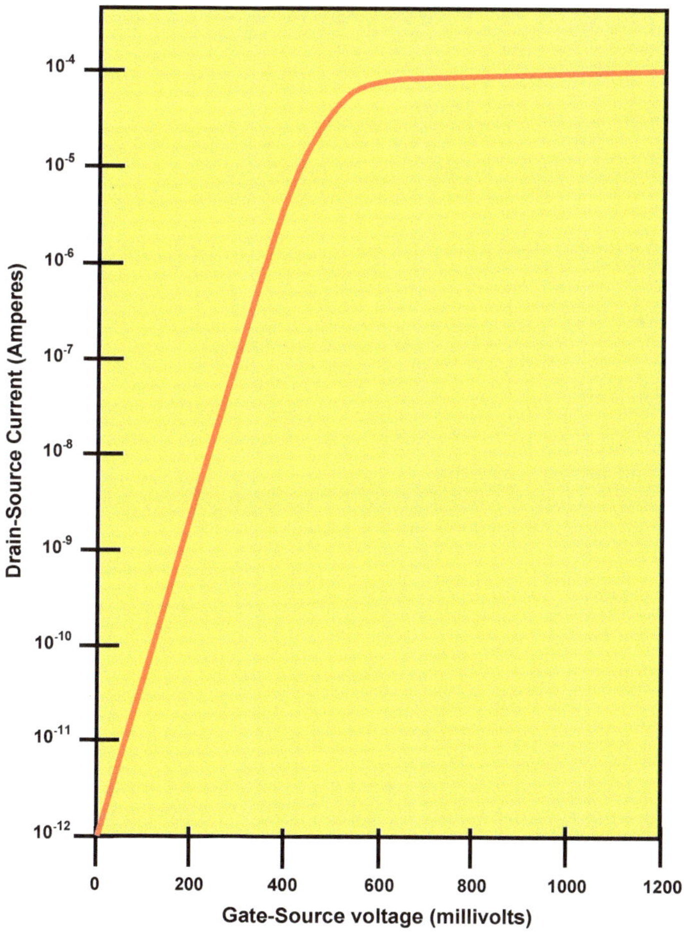

Fig. 4.11 Subthreshold characteristic of a MOSFET

$$I_{\text{Subth}} \propto \exp\left(\frac{qV_{\text{GS}}}{nk_BT}\right) \tag{4.23}$$

where q is the electronic charge, k_B is Boltzmann constant = 8.62×10^{-5} eVK^{-1} and T is temperature in Kelvin scale.

The above Eq. (4.23) can be recast as

$$\ln(I_{\text{Subth}}) \propto \frac{qV_{\text{GS}}}{nk_BT}$$

or,

$$2.3\log_{10}I_{\text{Subth}} \propto \frac{qV_{\text{GS}}}{nk_BT}$$

or,

$$\log_{10}I_{\text{Subth}} \propto \frac{1}{2.3}\left(\frac{qV_{\text{GS}}}{nk_BT}\right) \tag{4.24}$$

Hence, the partial derivative of $\log_{10}I_{\text{Subth}}$ with respect to V_{GS} yields a constant slope called the subthreshold slope (S)

$$S = \frac{\partial}{\partial V_{\text{GS}}}(\log_{10}I_{\text{Subth}}) = \frac{\partial}{\partial V_{\text{GS}}}\left\{\frac{1}{2.3}\left(\frac{qV_{\text{GS}}}{nk_BT}\right)\right\} = \frac{1}{2.3}\left(\frac{q}{nk_BT}\right) \tag{4.25}$$

Subthreshold slope of a MOSFET is the slope of the graph of $\log_{10}(I_{\text{DS}})$ against V_{GS} in the subthreshold region.

Subthreshold swing (SS) of a MOSFET is defined as the inverse of subthreshold slope. It is given by

$$\begin{aligned} SS &= \frac{\partial V_{\text{GS}}}{\partial(\log_{10}I_{\text{Subth}})} = 2.3 \times n\left(\frac{k_BT}{q}\right) = 2.3 \times 1 \times \left(8.62 \times 10^{-5} \times 300\right) \\ &= 0.0595\,\text{V} = 59.5\frac{\text{mV}}{\text{decade}} \approx 60\,\text{mV/decade} \end{aligned} \tag{4.26}$$

taking $n = 1$.

The SS parameter shows how abruptly the transistor turns off with decreasing gate voltage V_{GS}. In other words, it tells us how well the channel surface potential can be controlled by the gate voltage. In order to turn off the transistor effectively, SS must be designed to be as small as possible. For a typical device at room temperature T, SS is always greater than the best value ~60 mV/dec.

A small value of subthreshold swing improves the ratio between the on- and off-currents. This ratio = (current at $V_{\text{GS}} = 0$)/(current at $V_{\text{GS}} = V_{\text{Th}}$). Let us assume a typical subthreshold slope value and a worst-case threshold voltage variation.

That being so the required minimum threshold voltage must be >0.4 V [6]. For high-performance operation, the threshold voltage V_{Th} must be much lower than the power supply voltage V_{DD}. Then the device remains much of its time out of the subthreshold regime. For this condition, a good requirement is fulfilled by [7]

$$V_{\text{DD}} > 3\,V_{\text{Th}} > 1.2\,V \tag{4.27}$$

Thus high-performance condition and subthreshold leakage current restrict the limits to which the supply and threshold voltages can be downscaled. The threshold voltage lies between ground and the supply voltage.

4.6 Discussion and Conclusions

Microprocessor, memory, and digital signal processing chips benefitted from the exponential shrinkage of MOSFET dimensions. The constant field scaling law was pursued up to 2005, reaching a MOSFET density $\sim 10^6/\text{mm}^2$ and minimum feature size ~ 100 nm [8]. In accordance with this scaling law, the switching speed decreased by a factor of λ, the power dissipation by λ^2, and the power-delay product by λ^3. However, the reduction of supply and threshold voltages by λ was not achieved. The raison d'être was that it was impossible to scale the subthreshold slope to values <59.6 mV/decade. Thus after several years of unabated progress in scaling, the efforts seemed to reach a pinnacle. Further progress was braked down by the short-channel effects encountered in these small devices.

Review Exercises

4.1 State Moore's law regarding the growth of transistor density in integrated circuits. What was the value of the variable x in the original version of this law? What are other two suggested values of x?

4.2 Draw the cross-sectional diagram of a classical long-channel MOSFET. Indicate the different connection terminals. What is the role of the body/substrate terminal?

4.3 What is meant by scaling of MOSFET? What are the motivations for scaling MOSFET?

4.4 What is the basic idea of constant electric field scaling? How is this idea implemented to achieve constancy of field?

4.5 Show with relevant equations how does constant field affect the following parameters of a MOSFET: (i) linear drain-source current, (ii) saturation drain-source current, and (iii) intrinsic gate delay.

4.6 Power density is not affected by constant field scaling. Show mathematically.

4.7 From the equations for source-substrate and drain-substrate junctions, show that the donor and acceptor concentrations are multiplied by the scaling factor during constant field scaling of a MOSFET. Up to what limiting value is it possible to raise the doping density?
4.8 Discuss the problems associated with lowering the threshold voltage of a MOSFET beyond a certain limit.
4.9 Give arguments in support of the approach followed in constant voltage scaling. Why is it a more practical method than constant field scaling?
4.10 During constant voltage scaling of a MOSFET, by what factor is the power density changed? Is it favorable for device operation? How do you compare the effect of constant voltage scaling on power density with that in constant field scaling?
4.11 Write three harmful effects of constant voltage scaling on the operation of a MOSFET.
4.12 What is meant by the subthreshold current of a MOSFET? How is it related to the standby power dissipation of a device?
4.13 What is subthreshold slope of a MOSFET? How is it related with its subthreshold swing? What is the ideal value of subthreshold swing?
4.14 What is the constraint on reducing the threshold voltage with respect to the supply voltage? State any relationship prescribed for deciding the threshold voltage at a given supply voltage.

References

1. Moore's law (2015) Encyclopedia Britannica. http://www.britannica.com/EBchecked/topic/705881/Moores-law. Accessed 21 Feb 2015
2. Nishi Y , Doering R (2007) Handbook of semiconductor manufacturing technology. CRC Press, Boca Raton, pp 1–27
3. Antoniadis DA, Aberg I, Chléirigh CN et al (2006) Continuous MOSFET performance increase with device scaling: the role of strain and channel material innovations. IBM J Res Dev 50 (4/5):363–376
4. Dennard RH, Gaensslen FH, Yu H-N et al (1974) Design of ion-implanted MOSFETs with very small physical dimensions. IEEE J Solid-State Circuits SC-9:256–268
5. Kim Y-B (2009) Review paper: challenges for nanoscale MOSFETs and emerging nanoelectronics. Trans Electr Electron Mater 10(1):21
6. McFarland G, Flynn M (1995) Limits of scaling MOSFETs. Technical Report CSL-TR-95-662 (Revised), p 6
7. Pfiester JR, Shott JD, Meindl JD (1985) Performance limits of CMOS ULSI. IEEE J Solid-State Circuits SC-20(1):253–263
8. Ferain I, Colinge CA, Colinge J-P (2011) Multigate transistors as the future of classical metal–oxide–semiconductor field-effect transistors. Nature 479:310–316

Chapter 5
Short-Channel Effects in MOSFETs

Abstract Short-channel effects are a series of phenomena that take place when the channel length of the MOSFET becomes approximately equal to the space charge regions of source and drain junctions with the substrate. They lead to a series of issues including polysilicon gate depletion effect, threshold voltage roll-off, drain-induced barrier lowering (DIBL), velocity saturation, reverse leakage current rise, mobility reduction, hot carrier effects, and similar other annoyances. Mitigation of the problem posed by polysilicon gate depletion effect via restoration of metal gate structure is presented. Threshold voltage reduction makes it difficult to turn the transistor off completely. By DIBL effect, electrostatic coupling between the source and drain makes the gate ineffective. Velocity saturation decreases the current drive. The leakage current increases the power dissipation. Enhanced surface scattering degrades the mobility of charge carriers affecting the output current. Apart from these factors, impact ionization and hot carrier effects seriously impair the MOSFET performance and cause the device to diverge in behavior from long-channel ones. Notable solutions are the gate oxide thickness cutback, use of high-κ dielectrics, strain engineering, etc. Nevertheless, the various effects mentioned severely downgrade the performance of planar CMOS transistors at process nodes <90 nm.

5.1 Meaning of "Short Channel"

Gate length L_g represents the physical length of the gate. Actual length of the channel L is obtained by subtracting the sum total lateral diffusions of the source and drain junctions from L_g. The length $L_g > L$ and L tracks L_g but the difference $(L_g - L)$ cannot be quantified precisely. The channel length L is being continuously reduced to increase the operational speed and to accommodate more number of components per chip. At a certain channel length, the so-called short-channel effects arise. These effects are named so because they occur explicitly by virtue of the fact that the channel is short. A more concrete definition of short channel follows in the next paragraph.

V.K. Khanna, *Integrated Nanoelectronics*,
NanoScience and Technology, DOI 10.1007/978-81-322-3625-2_5

The decrease of channel length is associated with the enlargement of source and drain depletion regions and their incursion into the channel. This incursion may take place even without any bias. It may lead to trespassing of the depletion regions into such zones which usually fall under the jurisdiction of the gate. A MOSFET device is deemed to have a short channel when the channel length is reduced to a certain consented degree. This degree has been mutually decided to be attained when the channel length is of comparable magnitude to the depletion layer widths (x_{dD}, x_{dS}) surrounding the drain and source junctions. A short-channel effect is an effect which is produced only when the channel has become short, and which is not observed otherwise. As a consequence of this effect, a MOSFET of channel length $L \approx (x_{dD}, x_{dS})$, deviates in behavior from a long-channel MOSFET having $L \gg x_{dD}, x_{dS}$. Short-channel effects originate from a variety of reasons: (i) the production of high electric fields in the channel region. (ii) the two-dimensional potential distribution in this region. This distribution depends on the transverse field E_x controlled by the gate voltage as well as the bias applied to the back surface. Longitudinal field E_y due to the drain bias also plays its role. The two-dimensional potential degrades the threshold comportment of the device. It makes the threshold voltage dependent on the channel length and biasing voltages. Circuit designers will be scared if V_{Th} changes with device dimensions or biasing voltages.

5.2 Polysilicon Gate Depletion Effect

In the beginning stages of MOSFET development, the metal electrode of the gate was replaced by a heavily doped polysilicon electrode. But this polysilicon is a semiconductor, not a conductor like a metal. It can deplete when the electric fields become high. The finite depletion layer created inside the polysilicon becomes uncooperative with continued device scaling. The potential dropped across the polysilicon layer becomes a large fraction of the supply voltage. This mandates a further reduction of oxide thickness to maintain a constant electric field across the gate dielectric. But the oxide is already very thin and does not allow further thinning. The polysilicon depletion effect can also occur in long-channel devices but its probability increases at smaller dimensions because of the strong electric fields that are created. The obvious solution to polysilicon depletion effect is to increase the doping concentration of polysilicon as much as possible, but this is restricted practically to $\sim 1 \times 10^{20}\ \text{cm}^{-3}$ for N-type and $\sim 1 \times 10^{19}\ \text{cm}^{-3}$ for P-type polysilicon. Fully silicided (FUSI) polysilicon gates, e.g., cobalt silicide, nickel silicide, hafnium silicide, platinum silicide, and titanium silicide gates are the useful approach for highly scaled CMOS [1, 2]. This disruptive method was given up due to the problems faced in controlling the silicide phase to achieve low threshold voltages. The other alternative is to replace the polysilicon gates with metal gates of work functions 4.1 and 5 eV for N^+ and P^+ polysilicon, respectively [Nishi]. Aluminum gates were the bastion of MOS integrated circuits until the introduction of polysilicon gate-based self-aligned process. A comeback of metal gates is mandated to avoid depletion of the gate conductor.

Advantages offered by metal gates are: (i) Low gate resistance and hence smaller gate RC (resistance–capacitance) delay. (ii) Absence of boron penetration from the polysilicon gate into the channel across the thin gate oxide. (iii) Appropriate work function adjustment: 4.1–4.4 eV for N-channel MOSFET and 4.8–5.1 eV for P-channel MOSFET for bulk silicon MOSFET and partially-depleted SOI-MOSFET. A spectrum of work function values covering the silicon band gap is available for fully-depleted SOI-MOSFET and FINFET devices. (iv) Diminution in electrical thickness of gate insulator.

5.3 Gate-First or Gate-Last Fabrication Flow

This issue is connected with that of high-κ dielectrics to be discussed in the next chapter. Actually, the gate consists of a series of layers; hence it will be expedient to talk about metal gate/high-κ or MG/HK stack. Two different fabrication flows are possible to integrate metal gates with the process [3]: (i) Gate-first approach: Traditional CMOS process is followed (Fig. 5.1). Hence, this kind of gate is called metal inserted-polysilicon gate (MIPG). The polysilicon gate is deposited at a very early stage in the process. It serves as a mask for the subsequent source/drain implantation. The lattice damages created in the implantation step are annealed out at a high temperature. Deposition of the metal gate in the beginning stage exposes it to high-temperature steps. Thermal annealing can destroy the integrity of MG/HK stack, causing thermal instability and threshold voltage shifts, besides contaminating the front-end equipment, specially the furnaces. These complications have obstructed the use of gate-first method for high-performance CMOS. (ii) Gate-last approach: This is designed to circumvent the obstacles faced in gate-first method. Here a dummy/sacrificial gate is made for masking the implantation (Fig. 5.2). After source/drain junctions have been formed and no further thermal cycling is required, the dummy gate is etched away. The real gate stack is then constructed. By chemical mechanical polishing (CMP), the gate metal is thinned down to desired thickness. Looking at the above process flow, this type of gate is known as a replacement metal gate (RMG). Table 5.1 presents a comparison between the two processes [4].

Essentially, the gate-first and gate-last approaches differ in the relative sequence or order of gate metal deposition and thermal annealing steps. In the gate-first approach, the metal gate is deposited prior to carrying out high-temperature activation annealing. In the gate-last approach, the annealing steps are carried out at an early stage in the process, and metal gate deposition is done afterwards.

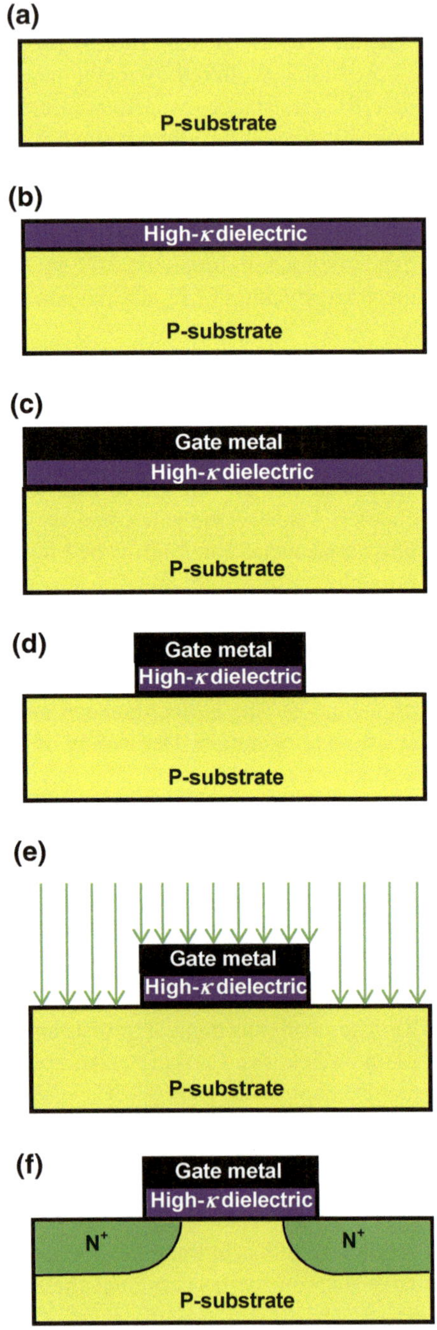

Fig. 5.1 Gate metal first process. **a** Starting silicon wafer. **b** Deposition of high-κ dielectric film. **c** Gate metal deposition. **d** Photolithography and etching of metal and dielectric films. **e** N^+ source/drain implant. **f** High-temperature annealing

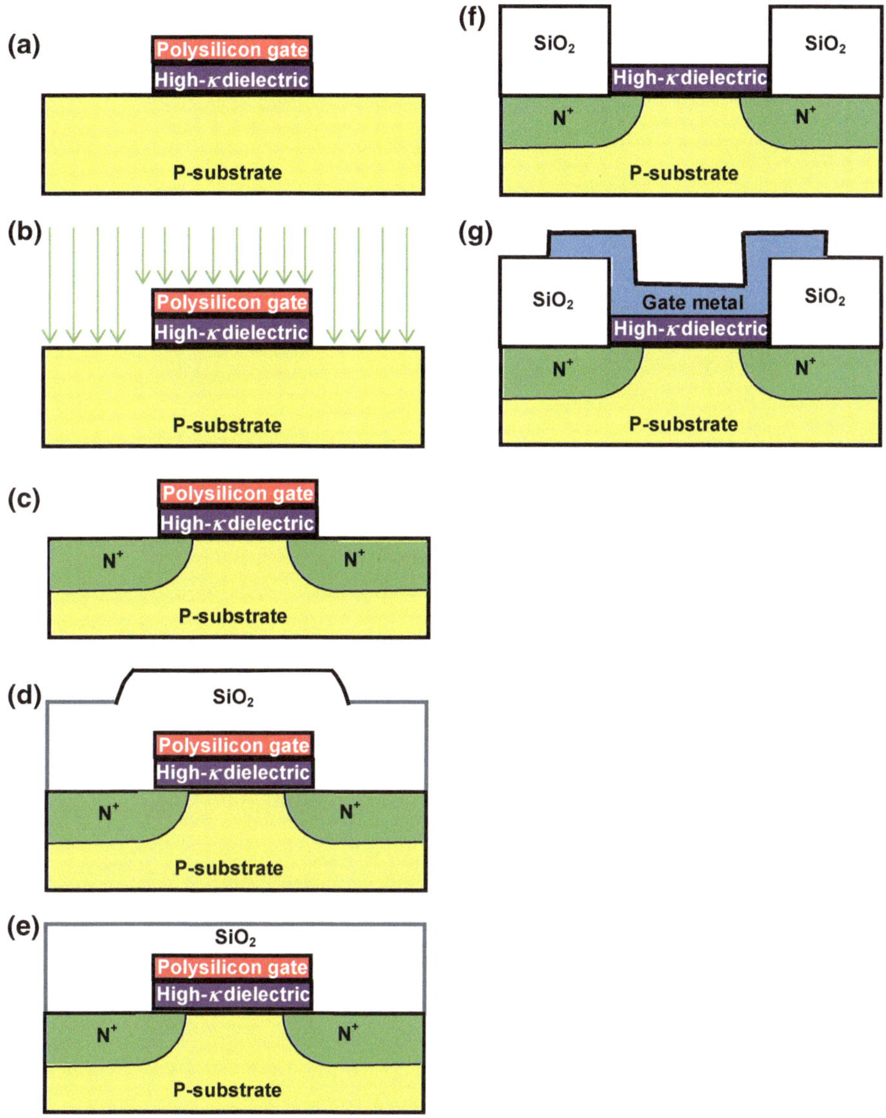

Fig. 5.2 Gate metal last process. **a** High-κ dielectric and polysilicon gate deposition. **b** N^+ source/drain implant. **c** High-temperature annealing. **d** Silicon dioxide coverage. **e** Chemical mechanical polishing. **f** Removal of dummy gate. **g** Replacement gate formation

Table 5.1 Comparison between gate-first and gate-last process flows

Feature	Gate-first	Gate-last
Gate insulator	First	First or last
Gate metal	First	Last
Process sequence	Conventional	Revised
Process complexity	Less	More
Thermal cycles given to gate metal	Large	Small
Reliability	Less because high-κ material has undergone temperature cycling	More
Applications	Low power DRAM where threshold voltage requirements are relaxed	High-performance applications

5.4 Threshold Voltage Roll-off and Drain-Induced Barrier Lowering (DIBL)

With the decrease in channel length of a MOSFET, the bulk charge terminating on the gate electrode decreases. The decrease in charge leads to a reduction of the threshold voltage. To explain this reduction, let us reiterate that the formation of an inversion layer underneath the gate dielectric is preceded by the depletion of this region up to a depth W_d. In depleting this region, the source and drain are copartners of the gate. Although the gate is responsible for a major fraction of this depletion, the source and drain junctions too contribute to depletion. Small portions of the depletion layer charge are balanced by the charges in the source and drain regions. Effectively, less gate charge is necessary for depletion than would be required if source and drain did not partake of the responsibility. This means that the threshold voltage is reduced by an amount ΔV_{Th} due to source and drain effects. For a long-channel MOSFET, ΔV_{Th} is negligibly small but for a short-channel device, it becomes conspicuous. Also, in the same wafer, any two transistors having different channel lengths will differ in V_{Th}. This is true even in the same die. The threshold voltage reduction due to the decreased channel length represents V_{Th} roll-off.

Drain-induced barrier lowering (DIBL) is the drain voltage-induced decrease in threshold voltage in a short-channel MOSFET at high drain voltages. It arises from electrostatic coupling between the drain and the source. In consequence to this coupling, the potential barrier of the source-to-channel junction is depressed. It is this depression of the potential barrier to current flow at the source under the influence of the drain voltage that is responsible for the reduction in V_{Th}.

DIBL occurs in a short-channel MOSFET due to the relatively pronounced charge sharing effect between the channel depletion region and source/drain depletion regions as compared to long-channel device case. If the barrier between the source and channel is decreased, electrons are more freely injected into the channel region. Consequent upon the injection of electrons, the transistor requires less gate voltage to deplete the substrate beneath the gate dielectric. Hence, V_{Th} is

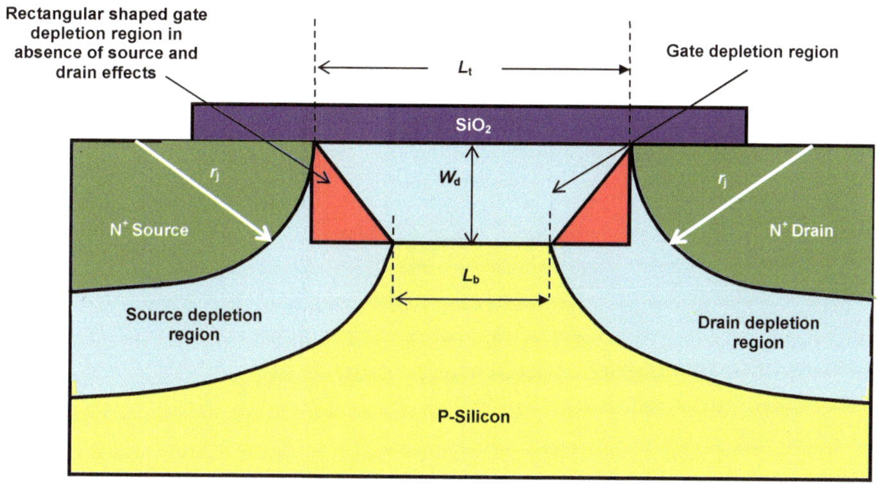

Fig. 5.3 Analysis of drain-induced barrier lowering

decreased and the transistor is switched on prematurely. Therefore, the threshold voltage is brought down and the gate has inferior control over the channel current. In a classical long-channel field-effect transistor, the barrier at the source end is situated far away from the drain contact. So, it is electrostatically shielded from the drain by the combination of the substrate and gate voltages. For this reason, the threshold voltage is independent of drain voltage.

Because of source and drain effects, the depletion region arising from the gate becomes trapezoidal in shape in place of the rectangular shape if these effects were absent (Fig. 5.3).

Let L_t, L_b denote the lengths of top and bottom parallel sides of the trapezium. Then the factor r by which the depletion charge is decreased from its value for the rectangle shape is given by

$$r = 1 - (L_t + L_b)/(2L_t) \tag{5.1}$$

If r_j is the radius of curvature of source/drain diffusion, it can be shown that

$$L_b = L_t - 2r_j\left(\sqrt{1 + (2W_d)/r_j} - 1\right) \tag{5.2}$$

where W_d is the depletion region width underneath the gate. For a first order analysis, the change in threshold voltage with respect to a long-channel device is

$$\Delta V_{\mathrm{Th}} \equiv |V_{\mathrm{Th}}| - \left|V_{\mathrm{Th(Longchannel)}}\right| = -\left(\frac{qN_A W_d}{C_{\mathrm{ox}}}\right)\left(\frac{r_j}{L_t}\right)\left(\sqrt{1 + (2W_d)/r_j} - 1\right) \tag{5.3}$$

where N_A is the doping concentration of the P-substrate and C_{ox} is gate oxide capacitance per unit area.

The change ΔV_{Th} is decreased by lowering N_A; by increasing C_{ox} with a thinner gate oxide; and by using a smaller r_j. For smaller r_j, the source and drain junctions can be made shallower but parasitic resistances R_{source}, R_{drain} of these regions will increase

$$R_{source}, R_{drain} \propto \rho/(Wr_j) \tag{5.4}$$

where ρ is the resistivity of source/drain region and W is the channel width. For r_j reduction, shallow extensions of the source/drain regions are formed. The accompanying increase in R_{source}, R_{drain} is considerably low. For providing shallow extensions, a dielectric spacer is included in the MOSFET structure.

At very high V_{DS} values, the depletion regions of the source and drain touch each other. When they mingle together, a high current flows from source to drain. It is irrepressible by the gate. This phenomenon is called punch-through. If L is the channel length, the punch-through voltage V_{PT} is given by

$$V_{PT} = qN_A L^2/(2\varepsilon_0\varepsilon_{Si}) \tag{5.5}$$

showing that the punch-through voltage decreases as L becomes smaller.

5.5 Velocity Saturation

As the gate length is decreased to smaller values, the longitudinal electric field E_x between the source and drain increases and becomes larger. The behavior of the carrier velocity v_d varies according to E_x. At low values of E_x, the carrier velocity v_d is proportional to E_x. But at higher values of E_x, the proportionality relationship is grossly violated. When E_x exceeds 3×10^4 V/cm for electrons and 10^5 V/cm for holes, the carrier velocity saturates at a value $v_{sat} = 10^7$ cm/s for electrons and $v_{sat} = 6 \times 10^6$ cm/s for holes. This saturation is explained by the increased scattering rate at the high electric fields. The saturation current I_{DS} is no longer a quadratic function of V_{GS}. It increases linearly with $(V_{GS} - V_{Th})$:

$$I_{DSsat} = v_{sat} W C_{ox} (V_{DS} - V_{Th}) \tag{5.6}$$

It is independent of channel length. It is lower than that for a long-channel MOSFET, resulting in a substantial decline in current drive. The short-channel MOSFET saturates at a lower V_{DS} value.

5.6 Carrier Mobility Degradation

5.6.1 Horizontal Field Effect

Mobility decreases as the carrier velocity saturates and becomes constant. Caughey-Thomas analytical equations for dependence of electron mobility $\mu_n(E)$ and hole mobility $\mu_p(E)$ on electric field E parallel to the current flow are [5]

$$\mu_n(E) = \mu_{n0}\left\{1+\left(\frac{\mu_{n0}E}{v_{\text{satn}}}\right)^{\beta_n}\right\}^{-1/\beta_n} \tag{5.7}$$

$$\mu_p(E) = \mu_{p0}\left\{1+\left(\frac{\mu_{p0}E}{v_{\text{satp}}}\right)^{\beta_p}\right\}^{-1/\beta_p} \tag{5.8}$$

$$v_{\text{satn}} = v_{\text{satp}} = \frac{2.4\times 10^7}{1+0.8\exp(T/600)}\text{cm/s} \tag{5.9}$$

where μ_{n0}, μ_{p0} are low-field electron and hole mobilities, respectively (μ_{n0} = 1375 cm^2/V s, μ_{p0} = 487 cm^2/V s); v_{satn}, v_{satp} are the corresponding saturation velocities; β_n, β_p are user-defined unit-less parameters determined experimentally ($\beta_n = 2$, $\beta_p = 1$); and T is the lattice temperature. These are empirical equations which are derived by the above authors from published experimental data.

5.6.2 Vertical Field Effect

Near the surface, enhanced scattering of carriers takes place due to surface acoustic phonons and surface roughness. Since the carrier transport is confined in the constricted MOSFET inversion layer near the silicon–silicon dioxide interface, the carriers experience great difficulty in moving parallel to the interface. Surface mobility is thereby decreased to $\leq$ half the bulk mobility. The mobility degradation by surface scattering is taken into account by writing the reciprocal of mobility as the sum of reciprocals of three terms according to Matthiesen's rule [6]:

$$1/\mu = 1/\mu_b + 1/\mu_{\text{ac}} + 1/\mu_{\text{sr}} \tag{5.10}$$

where μ_b is the carrier nobility in the bulk, μ_{ac} is the mobility limited by surface acoustic phonons, and μ_{sr} is the mobility limited by surface scattering. The μ_b term is obtained from the Klaassen model [7]. Lombardi model gives the μ_{ac} term [8]. Surface roughness scattering is treated in [8, 9].

5.7 Impact Ionization

In a short-channel NMOS transistor, the electrons accelerated to high velocities by the large longitudinal electric field bombard silicon atoms, liberating electrons from their outermost shells. The electrons thus released also acquire high velocities, taking part in further collisions, and generating electron–hole pairs. The drain attracts the ejected electrons while the holes move to the P-substrate. This multiplicative process produces an avalanche of free carriers. The N^+-source-P-substrate-N^+-drain acts like an NPN transistor. If the aforesaid holes are collected by the source, and the hole current produces a voltage drop in the substrate, the source-substrate junction will be forward-biased. Then electron injection will start from the source to the substrate. These electrons can move toward the drain, creating electron–hole pairs and aggravating the situation.

5.8 Hot Carrier Effects

The decreasing feature size of MOSFET devices is accompanied by an increase of the electric field in their channel regions. Hot carriers are charged particles, either electrons or holes. They include particles which have acquired very high kinetic energies upon acceleration by the large electric field prevalent across the channels of MOSFETs. These carriers have higher energies than those of carriers normally found in semiconductor devices. Due to their high energies, hot carriers may migrate into and roam around the unwelcome areas of the devices. Such areas are the gate dielectric and substrate of a transistor. They cause shifts in threshold voltage. Device transconductance is also degraded.

Hot carrier injection is more serious in N-channel MOSFETs than P-channel devices because of the higher mobility of electrons. Consequently, electrons acquire higher energies and become hotter than holes. Further, the energy barrier is lower for electrons than that for holes.

5.8.1 Substrate Hot Electron (SHE) Injection

When exceedingly high positive or negative positive or negative voltages are applied to the MOSFET substrate or body, SHE injection is triggered. Then the substrate field impels charge carriers of one type in the substrate toward the Si–SiO_2 interface. They gain high kinetic energy and are hurled into SiO_2.

5.8.2 Channel Hot Electron (CHE) Injection

When both V_{GS} and V_{DS} are very high, some electrons are driven toward the gate oxide.

5.8.3 Drain Avalanche Hot Carrier (DAHC) Injection

When $V_{DS} > V_{GS}$, the acceleration of charge carriers in the channel region causes impact ionization from atomic-level collisions near the drain. Electron–hole pairs are thus produced and further carrier multiplication ensues. The generated electron–hole pairs gain enough energy to break the barrier at Si–SiO_2 interface and penetrate into the oxide.

5.8.4 Charge Generation Inside SiO_2

(a) Negative charge buildup by trapping of hot electrons in the oxide near the drain. (b) Positive charge accumulation by injection of holes into the oxide. (c) Interface state generation at Si–SiO_2 interface. This happens because some Si–H, Si–Si and Si–O bonds need less energy to break. Taking advantage of this situation, any electron with energy >2 eV can release H_2 and create interface states.

5.9 Random Dopant Fluctuations (RDF)

These arise from statistical irregularities in dopant concentrations, which become pronounced when the MOSFET area decreases. As a consequence, threshold voltages of neighboring devices may differ.

5.10 Overcoming Short-Channel Effects in Classical MOSFETs

5.10.1 Avoiding DIBL Effect

DIBL effect is reduced by decreasing the gate oxide thickness. The thickness reduction makes the gate more effective in controlling the channel region.

5.10.2 Reducing Gate Leakage Current

Thermal silicon dioxide is amorphous in structure with dielectric constant 3.9. It is a three-dimensional network of tetrahedral cells. Atomic diameter of silicon is 0.236 nm, that of oxygen is 0.13 nm, average silicon–oxygen bond length = 0.162 nm, oxygen ion–oxygen ion distance = 0.262 nm, silicon–silicon bond distance = 0.31 nm [10]. Thermal SiO_2 has been the preferred MOSFET gate insulator. It was constantly scaled down in thickness up to the 130 nm technology node at the rate of $0.7x$ per MOSFET generation. But after reaching this node, the scaling became slower in pace particularly in sub-100 nm range. At 90 nm and 65 nm, it was considerably sluggish due to the resulting high leakage current. The ultimate limit at which bulk oxide could be used was that required in 70 nm technology node = 0.7 nm. It is about two atomic layers in thickness.

Now, gate oxide thickness = 1.2 nm is being used in MOSFET manufacturing. So, the present oxide thickness is hardly a single atomic layer thick [11]. The oxide thickness at which direct tunneling starts is 3 nm. With reduction of gate oxide thickness, direct quantum-mechanical tunneling of electrons from the gate across the gate oxide to the underlying silicon causes an increase in the gate leakage current. The gate leakage current density = 100 Acm^{-2} at 1 V for 1.2 nm thick oxide.

After reaching the 70 nm node, a desperate need was felt to use a high dielectric constant insulating material above the silicon dioxide layer to subdue the leakage current to ignorable proportions. The problem created by ultra-thin oxide has been tackled by using gate dielectric materials of high dielectric constant k such as zirconium oxide (ZrO_2) with $k = 25$, hafnium oxide (HfO_2) with $k = 30$, etc. For a high k dielectric, an equivalent oxide thickness (EOT) is defined as

$$\mathrm{EOT} = t_{\mathrm{ox}} = t_k \times (3.9/k) \tag{5.11}$$

where t_{ox} is thickness of oxide film and t_k that of the gate dielectric. Because $t_k \gg t_{\mathrm{ox}}$, the gate leakage current is significantly reduced by using such a dielectric.

Beyond 70 nm node, requirement of high-κ materials was essential. They were necessary despite the fact that many of these materials showed poor thermal stability and interface quality with silicon dioxide. Most of these materials are oxides of transition metals. They are deposited on silicon dioxide instead of being thermally grown. The methods commonly used are based on physical and chemical vapor deposition; sol-gel process is also used. Among physical vapor deposition (PVD) methods, thermal evaporation and reactive sputtering stand out prominently. Chemical vapor deposition methods include atomic layer deposition (ALD), plasma-enhanced atomic layer deposition (PEALD), plasma-enhanced chemical vapor deposition (PECVD), metal-organic chemical vapor deposition (MOCVD), and molecular beam epitaxy (MBE). Surface preparation prior to deposition plays a critical role in obtaining desired film quality and adhesion. Post deposition thermal treatments such as sintering or annealing too are decisive factors. Besides oxides, silicate films, notably $ZrSiO_x$ and $HfSiO_x$ are also used.

5.10.3 Strain Engineering for Enhancing Carrier Mobility

One could replace silicon (electron mobility $\leq$ 1400 cm^2V^{-1} s^{-1}) with high-mobility semiconductors GaAs (electron mobility $\leq$ 8500 cm^2V^{-1} s^{-1}) or InP (electron mobility $\leq$ 5400 cm^2V^{-1} s^{-1}). But for these semiconductors, the level of technological maturity for large-scale production of ICs is much lower.

Use of strained silicon increases mobility. In strained silicon, the silicon atoms are pulled apart from their normal positions in the lattice, increasing their inter-atomic distance by a small amount $\sim$1%. Strain engineering is a strategy employed in silicon IC manufacturing to increase the carrier mobility. By virtue of the increase in spacing between the atoms than for regular silicon, the electronic band structure of silicon is modified in such a manner that effective mass of charge carriers in silicon is reduced. Lower is the effective mass higher is the mobility. Also, carriers are diverted to regions of lower effective mass. Effectively, better transport properties are achieved [12].

Strain engineering is done in one of the two ways, either globally or locally. Uniaxial global strain is introduced by bonding the wafer to a substrate with cylindrical surface. Biaxial global strain is generated by epitaxial growth of a thin strained silicon layer on a thick relaxed SiGe virtual substrate.

Epitaxy is the process of growing a single crystal film on a crystalline substrate. In this growth process, the substrate serves as a template according to which the deposited atoms arrange themselves. Without this template, the growth will be either polycrystalline or amorphous. Since Si and Ge have identical crystal structure, a Si overlayer can be grown on a Ge substrate. But Ge atoms are larger in size than Si atoms. Moreover, interatomic spacing is 4.2% greater in Ge than in Si. Due to this reason, direct deposition of Si atoms on Ge lattice results in a high defect concentration in the form of dislocations. A favorable condition is provided by the fact that lattice constant of SiGe alloy defining the unit cell of the lattice has a value between the lattice constants of Si and Ge. So if a Si overlayer is deposited on a SiGe substrate, the silicon atoms attempt to be coincident with SiGe atoms below. As a result, the Si atoms are placed at larger distances than they would be in a normal Si lattice. So, the Si layer grown by epitaxy is stretched slightly as if like a rubber diaphragm from the normal Si lattice. This is equivalent to generation of a strain in the Si lattice similar to the strain in the extended diaphragm. But the crux of the problem is that it is difficult to obtain a SiGe template of high quality on which the epitaxial Si layer could be grown. A way to get out of this situation has been found out (Figs. 5.4 and 5.5). One starts with Si substrate instead of SiGe substrate. A buffer layer is grown over this substrate. This buffer layer contains Si and Ge. To begin with, the concentration of Ge is zero at the bottom of the buffer layer. As thickness increases, the concentration of Ge is slowly raised. The final

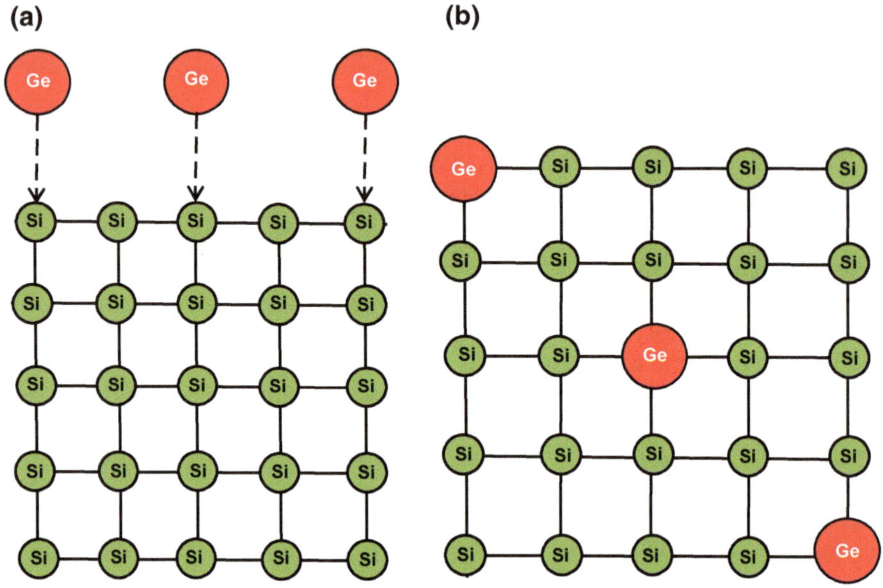

Fig. 5.4 First step in strained silicon process: **a** adding Ge atoms to surface layers of silicon crystal to produce **b** $Si_{1-x}Ge_x$ layer

concentration of Ge at the top of the buffer layer is 20%. After the buffer layer has been grown, a template layer of uniform SiGe concentration is grown. As this template layer maintains the concentration of the buffer layer, it is not strained. It is therefore a relaxed layer. Over this layer, the Si capping layer is grown. This capping layer made of Si only follows the structural arrangement of SiGe template layer below. Obviously, it is a strained layer since the lattice constant of Si differs from that of SiGe. But the lattice constant does not differ to the extent to be able to cause dislocations. Hence, it is said to be optimally strained.

In the 90 and 65 nm modes, process-induced stress generation was extensively applied: compressive stress for PMOS (Fig. 5.6) and tensile stress for NMOS (Fig. 5.7). Local compressive stress is created by selectively growing a thin epitaxial film of SiGe in the source/drain regions of P-channel MOSFET.

In case of N-channel MOSFET, growth of SiC film develops local tensile stress. Other local stress generation techniques include the formation of shallow trench isolation or a stressed silicon nitride capping layer [13].

Use of silicon nitride as a strain-inducing capping layer (Fig. 5.8) offers the flexibility of controlling the type of strain (compressive or tensile) as well as the degree of strain by proper choice of deposition conditions, mainly temperature. This approach is known as dual-stress liner method [14]. In a CMOS process, after the self-aligned silicide step, a tensile silicon nitride layer is deposited over the complete wafer. Photolithography is performed to selectively remove the silicon nitride from regions where P-channel MOSFET is located. Thereafter, a compressive

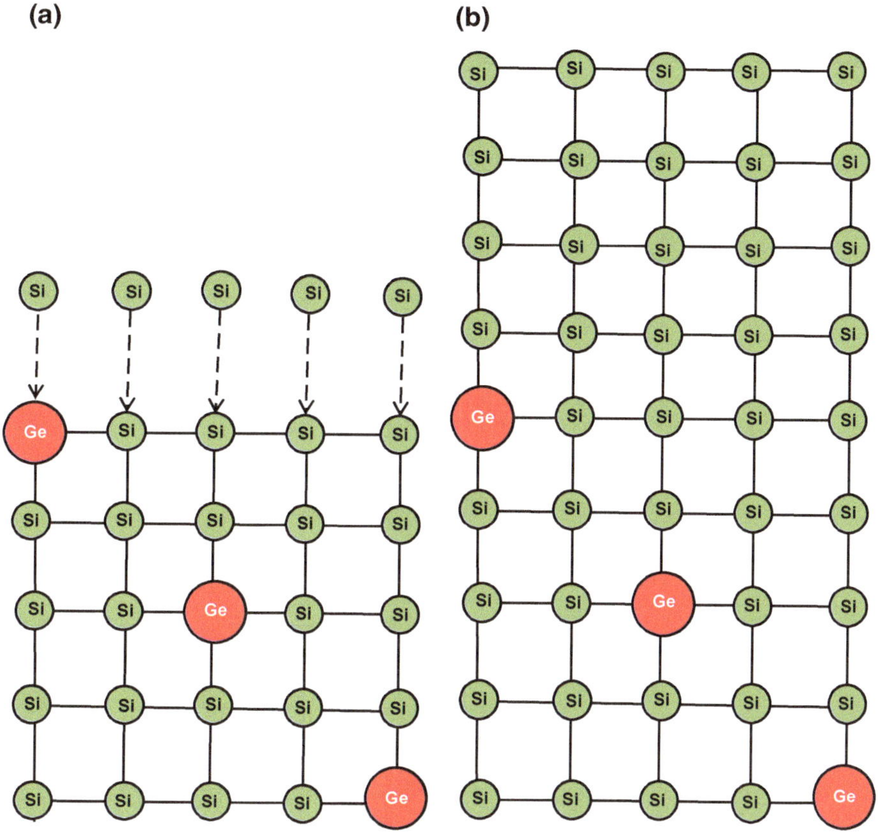

Fig. 5.5 Second step in strained silicon process: **a** epitaxial growth of silicon on $Si_{1-x}Ge_x$ layer, and **b** resulting strained silicon

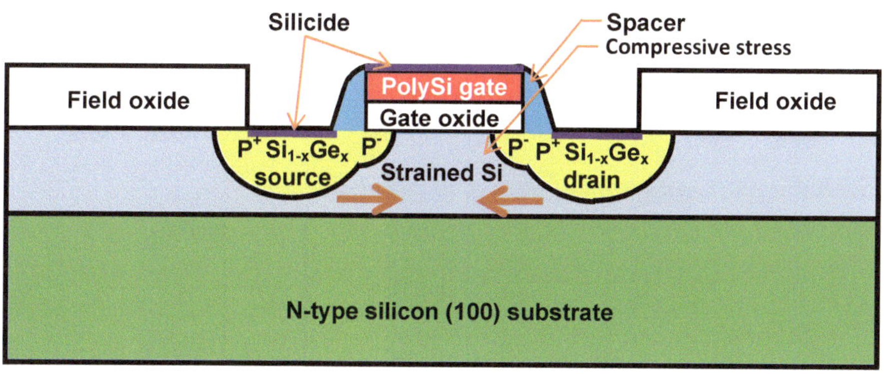

Fig. 5.6 Compressive strain induction in the silicon channel region of a P-channel MOSFET by $Si_{1-x}Ge_x$ source/drain regions

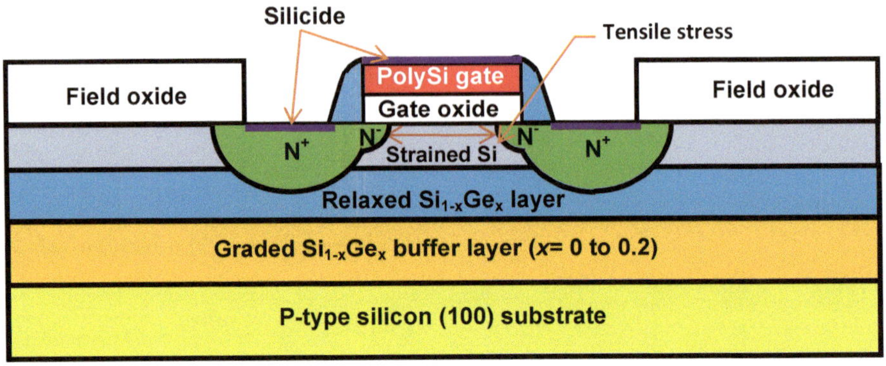

Fig. 5.7 Tensile strain generation in the silicon channel region of an N-channel MOSFET; the channel region is grown epitaxially on a $Si_{1-x}Ge_x$ layer over a P-silicon substrate

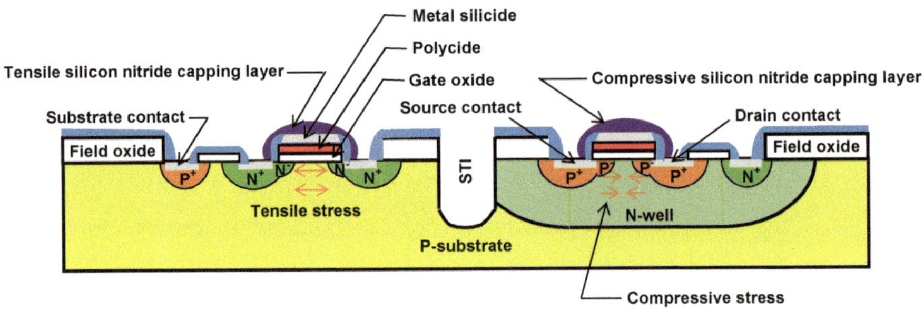

Fig. 5.8 Strained silicon produced by two types of silicon nitride stress liners formed under different deposition conditions

silicon nitride layer is deposited over the full wafer and etched from N-channel MOSFET regions. A significant increase in drive current was reported by simultaneous action of two different strain layers in the CMOS process flow without any SiGe layer.

5.10.4 Minimization of Hot Carrier Effects

Different techniques have been proposed for reduction of hot carrier effects, such as

(i) **Gate Oxide Thickness Reduction** By using a thinner oxide, the point of peak of electron injection is shifted to a greater extent toward the drain region. Hence, the stretch of damaged region overlying the channel is smaller in size.
(ii) **Lightly Doped Drain (LDD) Structure** This structure consists of two doping concentrations, one high and another low. Light doping, shallow implant is

done in a region abutting the channel, and therefore establishing contact with it. Small overlap of the gate with source or drain regions produces minimum overlap capacitance. Further, the reduced carrier concentration at the drain edge decreases the field between drain and channel regions. The field thus lowered brings down the amount of carrier injection into the oxide, impact ionization and other related effects. The heavy doping, deep implant is done after depositing silicon nitride sidewall spacers on both sides of the polysilicon gate. It covers most of the source and drain areas, creating low series resistance with the channel. Thus the combination of two implants not only provides small access resistance and overlap capacitance but also minimizes hot carrier injection. Figure 5.9 shows the main steps of this process.

Elaborating the mechanism of the LDD structure for hot carrier inhibition, it essentially performs a kind of drain engineering. In this drain engineering, the peak of the lateral electric field located near the edge of the drain is moderated and weakened by modification of doping profile through low dose implantation in the concerned regions. As the lightly doped regions created by low dose implantation look like extensions of actual heavily doped source and drain, they are referred to as source/drain extensions.

(iii) **Double-Diffused MOSFET Structure** It has deeper N-phosphorous profile than N^+As profile. The outcome is that the path of maximum current is located away from the position of the peak field. This helps to reduce the impact ionization, and thereby hot carrier generation.
(iv) **Incorporating Si_3N_4 as the Gate Oxide** The Si–N bonds require more energy to break than Si–H bonds.
(v) **Deuterium Post-metal Annealing** Post-metallization annealing in hydrogen ambient at a low temperature can reduce Si–SiO_2 interface traps.

5.10.5 Preventing Punch-Through

It can be avoided by increasing the channel doping. The higher doping shortens the source and drain depletion regions. However, higher channel doping decreases the mobility by enhanced scattering. Therefore, it will solve the punch-through problem at the expense of mobility reduction and consequent decrease in on-state current.

Halo or pocket implants (Fig. 5.10) are implants used to suppress punch-through between source and drain through the substrate. Hence, they are often called punch-through suppression implants. To achieve this objective without mobility degradation, the dopants are placed a little below the channel adjoining the source and drain regions. In this way, they can accurately adjust the background doping concentration of the well in the intended location. At the same time, carrier mobility is not decreased.

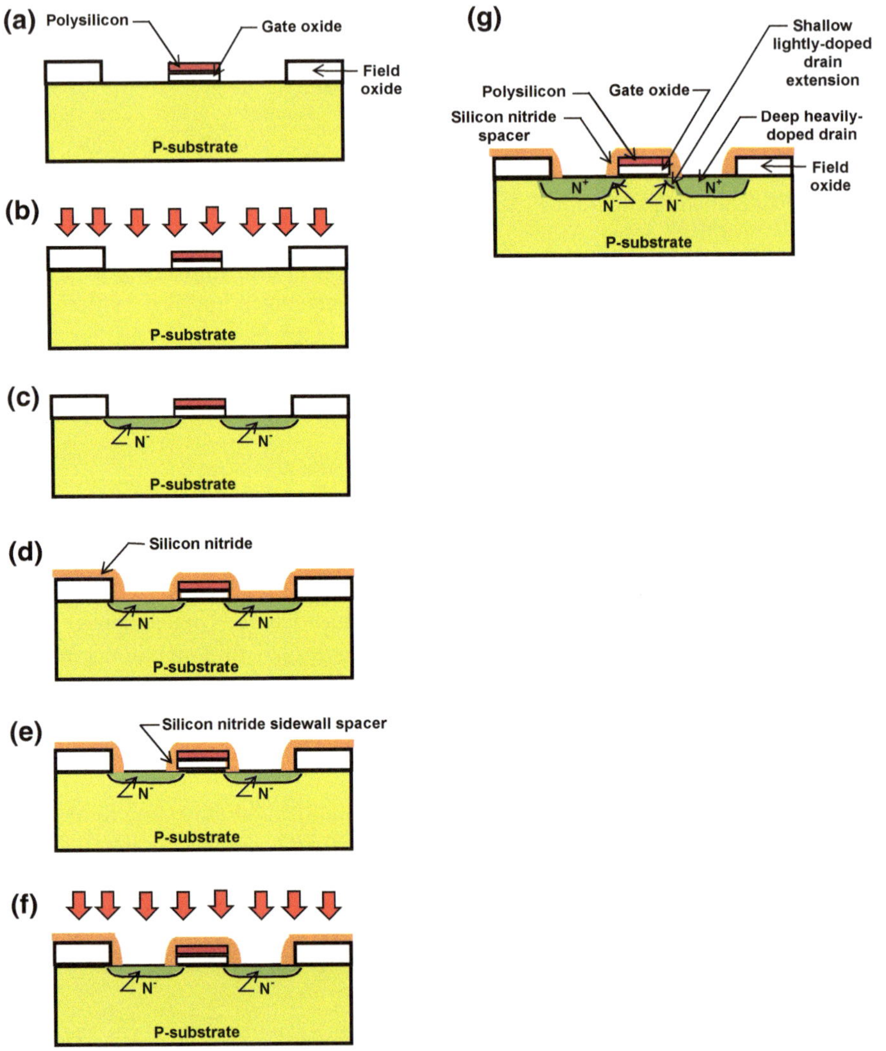

Fig. 5.9 Self-aligned process for fabrication of lightly doped source/drain structure. **a** Gate oxide and polysilicon etching. **b** N-source/drain extension implant. **c** High-temperature annealing. **d** Nitridation for sidewall spacer formation. **e** Nitride etching. **f** N^+ source/drain implant. **g** High-temperature annealing

Another strategy for avoidance of punch-through employs the super steep retrograde well (SSRW) and a thin intrinsic region in the channel region of the MOSFET. "Retrograde" means backwards. In a retrograde well, the doping concentration is lowest in the region near the gate insulator. It increases as one moves far away from the gate insulator and the channel region deep into the P-substrate. By doing so, the mobility in low-concentration zone close to the gate insulator is

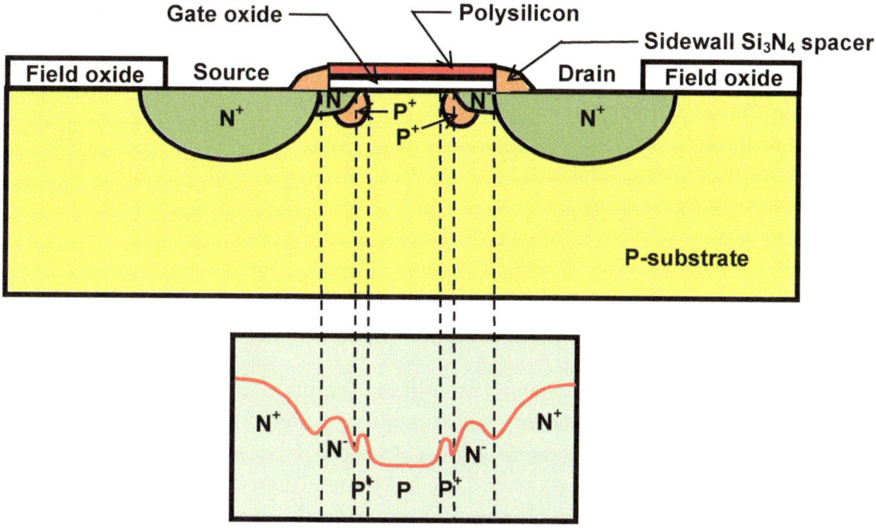

Fig. 5.10 Cross-sectional diagram of N-channel MOSFET with lightly doped drain structure and halo implant. Doping profile at the surface of the device is shown below the diagram

preserved at a high value. This happens because ionized impurity scattering in the zone is reduced. Hence, on-state current is increased. Also, the high-concentration zone buried deep into the substrate counteracts the substrate punch-through. The off-state current is minimized because the depletion depths on both sides of the channel are smaller due to the high doping. As a result, space charge generation current and thereby off-state current are diminished.

5.10.6 Innovative Structures Superseding Classical MOSFET

Evidently, short-channel effects are a manifestation of loss of control over the channel by the gate electrode due to interfering electric fields. The channel should be under tighter control of the gate. Then only such disturbances can be barred from exerting their influence.

5.11 Discussion and Conclusions

With strained silicon using SiGe, bulk MOSFET crossed the 100 nm signpost, reaching 90 nm technology node in 2003. Second generation SiGe led to 65 nm node devices in 2005. Metal gate and high-κ dielectric with gate-last approach provided 45 nm node in 2007 and 32 nm node in 2009 in second generation devices. Gate length is 70% less than the technology node, e.g., for 90 nm node, the gate length is <63 nm, … for 32 nm node, it is <22.4 nm. An important observation is that in all these efforts, either by using SiGe or other strain-inducing layers, or by reverting to metal gates with high-κ dielectrics, enhancements in properties of materials helped us a great deal in treading the Moore's curve. However, no major structural breakthroughs were necessary.

Review Exercises

5.1 When is a MOSFET channel said to be: (i) long, and (ii) short? Can a short channel be avoided during miniaturization of MOSFET?

5.2 What is polysilicon depletion effect? Up to what extent doping elevation can obviate this effect? Can it be avoided by using metal gates?

5.3 Do source and drain junctions play any role in depleting the MOSFET substrate? In what way does this contribution of source and drain junctions change in a short-channel device? What is threshold voltage roll-off?

5.4 Why is the threshold voltage of a long-channel MOSFET independent of drain voltage but this is not so in a short-channel device?

5.5 Explain drain-induced barrier lowering in a MOSFET. What is the effect on threshold voltage of the device?

5.6 How does the threshold voltage reduction in a short-channel MOSFET differ for source/substrate and drain/substrate junctions, which are: (i) deep, and (ii) shallow?

5.7 At high drain voltages, the source and drain junctions may touch each other. What is this phenomenon called? How is the voltage at which this phenomenon occurs related with the channel length of MOSFET?

5.8 What is velocity saturation? What is its effect on the current drive of a MOSFET?

5.9 Write the Caughey-Thomas equations for the dependence of mobility on electric field.

5.10 How are the contributions of bulk and surface effects in mobility of carriers in a MOSFET inversion layer expressed by Matthiesen's rule?

5.11 How is impact ionization produced in a MOSFET at high drain voltages? How the parasitic bipolar transistor makes the situation worse?

5.12 What is a hot carrier? Describe the different hot carrier effects in a MOSFET.

5.13 How is leakage through gate oxide avoided by using a high-κ dielectric? What is equivalent oxide thickness?

5.14 Write the equation relating the equivalent oxide thickness with thickness of high-k dielectric.

5.15 How does straining the silicon lattice increase the carrier mobility? Briefly explain the dual-stress liner approach.

5.16 How do the following structures help in avoiding hot carrier effects? (i) light-doped drain, and (ii) double-diffused MOSFET?

References

1. Maszara WP (2005) Fully silicided metal gates for high-performance CMOS technology: A Review. J Electrochem Soc 152(7):G550–G555
2. Kittl JA, Lauwers A, Mv Dal et al (2006) Ni, Pt and Yb based fully silicided (FUSI) gates for scaled CMOS technologies. ECS Trans 3(2):233–246
3. Moyer B (2011) Gate first vs last: a summary of the issue now that things should have settled down. Electronic Engineering Journal © 2003–2015 techfocus media, Inc. http://www.eejournal.com/archives/articles/20111114-gate/. Accessed 8 Oct 2015
4. Hoffmann TY (2015) Integrating high-κ/metal gates: gate-first or gate-last? Solid State Technology ©2015 Extension Media, http://electroiq.com/blog/2010/03/integrating-high-k/. Accessed 8 Oct 2015
5. Caughey DM, Thomas RE (1967) Carrier mobilities in silicon empirically related to doping and field. Proc IEEE 55:2192–2193
6. Darwish MN, Lentz JL, Pinto MR et al (1997) An improved electron and hole mobility model for general purpose device simulation. IEEE Trans Electron Devices 44(9):1529–1538
7. Klaassen DBM (1992) A unified mobility model for device simulation—Part I: model equations and concentration dependence. Solid State Electron 35(7):953–959
8. Lombardi C, Manzini S, Saporito A et al (1988) A physically based mobility model for numerical simulation of nonplanar devices. IEEE Trans Comput Aided Des 7:1164–1171
9. Hartstein A, Ning TH, Fowler AB (1976) Electron scattering in silicon inversion layers by oxide and surface roughness. Surf Sci 58:178–181
10. Silicon Dioxide Properties. http://www.iue.tuwien.ac.at/phd/filipovic/node26.html. Accessed 8 Oct 2015
11. Misra D, Iwai H, Wong H (2005) High-κ gate dielectrics. Electrochem Soc Interface Summer 2005:30–34
12. Intel: Strained Transistors. http://www.intel.com/pressroom/kits/advancedtech/doodle/ref_strain/strain.htm. Accessed 30 Aug 2015
13. Ungersboeck E, Sverdlov V, Kosina H et al (2006) Strain engineering for CMOS devices. In: 8th international conference on solid-state and integrated circuit technology (ICSICT'06), Shanghai, Oct 2006, pp 124–127
14. Yang HS, Malik R, Narasimha S et al (2004) Dual Stress liner for high performance sub-45 nm gate length SO1 CMOS manufacturing. In: IEEE international electron devices meeting, IEDM technical digest, 13–15 Dec 2004, pp 1075–1077

Chapter 6
SOI-MOSFETs

Abstract Continuing the onward advancement from where the classical MOSFET failed to meet the expectations of Moore's law, it was widely accepted that novel MOSFET structures were direly needed in order that the pace of the progress is not slackened. It was also evident that short-channel effects could only be obviated if the gate action could be strengthened so that the channel region is always under the solitary control of the gate. The advent of silicon-on-insulator technology came as a breakthrough to rescue the CMOS engineers. First partially-depleted silicon-on-insulator (SOI) MOSFETs entered the market followed by the fully-depleted MOSFET devices. The fully-depleted MOSFETs represent a cornerstone of technological transformation leading to downscaling to lower levels.

6.1 Introduction

Scaling down MOSFETs from 28 to 20 nm gate length has been vehemently opposed by short-channel effects. Economic viability of planar semiconductor processes has been questioned at ≤ 20 nm gate length. Process and design engineers have come up with two novel structures: the fully-depleted SOI-MOSFET and FINFET. In the present chapter, SOI-MOSFET will be comprehensively discussed. The next chapter will describe the FINFET.

The silicon-on-insulator (SOI)-MOSFET is a structural innovation, which brings the channel under stringent control of the gate. The novelty in the structure is the use of a buried oxide layer under the channel.

In SOI technology, transistors are fabricated in a "body region" as opposed to their formation in "bulk substrate." Electrical isolation is achieved through trench oxide. In the SOI-MOSFET, the drain-to-substrate capacitance is very low because the dielectric constant of oxide (3.9) is $\ll$ that of silicon (11.7). The technology provides an overall improved performance than bulk MOSFET, with reduction of parasitic effects. High chip density is obtained with lower power consumption. Other advantages include enhanced latchup immunity, better radiation hardening,

V.K. Khanna, *Integrated Nanoelectronics*,
NanoScience and Technology, DOI 10.1007/978-81-322-3625-2_6

Table 6.1 Bulk versus SOI-CMOS

Feature	Bulk CMOS	SOI-CMOS
Silicon used in electrical conduction	Only a diminutive depth of silicon in close proximity to the surface (0.1%) is exploited for device operation; residual silicon has a redundant role	Device operation is restricted to a thick/thin film of silicon, which constitutes the total silicon thickness leaving no unused portion of silicon
Parasitic effects	Interaction between the device and inactive regions causes detrimental parasitic effects	Active region is isolated from the substrate by buried oxide, thereby forestalling parasitic effects
Isolation	Inferior isolation through reverse-biased P–N junctions	Superior isolation through dielectric
Fabrication	Complex	Simple
Ground taps for isolation rings around transistors	Required	Not required by isolation trenches
Network of ground metal lines over whole chip	Required	Not required. Substrate biasing all around edge eliminates ground routing
Leakage current	Very high through isolation taps and substrate interface	Very low through buried oxide and isolation trenches
Power dissipation	High due to excessive leakage	Low owing to reduced leakage
Transistor area and packing density	Large area and low density due to large separation between drain/source region and isolation taps	Small area and high density because drain/source regions can touch isolation trenches
Necessity of well fabrication	N-well required for P-channel transistors	No well needed. N-body and P-body exist
Substrate capacitance	Large created by transistor formed in substrate; hence lower switching speed	Small due to buried oxide layer; hence higher switching speed
Backside gate	No	Yes
Latchup susceptibility	Yes. Parasitic NPN and PNP transistors create subsurface thyristors triggering latchup	No. Latchup-free operation is possible because N-channel and P-channel transistors reside in separate floating body tanks without any access to substrate due to the intervening buried oxide layer
Charge flow	Restrained because wells are biased to avoid forward biasing of P-N junctions	More liberty to flow of charge because of floating body regions, resulting in improved speed
Floating body effects	No	Yes
Short-channel effects	More	Less
Subthreshold swing	More	Less
Tolerance to harsh environments (high temperatures and high radiation dose rates)	Less tolerant	More tolerant
Self-heating	Less	More

less parameter variation with temperature, and capability of higher temperature operation, with more immunity to cross-talk effects. A side-by-side comparison between bulk CMOS and SOI-CMOS is made in Table 6.1.

6.2 SOI Wafer Manufacturing

SOI technology involves the growth of a thin layer of crystalline silicon. This thin layer is grown above an insulating film, usually made of silicon dioxide. Hence, this film is called buried oxide (BOX). Processes of SOI wafer manufacturing are elucidated in the following subsections.

6.2.1 Separation by Implanted Oxygen (SIMOX) Process

This is the most popular method for large-scale SOI wafer production. It consists of two main steps (Fig. 6.1): (*a*) *Oxygen ion implantation in silicon substrate* During oxygen ion implantation, the substrate is maintained at a temperature of 600 °C. The high temperature is necessary to prevent undesirable amorphous silicon overlayer formation. The implantation energy is 120–200 keV and dose (number of ions cm^{-2} of silicon wafer) is $3 \times 10^{17} - 1.8 \times 10^{18}$ cm^{-2}. (*b*) *Annealing at a high temperature* The high-temperature annealing is done at a temperature of 1300 °C for 6 h in an inert ambient to strip the silicon overlayer of implantation-induced defects. A uniform buried oxide layer is formed having distinct interfaces with silicon on both sides. Typical silicon overlayer thicknesses are: 50–250 nm and buried oxide layer thicknesses are: 100, 200, 500 nm, ….

6.2.2 Bond and Etch-Back SOI (BESOI) Process

This process consists of five steps [1] shown in Fig. 6.2: (*a*) *Thermal oxidation of wafer A* The buried oxide of required thickness is grown. (*b*) *Direct fusion bonding of wafer B on the oxidized wafer A* The wafers A and B are made hydrophilic by proper surface treatment. A preprocessing plasma treatment and wet cleaning by RCA (Radio Corporation of America) procedure is essential. Following surface preparation, they are brought into intimate contact in a clean environment and gently pressed together at low temperature (400 °C) in vacuum at one central point for pre-bonding. As the contact wave spreads, the hydrated wafers adhere together by van der Waal's forces of silanol groups (Si–OH) formed by water chemisorption on their surfaces: Si–OH + Si–OH → Si–O–Si + H_2O. (*c*) *Post-bonding annealing*

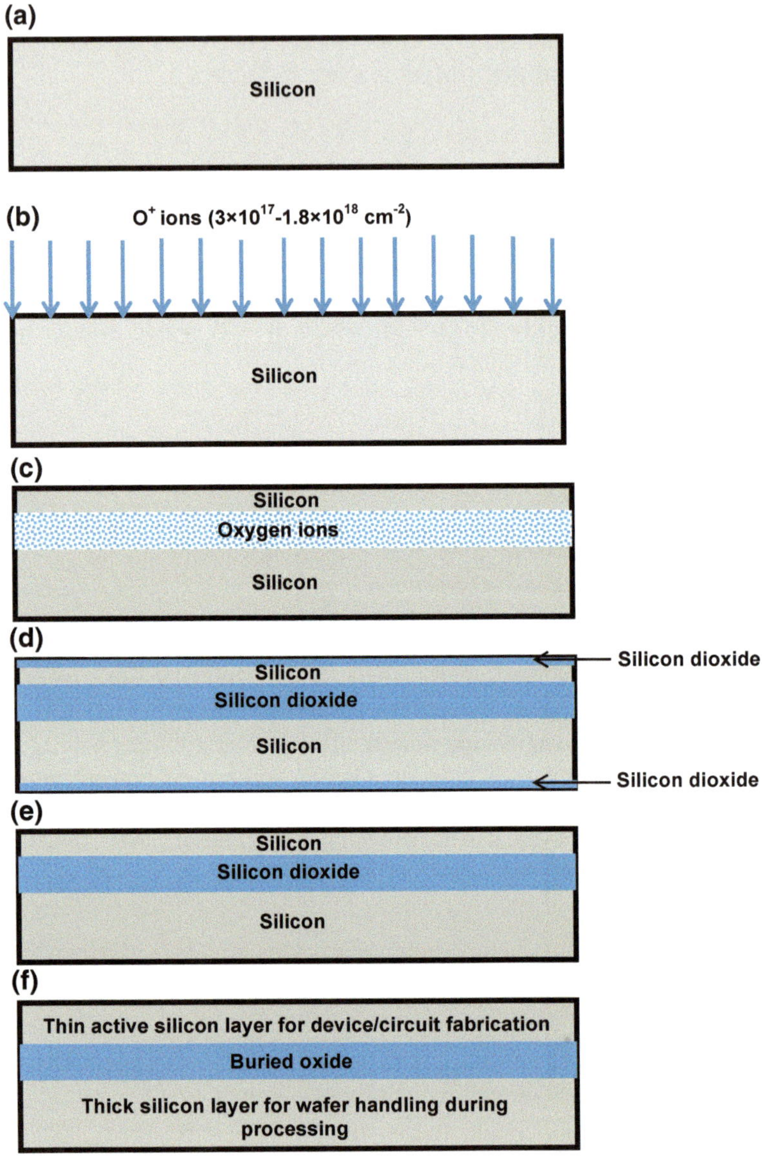

Fig. 6.1 SIMOX process steps for manufacturing SOI wafers. **a** Starting silicon wafer. **b** Oxygen ion implantation. **c** Cleaning. **d** Annealing at 1300 °C for oxidation. **e** Cleaning and surface oxide removal. **f** Epilayer Si growth if necessary

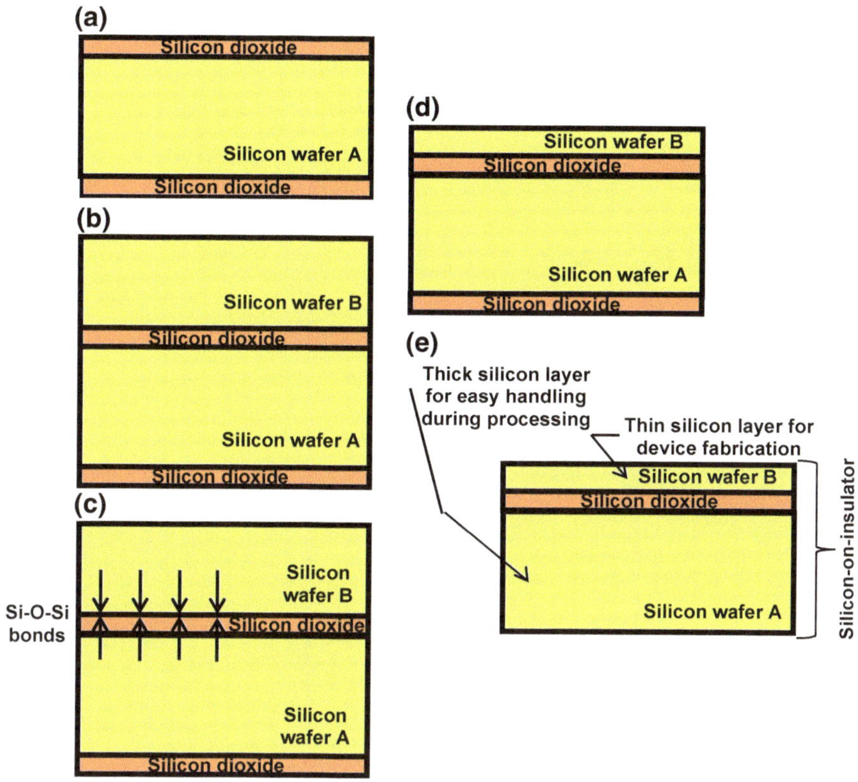

Fig. 6.2 SOI wafer fabrication using bond and etch-back process. **a** Thermally oxidize wafer A. **b** Bond wafer B to wafer A. **c** Anneal at high temperature. **d** Thinning of wafer B by etching. **e** Remove back oxide

The wafers are subjected to an annealing process at elevated temperature (1100 °C). During annealing, the water molecules diffuse away along the interface, forming strong Si–O–Si bonds. (*d*) *Etching back wafer B* The wafer B is thinned down to required thickness by etching silicon. (*e*) *Back oxide removal.*

6.2.3 Smart Cut® Process

The important steps of this process [2] are depicted in Fig. 6.3: (*a*) *Growth of silicon dioxide on wafer A* By thermal oxidation, silicon dioxide layer is grown. (*b*) *Hydrogen ion implantation* Hydrogen ions are implanted at a dose $\sim 5 \times 10^{16}$ ions cm^{-2} to form a Gaussian profile. The energy of hydrogen ions determines the distance of the peak of the Gaussian profile from the surface of the wafer. The distance corresponding to the peak gives the region of maximum lattice damage

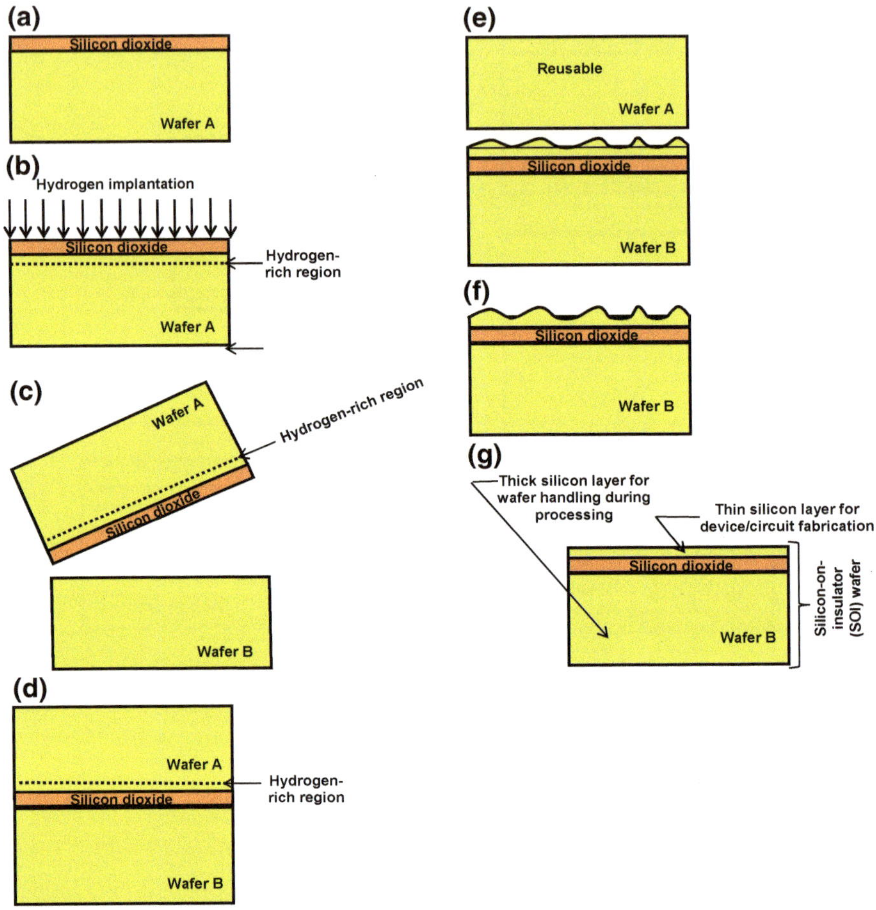

Fig. 6.3 Representation of SOI wafer fabrication by smart cut process using hydrogen ion implantation. **a** Thermally oxidize wafer A and remove oxide from one side. **b** Hydrogen ion implantation. **c** Invert wafer A on wafer B. **d** Wafer bonding. **e** Thermal splitting. **f** Annealing. **g** Chemical mechanical polishing

caused by hydrogen ion implantation, where the lattice bonds are significantly weakened. (*c*) *Direct bonding of handle wafer B to wafer A* This follows the same process as for BESOI case. (*d*) *Low-temperature splitting* Thermal processing at 400–600 °C separates wafer B from wafer A along the peak concentration of the profile of hydrogen implantation. (*e*) *Thermal annealing at high temperature* Annealing at 1100 °C strengthens the bond between implanted and handle wafers. (*f*) *Fine chemical mechanical polishing* (*CMP*) This removes roughness of the surface of SOI wafer. Wafer A is reusable.

The major aspects of aforementioned three SOI wafer manufacturing processes are itemized in Table 6.2.

Table 6.2 Comparison of SOI wafer manufacturing processes

SIMOX process	BESOI process	Smart cut® process
It is the most researched of the three methods. It has been commercially available since a long time > two decades. It is based on ion implantation. Hence, it offers consistent reproducibility in quality and SOI thickness	Silicon overlayer is not exposed to oxygen ion implantation at a high dose. Wide flexibility is available in thicknesses of SiO_2 and silicon overlayer. In absence of an etch stop, wide variability in SOI thickness is a serious drawback	It benefits from the repeatability of ion implantation process, as in SIMOX process. It also derives advantage from flexibility of BESOI process regarding the thicknesses of buried oxide and silicon overlayer. Recyclability of the wafer from which silicon overlayer is split enables the production of one SOI wafer from one silicon wafer

6.3 Classification of SOI-MOSFETs

A MOSFET having a thick layer of silicon, which is not fully depleted during operation, is called a partially-depleted (PD) MOSFET (Fig. 6.4a). It is often labeled as "thick film SOI device." A MOSFET fabricated with a thin layer of silicon having thickness < the depletion region thickness is known as a fully-depleted (FD) MOSFET (Fig. 6.4b). Frequently, it is called a "thin film SOI device." Both versions have their relative merits and demerits. The first partially depleted MOSFET dates back to 1964 while the first fully depleted device came in the early 1980s [3]. Please see Table 6.3 for comparison between partially- and fully-depleted SOI-MOSFETs [4, 5].

6.4 Floating Body Effects in SOI-MOSFET

6.4.1 *Kink Effects in Partially-Depleted SOI-MOSFET*

In SOI-MOSFETs, the body terminal is often left floating. Leaving the volume of silicon underneath the gate at a floating potential gives rise to some peculiar effects, which are unique to these devices, and are aptly called floating body effects [6]. The unbiased and undepleted portion of the body region can store charge. The charge Q thus accumulated produces a floating body potential V given by

$$V = Q/C \tag{6.1}$$

where C is the capacitance of the region.

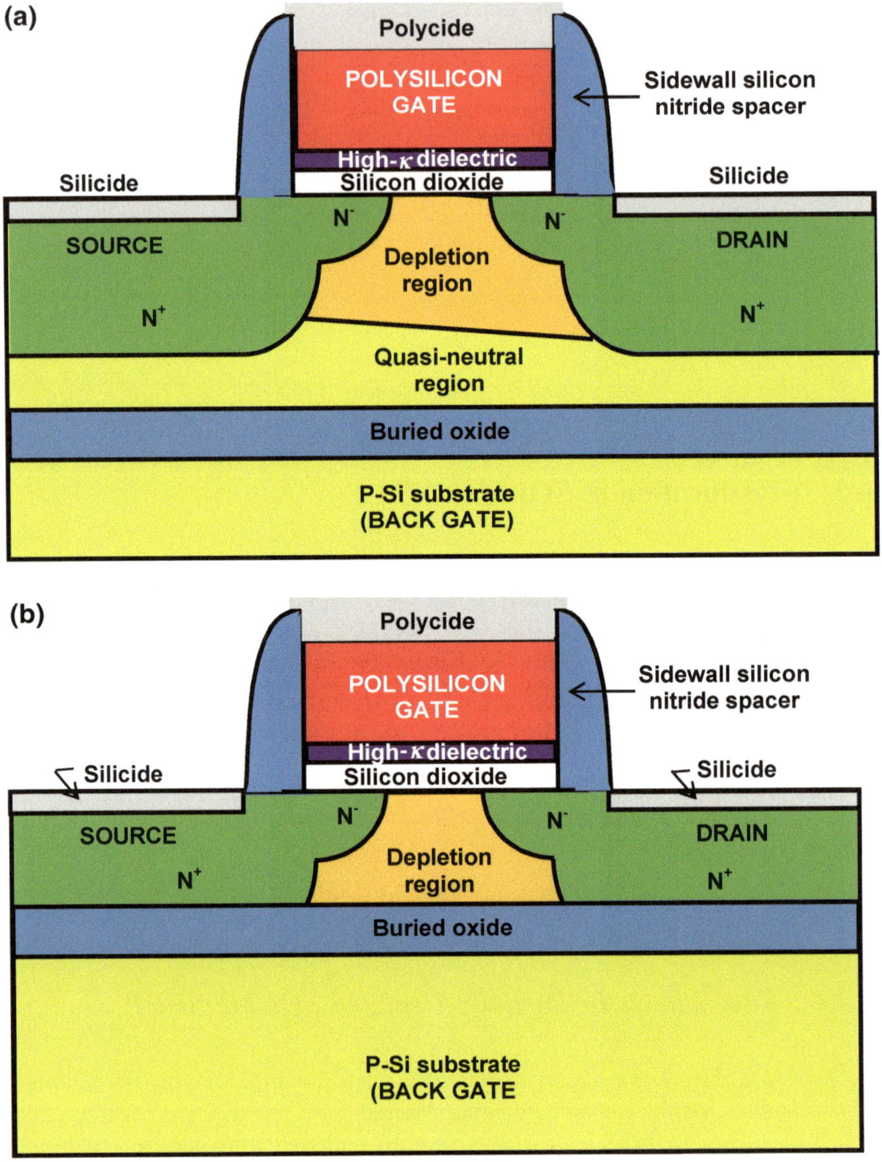

Fig. 6.4 Two types of SOI-MOSFET: **a** partially depleted and **b** fully depleted

Considering an N-channel SOI-MOSFET, the N^+-source, P-substrate and N^+-drain regions act as the N^+-emitter, P-base and N^+-collector of a parasitic NPN bipolar transistor. In a bulk silicon MOSFET, the P-substrate terminal is grounded but in an SOI-MOSFET, there is no such connection and this terminal is left floating. So, the parasitic transistor is completely isolated from the substrate leading to new

Table 6.3 Partially- and fully-depleted SOI-MOSFETs

Feature	Partially-depleted SOI-MOSFET	Fully-depleted SOI-MOSFET
Acronym	PD-SOI-MOSFET	FD-SOI-MOSFET
Typical buried oxide layer thickness (nm)	100–200	5–50
Typical silicon layer thickness (nm)	50–90	5–20
Channel doping	Yes	No (or light doping); hence avoids doping-induced process variability
MOSFET body	Thin	Ultra-thin
Transistor footprint	Large	Small
Body ties	Possible	Not required
Floating body effect	Yes. Charge accumulating at the drain due to impact ionization causes this effect	No such effect
Subthreshold swing	Low but more than FD-SOI-MOSFET	Extremely low < 65 mV/decade
Sensitivity to process variation	Less	More, e.g., variation in properties of thin silicon film causing fluctuation in threshold voltage
High threshold voltage	Easily achievable	Difficult because increase of silicon thickness for high V_{Th} makes the device PD; decrease of thickness for FD device, reduces V_{Th}
Threshold voltage variation with temperature	More	2–3 times less than PD-SOI-MOSFET
Multiple threshold voltage offering	Yes	No
Commercial introduction	First	Later
Disadvantages	Floating body effect	Challenging manufacturability. Requires advanced metrology for defect detection in thin layers
Advantages	Easy and established manufacturing technology. Can utilize floating body effect for memory applications	Absence of floating body effect. Easy controllability of short-channel effects. Non-susceptibility to random threshold voltage fluctuation in undoped channel devices. Extremely low leakage current and power consumption
Applications	Microprocessors and memory chips; automotive, aerospace and military equipment	Microprocessors, low- and ultra-low power electronics

phenomena, which are not observed in bulk silicon MOSFETs. When the drain voltage is high in such a floating body partially-depleted SOI-MOSFET, the electrons in the MOSFET channel attain high velocities due to the large electric field. These fast-moving electrons collide with silicon atoms in their paths, dislodging electrons, and creating electron–hole pairs. The electron–hole pairs thus generated trigger more collisions, and an avalanche of charge carriers builds up by impact ionization. The positive drain terminal attracts the electrons while the holes pile up in the floating body, raising the potential of the body. This positive potential is subtracted from the positive potential to be applied to the gate for inversion layer creation, leading to a decrease in threshold voltage of the device. The drain current therefore increases dramatically. The increase in drain current causes more impact ionization and further increases in the number of holes produced. As more holes gather in the body region, threshold voltage is again lowered. This cumulative process continues until the source-body diode becomes forward-biased and the accumulated holes can exit through this diode.

The positive feedback mechanism of generation of holes and their collection in the body of the MOSFET leads to an abrupt increase in drain–source current of the device. This sudden increase in drain–source current is observed as a kink in the output characteristics of the transistor. It is the first kink in the characteristics (Fig. 6.5). If the minority carrier lifetime in the silicon is high, the phenomenon receives a boost from the NPN transistor action. The base hole current is amplified by the bipolar gain. This amplification is often the cause of a second kink noticed in the output characteristics. The two kinks comprise the kink effects.

Elimination of the kink effects is achievable by provision of a body contact to the SOI-MOSFET. This contact will draw away any excessive majority carrier holes in the body, thus avoiding the associated increase in drain current. However, this will require additional area and so will have to be done at the expense of area.

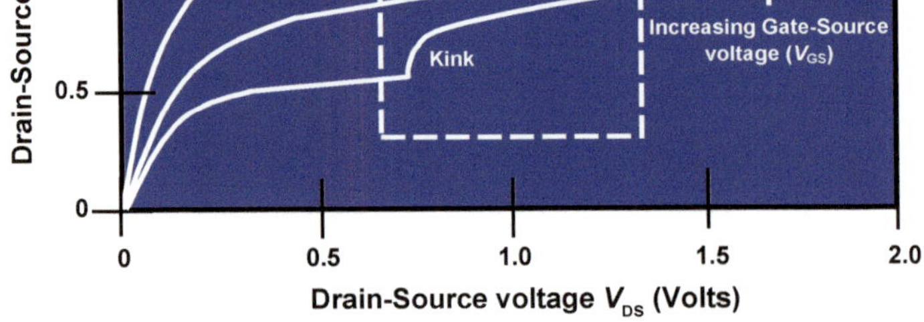

Fig. 6.5 Kink effect in output characteristics of a partially-depleted SOI-MOSFET

Moreover, these contacts will introduce complications of time constants for body charging and discharging. Another remedy is to decrease the minority carrier lifetime in the body. This is possible by implanting silicon at a high dose. The resultant lattice damage and amorphization of silicon will create defects which will serve as recombination centers facilitating the annihilation of holes by recombination. Accumulation of holes in the body will be greatly hampered, thus smothering the kink effect. But the single-crystal structure of silicon will also be disturbed in the channel region, reducing the free carrier density and hence the drive current. A better option available is to use a selective back oxide (SELBOX) structure in which the buried oxide does not fully close the body of the transistor at the bottom but has a window in the middle portion through which the holes migrate to the substrate and are removed. The size of the window or the gap in the buried oxide is carefully selected to derive the maximum advantages of bulk silicon MOSFET as well as SOI-MOSFET [7].

6.4.2 Absence of Kink Effects in Fully-Depleted SOI-MOSFET

These MOSFETs do not exhibit any floating body effects because no portion of their body is at neutral potential that can be charged and kept at a floating potential.

6.5 Disadvantage of SOI Technology: Self-heating Issue

The main shortcoming of SOI-MOSFET compared to bulk silicon MOSFET arises from the fact that the SOI-transistor is surrounded all throughout: on top, bottom, and all sides by silicon dioxide. Indeed, the transistor is housed inside a silicon dioxide enclosure except for the metal contact regions serving as heat dissipation conduits. The thermal conductivity of silicon dioxide is 100 times lower as compared to silicon. Due to this smaller thermal conductivity, heat produced cannot be readily conducted away. Bulk CMOS affords efficient heat transfer to substrate in the vertical direction whereas in SOI-CMOS, buried oxide layer impedes the vertical heat transfer. However, if trench isolation is used, lateral heat transfer and that through gate oxide are similar in the two cases.

Consequent upon the heat generated in SOI-CMOS, the mobility of carriers in the transistor is degraded and threshold voltage may be shifted. This effect is named as "self-heating effect." Particularly, the FD-SOI-MOSFET is greatly affected because the silicon film of this device is relatively thinner than in PD-SOI-MOSFET.

6.6 Double-Gate, Multiple-Gate, and Surround Gate MOSFETs

Double, triple, quadruple, and gate-all-around devices were introduced to achieve higher current drive capability and improved short-channel characteristics. The current drive of a multiple-gate MOSFET is proportional to the total gate width considering all its gates. As these devices offer more gate width per unit of silicon area, they provide much higher current drive per unit area than single-gate devices of the same gate width and gate length. The better control over the surface depletion region eliminates the short-channel effects [8].

6.7 Discussion and Conclusions

Buried oxide in SOI-MOSFETs is the focus of attention from reliability consideration. First is the self-heating issue. Second is the degradation of buried oxide by hot carriers created in the high electric field of short-channel devices. These hot carriers can interfere with device operation remembering that the wafers produced by the SIMOX process contain a high concentration of electron traps. Hot carriers may exert a deteriorating effect on the backgate and the back channel. By coupling effect, the overall performance of the MOSFET may be impaired.

Review Exercises

6.1 How does an SOI wafer differ from a bulk silicon wafer? Which of the two wafers is more expensive?
6.2 Which is prone to latch-up problem: Bulk CMOS or SOI-CMOS? Why?
6.3 Which is more resistant to radiation effects? Bulk CMOS or SOI-CMOS?
6.4 What are parasitic effects in bulk CMOS? Do they exist in SOI-CMOS?
6.5 Floating body effects occur in which of the following devices: (i) bulk silicon MOSFET (ii) Partially-depleted SOI-MOSFET, and (iii) fully-depleted SOI-MOSFET?
6.6 Which dissipates more power during operation? (i) bulk CMOS or (ii) SOI-MOSFET. Why?
6.7 Which has a greater subthreshold swing? (i) bulk CMOS or (ii) SOI-CMOS.
6.8 Write the full forms of the acronyms: (i) SIMOX and (ii) BESOI.
6.9 Describe the oxygen ion implantation-based method used for SOI wafer manufacturing? What are the principal steps involved? Why is thermal annealing at high temperatures necessary?
6.10 How are the roles of oxygen and hydrogen ion implantation in SOI wafer production different? Discuss.

6.11 Name two SOI manufacturing methods based on direct fusion bonding of wafers. Explain the main steps involved in both techniques.
6.12 Which of the two devices, PD-SOI-MOSFET or FD-SOI-MOSFET does not need doping of the body region during fabrication? How does elimination of this step help in avoiding process variations from device to device?
6.13 Why is a high threshold voltage not achievable in FD-SOI-MOSFET? Which has a superior subthreshold swing? PD-SOI-MOSFET or FD-SOI-MOSFET?
6.14 What is kink effect in PD-SOI-MOSFET? Explain the origin of first and second kinks in the output characteristics of the device? Why kink effect is not produced in a FD-SOI-MOSFET?
6.15 Describe different strategies for avoidance of kink effect in a PD-SOI-MOSFET without changing its structure? Mention any accompanying harmful effects of each scheme on device properties.
6.16 What is the reason of self-heating of an SOI-MOSFET? Which of the two devices is more affected by self-heating: (i) PD-SOI-MOSFET or (ii) FD-SOI-MOSFET? How does self-heating influence the electrical performance of the device?

References

1. Schmidt MA (1998) Wafer-to-wafer bonding for microstructure formation. Proc IEEE 86 (8):1575–1585
2. Silicon photonics (2015). http://homepage.ntu.edu.tw/~dwhuang/courses/sp/sp_05a.pdf. Accessed 4 Oct 2015
3. Colinge JP (2004) Multiple-gate SOI devices. Solid State Electron 48:897–905
4. Advanced substrate news: fully depleted (FD) vs. partially depleted (PD) SOI, May 14, 2008. http://www.advancedsubstratenews.com/2008/05/fully-depleted-fd-vs-partially-depleted-pd-soi/. Accessed 30 Aug 2015
5. Kim Y-B (2009) Review paper: Challenges for nanoscale MOSFETs and emerging nanoelectronics. Trans Electr Electron Mater 10(1):21
6. Vandana B (2013) Study of floating body effect in SOI technology. Int J Mod Eng Res (IJMER) 3(3):1817–1824
7. Narayanan M, Al-Nashash H, Mazhari B et al. (2012) Analysis of kink reduction in SOI MOSFET using selective back oxide structure. Act Passive Electron Compon 2012: Article ID 565827, 9 p
8. Gili E, Kunz VD, de Groot CH (2004) Single, double and surround gate vertical MOSFETs with reduced parasitic capacitance. Solid-State Electronics 48:511–519

Chapter 7
Trigate FETs and FINFETs

Abstract The ever-increasing leakage current with every successive generation of MOSFET urged the researchers to look for a revolutionary change in device architecture. The changeover to SOI-MOSFET, particularly the FD-SOI-MOSFET, succeeded to a large extent in meeting the challenges without any fundamental modification of the structure. Alternative choices proposed were trigate FET and FINFET structures, which marked the end of planar era and entailed a radical change from a planar device to a three-dimensional shape for rejuvenating the IC industry. This chapter explains how wrapping the gate insulator around the body region of a MOSFET is an effective way of increasing the capability of the gate to mitigate the various encumbrances faced with short-channel devices. A comparative study of FINFETs fabricated on SOI wafers and bulk silicon wafers is presented. The neck-to-neck battle between FINFET and FD-SOI-MOSFET to clinch the supreme position is described by pointing out their relative beneficial aspects and downsides.

7.1 Introduction

In multigate MOSFETs, placement of several gates in vicinity of the channel helps in screening the channel much more effectively from the drain voltage than a single gate. Therefore, for multiple-gate MOSFET structures, performance indices showing the alleviation of short-channel effects are much superior than for planar devices. These indices include subthreshold swing, DIBL effect, and V_{Th} roll off. An outcome of the improvement in these indices is that the degradation in V_{Th} due to scaling is relatively less, which means that the deterioration in off-state current with scaling is lowered.

The trigate FET and FINFET were introduced as welcome alternatives to the planar MOSFET. FINFET became more popular primarily due to its structural simplicity and easy fabrication. In this chapter, we attempt at the conceptual evolution of trigate FET/FINFET from the primeval planar MOSFET structure.

V.K. Khanna, *Integrated Nanoelectronics*,
NanoScience and Technology, DOI 10.1007/978-81-322-3625-2_7

7.2 Relooking at MOSFET Concept in Nanoscale

Our time-honored concept about a MOSFET has been that of a planar device in which a horizontal channel is established between two heavily doped source and drain diffusions. This primeval MOSFET was introduced as a long-channel structure. Gradually the channel was reduced in length. Many short-channel effects came into play. To mitigate these effects, the lattice was strained by incorporating germanium to augment the mobilities of carriers in the channel. Metal gate process and high-κ dielectrics brought further improvement. With the decrease in chip size, the gate began to lose its control over the channel. In short-channel MOSFETs, the drain began to dominate, overpower, and overtake the gate in its influence over the channel through drain-induced barrier lowering. Device engineers came up with the new structure of SOI-MOSFET. It was found that in a bulk MOSFET, a large portion of the silicon substrate was unnecessary. Not only redundant, this portion of silicon was responsible for parasitic effects, namely, the large drain-substrate capacitance, and the detrimental effects of in-built bipolar junction transistor. The gate could hardly exercise its influence in remote regions that were far away from it. To eliminate these obstacles, double-gated SOI-MOSFETs were suggested. The two variants of these devices, viz., partially depleted and fully depleted types could overcome many roadblocks. As technological expertise developed, the fully depleted device was mastered.

7.3 The Path of MOSFET Restructuring

Introspection of the novelties in all these advancements reveals that the efforts are moving in one straightforward direction, viz., making the gate electrode more influential and effective in controlling the phenomena taking place inside the device, whether it is by getting rid of the unwanted silicon in bulk MOSFET or by using two gates to be able to control the channel in a better way. The FINFET device has emerged as a consequence of these endeavors of building an architecture in which gate will be the master electrode under whose authoritarian control all device activities will be managed.

7.4 Rotating the SOI-MOSFET by 90° for Making Trigate FET

The SOI-MOSFET had already shown that two gates were far better than a single one. But still the MOSFET concept was that of a horizontal device in which there was a horizontal channel in the body because of the top gate and another horizontal channel in the body induced by the bottom gate. These gates were also called the front gate and backside gate. Comparing with the room of a house, one could name them as a gate over the roof and another under the floor. Then came the new idea. One could rotate the room by 90°. Then these gates will be located on two opposite walls. Besides the two gates over the opposite walls of the room, one could have another gate over the roof. Thus there will be three gates. These three gates could be connected with each other to form a single trilayer gate. Thus the notion of MOSFET being a planar device enlarged into one of a three-dimensional structure in which carriers flow in both horizontal and vertical planes. The three layers of gate, two on the two opposite walls, and one on the roof are producing inter-merged channels: two channels in the vertical plane parallel to the surfaces of the walls and one channel in the horizontal plane parallel to the roof. These channels originate from a source electrode lying in the vertical plane and terminating in a drain electrode, which is also vertically oriented. At the floor of this room is the buried oxide layer of the SOI wafer. Thus the body region of this MOSFET is surrounded by gate on three sides, instead of one side only in bulk silicon MOSFET or two sides only in an SOI-MOSFET. Gate control acts through three sides and is much stronger than from a single side. The total gate width W is obtained by adding together the heights of the two side gate layers (wall layers) $H_{\text{side-gate}}$ and length of the topmost layer (the roof layer) $L_{\text{roof-gate}}$

$$W = 2H_{\text{side-gate}} + L_{\text{roof-gate}} \tag{7.1}$$

The gate length is the distance between the source and drain diffusions.

Exerting gating action from three sides is only one side of the story. The body region of this MOSFET must be extremely thin. Then only the gates will be in proximity to all areas of the body and able to suppress the leakage current properly. A thicker body region will contain several distant leakage paths which the gates will not be able to handle. Inadequacy of gates in turning off the leakage current in these far-off zones will lead to the same situation as in bulk silicon MOSFET, which we are trying to evade.

The above MOSFET containing three gates is referred to as a trigate FET. A simpler version of this device without the top gate has been made. In the fabrication of this two-gate vertical device, the highly selective gate etching step to thin the top gate oxide is eliminated which significantly reduces the process complexity. This two-gate device called the FINFET is compared with a planar MOSFET in Fig. 7.1. It will be elaborated in the next section.

7.5 Advent of FINFET

The body region of the two-gate MOSFET described in the preceding section has the appearance of thin sheet in the vertical plane. The MOSFET consists of a thin vertical strip straddled by gates on the two opposite surfaces [1]. We know that thin membranous appendages protrude from the body of fishes or other aquatic animals.

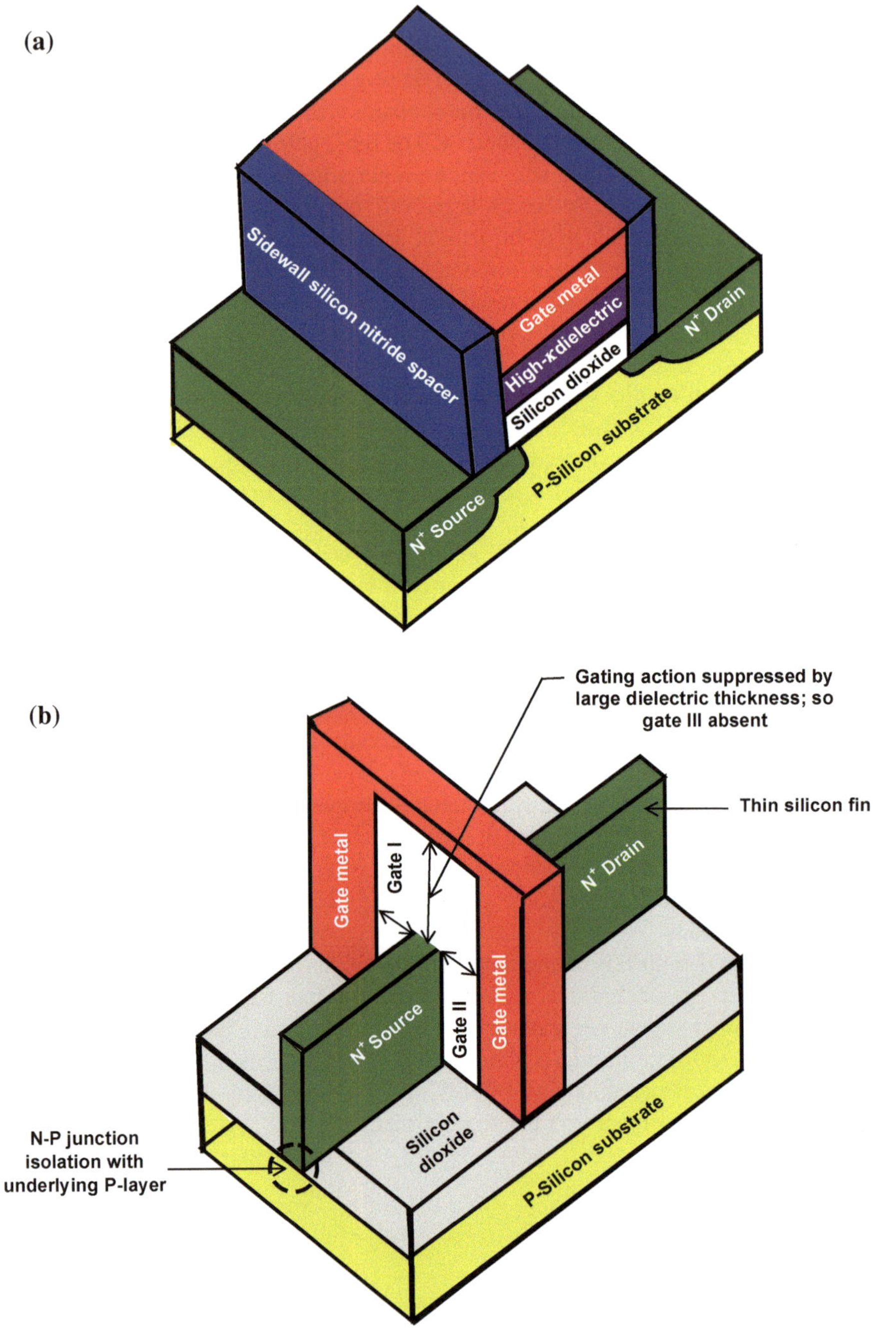

Fig. 7.1 Comparison between **a** planar MOSFET and **b** FINFET fabricated on bulk silicon wafers

These appendages help these animals in steering and stabilizing their motion in water. They are known as fins. The body regions of the new MOSFET device bear a striking similarity to the dorsal fin of a fish because it is a thin sheet extending from a large silicon block. To reiterate, the fin cannot be made thick because making it thicker will degrade its basic function. Like the large unwanted chunk of material in a bulk MOSFET, a thicker fin will contain only wasteful material of no use to device activities.

For the FINFET, the equation for gate width is written as

$$W = 2H_{\text{side-gate}} + L_{\text{roof-gate}} = 2H_{\text{side-gate}} + 0 = 2H_{\text{side-gate}} \tag{7.2}$$

If the current level of the MOSFET has to be increased, it is only possible to do so by connecting a large number of fins in parallel and providing a common gate electrode to these parallel-placed fin structures. The total gate width is an integral multiple of the width of a single gate. It can only be increased in discrete multiple values, not as a continuous variable. As the gate width determines the drive current available from the transistor, the FINFET provides quantized values of current. Arbitrary values of drive current are forbidden. This is known as width quantization. It means that drive current can only be increased during chip layout design for mask making. For a FINFET having n fins, the total gate width is

$$W = 2nH_{\text{side-gate}} \tag{7.3}$$

where n is an integer = 1, 2, 3, ….

It is easy to envisage that in this structure, the width of the gate is determined by the height of the fin. Therefore, a chip designer should pay due attention to fin thickness and fin height. These are critical parameters affecting the performance of the FINFET. If the fins selected are short in height, then for a given gate width, a larger number of fins are required to achieve the same current. If the fins are tall, less number of fins can serve the same purpose. But the structural stability is compromised with. The device becomes unstable. Generally accepted guideline is that fin height must be taken to be less than four times the fin thickness.

The new MOSFET is formed by wrapping or casing the silicon fin in a gate electrode. This special method of construction of the device gives it the name 'FINFET'. Profs. Chenming Hu, Tsu-Jae King-Liu and Jeffrey Bokor, at the University of California, Berkeley coined the term.

7.6 What About the Source and the Drain of FINFET?

In the discussion so far, attention was concentrated on the gate of the FINFET. The question might be asked: In what plane are the source and drain of FINFET located? Is it the vertical plane or horizontal plane? Yes. The source and drain are also diffused on the thin fin in the vertical plane. This dramatically changes the concept of MOSFET.

7.7 FINFET Versus Trigate FET

The FINFET appears to be in a slightly disadvantageous position with respect to the trigate FET because in the trigate FET, the top roof gate provides additional gate width (Fig. 7.2). So the gate width of the FINFET is a little lower than that of the trigate FET, providing smaller drive current. Further, the current flowing through the top surface adds to current conduction when the MOSFET is on, and leads to lower gate–source capacitance for the trigate device. Hence, this capacitance is a less severe setback in trigate FET than FINFET. But the extra path of current also increases the parasitic resistance.

On the whole, it has been shown that performance metrics characterizing the short-channel effects are superior for FINFETs than those for the trigate FETs. To bring the metrics of trigate FETs at par with FINFET, appreciable chip area is wasted [2].

7.8 FINFET Fabrication

A crucial part of FINFET fabrication is the fin patterning and etching [3].

7.9 FINFET on SOI or Bulk Silicon Wafers?

Although in the above discussion, attention was focused on FINFETs made on SOI wafers, and the FINFETs were launched initially on SOI platform, the reader should not acquire the notion that FINFET is an SOI-MOSFET. It can be made on bulk silicon wafers as well [4], Fig. 7.3.

Making FINFETs on SOI wafers is only a matter of technological convenience (Fig. 7.4). One starts by etching the silicon wafer to produce the fin. The etching must stop at a certain depth after the required fin height has been achieved. SOI wafer provides the way of stopping the etching because when silicon has been etched up to the level of buried oxide, the etching automatically ceases. In this manner, the SOI wafers can provide fins of exact heights. They impart reproducibility to the process through the buried oxide layer serving as an etch stop. Figure 7.5 shows the process steps involved in fin fabrication on SOI wafers.

After fin fabrication, the FINFET fabrication process is similar to standard process consisting of source and drain implants, followed by gate stack deposition. Alternatively, doped polysilicon may be deposited as source/drain, patterned and then nitride deposition is done with spacer etch, and ensuing steps of gate formation.

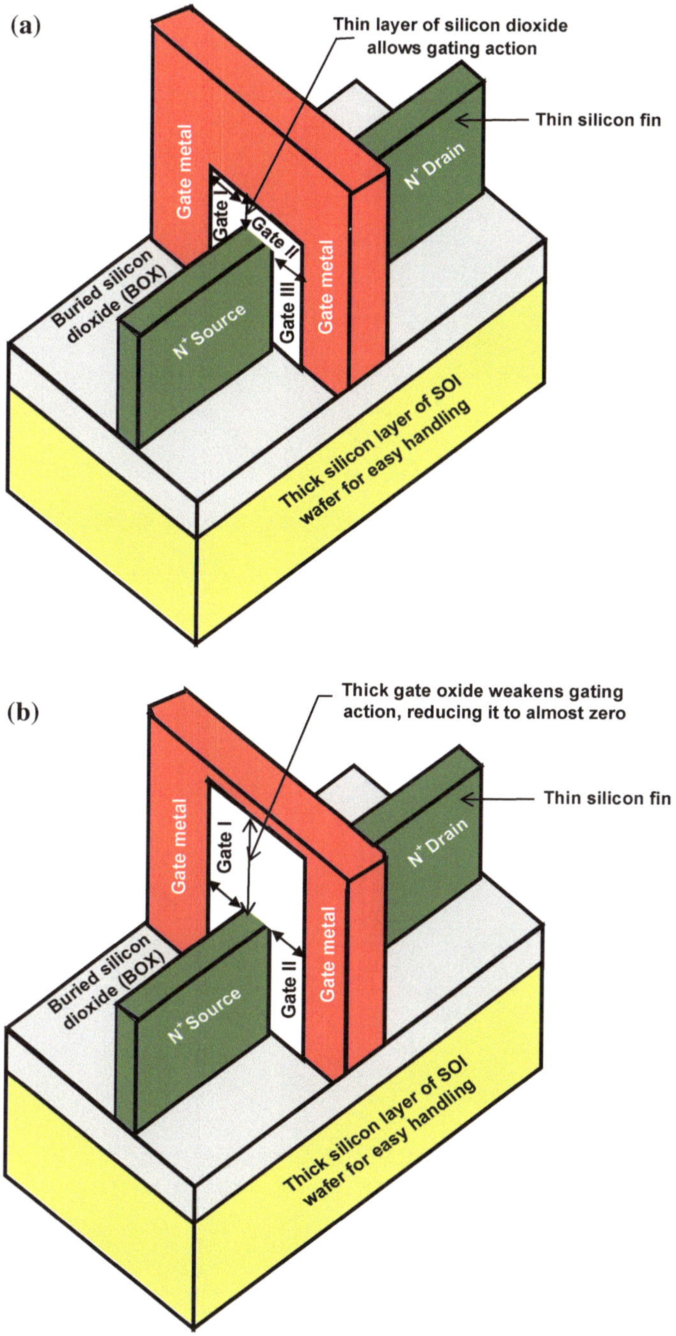

Fig. 7.2 Comparison of **a** a tri gate FET with **b** a FINFET

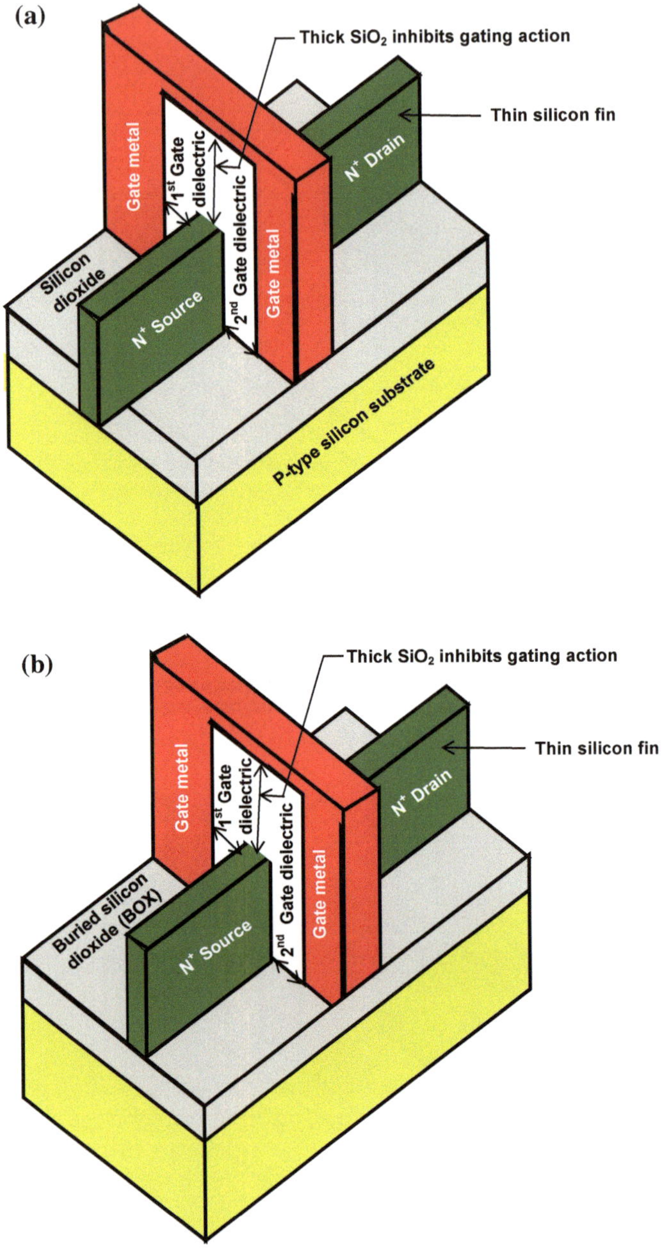

Fig. 7.3 Comparison of **a** FINFET fabricated on bulk silicon wafer with **b** a FINFET on SOI wafer

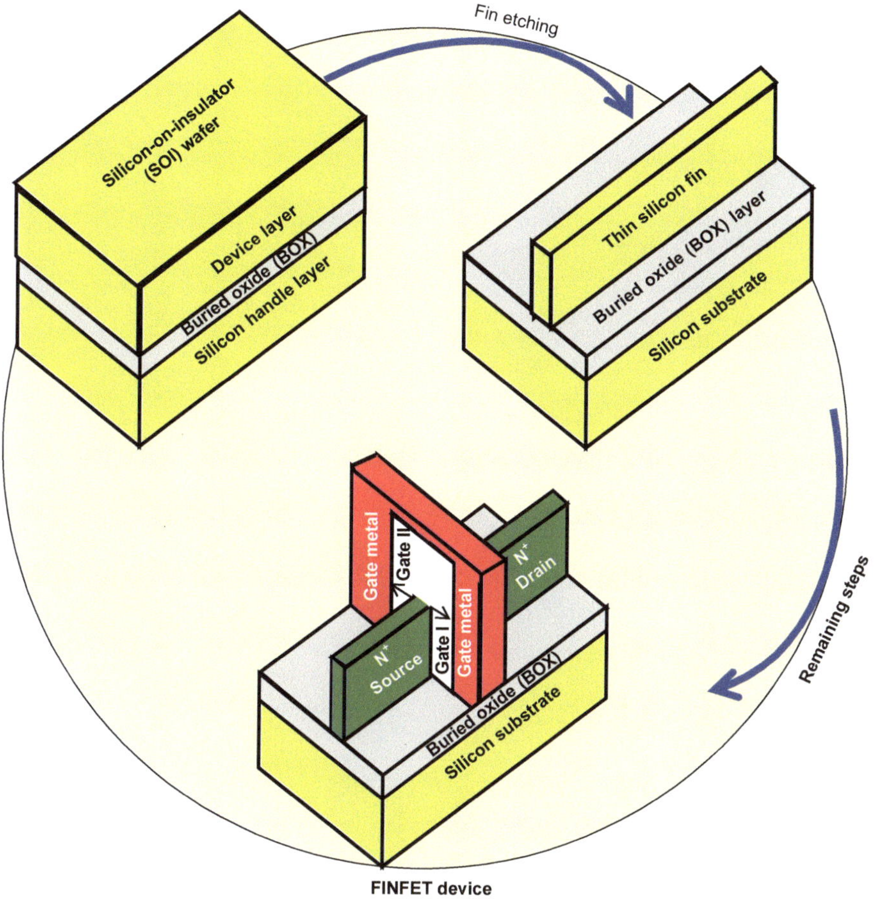

Fig. 7.4 Fabrication process of FINFET on SOI wafers

In a bulk silicon wafer (Figs. 7.6 and 7.7), the etching has to be controlled by reckoning the etching time. As etching depends on several variables, and the environmental parameters cannot be exactly repeated, process-induced variations are likely to yield fins of different heights both in the same batch or from batch to batch. Further to provide P–N junction isolation in place of the oxide isolation at the base of the fins, a high dose junction implant is essential.

The FINFET made in bulk silicon suffers from the leakage current flowing through the NPN parasitic bipolar junction transistor consisting of N^+-source, P-substrate and N^+-drain in case of N-channel FINFET. As the gate is at a large distance from this parasitic BJT, this transistor in parallel with the channel causes leakage current flow. The BJT transistor action can be considerably weakened by

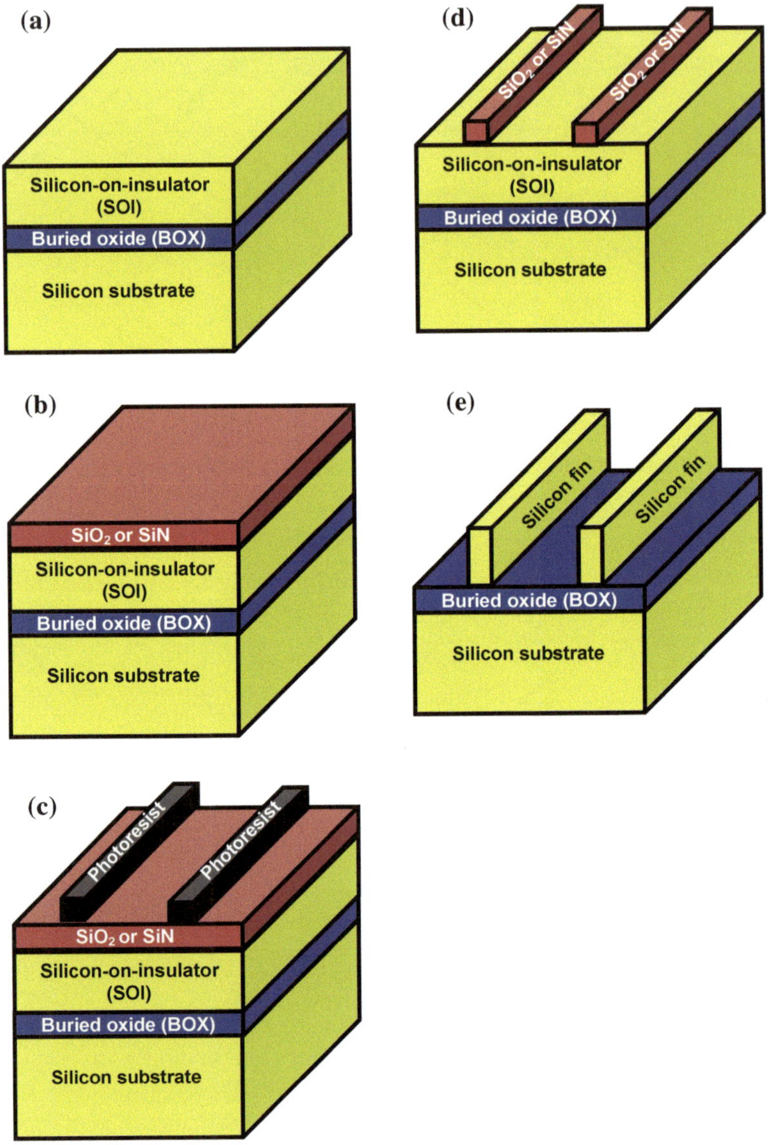

Fig. 7.5 Fin fabrication on silicon-on-insulator wafers. **a** Silicon on insulator wafer. **b** Hard mask (SiO_2 or SiN). **c** Photolithography for fin definition. **d** Etching of SiO_2 or SiN. **e** Fin etching

heavily doping the P-substrate to reduce the injection efficiency of the transistor because the P-substrate is the base of this NPN transistor. This doping is done by ion implantation, which is known as channel stop implantation.

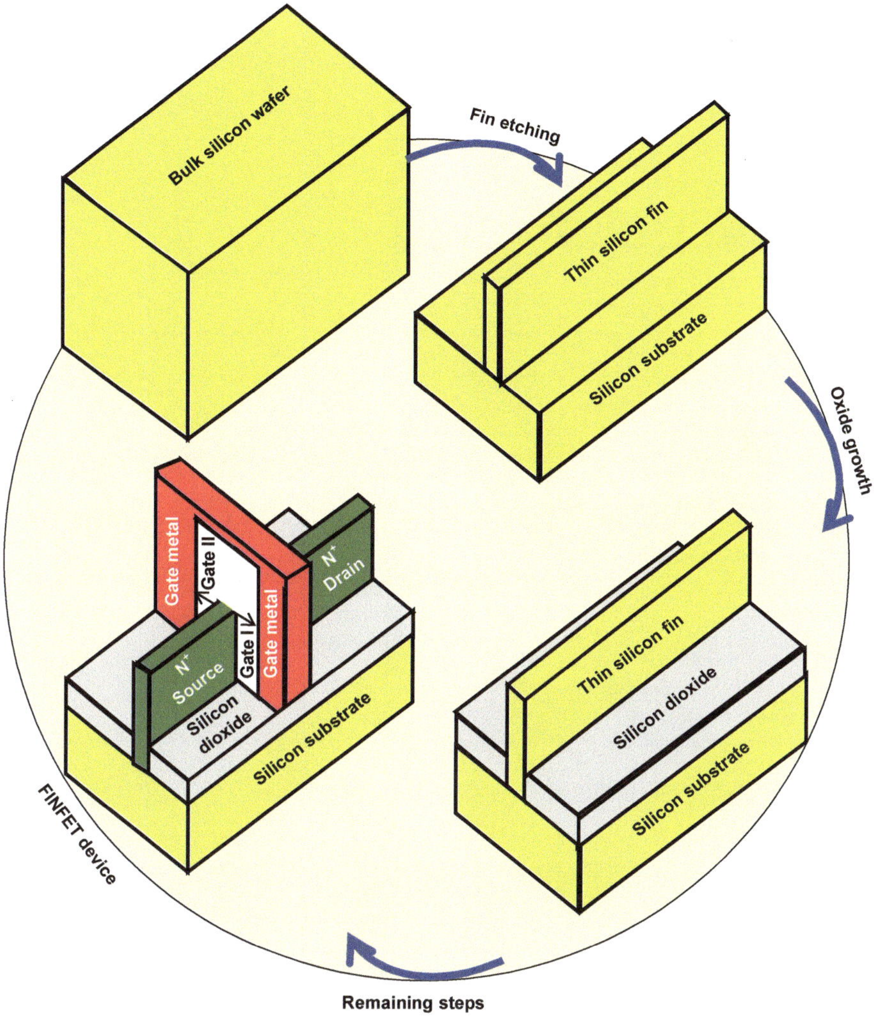

Fig. 7.6 Fabrication process of FINFET on bulk silicon wafers

Another problem with bulk silicon process is that the shallow trench oxide required for isolation of one fin from another cannot be planarized. Naturally, the heights of the fins above the oxide will differ. SOI wafers give fins of more regular rectangular shapes and precise heights as compared to bulk silicon.

The etch control difficulty seems to negate three vital advantages of the bulk silicon process. First, bulk silicon wafers are comparatively cheaper than SOI wafers. So, the use of bulk silicon wafers is far less expensive than SOI wafers and can drastically reduce the cost of the FINFET. Second, from a scientific viewpoint, as we had seen for the SOI-MOSFET, the buried oxide is a thermal insulator.

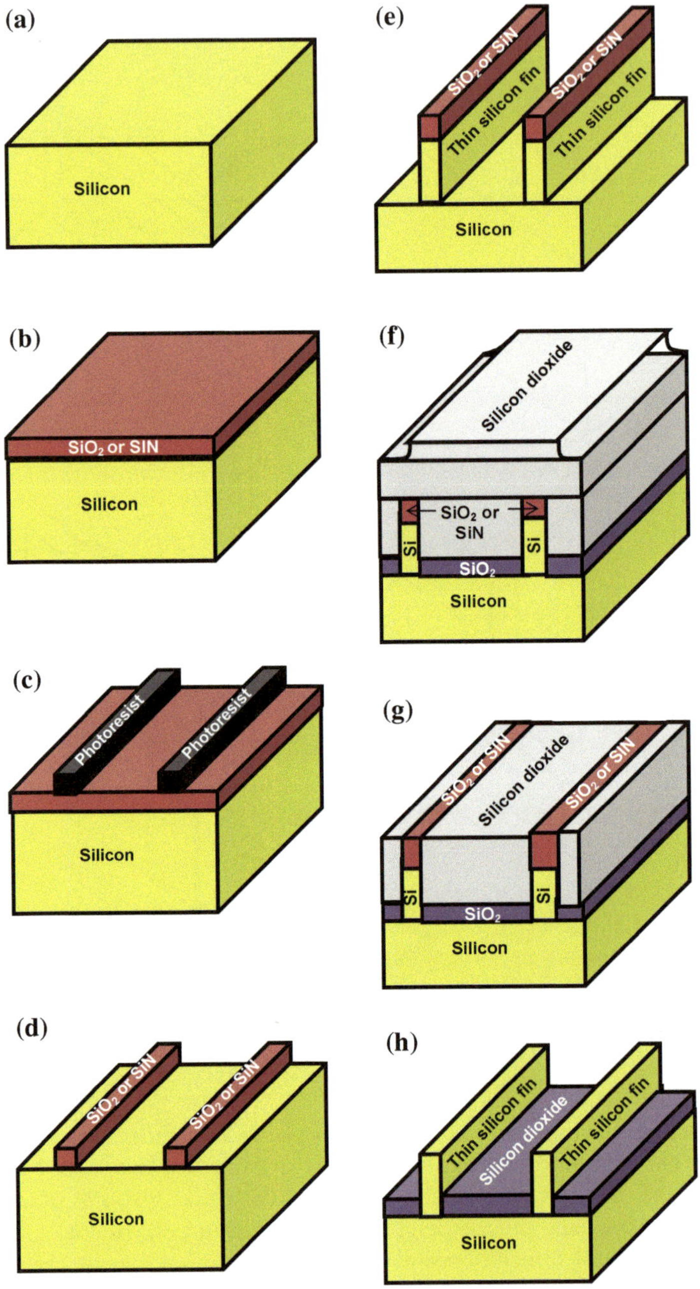

Fig. 7.7 Fin realization on bulk silicon wafers. **a** Bulk silicon wafer. **b** Hard mask (SiO_2 or Si_3N_4). **c** Photolithography for fin definition. **d** Hard mask etching. **e** Etching for desired fin height. **f** Oxide filling. **g** Surface planarization by chemical mechanical polishing (CMP). **h** Time-controlled etching

Silicon has a much higher thermal conductivity than silicon dioxide. Therefore, the FINFET device made in an SOI-wafer suffers from heat dissipation problem. The bulk silicon FINFET is a cooler device during operation. Generation of heat in an SOI-FINFET is likely to bring in the same problems of carrier mobility degradation, as encountered in SOI-MOSFET. This discourages the use of SOI wafers. Third, for mobility enhancement, SiGe source/drain layers are grown epitaxially. In case of the bulk wafer, the silicon can be removed from source/drain regions and epitaxial layers can be grown on the template layer. The same is not true for the SOI wafer because after etching silicon, the BOX layer is exposed and no template is available for epitaxial growth. Once again, the SOI wafers are in unfavorable position.

On the whole, both bulk silicon and SOI have their relative merits and demerits. FINFETs are being fabricated on both types of wafers. Either technology may win in the long run. Both technologies face the technological challenges of etching uniformity to obtain straight, tall fin structures, uniformly doping the three-dimensional shapes and achieving conformability in depositing the different films in the gate stacks.

7.10 FINFET Comparison with Fully-Depleted SOI-MOSFET

The planar fully-depleted SOI-MOSFET was also proposed as a remedy for short-channel effects. It is therefore interesting to compare FINFET with fully depleted SOI-MOSFET to bring out their relative strengths and weaknesses [5]. Table 7.1 presents a comparative chart of these devices.

7.11 Classification of FINFETs

Broadly, FINFETs are subdivided into two main classes (Fig. 7.8): (i) Shorted gate FINFET: In these FINFETs, acronym SG-FINFET, the two gates are connected together. They are three-terminal devices. (ii) Isolated gate FINFET: The two gates of these FINFETs, abbreviated as IG-FINFETs, are not mutually connected. These FINFETs are four-terminal devices.

Because in a SG-FINFET, the channel is subjected to the controlling action of both the gates acting together, the SG-FINFET provides higher on-state and off-state currents than the IG-FINFET. On the other side, the IG-FINFET offers the advantage that both the gates are capable of independent operation. As a result, different voltage signals can be applied to the two gates depending on the application at hand, e.g., the threshold voltage of the front gate can be modulated in a

Table 7.1 FINFET and FD-SOI-MOSFET

Feature	FD-SOI-MOSFET	FINFET
Power consumption	Low	Low
Switching speed	High	High
Wafer on which fabricated	SOI	SOI or bulk silicon wafer
SOI wafer cost	A primary disadvantage	Avoidable using bulk wafer
Strain engineering	No	Yes
Manufacturing	Easy and standard	Extra complexity and expenditure
Drive current quantization	No	Quantized in terms of number of fins
Analog applications	Better controllability	Less flexibility due to restricted design choices
Threshold voltage variability	Caused by differences in thickness of silicon film	Arises from variations in fin width and quality of edges
Multi-V_{Th} implementation	Complex	Provides effective trade-off between speed and power with multi-V_{Th}
Designing	Compatible with existing libraries of planar, bulk devices	Circuits migrated from planar processes need to be rechecked by modeling and simulation

linear fashion by using the back gate. At the same time, it must not be forgotten that two isolated gate contacts need to be placed in IG-FINFET, instead of a single contact for SG-FINFET. Two gate contacts of IG-FINFET are considerably extravagant from the viewpoint of silicon area consumption.

A further classification of SG-FINFET is based on asymmetries of different kinds (Fig. 7.9). One kind of asymmetry results from the fabrication of polysilicon gates of different work functions by selectively doping these gates at different carrier concentrations [6]. The FINFET thus fabricated with unequal work functions for the two gates is an asymmetric gate-workfunction (ASG-workfunction) FINFET (Fig. 7.9b). The symmetric gate-workfunction FINFET (SG-workfunction FINFET) (Fig. 7.9a) is one in which the work functions of the two gates are equal, as usually done.

The advantage of fabricating an ASG-workfunction FINFET is the improvement in short-channel behavior. Compared to the SG-workfunction FINFET, the ASG-workfunction FINFET has lower off-state current by as much as two orders of magnitude while its on-state current is only marginally lower than that of the SG-workfunction FINFET. Another variety of asymmetric structure is the asymmetric drain-spacer-extended (ADSE) FINFET. It has improved short-channel

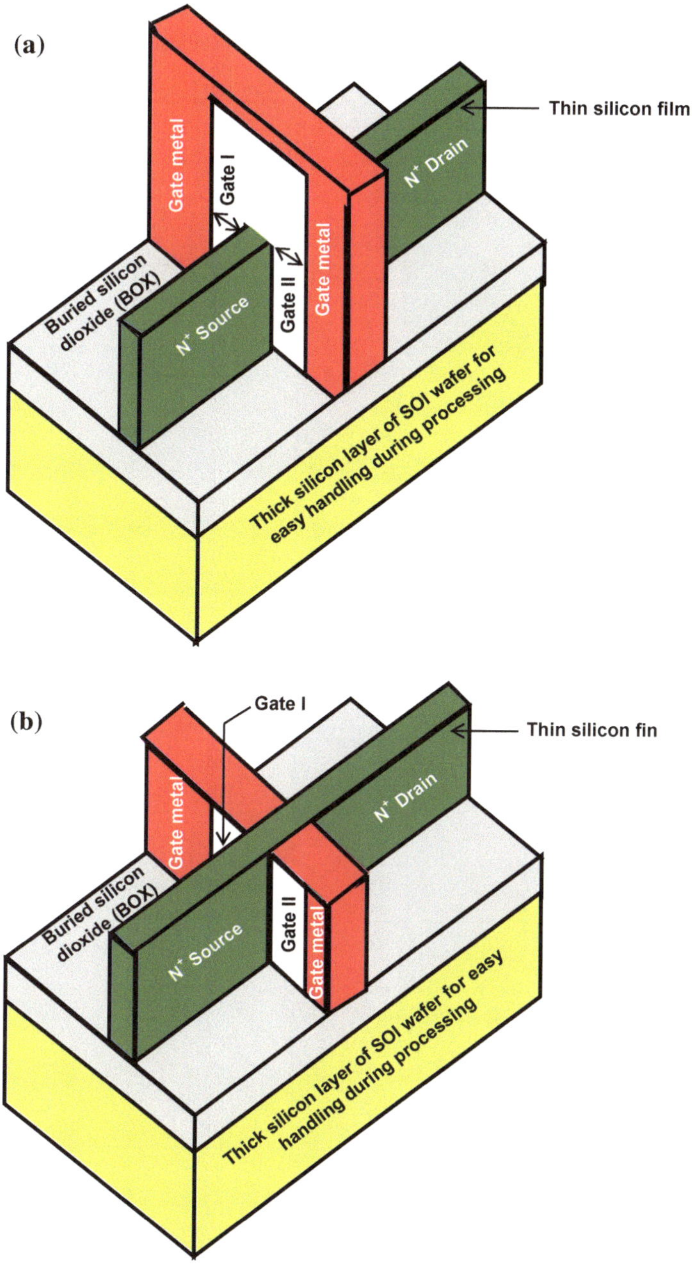

Fig. 7.8 Comparison of **a** an SG-FINFET with **b** an IG-FINFET

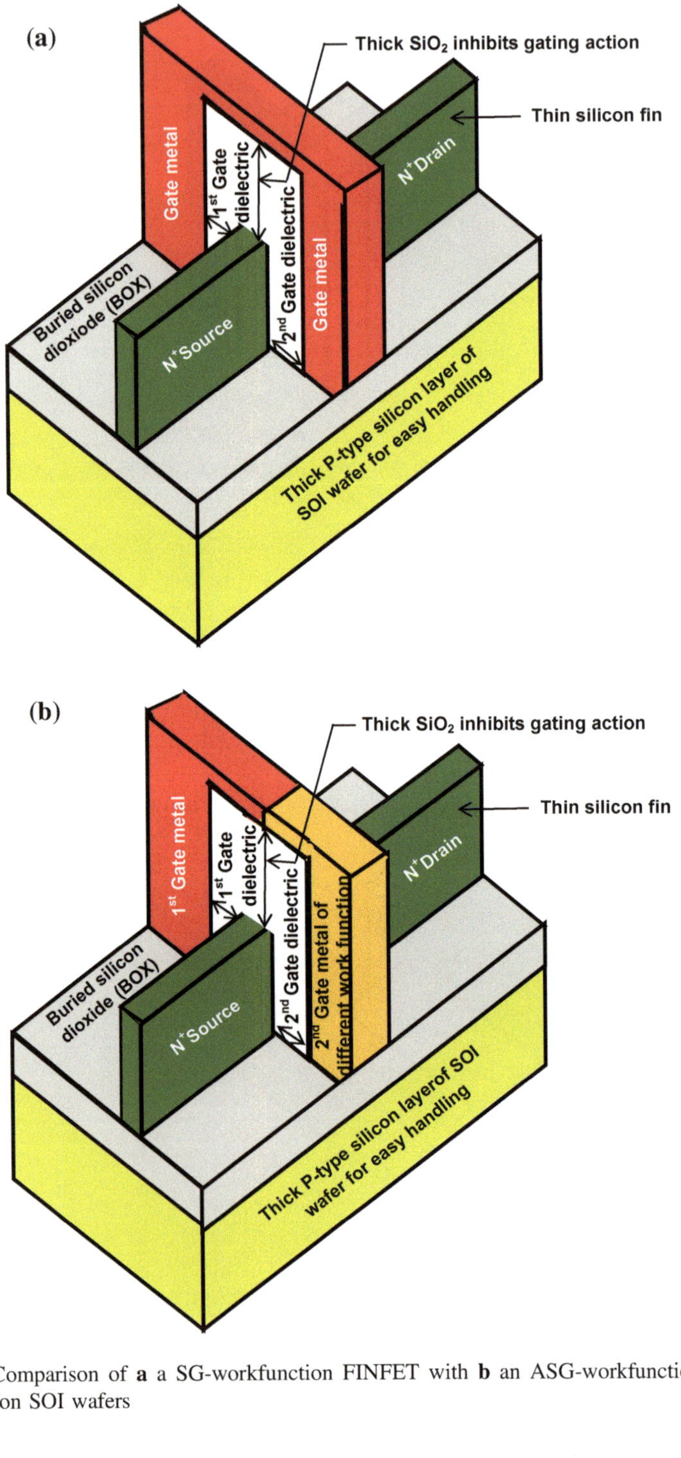

Fig. 7.9 Comparison of **a** a SG-workfunction FINFET with **b** an ASG-workfunction FINFET fabricated on SOI wafers

characteristics but at the expense of chip area. If this structure is adopted, the interchangeability of source and drain in CMOS structure no longer exists. A third type of asymmetric structure is asymmetric drain-source doped (AD) FINFET. In this class of FINFETs, the source and drain doping concentrations differ by an order of magnitude. This structure also shows good short-channel behavior but the symmetry between I_{DS} and I_{SD} is violated.

7.12 Impact of Random Doping Effects and Other Process Variations on FINFETs

To counter short-channel effects, a high concentration of dopant is required in the channel region of a planar MOSFET. When such a MOSFET device is intensively scaled down, the channel doping becomes erratic. This randomness is evident from the wide variation in threshold voltage. Unlike, the planar MOSFET, the FINFET can operate with much lower doping concentration in the channel region $\sim 10^{15}$ cm^{-3} inside the thin fin. FINFET operation is possible at this low doping level. In FINFET, a high threshold voltage is realized because the second gate ensures suppression of short-channel effects. Instead of keeping a high doping concentration in the channel, the FINFET adjusts the threshold voltage by changing the work function of the gate material. Lightly doped channel region also results in higher carrier mobility in the channel, raising the on-current. By virtue of this light doping, the FINFET overcomes process-induced random doping fluctuations. Nonetheless, the FINFET is susceptible to physical parameter fluctuations such as variations in gate oxide thickness, fin thickness, gate length and other parameters [7].

7.13 Discussion and Conclusions

As process technologies marched toward 20 nm, scaling started to look skeptical, and the need for a drastic change was realized. FINFETs represent a new lineage of devices, which can cross the 20 nm barricade previously considered to be insurmountable. FINFETs are three-dimensional structures. They rise above the plane of the substrate as opposed to planar MOSFETs which are restricted to two dimensions. For the same plane area, the three-dimensional structure gives them more volume than a planar gate. Their operating speed is 30% faster than competing devices. Also, static leakage is decreased by 90% than these devices [8]. They can be run at higher speeds than planar transistors using the same power. They can also be run at the same speed as planar transistors using less power.

Review Exercises

7.1 Why is a large portion of substrate of no use to device operation in bulk silicon MOSFET? What harmful effects does this unutilized portion of substrate produce?
7.2 Elaborate the effects of thickness of the body region of a nanoscale MOSFET on its performance.
7.3 How does the deployment of multiple gates affect the operation of a MOSFET?
7.4 Explain the origin of the name of FINFET device from its construction.
7.5 Is FINFET an SOI-MOSFET? Give reasons for your answer.
7.6 How does the substrate of a FINFET made from bulk silicon wafer impair its functioning? What remedy do you suggest?
7.7 Which of the two devices is susceptible to self-heating? FINFET or FD-SOI-MOSFET? Why?
7.8 Why is the drive current of a FINFET said to be quantized?
7.9 Why cannot a single FINFET device made of a thicker body region supply a larger drive current? Why is it necessary to connect several fins in parallel?
7.10 Point out the pros and cons of FINFET with respect to FD-SOI-MOSFET.
7.11 How does a shorted gate FINFET differ from an isolated gate FINFET? Discuss their relative merits and demerits.
7.12 Name three types of FINFETs classified on the basis of asymmetry. How do these classifications relate to interchangeability of source and drain?
7.13 Why is FINFET less vulnerable to random doping effects than planar MOSFET? What type of variations cannot be avoided with FINFET?

References

1. Kedzierski J, Ieong M, Kanarsky T et al (2004) Fabrication of metal gated FinFETs through complete gate silicidation with Ni. IEEE Trans Electron Devices 51(12):2115–2120
2. Yang J-W, Fossum JG (2005) On the feasibility of nanoscale triple-gate CMOS transistors. IEEE Trans Electron Devices 52(6):1159–1164
3. Hisamoto D, Lee Wen-Chin, Kedzierski J et al (2000) A self-aligned double-gate MOSFET scalable to 20 nm. IEEE Trans Electron Devices 47(12):2320–2325
4. Jurczak M, Collaert N, Veloso A et al (2009) Review of FINFET technology, SOI Conference, 2009 IEEE International, 5–8 Oct 2009, Foster City, CA, pp 1–4
5. Swinnen M, Duncan R (2013) Physical verification of FINFET and FD-SOI devices. © 2012 The Curation Company. http://www.techdesignforums.com/practice/technique/physical-verification-design-finfet-fd-soi/Accessed. Accessed 7 Oct 2015
6. Kedzierski J, Fried DM, Nowak EJ et al (2001) High-performance symmetric-gate and CMOS-compatible Vt asymmetric-gate FinFET devices. In: Proceedings of the IEEE international electron devices meeting (IEDM'01), pp 437–440

7. Bhattacharya D, Jha NK (2014) FinFETs: From devices to architectures. Adv Electron 2014, Article ID 365689, 21 p
8. Poole I (2015) Radio-Electronics.com: FinFET technology & basics © Adrio Communications Ltd. http://www.radio-electronics.com/info/data/semicond/fet-field-effect-transistor/finfet-technology-basics.php. Accessed 7 Oct 2015

Part III
CMOS-Supportive Nanotechnologies

Chapter 8
Nanophotonics

Abstract Two subbranches of nanophotonics are distinguished based on far-field propagating light and near-field non-propagating light. These subbranches are known as diffraction-limited and beyond-diffraction-limit nanophotonics, respectively; Japanese researcher Ohtsu proposed the later. Under the diffraction-limited nanophotonics fall plasmonics, photonic crystals, quantum dot lasers, and silicon nanophotonics. These utilize conventional light waves for transmission of signals. In the beyond-diffraction-limit nanophotonics, prototype AND and NOT gate arrangements are presented. These work on near-field energy transfer between quantum dots. They use optical near field for conveying signals. Fundamentally different criteria are to be evolved for designing nanophotonic devices exploiting conventional and near-field approaches. The near-field approach may render possible the development of novel photonic systems.

8.1 Introduction

Nanophotonics is the merger of photonics with nanotechnology. It applies photonics at nanoscale dimensions for enabling the transmission, manipulation, and detection of light to derive significantly superior performance from previously existing applications or to facilitate provision of hitherto unknown functionalities, resulting in new applications. Therefore, apart from research areas where the impact of nanophotonics is clearly visible, it is also necessary to explore new areas.

Nanophotonics has two subdivisions: (i) Diffraction-limited nanophotonics employing free-propagating light. It covers plasmonics, photonic crystals, quantum dot lasers, and silicon nanophotonics, which rely on the services of conventional light waves for transmitting signals. (ii) Nanophotonics beyond the diffraction limit is based on the concept of a non-propagating optical near field [1–3], which is localized to the source of optical radiation and shows remarkably different properties from the free-propagating light. An optical near field can localize optical energy to smaller lengths than the diffraction limit which is approximately half the

V.K. Khanna, *Integrated Nanoelectronics*,
NanoScience and Technology, DOI 10.1007/978-81-322-3625-2_8

wavelength of light. Nanophotonics beyond the diffraction limit is defined as the technology based on the local electromagnetic interactions of a nanometer-size element with an optical near field.

8.2 Diffraction-Limited Nanophotonics

8.2.1 Plasmonics

When an electromagnetic wave such as light strikes the surface of a metal, the free electrons in the metal are disturbed from their equilibrium positions with respect to the positive nuclei. It is known that the metal contains a large number of free electrons called the free electron gas. Therefore, the situation is imagined as the creation of ripples in the sea of free electrons of the metal by the impact of the light beam. This interaction between light and electrons occurs due to the presence of electric and magnetic fields in the light wave. The electric field of the light interacts with the electrons.

The free electron gas of the metal is referred to as electron plasma and the collective oscillations of the density of free electron gas are called plasma oscillations. These plasma oscillations propagate in the form of waves of electron density on the surface of the metal. As air is a dielectric medium, the waves are described as moving along the metal-dielectric interface, sticking to the metal surface as if they were glued to it.

A quantum of the collective oscillations of density of electrons in a metal is known as a plasmon. This nomenclature is similar to calling of quantum of mechanical vibration by the name, "phonon." Therefore, the plasmon is a quantum of plasma oscillations. It is a bosonic quasi-particle. Essentially, a plasma oscillation consists of plasmons. Surface plasmon polaritons are electromagnetic waves. These waves propagate on a metal-air or metal-dielectric interface. The connotation of the term "surface plasmon polariton" is that: (i) the wave is advancing on a surface; (ii) movement of charges is taking place in the metal (plasmon); (iii) an electromagnetic wave is traveling in air or dielectric (polariton). A polariton is a bosonic quasi-particle. It is obtained by strong coupling of electromagnetic waves with an electric/magnetic dipole-carrying excitation.

Plasmonics is the study of interactions of electromagnetic field with the free electrons of a metal, which lead to the production of plasmons. It is the investigation of coupling of light with the electrons of the metal.

8.2.1.1 Plasmonics for Data Transference Through Optical Interconnections

It may be recalled that photonics is the globally proven technology for data communications. It has already substituted electronics in long-distance data communication, as exemplified by the deployment of optical fibers in the Atlantic and Pacific oceans. In local area networks too, optical fiber communication links have rendered the older metallic conductor links obsolete. The optical fiber links provide superfast communication at the speed of light and offer a larger bandwidth too. However, these fibers are bulky and space consuming.

Plasmonics has the potential to enable extremely fast transference of data along the surface of a wire. Hence, it is often called, "light on a wire." But presently the problem faced is that the plasmons tend to lose energy after traveling a short distance of a few μm only. Plasmonics is a consequence of the wave nature of light. Hence, it is influenced by diffraction phenomenon in which light bends on passing through narrow apertures. The diffraction is exhibited as a spreading of light beam. Meaningful distances range from a few microns intra-component and inter-component separation in computer chips to thousands of kilometers in global communications. Therefore, initially scientists are not talking about long-distance communication. Researchers are first aiming at developing techniques to apply surface plasmons in integrated circuits used in computers, where shorter distances are required to be traversed.

In present-day computers, data moves in the form of an electronic current along metallic wires. This method of data transmission is slow. Plasmon polaritons can serve as the panacea for this difficulty. They can act metal waveguides which are as small in size as the metal conductor lines. So, the signals can travel at the speed of light using metal lines. Data will be conveyed by light-like waves, and not due to motion of electrons.

By replacing the metallic nanowires by plasmonic nanowires, the speed of data transmission will become quite faster much like fiber optics. It can be argued that optical fibers could be used for this purpose in computers. But their size is determined by the wavelength of light and they are of quite larger size than electronic conductors. Hence their usage is impractical. Plasmonic nanowires are comparatively much thinner and lightweight than optical fibers. The plasmonic nanowires in the form of nanometer thick metallic films can act as high bandwidth pathways for signal transmission. They may be looked upon as the nanophotonic counterparts of the optical fibers. They may be successful in replacing the metallic nanowires in computers, though over relatively shorter distances, leading to ultra-lightweight information technology.

To effectively utilize plasmonic nanowires, the diffraction problem associated with them must be eliminated. Otherwise, the quality of the signal will deteriorate as the distance increases. Lin et al. [4, 5] introduced a new surface wave. This surface wave is called a cosine-Gauss beam. To generate this beam, they fabricated a plasmon launcher consisting of two intersecting gratings made from metal. They made two sets of small grooves ~ 10 μm in length in a thin Au layer adhering to a glass backplate, and inclined the grooves by a small angle to construct a *V*-shaped pattern. Upon illuminating the pattern with an infrared source, two surface polaritons were produced. By their convergence and through constructive interference, a focused beam resulted with different degrees of confinement transversally and varying directions. This beam could cover a distance up to 80 μm, moving rectilinearly and tightly bound to the gold film without experiencing any spreading or diffraction. A near-field scanning optical microscope was used to follow the beam moving on the gold surface. The cosine-Gauss beam is the stepping-stone toward the realization of optical interconnections using surface plasmons.

8.2.1.2 Nanoparticle-Enhanced Surface Plasmon Resonance (SPR)-Based Biosensors

Surface plasmon biosensors commonly employ the Kretschmann configuration [6] shown in Fig. 8.1. In this configuration, a prism made of high refractive index glass

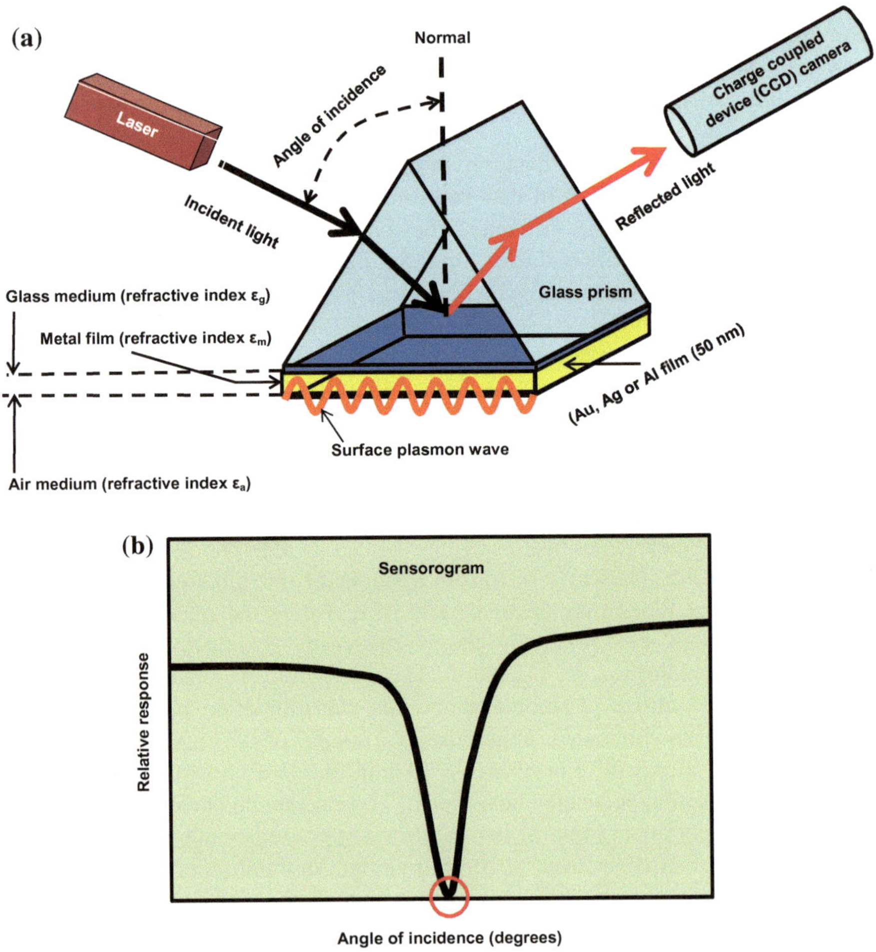

Fig. 8.1 Surface plasmon resonance: **a** Kretschmann configuration, and **b** the obtained sensorogram. The location of the minimum of reflectance analysis curve is proportional to the refractive index of the dielectric in contact with the gold film surface

is used. One face of the prism is coated with a thin gold film. Monochromatic, polarized light falls on the sensor chip, as shown in the figure. It is partly refracted and partly reflected. The angle of incidence is varied. At the critical angle, total internal reflection takes place. At angles greater than this critical angle, all the light suffers reflection at the glass–metal interface. The reflected light is received in the detector which measures its intensity.

The light falling on the Au film excites plasma oscillations on its surface. These oscillations lead to the production of surface plasmons. The electric field at the Au surface pierces through up to a shallow depth of 100 nm inside. Thus penetrating, it decays exponentially with distance, and propagates along the Au-glass interface. This electric field is known as the evanescent field.

Within total internal reflection, at a particular angle of incidence, the phenomenon of surface plasmon resonance takes place. At this angle, the momentum of photons of the light beam perfectly matches the momentum of surface plasmons in magnitude and direction. The energy of light beam is fully coupled into the Au-metal surface. The incident photons are fully utilized for plasmon generation so that there are no photons left in the reflected beam. Then the intensity of reflected light measured in the detector is lowest. Thus the onset of surface plasmon resonance is identified by the dip observed in the curve showing the intensity of reflected light as a function of angle of incidence.

To exploit the SPR resonance for biosensing (Fig. 8.2), the Au film is coated with suitable receptor molecules. The angle of incidence for SPR resonance is measured using Au-receptor molecule film. Next, this film is exposed to an analyte solution containing the target molecules. During exposure, the target molecules are bound to the receptors. The (Au+receptor+target) composite layer has different optical properties than (Au+receptor) layer. As a result, the incidence angle for SPR resonance is shifted in accordance with the concentration of target molecules. So, this shift in incidence angle provides an accurate estimation about target molecules. Thus SPR sensing operates on the principle of change in refractive index by the adsorption of target molecules which is seen as a variation in incidence angle for resonance.

To amplify the SPR signals and thereby improve the sensitivity of SPR-biosensors, gold nanoparticles have been extensively used. Incorporation of Au nanoparticles has been shown to increase the sensitivity of these biosensors by as much as a factor of 25. By using the metal nanoparticles, localized SPR (LSPR) resonance is found to occur (Fig. 8.3). The enhancement in SPR effect is attributed to an interaction between SPR and LSPR. LSPR resonance is absent when metals are used in bulk form. The gold nanoparticles are directly immobilized on the surface of the gold film using a self- assembled monolayer (SAM) of a dithiol. The dithiol contains two thiol groups; the thiol group is −SH. It is an organo-sulfur compound.

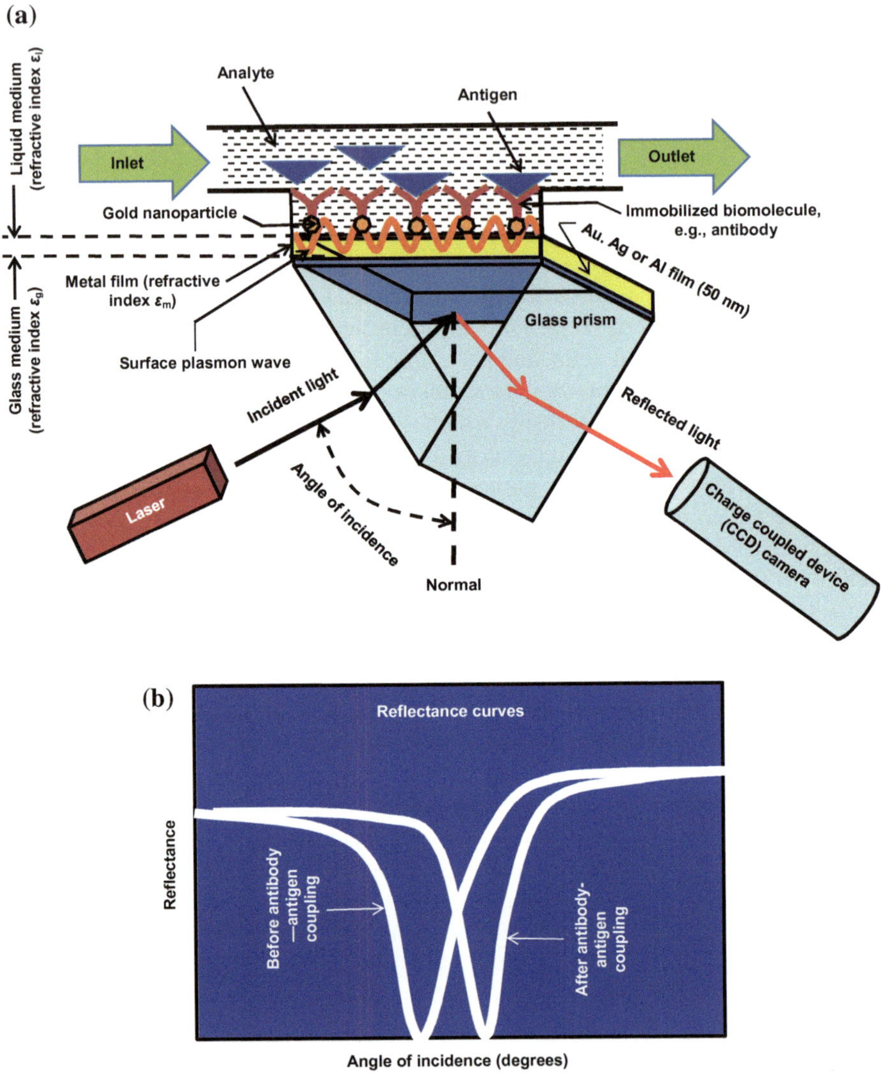

Fig. 8.2 Surface plasmon resonance: **a** biosensor based on Kretschmann configuration, and **b** its reflectance curves before and after antibody–antigen coupling

8.2.2 Photonic Crystals

These are periodic inhomogeneous micro/nanostructures made of dielectrics, metallo-dielectrics or superconductors (Fig. 8.4). They are engineered by using two materials of different dielectric constants. They are called semiconductors of light because they exert the same influence on light as a semiconductor crystal has on

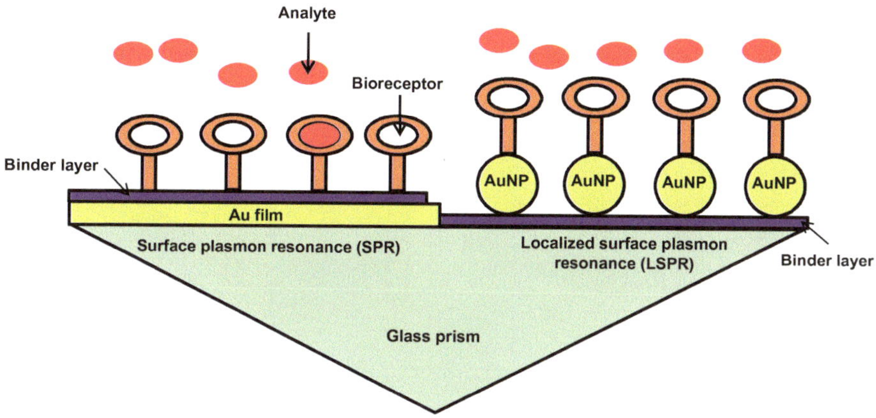

Fig. 8.3 Surface plasmon resonance and localized surface plasmon resonance

electron motion. In semiconductor theory, a periodic lattice defines allowed and forbidden energy bands. In the same way, in optical theory, a repetitive structure containing low and high dielectric constants, defines allowed and disallowed wavelength bands. Wavelengths that can pass through the photonic crystal structure are called modes. Groups of modes are known as bands. Wavelengths prohibited from passing through the structure constitute photonic band gap.

As already mentioned, modern data networks contain a mixture of optical and electronic sections. Optical section includes optical fibers to support signal transmission whereas electronic section consists of components and circuits to control the routing and queuing of signal traffic. The overall performance of the network is limited by the electronic section which has the utmost speed of gigabits/s. Photonic crystals offer a convenient technology to build an all-optical circuit. As an illustration, they can enhance the functionality of computer chips by serving as a filter device. This filter device permits the desired wavelengths to travel from one chip to another while stopping the undesired ones. By introducing an irregularity or defect into the structure, localized defect states are formed. The states obtained in this manner can help in guiding light on a sub-wavelength scale. Thus a low-loss waveguide is formed. More complex functions can be executed with other forms of disorder to make an optical integrated circuit [7]. Conventional design methodology starts with a proposed structure. The required parameters are found by playing around this structure. Alternatively, the circuit functions may be defined in the beginning. An inverse design approach may be followed to obtain the topology fulfilling these functions.

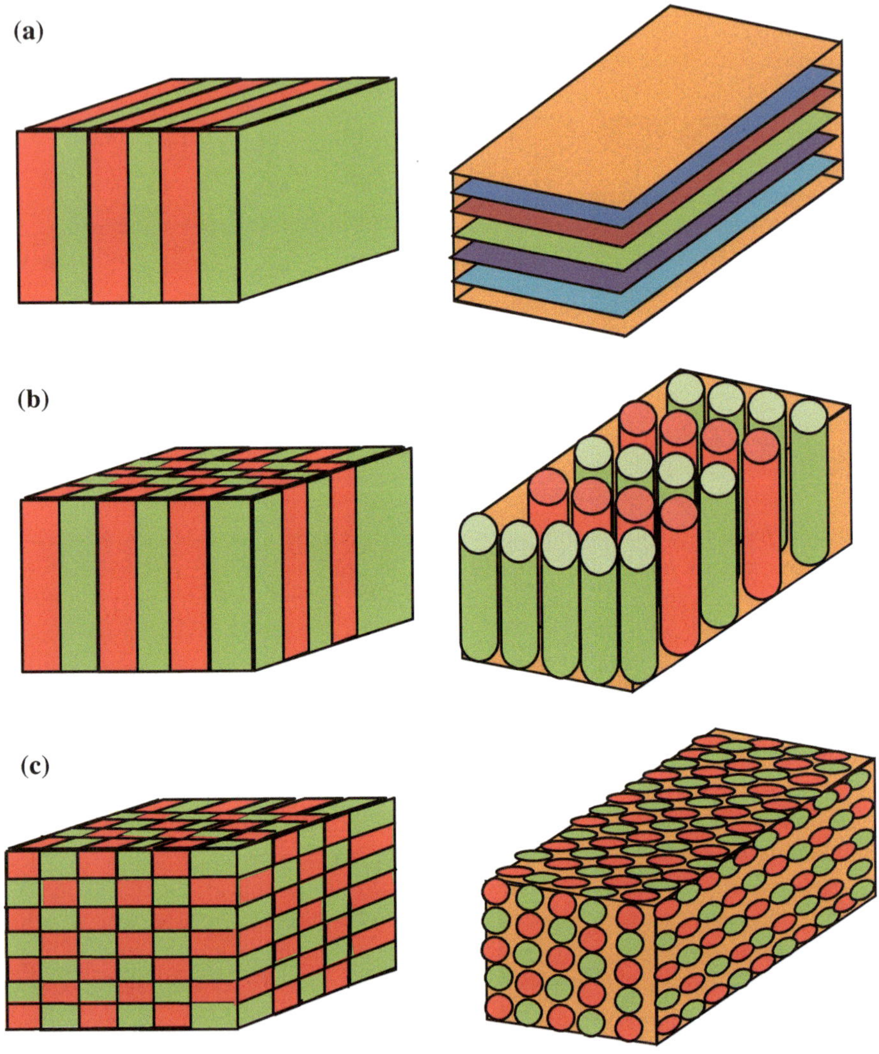

Fig. 8.4 Photonic crystals: **a** 1-D, **b** 2-D and **c** 3-D. Various colors signify materials of different relative permittivities

8.2.3 Quantum Dot Lasers

These lasers constitute a premier semiconductor-based technology containing high-quality quantum dots formed by self-assembly process in the active layer. They obviate the stability problem faced with existing semiconductor-based data communication systems causing excessive power consumption. Lower sensitivity to thermal fluctuations provided by them makes the use of temperature controllers

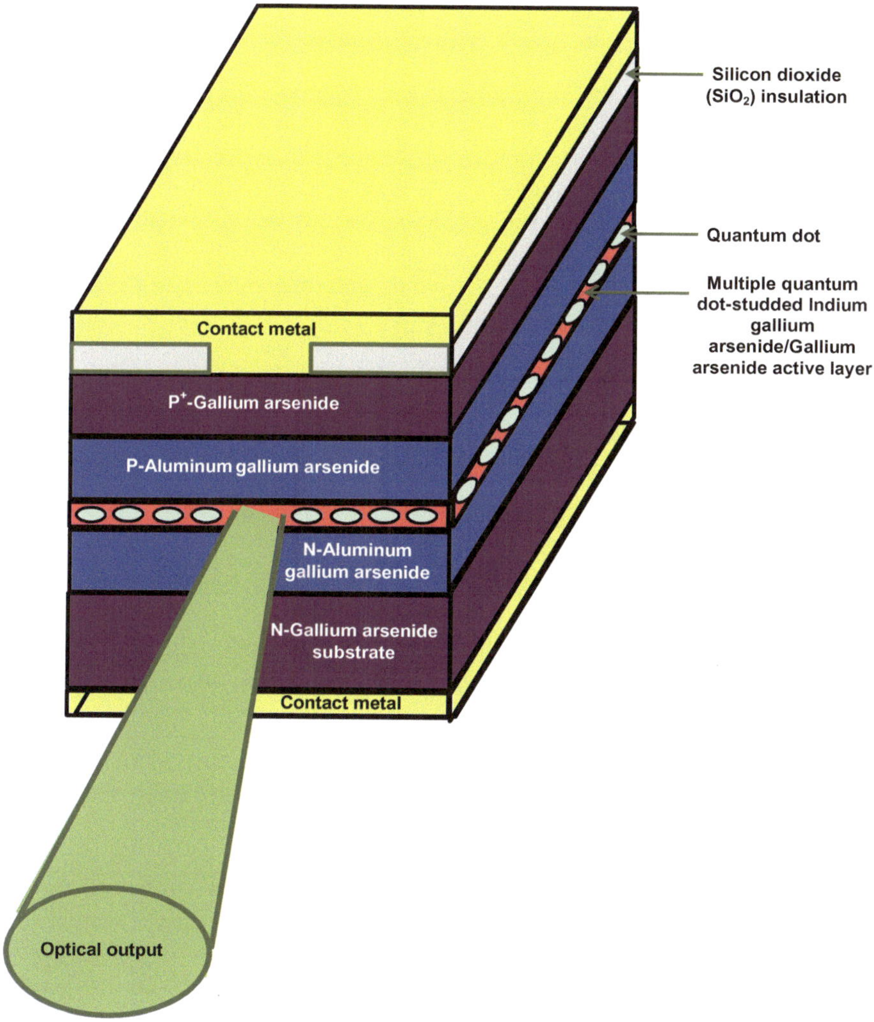

Fig. 8.5 Integration of several quantum dots in the active region of an edge-emitting semiconductor laser diode

unnecessary, thereby substantially cutting costs. Reduced power consumption, less distortion, and higher speed are additional benefits. The data transfer speed of 10 Gbps was raised to 25 Gbps by forming high-density indium antimonide quantum dots on a gallium arsenide substrate, with twice the number of dots per unit area and stacking more layers of quantum dots, increasing from 5 to 8 [8]. Figure 8.5 shows a laser diode with capability augmented with the help of quantum dots.

8.2.4 Silicon Nanophotonics

Silicon is the key element for fabrication of electronic devices and circuits. So, realization of optical components in silicon technology paves the way toward the goal of all-silicon optical/electronic integrated circuits [9]. Combination of silicon nano-optical devices, e.g., modulators, Ge photodetectors and wavelength-division multiplexers, with analog and digital CMOS circuits has been demonstrated on a single chip of silicon in 90-nm semiconductor fabrication, leading to optical communication transceivers with >25 Gbps data rates and able to feed several parallel output data streams into a single fiber [10].

8.3 Nanophotonics Beyond the Diffraction Limit

This optical technology was proposed by Prof. Ohtsu at University of Tokyo, Tokyo. The reader is encouraged to peruse the original papers for a comprehensive treatment [1–3]. A succinct description follows.

8.3.1 Near Field, Dressed Photons, and Nanophotonics

On illuminating a nanoparticle with conventional propagating light, free photons are emitted from the orbital electrons in the nanoparticle. These free photons moving away from the nanoparticle constitute the scattered light or the far field. Along with the free photons, another set of photons is emitted from the electrons in the nanoparticle. These photons do not move away but remain confined in the close vicinity of the nanoparticle. They are reabsorbed by the nanoparticle within a short interval of time. They can couple or team up with the electrons of the nanoparticle and are called dressed photons. It is these dressed photons which constitute the non-propagating optical near field. The optical near field is the elementary surface excitation on the nanoparticle, i.e., the dressed photons which are the carriers of energy of the material. It is an electromagnetic field acting as an intermediary or moderator for interactions between nanometric objects located proximate to the field. This optical near field enjoys freedom from diffraction of light by virtue of its confinement contingent upon size and through its resonance characteristics.

Nanophotonics is the optical nanotechnology dealing with the local electromagnetic interactions taking place between the optical near field and nanometric objects. Novel nanoscale devices can be fabricated, operated, and integrated by nanophotonics based on the new phenomena resulting from near field interactions. These phenomena cannot be observed with propagating light.

8.3.2 *Relevance of Plasmonics*

In the context of the above definition of nanophotonics, it is necessary to relook into the utilization of plasmonics for dimensional reduction of photonic devices. In the word,"plasmon," the letters, "on" signify that the plasma oscillation of free electrons in the metal is a quantum-mechanical phenomenon. Consider shining a light beam consisting of quanta called photons on a metal. Transformation of light energy into plasma oscillations of electrons takes place with a brief relaxation period. The moment this transformation occurs, the quantum-mechanical property is lost. In order to decrease the device dimensions and the heat produced, it is inadequate to introduce the idea of quanta of plasma oscillations. The insufficiency arises because from uncertainty principle, the position of the photon is defined in a space larger than the wavelength of light. Hence the wavefunction of the photon is undefined in the sub-wavelength space. On the other hand, for light absorption by a sub-wavelength size particle of nanoscale dimensions, the nanoparticle is a photodetector. The incident photon is perceived and its position is accurately obtained from the size of the particle. In a nutshell, reduction of the device size beyond diffraction limit warrants the involvement of a localized interaction between the nanoparticle and the photon. Moreover, energy dissipation is essential in the nanoparticle or surrounding macroscopic material. Inasmuch as plasmonics does not deal with local dissipation of energy, its pertinence for miniaturizing photonic devices is questionable. Plasmonics is founded on the classical wave optical formulation wherein notions like refractive index and wave number are used.

8.3.3 *Exciton-Polariton Exchanges*

An exciton is a quasi-neutral particle representing the bound state of an electron and an electron hole (the place from which the electron is removed), experiencing mutual attraction by Coulomb's law. Recalling the definition of polariton in Sect. 8.2.1, a strong coupling of exciton with photon yields the quasi-particle exciton-polariton.

In the local electromagnetic interaction between two nanoscale objects, virtual as well as real exciton-polariton exchanges take place. Virtual exciton-polariton exchange corresponds to non-resonant interaction through near-field non-propagating light represented by a Yukawa function with the optical near field localized around the nanoscale object(s). Real exciton-polariton exchange correlates with resonant interaction via the far-field scattered propagating light described by the spherical wave function.

8.3.4 Nanophotonic Devices

Optical near-field interaction is a short-range interaction. It is arbitrated by an optical electromagnetic field. In this interaction, excitation energy is transferred. Interactions among quantum dots, e.g., electron tunneling can be utilized to realize nanophotonic devices. Optical near-field interaction between quantum dots can be described as an exciton-polariton tunneling process. The quantized exciton energy level is specified by the quantum numbers (n_x, n_y, n_z).

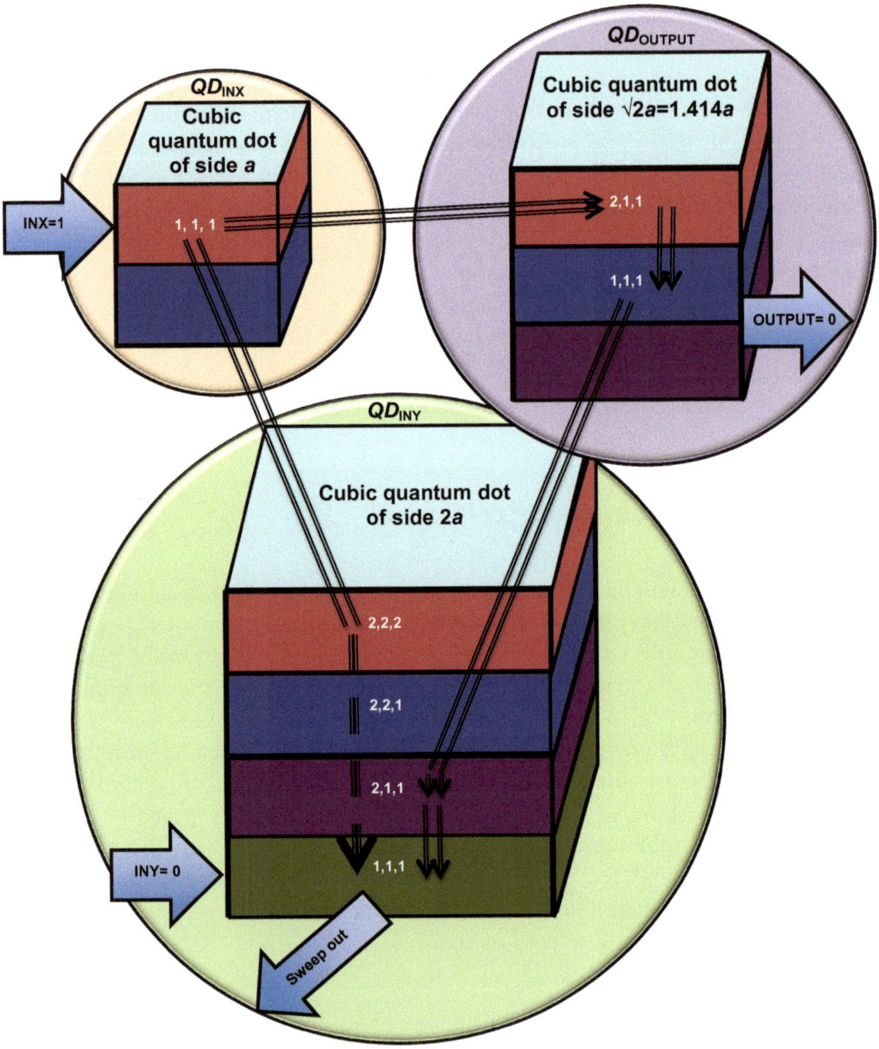

Fig. 8.6 Operation of a nanophotonic AND gate when INX = 1, INY = 0 and OUTPUT = 0

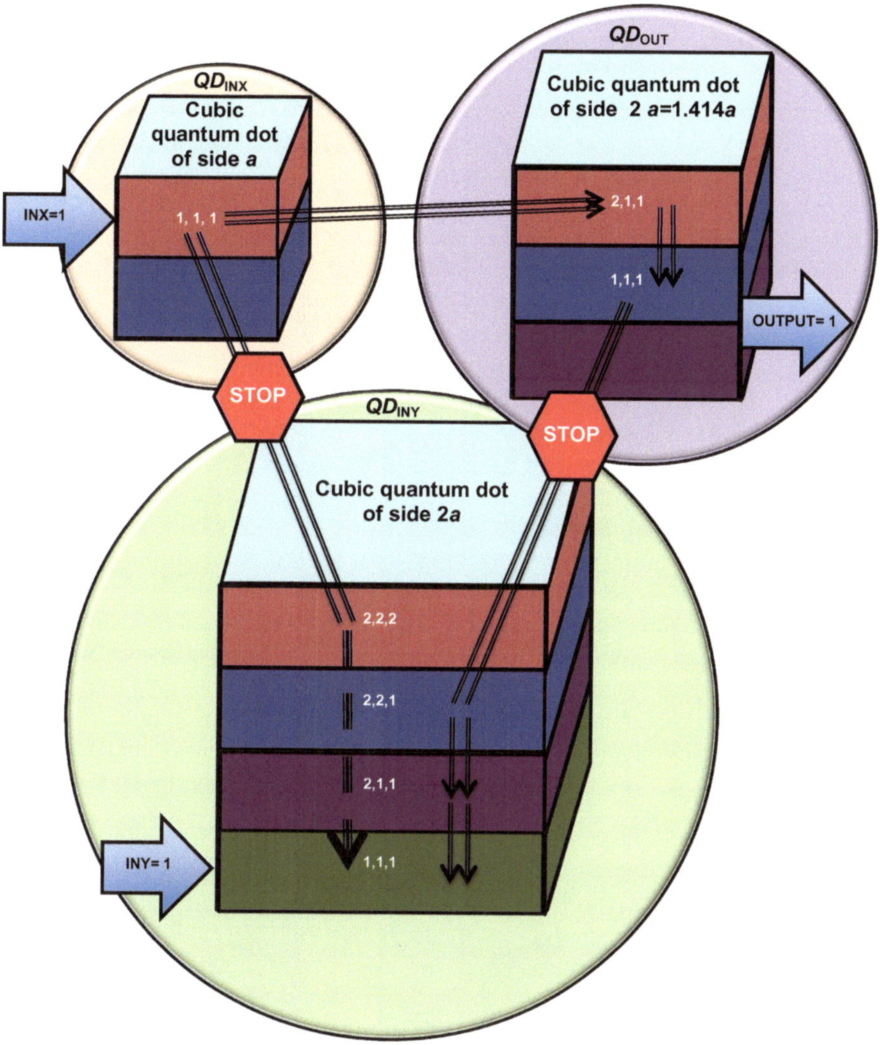

Fig. 8.7 Functioning of a nanophotonic AND gate when INX = 1, INY = 1 and OUTPUT = 1

Figures 8.6 and 8.7 show a nanophotonic AND gate. It consists of three closely located cubic quantum dots. Of these, two quantum dots are used for feeding the input signals, and one quantum dot is used for extracting the output signal. The two input quantum dots with inputs *X*, *Y* are labeled as QD_{INX} and QD_{INY}, respectively; and one output quantum dot is denoted by QD_{OUT}. The ratio of sizes of the three quantum dots is presupposed as Size of QD_{INX}: Size of QD_{INY}: Size of QD_{OUT} is $1{:}\sqrt{2}{:}2$. The exciton energy levels (1, 1, 1) in QD_{INX}, (2, 2, 2) in QD_{INY}, and (2, 1, 1) in QD_{OUT} resonate with each other. So also do the energy levels (2, 1, 1) in QD_{INY} and (1, 1, 1) in QD_{OUT}.

On application of a single input signal to the AND gate (Fig. 8.6), i.e., for INX = 1, INY = 0, the exciton energy in QD_{INX} is completely transported to the (1, 1, 1) level in QD_{INY}. Then QD_{OUT} = 0.

When both the input signals are applied to the AND gate (Fig. 8.7), i.e., for INX = 1, INY = 1, the second input signal INY impedes the transfer of energy to QD_{INY}. This is because of the state filling in QD_{INY}. Consequently, an output optical signal is produced in the quantum dot QD_{OUT}. Thus, the output signal QD_{OUT} = 1 when both INX = 1 and INY = 1.

Similarly, for INX = 0, INY = 1, QD_{OUT} = 0. Also, for INX = 0, INY = 0, QD_{OUT} = 0. Thus this arrangement of quantum dots serves as an AND gate.

The operation of a NOT gate is expounded with reference to Figs. 8.8 and 8.9. The NOT gate consists of two quantum dots. One quantum dot for feeding the input signal IN is labeled as QD_{IN}. The other quantum dot for taking out the output signal OUT is labeled as QD_{OUT}. A presumption is made regarding the ratio of the sizes of the two quantum dots: Size of QD_{IN}: Size of QD_{OUT} is $(1 + \xi):\sqrt{2}$ where $\xi \ll 1$. Then the energy levels of the exciton: (2, 1, 1) in QD_{IN} and (1, 1, 1) in QD_{OUT} are off-resonant by a small amount.

In the absence of an input signal (Fig. 8.8), IN = 0, the optical power source produces an exciton in the quantum dot QD_{OUT}. The exciton vanishes after emission of a photon. This happens because off-resonance condition leads to non-availability of any pathway for transferring energy. The emitted photon is obtained as the output signal. Thus when IN = 0, OUT = 1.

Upon application of an input signal, IN = 1 (Fig. 8.9). Then an optical near-field interaction enables the transference of exciton energy in quantum dot QD_{OUT} to the quantum dot QD_{IN}. This transference happens because the energy level (2,1,1) in

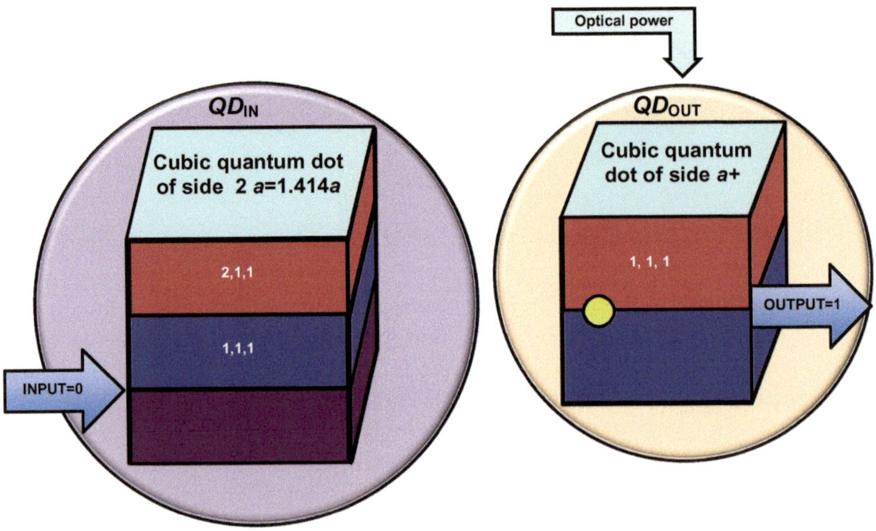

Fig. 8.8 Nanophotonic NOT gate operation when input = 0

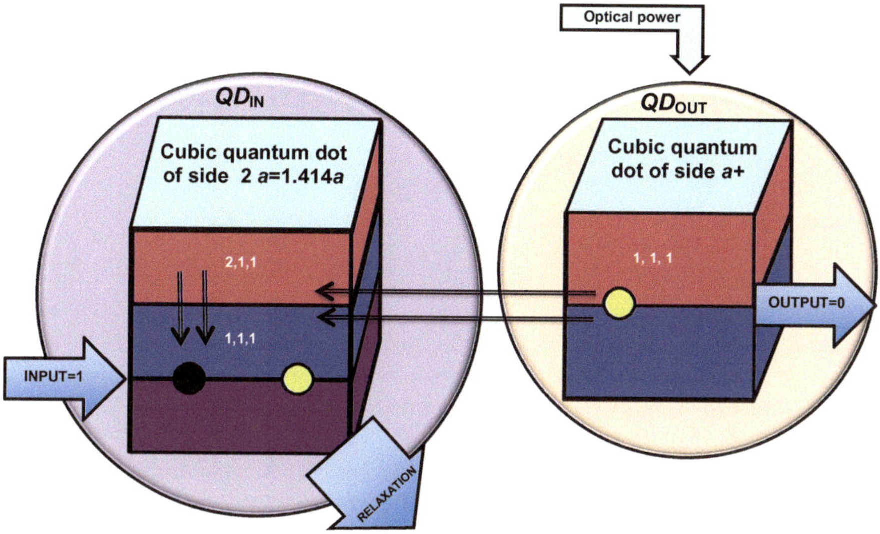

Fig. 8.9 Nanophotonic NOT gate operation when input = 1

QD_{IN} is shifted to become aligned with the energy level (1,1,1) in QD_{OUT}. One reason of the shift is the decrease in phase relaxation time of the carriers. Another reason is the carrier-carrier scattering. Consequent upon this resonance mechanism, output signal OUT = 0. In summary, the gate performs the following actions: If IN = 0, OUT = 1; when IN = 1, OUT = 0, which is essentially NOT gate function.

The above AND and NOT gates are used as basic building blocks which can be applied by structural modifications and in different combinations to fabricate the complete family of logic gates in nanophotonics.

8.4 Discussion and Conclusions

Plasmonics is promising to revolutionize optical interconnects within computer chips. Photonic crystals can be used for filtering and waveguiding. Quantum dot lasers provide temperature-independent output for speedy data communication. Silicon nanophotonics helps in availing the facilities of silicon technology for fabricating optical components so that both optical and electronic components are made using silicon. Near-field optical interactions are useful for building quantum dot-based logic gates, and hence nanophotonic integrated circuits.

Review Exercises

8.1 Name the two subbranches of nanophotonics? Describe their aims and scope.

8.2 How are oscillations produced in the density of free electron gas in a metal? What are these oscillations called?

8.3 Define the following terms: (i) plasmon, (ii) polariton, and (iii) surface plasmon polariton.

8.4 What is plasmonics? Why is it susceptible to diffraction effects?

8.5 Photonics is an accepted technology for long-distance communication networks. Then what is the shortcoming of photonics, which plasmonics could solve? Can it provide the solution for large distances? If not, for what order of distances is it likely to be utilized and for what applications?

8.6 Data can be transported inside computer chips by waves similar to light waves? Explain how is this possible?

8.7 What benefit could be derived by using plasmonic nanowires in computers? Why cannot the conventional optical fibers be used here?

8.8 How is a cosine-Gauss beam generated? What are its special features? Discuss its use as an optical interconnect in computer chips.

8.9 Describe the widely used prism coupling arrangement for surface plasmon resonance experiments. How is the occurrence of resonance recognized by measuring the intensity of reflected beam?

8.10 Explain the effect of gold incorporating nanoparticles on the sensitivity of SPR-biosensors.

8.11 What are photonic crystals? How are they engineered?

8.12 Why are photonic crystals called "semiconductors of light"? Explain.

8.13 How is the performance of a data network restricted by its electronic section? How can photonic crystals be used to construct an all-optical data network? Discuss.

8.14 What specific problem faced with conventional lasers is overcome by using quantum dot lasers? Up to what extent data transfer speed has been elevated by using these lasers?

8.15 What is the advantage of fabricating optical components in silicon technology? Name three components which have been fabricated in silicon.

8.16 What is meant by 'near field' and 'far field' in optics? Define nanophotonics using the definition of near field interactions. What is a dressed photon?

8.17 Why plasmonics is considered irrelevant for reduction of size of photonic devices? Justify with arguments.

8.18 What is an exciton-polariton? Explain the exciton-polariton exchanges that take place between nanoscale objects. What are the mathematical functions used to describe these exchanges?

8.19 On the basis of optical near-field interactions describe the implementation of following gates in nanophotonics with labeled diagrams: (i) AND gate and (b) NOT gate.

References

1. Ohtsu M, Kobayashi K, Kawazoe T et al (2002) Nanophotonics: design, fabrication, and operation of nanometric devices using optical near fields. IEEE J Sel Top Quantum Electron 8(4):839–862
2. Ohtsu M, Kawazoe T, Yatsui T et al (2008) Nanophotonics: application of dressed photons to novel photonic devices and systems. IEEE J Sel Top Quantum Electron 14(6):1404–1417
3. Ohtsu M, Kobayashi W, Kawazoe T et al (2008) Principles of nanophotonics. CRC Press, Boca Raton, 248 pp
4. Lin J, Dellinger J, Genevet P et al (2012) Cosine-Gauss plasmon beam: a localized long-range nondiffracting surface wave. Phys Rev Lett 109(9):093904 (5 pp)
5. The Agency for Science, Technology and Research (A*STAR). (2013). Plasmonics: a wave without diffraction. ScienceDaily. www.sciencedaily.com/releases/2013/05/130522131024.htm. Accessed 12 Sept 2015
6. Bedford EE, Spadavecchia J, Pradier C-M et al (2012) Surface plasmon resonance biosensors incorporating gold nanoparticles. Macromol Biosci 12:724–739
7. Andonegui I and Garcia-Adeva AJ (2013) Designing integrated circuitry in nanoscale photonic crystals. SPIE Newsroom. doi:10.1117/2.1201311.005035, http://spie.org/x104683.xml. Accessed 12 Sept 2015
8. Borghino D (2010) World's first 25 Gbps data communication using quantum dot laser achieved. http://www.gizmag.com/25gbps-communication-with-quantum-dot-laser/15310/. Accessed 12 Sept 2015
9. Khriachtchev L (2008) Silicon nanophotonics: basic principles, current status and perspectives. Pan Stanford Publishing, 472 pp
10. Silicon Integrated nanophotonics technology: from the lab to the fab. http://researcher.watson.ibm.com/researcher/view_group.php?id=2757. Accessed 12 Sept 2015

Chapter 9
Nanoelectromechanical Systems (NEMS)

Abstract NEMS consist of electronic and nonelectronic components and functions on the nanoscale. These components and functions include sensing, actuation, signal acquisition, and processing. Sometimes display, control, interfacing, and ability to perform chemical and biochemical interactions are also included. NEMS follow both approaches: downscaling previous MEMs components to nanodimensions, and introducing new concepts based on phenomena that are exclusive to nano-regime. Limitations in downscaling are pointed out as well as novel sensing/actuation techniques are presented. NEMS play a critical role in medical diagnostics, displays, energy harvesting, nonvolatile memory, and providing ultra-sharp tips for atomic force microscopy.

9.1 Introduction

Parallel to the relentless march of nanoelectronics toward miniaturization in the "More Moore" sub-domain, the sensing devices too are moving ahead in the "More-than-Moore" roadmap to nanoscale sizes, aiming at the ultimate limits of atoms and molecules. Nanoelectromechanical systems (NEMS) include man-made mechanical elements, sensors, actuators, and signal processing circuits having critical feature sizes between 100 and 1 nm. In NEMS, the mass, thermal capacity, and power consumption decrease as the critical dimension becomes smaller. Contrarily, the fundamental frequency, mass/force sensitivity, and quality or Q-factor increase as the critical dimension is reduced. Present technology has enabled the fabrication of NEMS of masses $\sim 10^{-18}$ g and cross-sectional area ~ 10 nm $\times$ 10 nm. By virtue of infinitesimally small sizes, NEMS have achieved frequencies ~ 100 GHz, Q-factors $\sim 10^3$–10^5 in moderate vacuum [1], force sensitivities $\sim 10^{-18}$ N, thermal capacities $\sim 10^{-24}$ cal, power consumption $\sim 10^{-18}$ W, integration density $\sim 10^{12}$ elements cm^{-2}, and ultra-low mass sensitivities up to molecular level [2]. These unique characteristics of NEMS strikingly differ from those of their predecessor MEMS. In comparison to MEMS, NEMS combine smaller mass with

V.K. Khanna, *Integrated Nanoelectronics*,
NanoScience and Technology, DOI 10.1007/978-81-322-3625-2_9

higher surface-area-to-volume ratio to achieve better sensitivities. Constructional materials for NEMS are silicon, silicon carbide, CNTs, Au, Pt, etc.

9.2 NEMS Sensor Classification

NEMS sensors are compartmentalized into two parts: (i) MEMS sensors that have been downscaled to nano–dimensions; and (ii) novel nanosensors and systems that have been developed especially for nanoscale, e.g., CNT-based. These sensors are unique to the nanodomain. They are required because several MEMS sensors cannot be downscaled due to restrictions imposed by noise and sensitivity. So, they cannot provide the desired resolution in their nano-versions. Further, many microsensors cannot be fabricated in nanosensor form due to technological problems. Therefore, downscaling fails. The metric used for measuring the performance of downscaled sensor is the ratio of the range of measurand to the resolution of the readings. This ratio is known as the dynamic range of the sensor [3].

9.3 MEMS Sensors Downscalable to NEMS Version

9.3.1 Piezoresistive Sensors

Piezoresistance is a basic material property in the toolbox of sensor design engineers. Piezoresistive sensors are used to measure displacement/force/pressure from the change in resistance of a piezoresistor formed on a cantilever beam [4] at the point of flexure. The cantilever, a beam fixed at one end and free at the other end, is a ubiquitous sensing element in the realm of micromachined devices (Fig. 9.1).

The piezoresistance change is caused by the surface stress generated in the cantilever beam on bending. The change in surface stress Δg resulting from deflection Δh of the cantilever is expressed by Stoney's formula [5]

$$\Delta g = [E\Delta h/\{4(1 - v)\}](t/L)^2 \tag{9.1}$$

where E and v are the Young's modulus and Poisson's ratio of cantilever material; t and L denote the thickness and length of the cantilever beam.

Piezoresistive sensors are also used to determine pressure from resistance changes of a piezoresistor made in monocrystalline silicon or polysilicon film deposited at the peripheral regions on a circular or rectangular diaphragm subjected to pressure [6], Fig. 9.2. By anisotropically etching silicon from the backside, a cavity is formed terminating near the surface of the silicon to form a circle-/rectangle-shaped

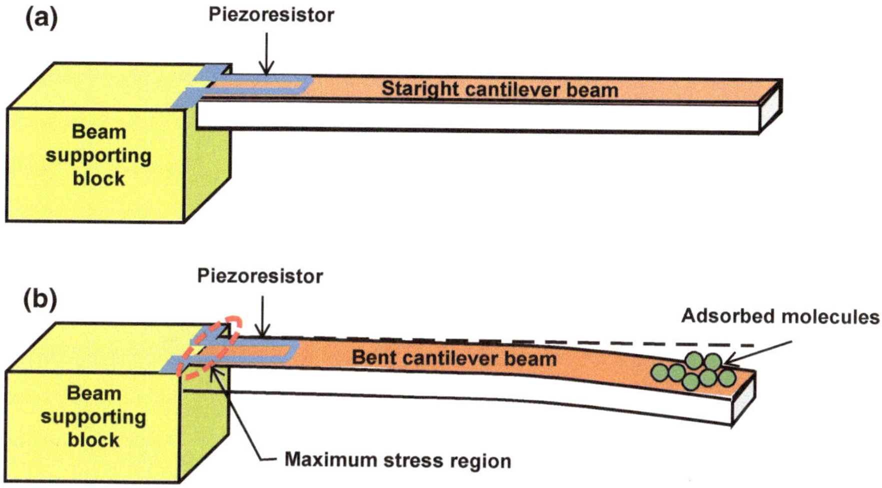

Fig. 9.1 Nanocantilevers: **a** straight and **b** bent by the mass of adsorption of molecules

diaphragm of required thickness [7]. A pressure applied on the diaphragm such as by a fluid produces mechanical tension at its edges. Piezoresistors are positioned at the locations of highest tension to achieve maximum sensitivity. Two resistors are placed such that current flows in a direction parallel to that of tension. The other two resistors are placed in such a way that current flows perpendicular to the direction of tension. The resistor areas are doped with ion implantation to get the desired resistance values. Through aluminum metallization film, the piezoresistors are connected together in a Wheatstone bridge configuration.

The sensor is susceptible to Flicker noise and Johnson noise. When it is decreased in size, flicker noise becomes dominant. The dynamic range is restricted to <60 dB.

9.3.2 *Tunneling Sensors*

They comprise a sharp tip in vicinity of a moving surface. As the tip is raster scanned over the surface, the tunneling current varies with distance between the tip and the surface providing high displacement sensitivity. The variation of tunneling current i with the distance d separating the tip and the surface is given by

$$i = \rho_{\mathrm{S}}(E_{\mathrm{F}})V\exp(-2\pi\lambda d) \tag{9.2}$$

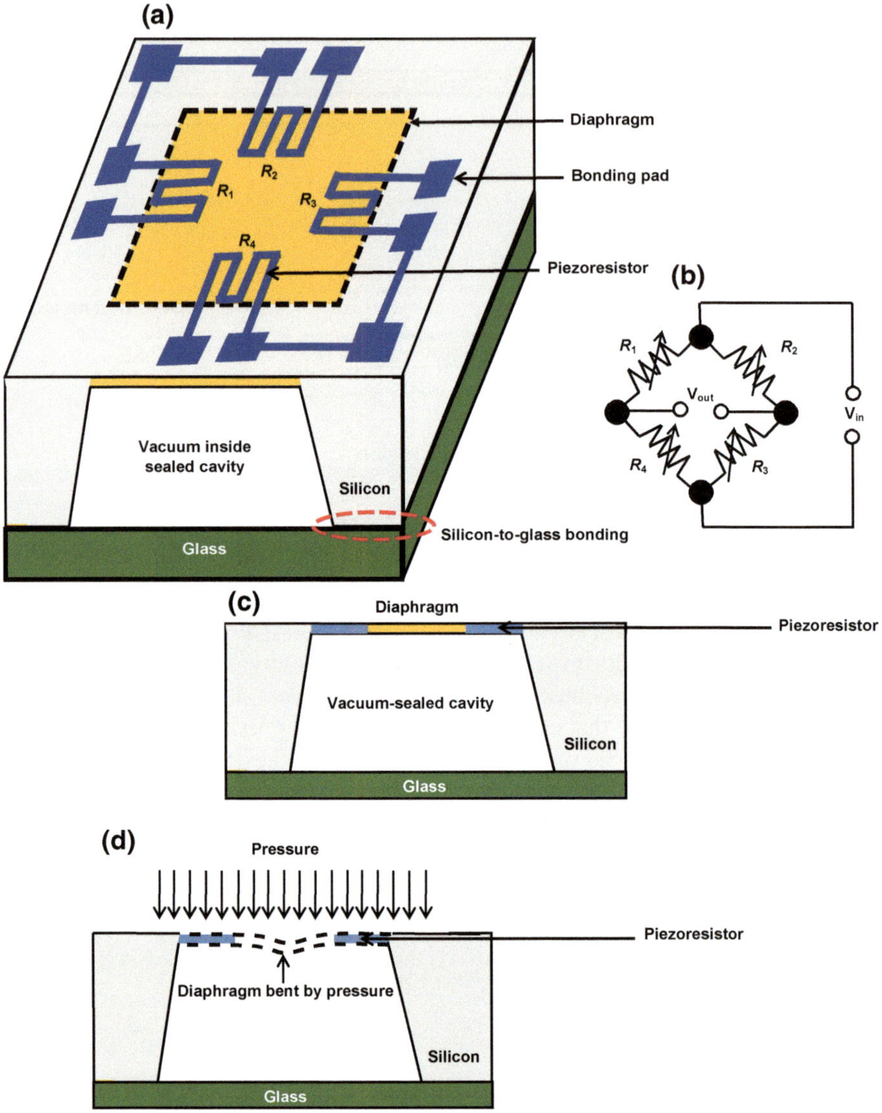

Fig. 9.2 Absolute pressure sensor: **a** 3-D view of pressure sensor chip bonded on glass; **b** wheatstone bridge connection of piezoresistors on the diaphragm; **c**, **d** cross-sectional diagrams showing straight and bent diaphragms

where V is the applied DC voltage, $\rho_S(E_F)$ is the density states of electrons localized around the tunneling region, E_F is the Fermi level, and λ is the decay constant of the electron wave function in the tunneling gap.

Their range and resolution do not downscale as the size decreases, making them ideally suited to nanorange.

9.4 MEMS Sensors Not Downscalable to NEMS Version

Piezoelectric sensors are active sensors, which use a film of a piezoelectric material such as ZnO or PZT deposited on a diaphragm or a clamped–clamped beam to measure vibrations or accelerations. They show a diminution in dynamic range with sensor size. Capacitive sensors use parallel-plate or comb drive configurations to detect out-of-plane/in-plane motion in accelerometers or gyroscopes. They do not work at all at nanoscale because their dynamic range decreases with size. Problem with Hall effect sensors used for measuring displacements is the interference from stray magnetic fields.

9.5 CNT-Based Piezoresistive Nanosensors

These sensors have intrinsically nanosizes. They provide large gauge factors and hence are capable of >60 dB dynamic range at a footprint of 1 μm × 1 μm. CNT sorting technique needs to be perfected so that CNTs of highest gauge factor can be employed for sensor fabrication.

Figure 9.3 shows the schematic diagram of a pressure sensor using SWCNT as the sensitive element. An SWCNT was adsorbed on the surface of an alumina membrane of thickness 100 nm [8]. Source and drain leads were connected to the SWCNT for electrical measurements. The membrane was circular in shape with a diameter of 50–100 μm. It was formed by bulk micromachining. For SWCNTs with metallic behavior, the piezoelectric gauge factor was found to be 210.

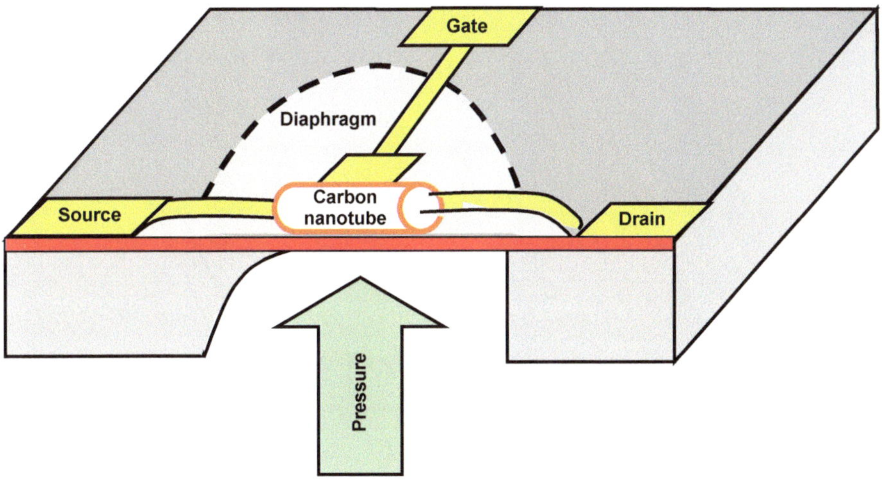

Fig. 9.3 CNT pressure sensor

9.6 NEMS Resonators

These are cantilevers or doubly clamped beams (Fig. 9.4). They have a miniscule mass. They dissipate very little power leading to extremely high quality factors. Their resonance frequency is inversely proportional to mass and to square of length of the device. It is very high >1 GHz, thereby enhancing the sensitivity as well as speed of signal processing.

Graphene nanoribbon has been used as a doubly clamped beam resonator (Fig. 9.5).

9.6.1 *Resonator-Based Mass Sensors*

Addition of a small mass to the resonator cases a shift in its resonance frequency, which is directly proportional to the incremental mass. The sensitivity of detection is given by

$$\Delta f_{\mathrm{r}}/\Delta m = (1/2m)f_{\mathrm{r}} \tag{9.3}$$

where Δf_{r} is the change in resonance frequency f_{r} when an extra mass Δm is added to the original mass m. Here, $\Delta m \ll m$. This equation shows that the sensitivity

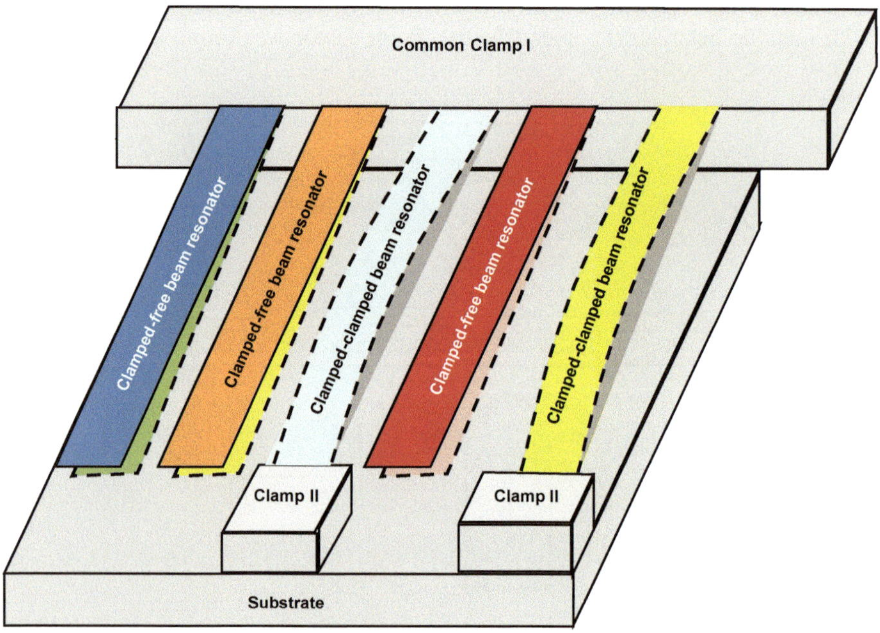

Fig. 9.4 Two types of resonators: singly clamped beam (cantilever) and doubly clamped beam

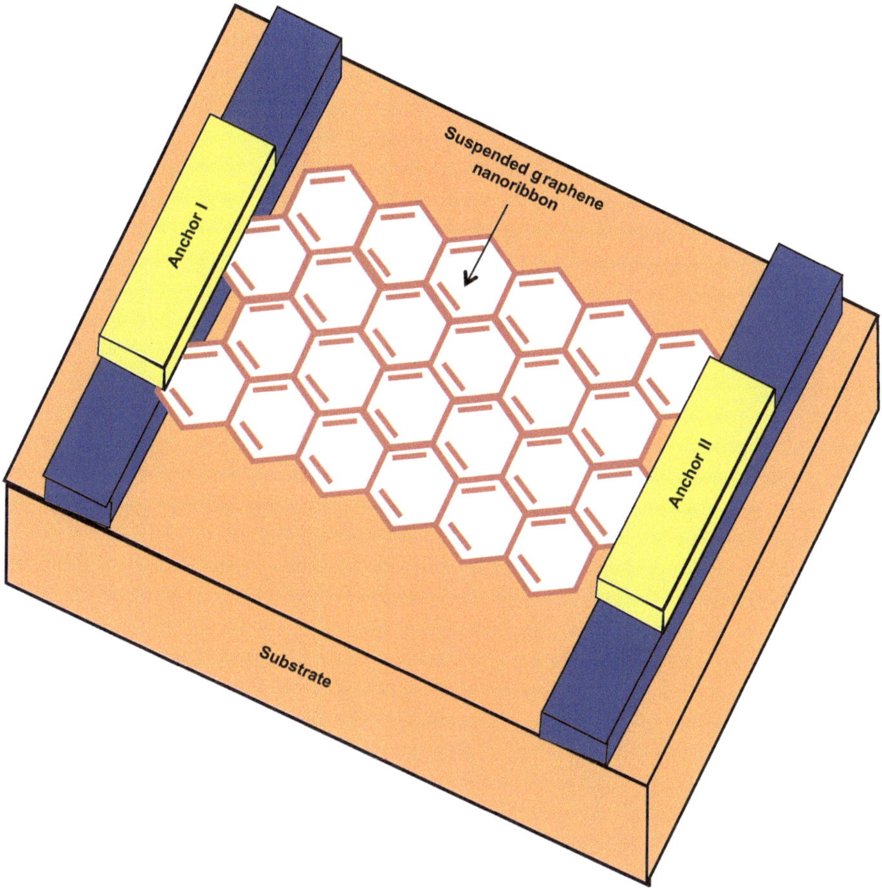

Fig. 9.5 Graphene nanoribbon as a NEMS resonator

increases with increase in resonance frequency. Hence, ultrahigh frequencies are needed for detecting very small masses. NEMS-based mass spectrometry has rendered possible detection of protein molecules and nanoparticles in real time immediately on their adsorption on the sensor [9].

Masses up to 10^{-18} g have been measured using silicon nanocantilevers or beams clamped on both sides. CNT-based resonators are capable of 10^{-21} g resolution. The NEMS resonators are prone to thermomechanical noise, limiting their dynamic range to 60 dB.

The resonant sensors have been utilized as a platform whose surface is functionalized for detection of gases or biomolecules. CNTs with carboxylic chloride groups were used for NO_2 detection [10].

9.6.2 Resonator-Based Strain Sensors

On subjecting a clamped–clamped flexure beam to tension, the beam increases in length. Tension makes the beam stiff and unbending. The stiffness of the beam causes a change in its natural frequency. This change in frequency is correlated to the strain. CNTs are very useful in these devices due to their high modulus of elasticity and large strength [11]. For similar reasons, graphene-based NEMS resonators are also promising [12].

9.7 NEMS Actuators

In NEMS applications, particularly for optical and RF systems, MEMS actuation reduces both speed and accuracy. So, suitable NEMS actuators offering nanoscale precision must be provided. Nanoactuation is actuation of a function using a nanosize object.

9.7.1 CNT Nanotweezers

Nanotweezers consist of two arms of CNTs [13]. These arms are fixed on a silicon tip with metal electrodes. For this fixation, viewing was done through SEM. On applying a DC voltage between the two CNTs, they move toward each other. The nanotweezers are operated electromechanically for manipulating nanomaterials in scanning probe microscopes such as STM (scanning tunneling microscope) and AFM (atomic force microscope).

9.7.2 Nanogrippers

A nanogripper or 'robotic hand for ultra-small objects' is a 'pick-and-place' device with two end effectors consisting of a nanotip made from a tungsten tip and CNT [14]. These end effectors can be used to apply forces acting in opposite directions on nanoscale objects to grasp them, lift them up from their positions, hold them, and place them, as desired. Nanogrippers have been fabricated whose arms can be opened/closed by electrostatic actuation that can also measure the size of the specimen [15]. Electrothermal actuation has been exploited in silicon nanogrippers to make them expand [16].

9.7.3 Magnetic Bead Nanoactuator

A titanium cantilever carries a bead made of a superparamagnetic material at its front end. The deflection of the cantilever can be controlled by an externally applied magnetic field gradient [17]. Superparamagnetism is a type of magnetism found in ferromagnetic/ferrimagnetic nanoparticles. It is the property by which the direction of magnetic moments changes in some materials at nanoscale and they behave like a paramagnet even below Curie temperature without applying any magnetic field. At the same time they show a high magnetic susceptibility like a ferromagnetic material, i.e., exhibit a high degree of response to an applied magnetic field.

9.7.4 Nanoactuation by Magnetic Nanoparticles and AC Fields

22-nm-size magnetic nanoparticles were injected into the brain tissue [18]. The neurons were excited by triggering thermally sensitive capsaicin receptors by applying an alternating electric field externally. By this method, calcium ions were introduced into neurons.

9.7.5 Ferroelectric Switching-Based Nanoactuator

When the electric field is absent, the nanoactuator is in the ground state in which the domains are polarized [19]. On applying an electric field, the polarization undergoes reorientation. Strain is thus produced. As soon as the electric field is withdrawn, the nanoactuator once again reverts back to the ground state. Thus a repeatable actuation cycle is completed.

9.7.6 Optical Gradient Force-Driven NEMS Actuator

Optical gradient force can produce mechanical deformation in the nano-regime [20]. This force is appreciably stronger in evanescently coupled waveguides owing to the larger value of the gradient of optical field intensity. Use of ring resonator of high quality factor increases the optical field intensity by many orders of magnitude, enhancing the force still further, and enabling its exploitation for nanoactuation. The nanoactuator, drawn according to Dong, consists of three main components (Fig. 9.6): (i) An actuation ring resonator: Its Q-factor is controlled by P–I–N electro-optics modulator. The rib waveguide forms a part of this resonator. The slab on one side is doped with P-type and that on the other side is doped with N-type

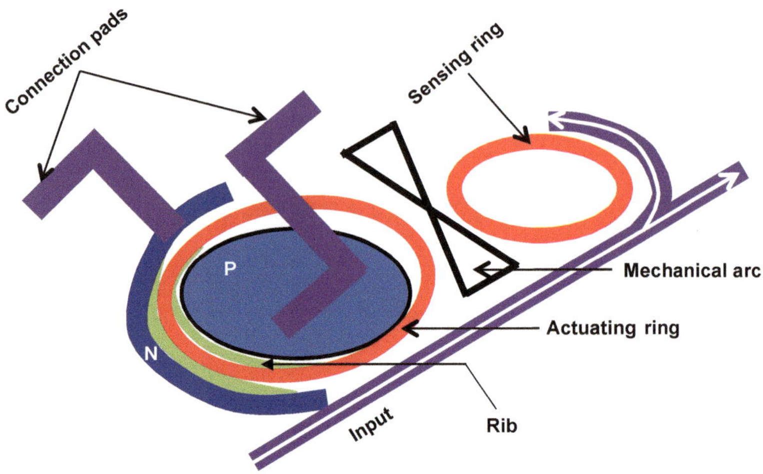

Fig. 9.6 Components of optical gradient actuator

constituting a P–I–N junction. The modulator induces free carriers into the actuation resonator to alter its absorption coefficient for changing the Q-factor. (ii) A sensing ring resonator: It measures the actuation displacement. It is optically linked with the actuation resonator. (iii) A mechanical arc: Optical gradient force is produced between the actuation resonator and the mechanical arc.

On changing the Q-factor of the ring resonator from 15000 to 6000, the actuation displacement reached 14 nm. The actuator can be applied to manipulating single molecules and sensors of high sensitivity.

9.8 NEMS Memories

These are memories utilizing NEMS switches to reduce leakage currents [21]. They provide high ON/OFF ratios $\sim 10^5$. Further, they can function in hostile environments such as at elevated temperatures and in radiation-polluted spaces. Different architectures consisting of two or three electrodes have been conceived. One common two-electrode architecture consists of a TiN cantilever/a hanging bridge and a stationary TiN/W electrode at a separation of 15–20 nm (Fig. 9.7a). Voltage applied between the two electrodes activates the switching action for closing or opening of the switch leading to its on and off states. In the suspended bridge structure, the top electrode anchored on an insulating support is mobile while the bottom electrode is static (Fig. 9.7b).

Another electrode architecture comprises crossed pairs of CNTs (Fig. 9.8). One pair of parallel CNTs running on an Si/SiO_2 substrate is connected to metal electrodes A and B. A second pair of parallel CNTs moves perpendicular to the first

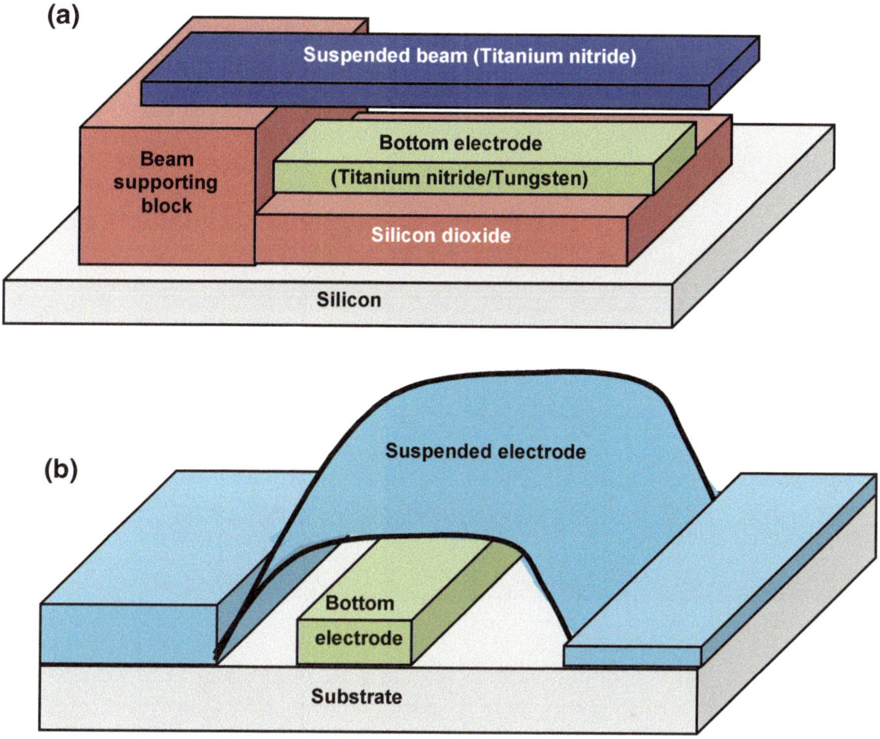

Fig. 9.7 Two electrode architectures using **a** cantilever and **b** suspended bridge

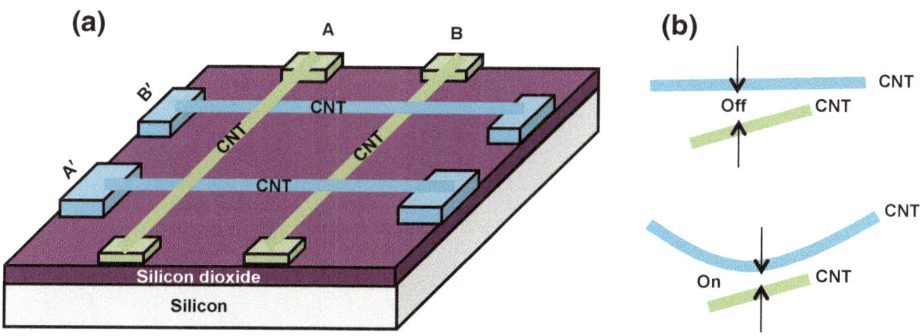

Fig. 9.8 CNT arrangement and disposition: **a** perpendicularly placed pairs of CNTs, **b** their disposition in off and on states

pair, and is connected to metal dots A′ and B′. The gap intervening the two CNT pairs at their intersection is very small ~1 nm or less. The on and off states are distinguished from the magnitude of tunneling currents between the two pairs.

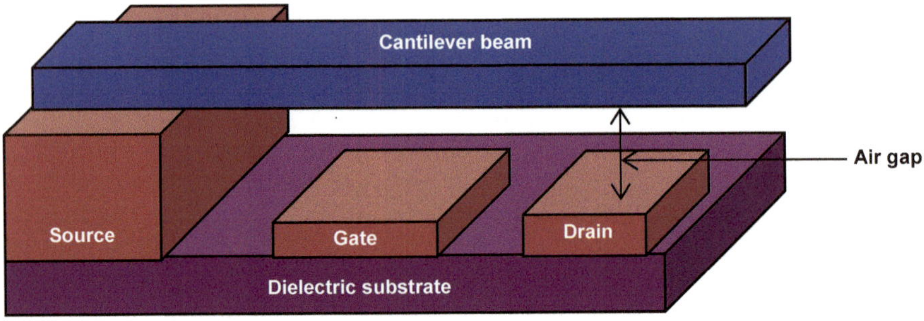

Fig. 9.9 Capacitive three-electrode structure

In the three-electrode architecture, a third motionless top electrode controls the motion of the cantilever/suspended bridge like the gate electrode of a field-effect transistor (FET), Fig. 9.9.

9.9 Discussion and Conclusions

A plethora of nanosensors and nanoactuators were described. NEMS components outperform their MEMS competitors in speed and resolution. They have established superiority due to their small sizes and abilities to execute tasks, which are impossible for MEMs devices. Such tasks include their placement inside the human body to provide medical diagnostic information in real time.

Review Exercises

9.1. What are NEMS? Name some materials commonly used in fabrication of NEMS. Compare NEMS with MEMs.

9.2 How does the reduction of critical feature size affect the following: (i) power consumption, and (ii) Q-factor?

9.3 What are the two sub-classes of NEMs sensors? What is the basis of this classification? Name and define the figure of merit to assess the performance of a downscaled sensor.

9.4 Write Stoney's formula of surface stress. Explain the symbols used.

9.5 At what locations are the piezoresistors placed for maximum sensitivity on (i) a cantilever, and (ii) a diaphragm? Explain the working of a piezoresistive pressure sensor.

9.6 What types of noises is a piezoresistor susceptible to? How do these noises impact the performances of piezoresistive sensors when their sizes are reduced to nanoscales?

9.7 How does the tunneling current vary as a function of distance between the electrode tip and the scanned surface? What makes tunneling sensors ideal for nanoscale?
9.8 Name three categories of MEMs sensors which do not perform well at nanodimensions. Enumerate the reasons for this behavior.
9.9 What are the merits of CNT-based piezoresistive sensors? Explain with a diagram the operation of a pressure sensor using SWCNT.
9.10 What is a NEMS resonator? How does its sensitivity vary with its resonance frequency?
9.11 Write the equation relating the incremental mass placed on a NEMS resonator and the shift in its resonance frequency? Can this device be applied to mass spectrometry?
9.12 What is a nanotweezer? How is it made and how is it used to manipulate nanomaterials?
9.13 What is a magnetic bead nanoactuator? How does it function?
9.14 Explain with a diagram the operation of an optical gradient force-driven nanoactuator? What is the typical range of displacement achieved by Q-factor modulation?
9.15 What are the advantages of NEMS memories? Describe the different two-electrode architectures used for switching between on and off states? How does the three-electrode architecture resemble an FET?

References

1. Roukes ML (2000) Nanoelectromechanical systems. Technical digest of the 2000 solid-state sensor and actuator workshop, Hilton Head Isl., SC, 6/4-8/2000, pp 1–10
2. Mukherjee S, Aluru NR (2006) Preface: Applications in micro- and nanoelectromechanical systems. Eng Anal Boundary Elem 30:909
3. Cullinan MA, Panas RM, DiBiasio CM et al (2012) Scaling electromechanical sensors down to the nanoscale. Sens Actuators, A 187:162–173
4. Li M, Tang HX, Roukes ML (2007) Ultra-sensitive NEMS-based cantilevers for sensing, scanned probe and very high-frequency applications. Nat Nanotechnol 2:114–120
5. Datar R, Kim S, Jeon S et al (2009) Cantilever sensors: Nanomechanical tools for diagnostics. MRS Bull 34:449–454
6. Barlian AA, Park W-T, Mallon JR Jr et al (2009) Review: semiconductor piezoresistance for microsystems. Proc IEEE Inst Electr Electron Eng 97(3):513–552
7. Maxim integrated: Application Note 871, Demystifying piezoresistive pressure sensors, Jul 17, 2002, pp 1–12. http://pdfserv.maximintegrated.com/en/an/AN871.pdf. Accessed 19 Sept 2015
8. Stampfer C, Helbling T, Obergfell D et al (2006) Fabrication of single-walled carbon-nanotube-based pressure sensors. Nano Lett 6(2):233–237
9. Naik AK, Hanay MS, Hiebert WK et al (2009) Towards single-molecule nanomechanical mass spectrometry. Nat Nanotechnol 4:445–450
10. Cobiano C, Serban B, Petrescu V et al (2010) Towards nanoscale resonant gas sensors. Ann Acad Rom Scientists 3(2):39–60
11. Cao G, Chen X, Kysar J (2005) Strain sensing of carbon nanotubes: numerical analysis of the vibrational frequency of deformed single-wall carbon nanotubes. Phys Rev B 72:195412

12. Lee C, Wei X, Kysar JW et al (2008) Measurement of the elastic properties and intrinsic strength of monolayer graphene. Science 321:385–388 (New York)
13. Akita S, Nakayama Y (2002) Manipulation of nanomaterial by carbon nanotube nanotweezers in scanning probe microscope. Jpn J Appl Phys 41:4242–4245
14. Lee J, Kwon S, Choi J et al (2004) Nanogripper using carbon nanotube. NSTI-Nanotech 3:180–182
15. Konno T, Hayashi H, Tan T et al (2006) US 20060220659 A1: Nanogripper device having length measuring function and method for length measurement executed with nanogripper device having length measuring function
16. Berger M (2008) Nanotechnology gets a grip Copyright Nanowerk LLC. http://www.nanowerk.com/spotlight/spotid=8390.php. Accessed 20 Sept 2015
17. Hartbaum J, Jakobs P, Wohlgemuth J et al (2012) Magnetic bead nanoactuator. Microelectron Eng 98:582–586
18. Nanotherics: nano actuation: using magnetic nanoparticles and AC magnetic fields in neurological and biomedical applications, Updated September 10, 2015. http://www.azonano.com/article.aspx?ArticleID=4130. Accessed 20 Sept 2015
19. Balakrishna AR, Huber JE, Landis CM (2014) Nano-actuator concepts based on ferroelectric switching. Smart Mater Struct 23(8):085016 (8 pp)
20. Dong B, Cai H, Ng GI et al (2013) A nanoelectromechanical systems actuator driven and controlled by Q-factor attenuation of ring resonator. Appl Phys Lett 103:181105-1–181105-5
21. Lacaze PC, Lacroix J-C (2014) Volatile and non-volatile memories based on NEMS, in non-volatile memories. John Wiley & Sons, Inc., Hoboken, NJ, USA. doi: 10.1002/9781118789988.ch4

Chapter 10
Nanobiosensors

Abstract A new generation of extra-ordinarily sensitive, fast-response devices utilizing the distinctive properties of nanomaterials or modulation of characteristics of nanoelectronic devices is described. Besides their ultrahigh sensitivities, these nanosensors exhibit much lower detection limits than their microscopic competitors. In these nanomaterials and nanosize sensing devices, the relevant analyte molecules bind with the functionalized surfaces of the concerned nanostructures, producing changes in the properties of materials or altering the characteristics of devices in accordance with the concentration of the target biomolecules. These molecular bindings constitute the basis for specificity in the detection of biological and chemical species. Perspectives of nanosensors based on gold nanoparticles, magnetic nanoparticles, quantum dots, carbon nanotubes, silicon nanowires, and nanocantilevers are sketched.

10.1 Introduction

Nanobiosensors (or any biosensor) can be classified on the basis of transformation of energy of the stimulus signal into electrical form, e.g., mechanical, thermal, optical, chemical, etc. Another approach to classification is from the viewpoint of biomolecule used, e.g., one using enzymatic reaction, or detecting by immune response or through microbial action. From nanotechnology standpoint, nanobiosensors can be subdivided into classes according to the type of nanomaterial or nanoelectronic device participating in the sensing mechanism, e.g., a metallic nanoparticle, quantum dot, nanotube or nanowire, field-effect transistor, cantilever, and so forth. Nanotechnological classification will be as follows.

10.2 Gold Nanoparticle (GNP) Biosensors

Gold nanoparticles are biocompatible, i.e., they can coexist with living matter without causing any toxic effects. They offer an environment very similar to the natural environment on which enzymes can be immobilized. They are easily

V.K. Khanna, *Integrated Nanoelectronics*,
NanoScience and Technology, DOI 10.1007/978-81-322-3625-2_10

produced by chemical route [1]. Their surfaces are easily modified by attaching biomolecules. Therefore, they are the favorites of nanobiosensor designers.

10.2.1 Gold Nanoparticle-Enhanced Surface Plasmon Resonance (SPR) Biosensor

Surface plasmon resonance (SPR) biosensor works on the SPR phenomenon in which the conduction band electrons of a metallic film experience collective coherent oscillations upon excitation with an impinging electromagnetic radiation. The frequency of oscillation of surface plasmons changes with the concentration of specific molecules bound to the surface of the metallic film; the metallic film along with bound molecules is called the sensing film. Hence, the molecular concentration can be determined from the change in oscillation frequency. The aforesaid frequency change occurs due to the variation of dielectric constant of the surface environment with the molecular concentration.

The sensitivity of the SPR biosensor has been found to dramatically increase on modifying the sensing film with gold nanoparticles [2]. This helps us to build ultra-sensitive nanobiosensors. Such Au nanoparticle-enhanced SPR originates from the coupling of the effects of LSPR of Au nanoparticles with the SPR of the sensing film [3]. Unification of LSPR with SPR provides the augmented response. The extinction coefficient or optical density of gold nanoparticles represents its LSPR. It determines the extent of absorption of light by the nanoparticles at a given wavelength. The extinction coefficient depends on the absorption and scattering efficiency. The contribution of absorption and scattering efficiency varies with the size of the nanoparticles. Hence, particles of different sizes differ in exerting their coupling effects in operation of the sensor.

For investigating the coupling effects of Au nanoparticles to SPR signal [2], nanoparticles having diameters of 40, 60, 70, and 80 nm were immobilized on the SPR sensing film with a 5-nm-thick dithiothreitol (DTT) spacer separating the sensing film from the nanoparticles (Fig. 10.1). A finite element simulation was performed to understand how the evanescent field in the region between the Au nanoparticles and the sensing film is perturbed by the presence of the nanoparticles. In the absence of nanoparticles, the simulations showed a clear evanescent field during SPR. But the local field was significantly increased when nanoparticles were placed nearby. The largest evanescent field perturbation was produced by 40-nm-diameter particles with the gap between nanoparticles and sensing film fixed at 5 nm. These theoretical predictions were in conformity with differential-phase SPR measurements. Therefore, the 40-nm-diameter particles provided the best amplification factor for SPR biosensing.

SPR biosensor to detect DNA hybridization was developed by conjugation of oligonucleotide probes with gold nanoparticles [4] (Fig. 10.2). The nanoparticles were used for amplification of the SPR response. Target oligonucleotides were

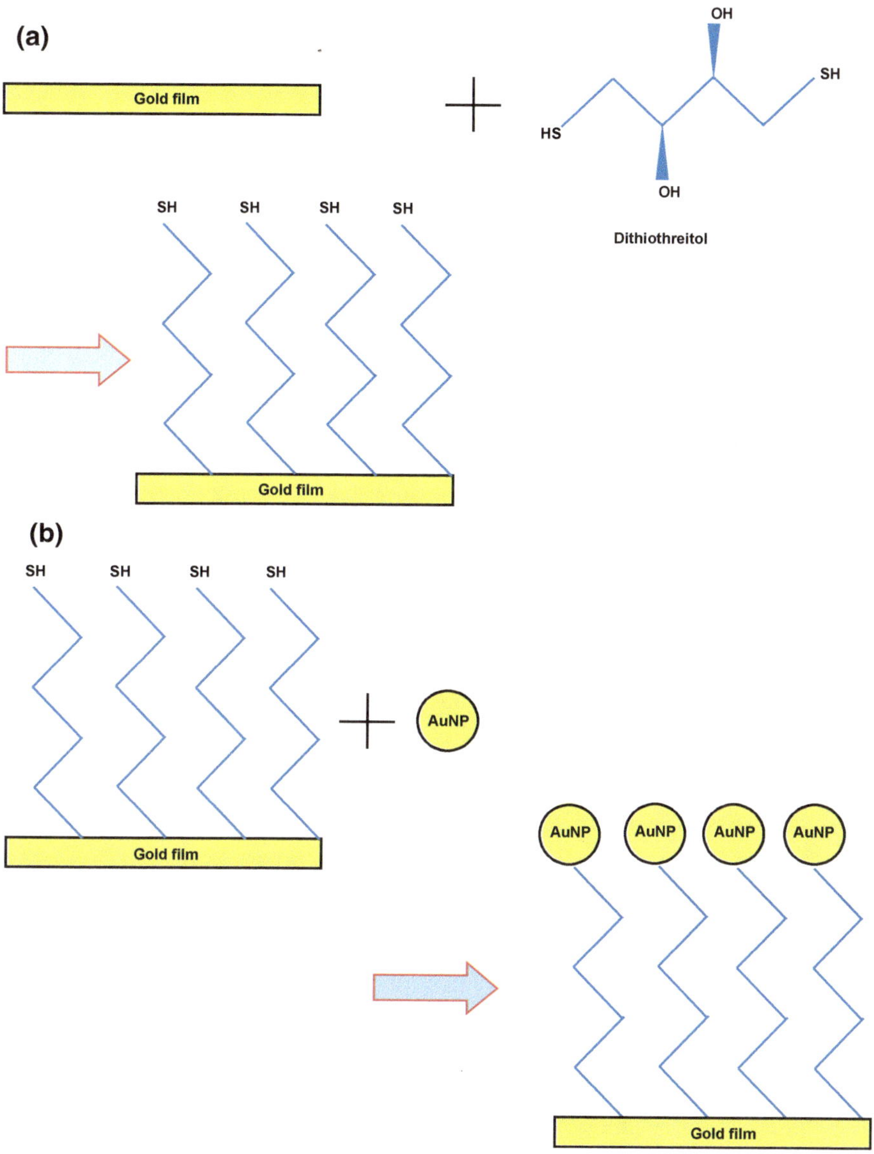

Fig. 10.1 Immobilization of gold nanoparticles on SPR sensing gold film through dithiothreitol layer: **a** binding of one end of dithiothreitol with gold film, and **b** binding of gold nanoparticles to the other end of dithiothreitol

bound to the probes by hybridization (Fig. 10.3). The response of Au nanoparticles is coupled with that of the Au film in SPR setup to greatly improve the detection sensitivity by as much as 1000 times of that when nanoparticles are absent, and to lower down the quantification limit to 10 pM.

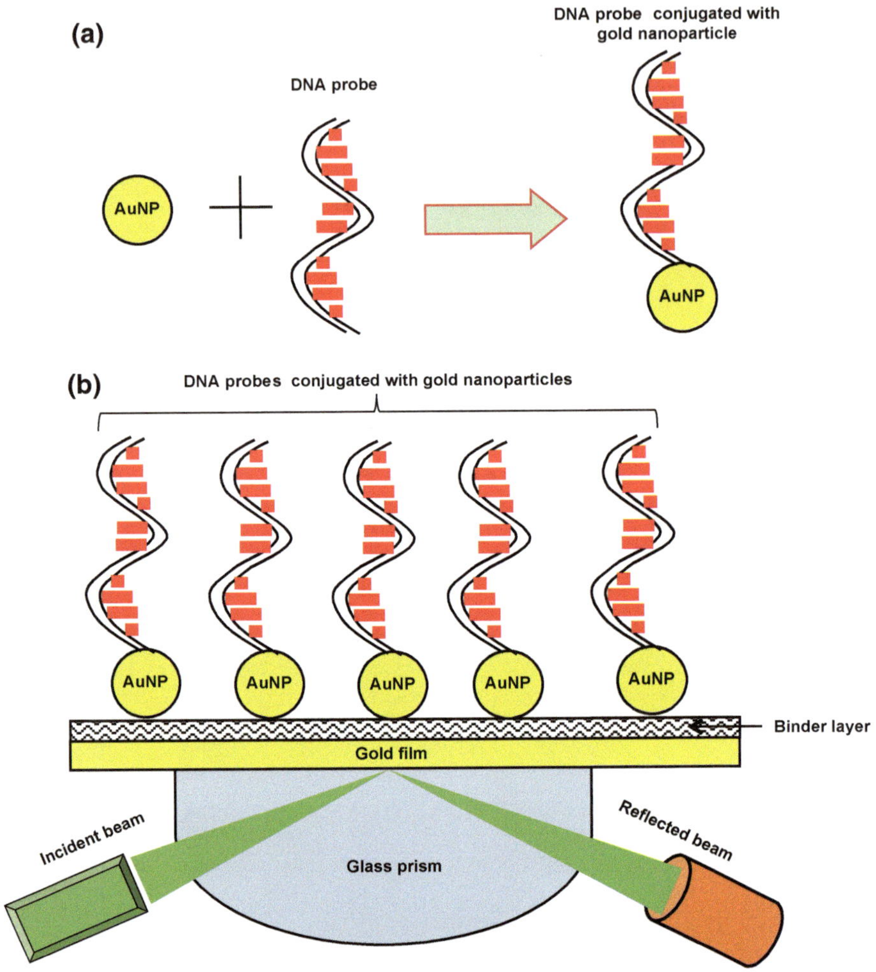

Fig. 10.2 Surface plasmon experiment with probe DNA: **a** preparation for experiment by binding probe oligonucleotides with gold nanoparticles and **b** SPR angle measurement

10.2.2 Gold Nanoparticle LSPR Biosensor

This biosensor differs from the conventional SPR transduction because it works on the localized change in optical properties of gold nanoparticles in accordance with molecular concentration. The primary condition for LSPR to take place is that the size of the Au nanoparticles must be $\ll$ the wavelength of incident electromagnetic wave. During resonant oscillation of surface plasmons, the electron density is displaced to one side of the nanoparticle because the oscillating electrons cannot propagate along the surface. This polarization of electron density depends on the size of the nanoparticle. It is also governed by the nanoparticle shape. The dielectric

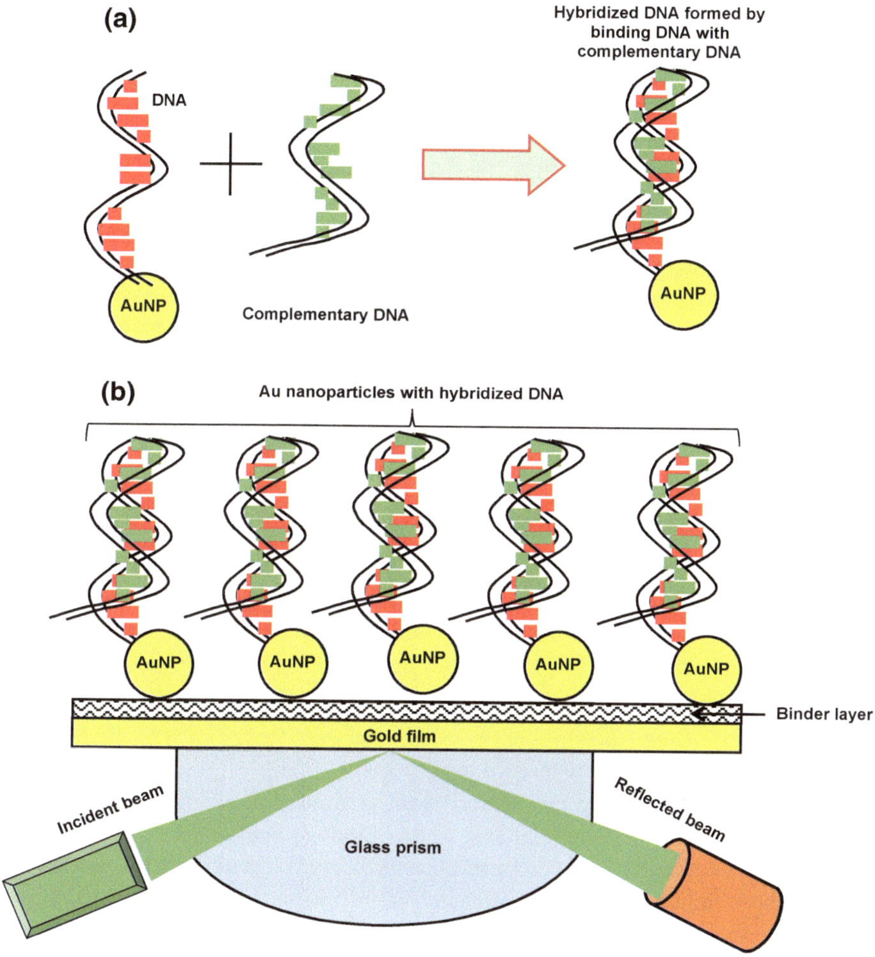

Fig. 10.3 DNA hybridization detection by surface plasmon resonance: **a** Hybridization of DNA probes by binding with complementary oligonucleotides, and **b** SPR angle measurement after hybridization

constant of the environment too impacts the polarization. Consequently, the frequency of oscillation of nanoparticles is altered. This change is perceived as change in their color and other optical properties. As an example, the SPR wavelength of Au nanoparticles coated with a monoclonal antibody is shifted toward the red color when the specific ligand binds with the antibody [5]. The underlying cause of this change in color was confirmed to be neither agglutination nor aggregation of the nanoparticles. It arises from the change in refractive index of the restricted medium surrounding the nanoparticles, and hence the localized SPR (LSPR).

Organophosphorous pesticide hampers the activity of a key enzyme acetylcholinesterase (AchE) in the nervous system of animals catalyzing the hydrolysis of acetylcholine (Ach), a neurotransmitter. A fibre-optic LSPR biosensor for this pesticide was fabricated by immobilizing the enzyme acetylcholinesterase through covalent coupling on a layer of gold nanoparticles [6]. Depending on the concentration of the pesticide, the catalytic activity of AchE for breaking down Ach is slowed down. The higher the pesticide concentration, the greater is the extent of slowing down. Accordingly, the light in the fibre is attenuated due to locally induced refractive index change. Hence, the degree of attenuation related to the degree of inhibition of enzyme activity helps to ascertain the concentration of pesticide and therefore the paraoxon which is a product of metabolism of the organophosphorous pesticide. The degree of inhibition D is given by

$$D = \{(I_0 - I_1)/I_0\} \times 100\% \tag{10.1}$$

where I_0 is the intensity of light in the beginning without inhibition by the paraoxon and I_1 is the final intensity after paraoxon inhibition. This biosensor could detect 1–100 ppb of paraoxon.

10.2.3 Gold Nanoparticle-Wired Electrochemical Biosensor

Oxidoreductases are a group of enzymes which catalyze oxidation/reduction reactions. They do so by facilitating the transference of electrons from the electron-donating reductant molecule to the electron accepting oxidant molecule. Regrettably, thick protein shells enclose the active centres of most oxidoreductases. These shells are insulating and hence obstruct the electron transfer between the electrodes and the active centres. Due to this blocking, electrochemical biosensors give a poor response. To overcome this hurdle, gold nanoparticles are used as mediators or shuttles for electron transfer. These nanoparticles serve as tiny conduction centres forming electron wires to support transference of charge. Using nanoparticles, the electrochemical signal is appreciably increased.

Efficient electron wiring was established in an H_2O_2 biosensor employing a nanocomposite made of *O*-carboxymethyl chitosan (CMCS) (Fig. 10.4) + Au nanoparticles [7]. The nanoparticles are stabilized by the $-NH_2$ ligand of the chitosan. Horseradish peroxidise (HRP) enzyme was immobilized in a silica sol-gel matrix in which the nanocomposite was embedded.

Direct electron transfer through HRP is facilitated by the nanocomposite. It provides a favourable surrounding for the enzyme to orient in desirable conformations which are suited to electron transfer. When the multiple functional groups in chitosan are exposed to solution, they act as adsorption sites assisting in electron transfer. The conducting channels built by Au nanoparticles act as stepping stones falling within reachable distances between the electrode and the enzyme. By shortening of the charge transfer spacing, the charge transfer is rendered more effective. This promotion of electron transport enabled a speedy amperometric response. A linear characteristic of the biosensor was obtained in the concentration

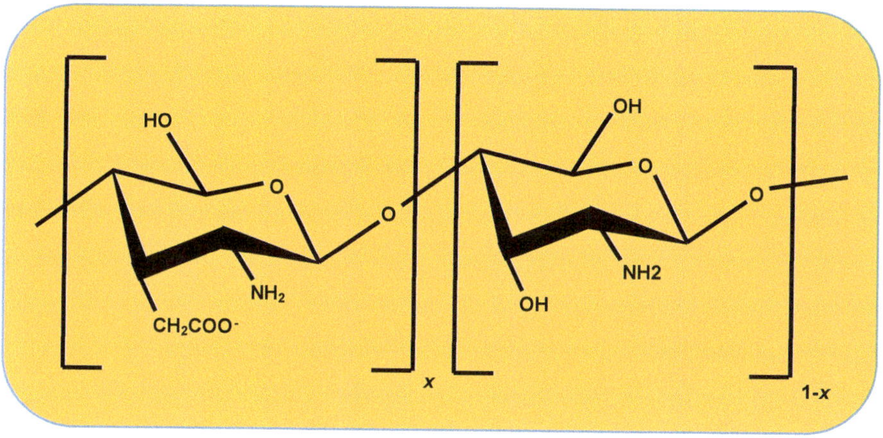

Fig. 10.4 *O*-carboxymethyl chitosan (CMCS)

range from 5.0×10^{-6} to 1.4×10^{-3} M. The limit of detection was pushed down to 4.01×10^{-7} M.

10.3 Magnetic Nanoparticle Biosensors

Magnetic nanoparticles exhibit a different behavior from bulk material. The direction of magnetization of magnetic nanoparticles flips randomly under thermal agitation, leading to zero average magnetization in the absence of an external magnetic field. This flipping is similar to that observed in a paramagnetic material in which the directions of magnetic moments of atoms change due to thermal effects.

When an external magnetic field is applied, the magnetic moments of nanoparticles are aligned parallel to the field, but the magnetization of magnetic nanoparticles is much higher than that of paramagnets. The high magnetization distinguishes then from paramagnets. Hence, this property of nanoparticles is called superparamagnetism. By virtue of this property, the magnetic nanoparticles do not exert mutual forces when magnetic field is absent but display intense magnetization when a field is applied.

Magnetic nanoparticles can be easily produced at nominal cost. They are physically and chemically stable. They are compatible with biological environment. Biological specimens are intrinsically nonmagnetic. They do not disturb the measurements performed with magnetic nanoparticles. Interference or noise generated from the sample toward magnetic nanoparticles is therefore a nonissue. Turbidity of the samples does not obscure the magnetic signals, as it does for the optical signal.

Magnetic nanoparticle biosensors have been constructed for the detection of various biomolecules such as DNA, proteins, tumor cells, etc. In these biosensors, the magnetic nanoparticles are used as proximity sensors to vary the spin–spin relaxation time of nearby water molecules [8]. This variation is quantifiable with clinical MRI scanners. This technique is called diagnostic magnetic resonance (DMR) in analogy to MRI.

DMR sensors consist of magnetic nanoparticles. These nanoparticles are conjugated with affinity molecules. The affinity molecules attach with and bind to target analyte molecules. As a result, there is a change in the rate of proton relaxation. In a subsequent washing step, the unbound nanoparticles are removed.

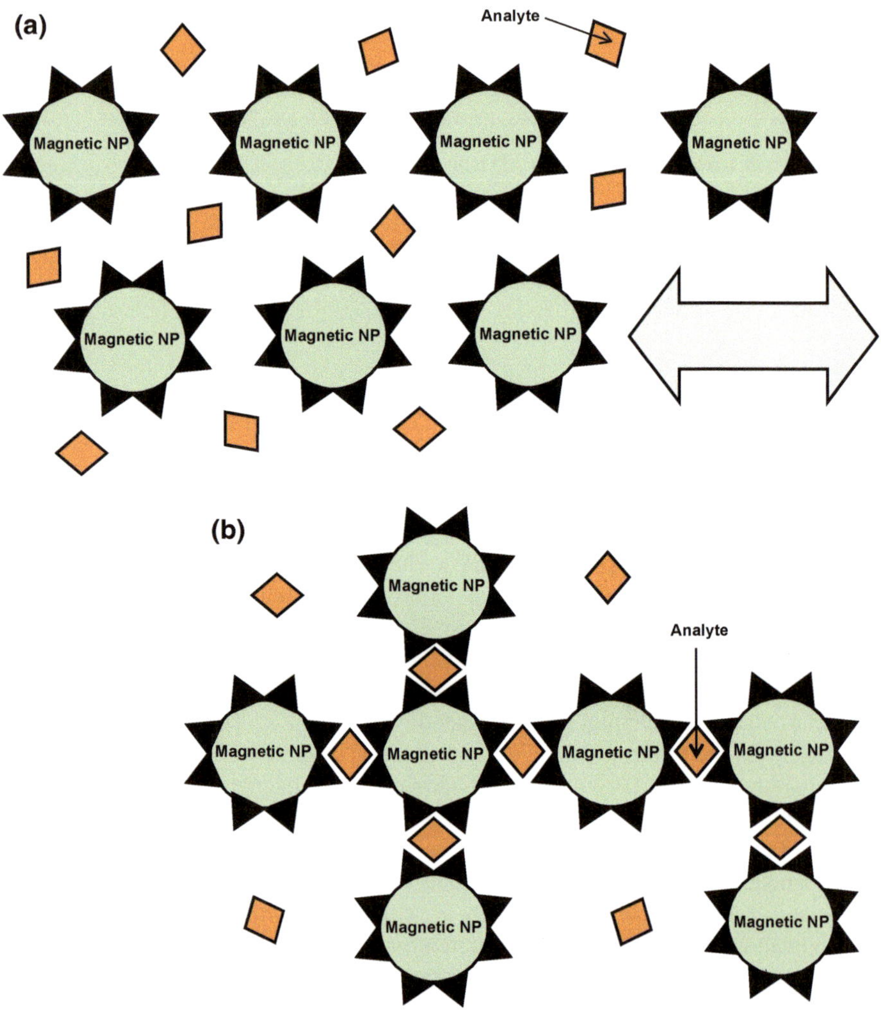

Fig. 10.5 Cartoon showing the working of magnetic relaxation switch: **a** magnetic nanoparticles in dispersed state, and **b** aggregation of magnetic nanoparticles on binding with target analytes. The aggregated nanoparticles dephase the spins of protons in the surrounding water molecules, decreasing the spin–spin relaxation time

Another application modality of magnetic nanosensor is in the form of magnetic resonance switches (MRSs), Fig. 10.5. In this mode, magnetic nanosensors were shown to detect molecular interactions in the rescindable self-congregation of individual magnetic nanoparticles into firm nanoclusters for different schemes, such as DNA–DNA, protein–protein, etc. [9]. Detection of these interactions is carried out by magnetic relaxation measurements including MRI.

Magnetic nanoparticles used in DMR sensors must possess a strong magnetic moment. Smaller size particles are preferred because of their higher stability in solution. The particles are often passivated with a hydrophilic, biocompatible layer. Widely used particles are cross-linked iron oxide (CLIO) nanoparticles. By doping the magnetite crystal with metal ions, e.g., Mn, Co, and Ni, the magnetic moment of the nanoparticles is noticeably strengthened.

10.4 Quantum Dot (QD) Biosensors

The most popular quantum dots are those made of nanosize crystals of cadmium chalcogenides (S, Se, Te). These quantum dots provide a broad absorption spectrum with relatively narrow emission spectrum. The emission spectrum is size-tunable. For different sizes of nanocrystals, the bandgaps of quantum dots vary, resulting in distinct emission wavelengths. Opportunities for multiplexed analysis are offered by the emission wavelengths thus obtained. Trapping of the excited electrons by defects in the crystal lattice leads to non-radiative relaxation. To increase the quantum yield and enhance the photostability, the quantum dots are fabricated in a core/shell structure. In this structure, a shell of wider bandgap material such as ZnS encloses the quantum dot made of smaller bandgap material. The surface defects are thus passivated. For avoidance of toxicity and making provision for attaching biomolecules, the quantum dots are coated with biocompatible materials. The biocompatible coatings contain functional groups on which the bioreceptors are immobilized. Modulation of the luminescence of QDs as a function of analyte concentration enables their application as optical biosensors. The main phenomena used in biosensing are as follows: (i) Förster (fluorescence) resonance energy transfer (FRET): It is a radiationless transference of energy from a QD donor in excited state to a suitable acceptor called the quencher. The quencher can be an organic fluorophore such as a dye. A fluorophore is a molecule or functional group which emits light after absorbing light or other form of electromagnetic radiation. FRET takes place only when the distance of separation between the donor and the acceptor is less than 10 nm. This distance is comparable with the macromolecular dimensions in biological samples. (ii) Bioluminescence resonance energy transfer (BRET): BRET takes place between a luminescent donor and fluorescent acceptor. In BRET, a light-emitting protein molecule transmits energy to the quantum dot. Hence, no external light source is needed for excitation. (iii) Fluorescence quenching via charge transfer: Here, the fluorescence intensity is reduced by a charge transfer reaction. Charge is transferred from an atom or molecule to an ion in the form of one or more electrons. The ion is promoted to an excited state. When it undergoes de-excitation, the energy is released as electromagnetic radiation. (iv) Chemiluminescence resonance energy transfer (CRET): Here, the light emitted from a chemical reaction serves as the source of light, e.g., CRET occurs between luminol and CdTe quantum dots during the chemiluminescence reaction between horseradish peroxidase (HRP)–QD conjugates and the luminol ($C_8H_7N_3O_2$)/H_2O_2 [10].

10.4.1 QD FRET Biosensor

A quantum dot-based DNA nanosensor [11], shown in Figs. 10.6 and 10.7, consists of DNA probes attached to quantum dots. These QD-appended DNA probes seize DNA targets. The target DNA strand fastens to a reporter strand labeled with fluorescent dye. Thus a donor–acceptor pair is constituted between the QD and the dye. A clear FRET signal is generated when the QD-appended DNA probes capture a small quantity of target DNA. In the absence of this capturing, the background signal is close to zero. Amplification of the signal is observed due to confinement of a large number of DNA targets by the QD-appended DNA probes in a small region of nanoscale dimensions.

10.4.2 QD BRET Biosensor

QD BRET biosensor is applied to monitor breakdown of proteases into smaller molecules [12]. Attachment of C-terminus of the protein Luc8 to the QDs yields an efficient biosensor to study protease activity. The Luc8 is a bioluminescent protein. The protein QD linkage is established through a protease peptide substrate (Fig. 10.8). Any splitting of protease disrupts the BRET process (Fig. 10.9). The discontinuity occurs due to liberation of the bioluminescent protein from the QDs. Consequently, the BRET ratio diminishes. The activity of protease can be accessed from the BRET ratio changes.

10.4.3 QD Charge Transfer-Coupled Biosensor

Charge transfer processes affect the optical properties of QDs. QD–dopamine–peptide bioconjugates serve as charge transfer-induced pH sensors [13]. Dopamine is estimated by oxidation of hydroquinone to quinone by oxygen at alkaline pH. Quinone is an electron acceptor. It quenches the fluorescence of QD in a pH-dependent manner. The pH dependence of QD quenching is used as a pH sensor to determine cytoplasmic pH variations.

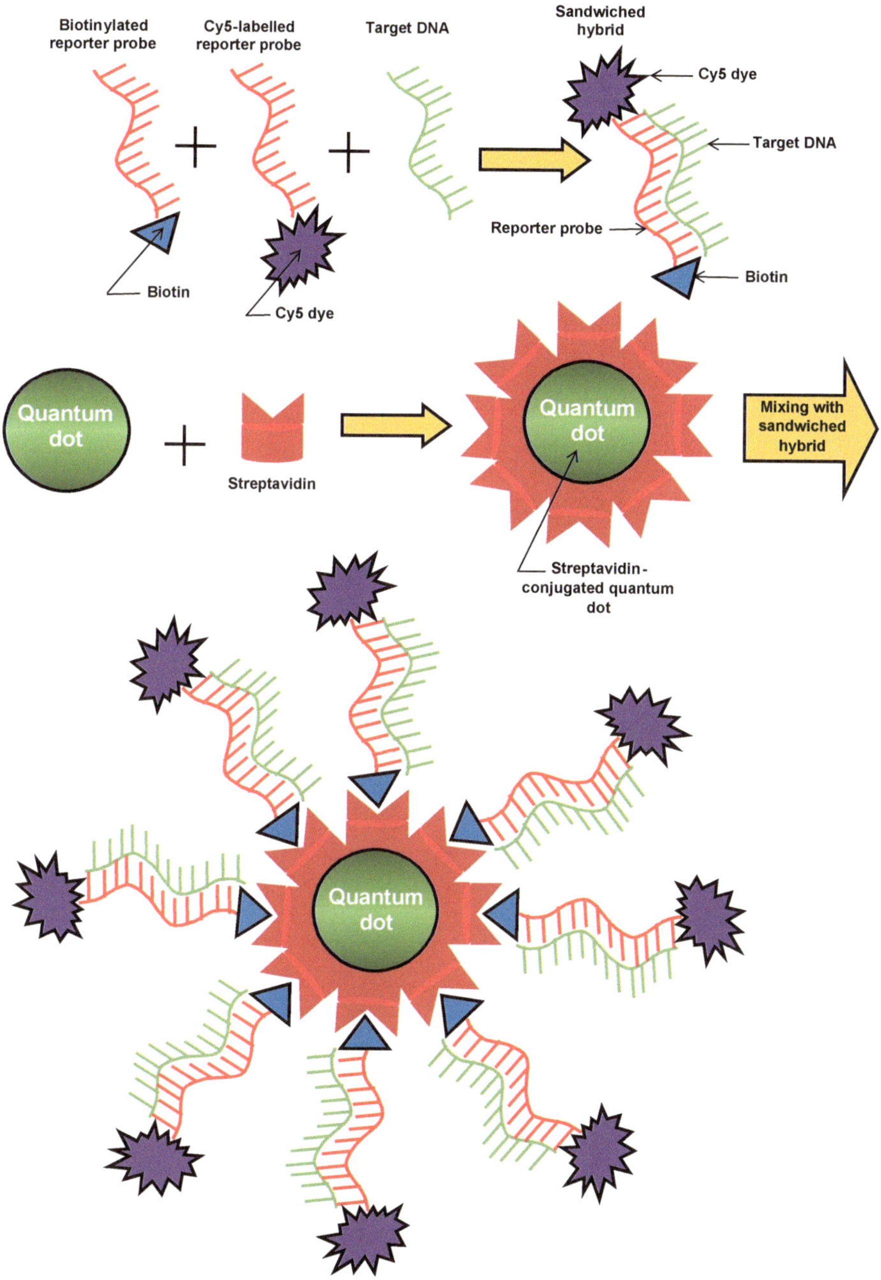

Fig. 10.6 Assembly of nanosensor in presence of targets

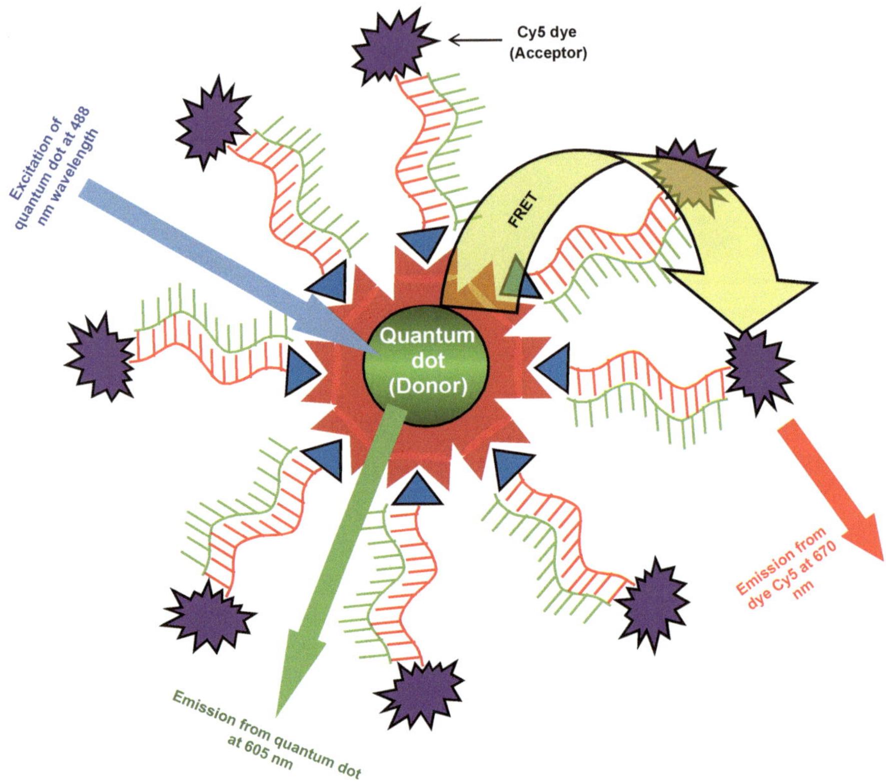

Fig. 10.7 Illumination of quantum dot, FRET between quantum dot donor and Cy5 acceptor, and resultant fluorescent emission from the Cy5 dye

10.4.4 QD CRET Biosensor

CRET aptamer sensors for thrombin and ATP were developed using catalytic hemin/G-quadruplexes [14]. An aptamer is a single-stranded DNA or RNA (ssDNA or ssRNA) molecule. It can attach to pre-chosen targets with high affinity.

Thrombin detection Hemin is incorporated into the thrombin/G-quadruplex aptamer. By the incorporation of hemin, active DNAzyme is formed. The DNAzyme promotes the production of chemiluminescence. For thrombin detection, antithrombin aptamers are conjugated to CdSe/ZnS QDs. Thrombin is sensed from the luminescence of the QDs. The luminescence occurs by CRET process, which is excited by the reaction between hemin/G-quadruplex/thrombin complex with luminol/H_2O_2. Detection limit of thrombin is 200 pM.

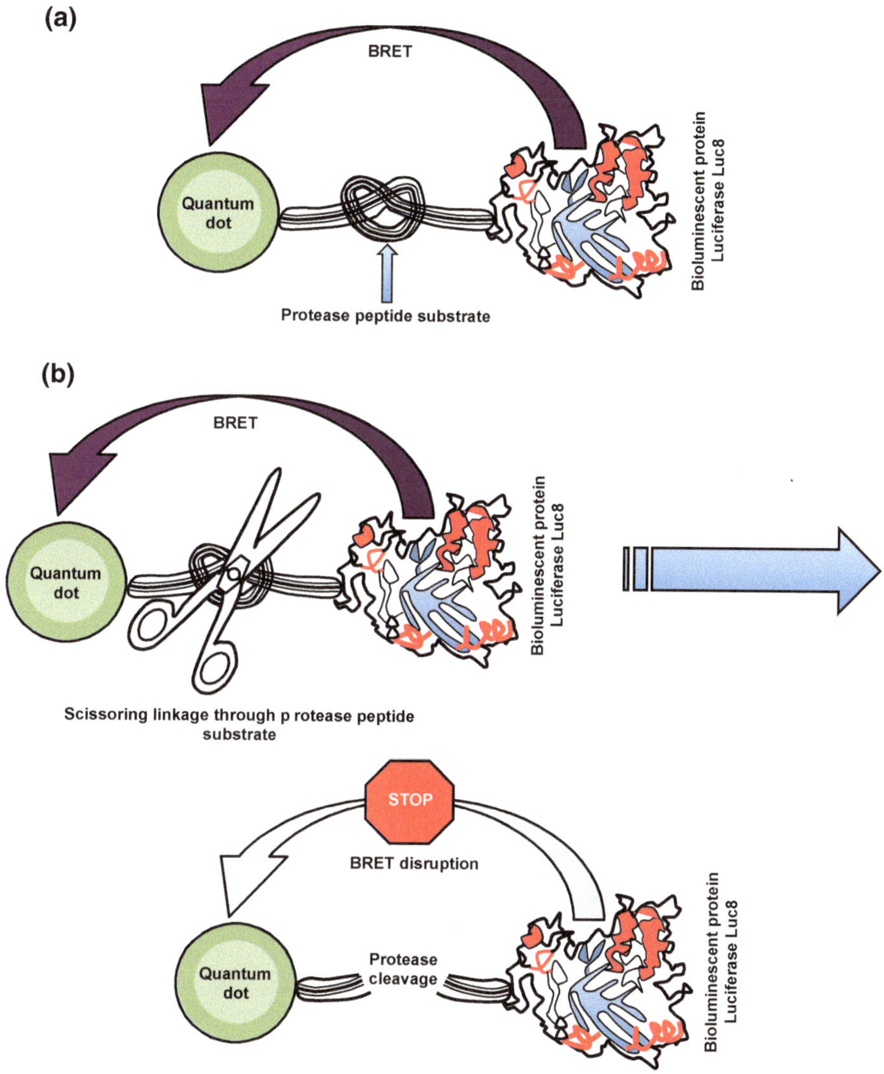

Fig. 10.8 Protease sensing: **a** Occurrence of BRET when the quantum dot is linked with Luc8 through protease peptide, and **b** Cessation of BRET when protease peptide linkage is broken

ATP Detection For ATP detection, anti-ATP aptamers are conjugated to CdSe/ZnS QDs. As before, luminescence of QDs helps to sense ATP. Reaction between hemin/G-quadruplex/ATP nanostructure and luminol/H_2O_2 excites the CRET. Detection limit of ATP is 10 μM.

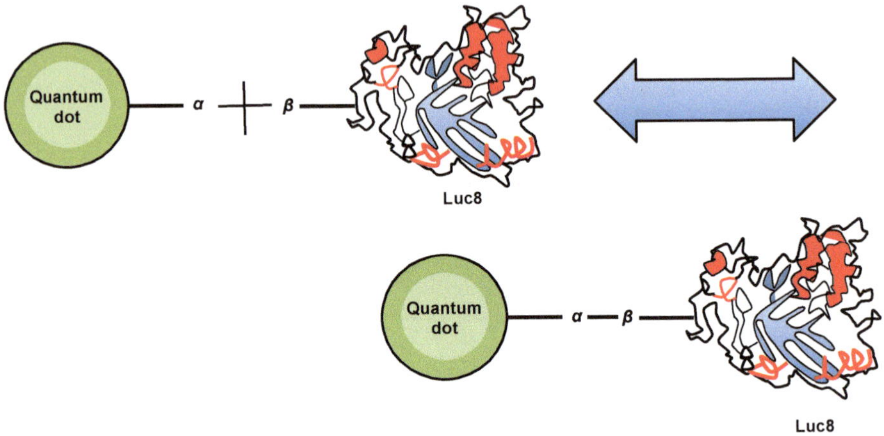

Fig. 10.9 Generalized quantum dot-BRET biosensing system in which α and β are two interactive associates linked to quantum dot and luciferase Luc 8, respectively

10.5 Carbon Nanotube (CNT) Biosensors

CNTs possess unique mechanical and electrical properties. CNTs can reach the active centers of redox enzymes. They act as wires connecting the active centers with the electrode. In this way, they make the electron transfer between the active centers and the electrode easier, faster, and more efficient. Besides mediating electron transfer, CNT films provide a large electro-active surface area. Their 3D networks offer a high density of docking sites to which bio-receptor molecules can be affixed.

To ensure fast electron transfer, a solution of CNTs is casted on the glassy carbon electrode (GCE) to modify the GCE. For mechanically strengthening the connection of CNTs with GCE, a protective coating of Nafion is formed. On a glassy carbon electrode modified with CNTs, the enzyme organophosphorous hydrolase (OPH) is immobilized to construct an amperometric biosensor for organophosphorous compounds, used as pesticides [15]. Using CNT-modified electrode, the p-nitrophenol produced is anodically detected with great precision. Up to 0.15 μM paraoxon could be detected by this biosensor.

Enzyme immobilization by direct absorption on CNT-modified GCE suffers from several shortcomings such as small amount of enzyme immobilized, and subsequent leaching out of the enzyme from the sensor degrading its stability. These shortfalls are overcome by decorating the surfaces of CNT-modified GCE with metallic nanoparticles. The CNTs are grown over graphite electrodes [16]. These electrodes are modified with platinum nanoparticles. Glucose oxidase (GOD) enzyme is immobilized on the nanoparticle-modified surface. The electrode thus prepared is covered with a thin film of Nafion to prevent the loss of enzyme. The resulting Nafion/GOD/Pt/CNT/Graphite electrodes were used as glucose biosensors. The sensors showed a high sensitivity of 91 mA/M/cm^2 and large detection range of 0.1–13.5 mM.

10.6 Si Nanowire (SiNW) Biosensors

Large surface area-to-volume ratio of silicon nanowires helps in providing high sensitivity along with low limit of detection ~fM concentrations. The dimensions of nanowires in the 1–100 nm range make them compatible with the scale of biological species.

10.6.1 SiNW Electrochemical Biosensor

Modified glassy carbon electrodes (GCEs) having the structure Nafion/Acetylcholinesterase(AChE)/Gold nanoparticles (AuNPs)/Silicon nanowires (SiNWs)/GCE served as excellent nanosensors for acetylcholine in the range of 1.0 μM–1.0 mM [17]. This sensor could detect up to 8 ng/L of an organophosphate pesticide dichlorvos or 2, 2-dichlorovinyl dimethyl phosphate ($C_4H_7Cl_2O_4P$).

Nickel hydroxide [$Ni(OH)_2$]/Silicon nanowires (SiNWs) electrode was used as a nonenzymatic H_2O_2 sensor [18]. To fabricate the electrode, the SiNWs were prepared by chemical etching while the $Ni(OH)_2$ film was deposited by electroless plating. Sensitivity of the developed electrode toward H_2O_2 detection was 3.31 mA/mM/cm^2. The limit of detection was 3.2 μM.

10.6.2 SiNW Field-Effect Transistor (FET) Biosensor

The SiNW is the gate of the FET structure. Considering an N-type SiNW, binding of negatively charged molecules on the surface of the nanowire repels the electrons away from the surface, depleting the nanowire of charge carriers, and thereby increasing the nanowire resistance. SiNW nucleic acid biosensor utilizes the change in charge concentration at the nanowire surface and therefore resistance of nanowire after hybridization of DNA probe attached to nanowire with complementary target DNA (Fig. 10.10). Up to 0.1 fM DNA was detectable [19].

Significantly, improvement was shown in DNA detection using a synthesized DNA probe [20]. This probe is ethylated DNA. In short, it is called E-DNA. The E-DNA is an analog of DNA. The advantage of this E-DNA probe is that it is electrically neutral. By virtue of its neutrality, it avoids the influence of charge noise on the measurements. It decreases the electrical repulsive force and shielding of charge carriers inside the SiNW. Therefore, as compared to regular DNA probe

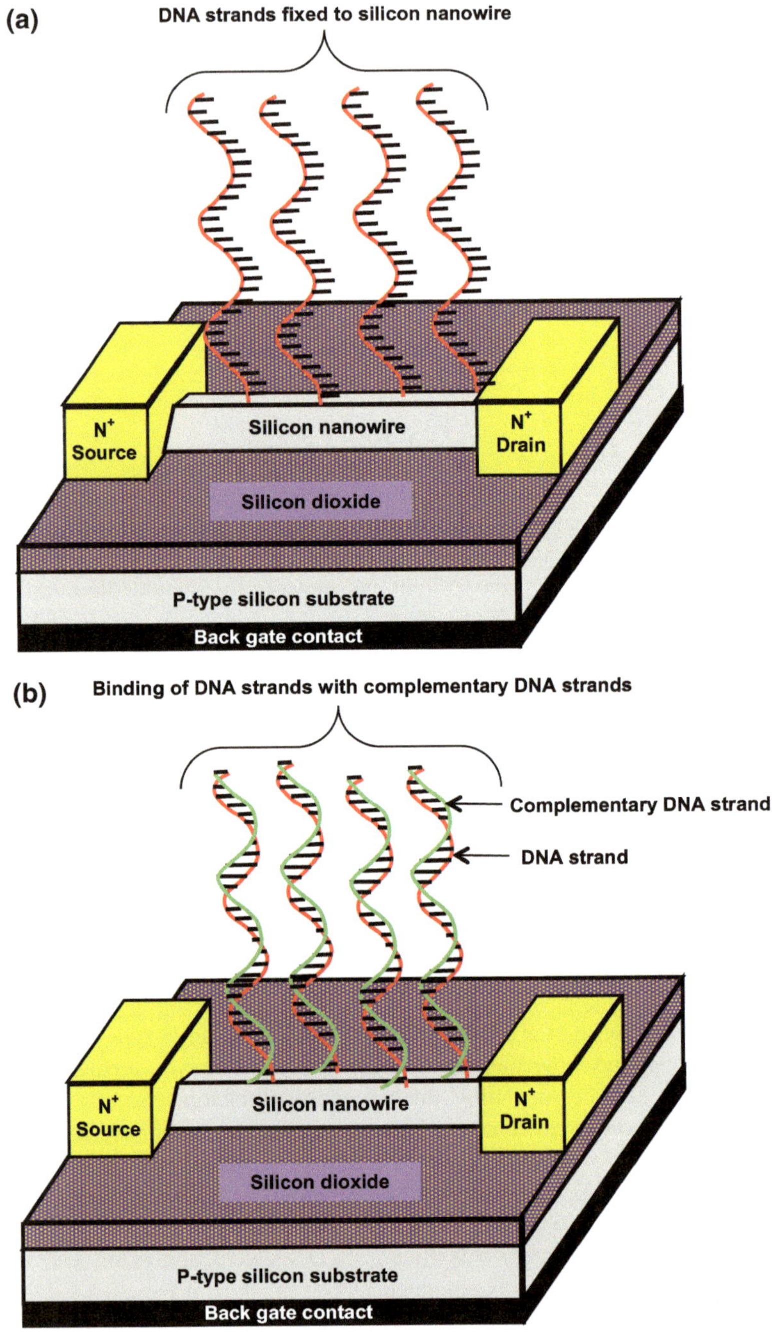

Fig. 10.10 Silicon nanowire FET: **a** before DNA hybridization and **b** after DNA hybridization

SiNW biosensor, the sensor with E-DNA probes exhibited a much higher response signal. It recorded an order of magnitude better signal-to-noise ratio. Other reasons for this ameliorated performance were the larger quantity of E-DNA probes immobilized together with the higher hybridization efficiency of E-DNA than regular DNA probes. The above was corroborated by surface plasmon resonance (SPR) examination.

10.6.3 SiNW Fluorescence Biosensor

SiNWs are used to fabricate molecular beacons (MBs) [21]. SiNWs decorated with gold nanoparticles act as efficient fluorescence quenchers, with efficiency >90% for different fluorophores such as FAM, Cy5, etc. They are stable in a wide range of salt concentrations from 0.01 to 0.1 M and over a temperature range of 10–80 °C. The nanoMBs formed from SiNWs are used for detection of DNA targets.

Small surface area is available on a planar silicon substrate. As a result, proteins can be attached in limited amount. To increase the capacity for protein loading, protein micro-patterns are created on substrates made of 8 μm high SiNWs of 150 nm diameter [22], Fig. 10.11. The nanowires were modified with (3-Aminopropyl)triethoxysilane (APTES) and glutaraldehyde for covalent immobilization of proteins. Using these substrates, immunobinding assays are carried out. These immunoassays are based on IgG (Immunoglobulin G) and anti-IgG. The IgGs are smallest but most widespread antibodies in body fluids, which combat bacterial and viral infections. Much higher intensity of fluorescent signals is recorded on micro-patterned SiNWs with respect to planar substrates. This is because the SiNW micro-patterns have ten times capacity than the planar surfaces.

10.6.4 SiNW Surface-Enhanced Raman Spectroscopy (SERS) Biosensor

Surface-enhanced Raman spectroscopy is a type of Raman spectroscopic technique, which provides considerably improved Raman signal than regular Raman spectroscopy, to the extent of 10^4–10^6 and even 10^8–10^{14} in some cases, from molecules adsorbed on metal nanostructures. In dissimilarity to Raman spectroscopy, which is not useful for surface studies, SERS affords selectivity of surface signal together with high sensitivity, up to single molecule. SERS yields a unique identification of the analyte molecule.

SiNW arrays made by chemical etching were used as templates for the fabrication of SERS-active Ag-coated SiNW arrays [23]. The morphology of the SiNW arrays and the Ag-plating solution used determined the degree of enhancement of

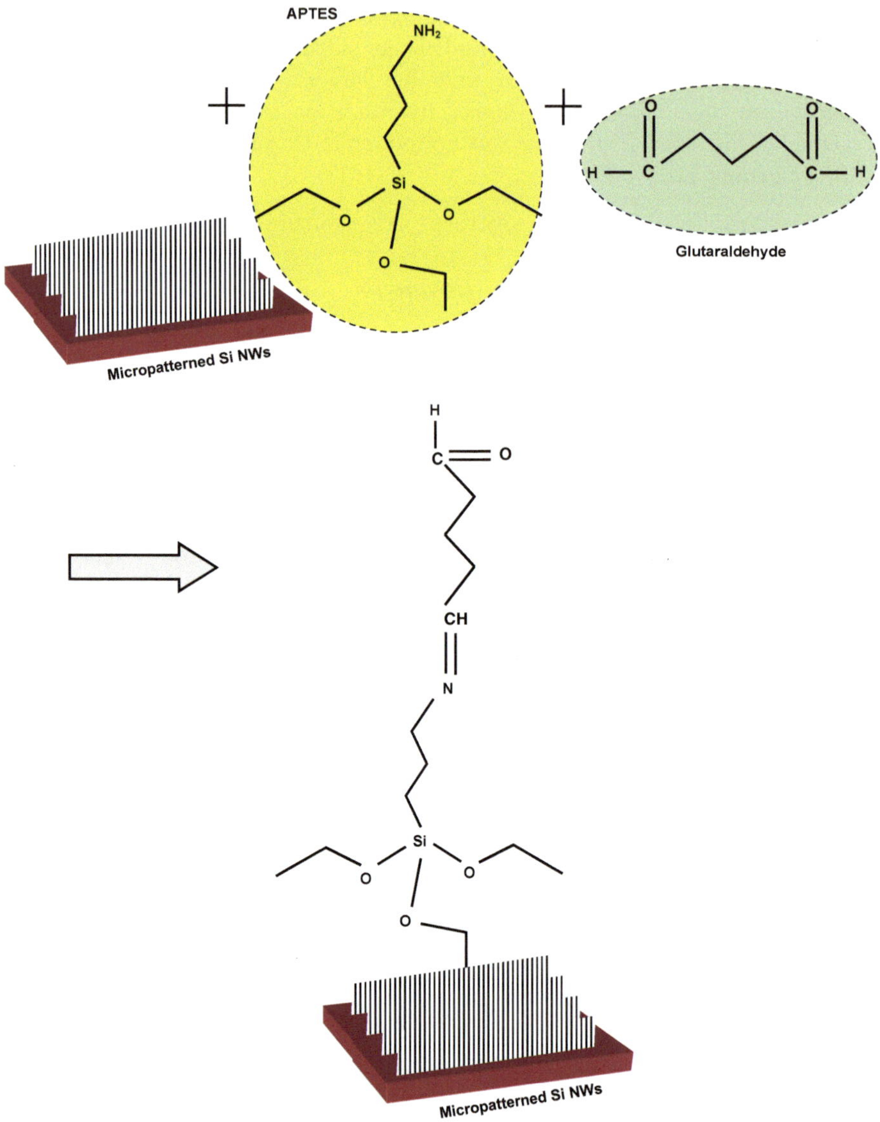

Fig. 10.11 Illustrating the preparation of micro-patterned Si nanowires for covalent immobilization of proteins

the SERS signal. The silver-coated SiNW arrays were used for detecting calcium dipicolinate (CaDPA), which is a biomarker for *B. anthracis* spores. The biomarker could be detected up to 4×10^{-6} M, which is 0.067th fraction of infectious dose.

10.7 Nanocantilever Biosensor

A nanocantilever is coated with a binding probe. This probe has an affinity for a specific amino acid or protein. When the target biomolecule is attracted toward the binding probe, it is fixed to the probe [24]. The cantilever biosensor is used in one of the two modes of detection: static and dynamic. In the static mode, the property used for detection is bending of the cantilever induced by surface stress caused by adsorption of the biomolecules; it is a stress sensor. In the dynamic mode, the change in resonance frequency due to added mass of the captured target molecules is the basis of detection; it is a mass sensor.

The surface of an AFM cantilever tip is treated with aminopropylsilatrane (APS) followed by adsorption of the lectin concanavalin A (Con A) [25]. Con A is a glucose-specific lectin. The resulting Con A/APS-modified cantilever tip is found to function as a reliable glucose biosensor.

10.8 Discussion and Conclusions

Unification of LSPR signal of Au nanoparticles with the SPR signal boosts up the overall surface plasmon signal, thereby raising the sensitivity of nanoparticle-based SPR biosensor much above the level of routine SPR. The incorporation of Au nanoparticles upgrades the response of an electrochemical biosensor by acting as electron wires or vehicles transporting charge carriers from active centers of biomolecules to the electrode to enable efficient detection. Magnetic nanoparticles are used as proximity sensors in diagnostic magnetic resonance technology. The extent of quenching of the fluorescence of quantum dots by resonance energy transfer mechanisms is used for quantitative estimation of several species. CNTs can act as electron shuttles much like Au nanoparticles. Si nanowires are the active constituents of both electrical and optical transduction devices. Besides, SiNW field-effect transistors and nanocantilevers, which are, respectively, nanoelectronic and NEMS devices, several other devices, are used as nanosensor platforms.

Review Exercises

10.1 Classify nanobiosensors: (a) from the viewpoint of type of excitation signal converted into electrical signal, (b) the biomolecule immobilized on the electronic platform, and (c) from nanotechnology perspective.

10.2 Why are gold nanoparticles attractive to biosensor designers?

10.3 The Kretschmann SPR configuration consists of a source of light, a prism, gold film, and detector. What role is played by gold nanoparticles in fabricating biosensors utilizing SPR?

10.4 Distinguish between SPR and LSPR. How is the sensitivity of an SPR biosensor for detecting DNA hybridization affected by conjugating oligonucleotide probes with Au nanoparticles?
10.5 Describe the operation of a fibre-optic LSPR biosensor for organophosphorous pesticide.
10.6 In what ways does the behavior of magnetic nanoparticles differ from that of bulk particles? What is superparamagnetism? What special properties of magnetic nanoparticles make them suitable for biological applications?
10.7 What is diagnostic magnetic resonance? Name a material from which the magnetic nanoparticles for DMR sensors are made.
10.8 Name a material commonly used for making quantum dots. How can quantum dots be utilized for multiplexed analysis?
10.9 What is the benefit of fabricating quantum dots in a core/shell structure?
10.10 Define the following terms: (a) FRET, (b) BRET, (c) quenching by charge transfer, and (d) CRET.
10.11 Describe the working of a QD FRET-based DNA nanosensor. How is the signal amplified by this structure?
10.12 How is a BRET-based QD biosensor used for studying protease activity?
10.13 Give an example of QD charge transfer-coupled biosensor.
10.14 Explain the operation of CRET aptamer sensors for thrombin and ATP.
10.15 How does the use of CNTs affect the sensitivity of biosensors? Describe a paraoxon sensor utilizing CNTs.
10.16 What is the advantage obtained by decorating the surfaces of CNT-modified glassy carbon electrodes with metallic nanoparticles? Describe the operation of a glucose biosensor based on this structure.
10.17 How a non-enzymatic sensor for H_2O_2 is constructed using silicon nanowires?
10.18 How does a silicon nanowire field-effect transistor sensor work? Explain the operation of a DNA sensor using this structure. What is E-DNA? In what way is the use of E-DNA probes better than regular DNA probes?
10.19 In what respects are SiNWs substrates with protein micro-patterns better than planar substrates for immunoassays based on IgG?
10.20 Explain the role played by silver-coated SiNW arrays in SERS biosensor.
10.21 What are two operational modes of a cantilever biosensor? Explain the operation of a cantilever-based glucose biosensor.

References

1. Hao E, Schatz GC, Hupp JT (2004) Synthesis and optical properties of anisotropic metal nanoparticles. J Fluoresc 14(4):331–341
2. Zeng S, Yu X, Law W-C et al (2013) Size dependence of Au NP-enhanced surface plasmon resonance based on differential phase measurement. Sens Actuators, B 176:1128–1133

3. Wang JL Munir A, Li ZH et al (2009) Aptamer-Au NPs conjugates-enhanced SPR sensing for the ultrasensitive sandwich immunoassay. Biosens Bioelectron 25(1):124–129
4. He L, Musick MD, Nicewarner SR et al (2000) Colloidal Au-enhanced surface plasmon resonance for ultrasensitive detection of DNA hybridization. J Am Chem Soc 122(38):9071–9077
5. Englebienne P (1998) Use of colloidal gold surface plasmon resonance peak shift to infer affinity constants from the interactions between protein antigens and antibodies specific for single or multiple epitopes. Analyst 123(7):1599–1603
6. Lin T-J, Huang K-T, Liu C-Y (2006) Determination of organophosphorous pesticides by a novel biosensor based on localized surface plasmon resonance. Biosens Bioelectron 22:513–518
7. Xu Q, Mao C, Liu N-N et al (2006) Direct electrochemistry of horseradish peroxidase based on biocompatible carboxymethyl chitosan–gold nanoparticle nanocomposite. Biosens Bioelectron 22:768–773
8. Haun JB, Yoon T-J, Lee H et al (2010) Magnetic nanoparticle biosensors, WIREs Nanomedicine and Nanobiotechnology, © 2010 John Wiley & Sons, Inc., 15 pp
9. Perez M, Josephson L, O'Loughlin T et al (2002) Magnetic relaxation switches capable of sensing molecular interactions. Nat Biotechnol 20:816–820
10. Huang X, Li L, Qian H et al (2006) A resonance energy transfer between chemiluminescent donors and luminescent quantum-dots as acceptors (CRET). Angew Chem 45(31):5140–5143 Dong
11. Zhang C-Y, Yeh H-C, Kuroki MT et al (2005) Single-quantum-dot-based DNA nanosensor. Nat Mater 4:826–831
12. Xia Z, Rao J (2009) Biosensing and imaging based on bioluminescence resonance energy transfer. Curr Opin Biotechnol 20:1–8
13. Medintz IL, Stewart MH, Trammell SA et al (2010) Quantum-dot/dopamine bioconjugates function as redox coupled assemblies for in vitro and intracellular pH sensing. Nat Mater 9:676–684
14. Liu X, Freeman R, Golub E et al (2011) Chemiluminescence and Chemiluminescence resonance energy transfer (CRET) aptamer sensors using catalytic hemin/G-quadruplexes. ACS Nano 5(9):7648–7655
15. Deo RP, Wang J, Block I et al (2005) Determination of organophosphate pesticides at a carbon nanotube/organophosphorus hydrolase electrochemical biosensor. Anal Chim Acta 530(2):185–189
16. Tang H, Chen J, Yao S et al (2004) Amperometric glucose biosensor based on adsorption of glucose oxidase at platinum nanoparticle-modified carbon nanotube electrode. Anal Biochem 331(1):89–97
17. Su S, He Y, Zhang M et al (2008) High-sensitivity pesticide detection via silicon nanowires-supported acetylcholinesterase-based electrochemical sensors. Appl Phys Lett 93 (2), Article ID 023113
18. Yan Q, Wang Z, Zhang J et al (2012) Nickel hydroxide modified silicon nanowires electrode for hydrogen peroxide sensor applications. Electrochim Acta 61:148–153
19. Gao A, Lu N, Wang Y et al (2012) Enhanced sensing of nucleic acids with silicon nanowire field effect transistor biosensors. Nano Lett 12(10):5262–5268
20. Chen W-Y, Chen H-C, Yang Y-S et al (2013) Improved DNA detection by utilizing electrically neutral DNA probe in field-effect transistor measurements as evidenced by surface plasmon resonance imaging. Biosens Bioelectron 41:795–801
21. Su S, Wei X, Zhong Y et al (2012) Silicon nanowire-based molecular beacons for high-sensitivity and sequence-specific DNA multiplexed analysis. ACS Nano 6(3):2582–2590
22. Han SW, Lee S, Hong J et al (2013) Mutiscale substrates based on hydrogel-incorporated silicon nanowires for protein patterning and microarray-based immunoassays. Biosens Bioelectron 45:129–135

23. Zhang B, Wang H, Lu L et al (2008) Large-area silver-coated silicon nanowire arrays for molecular sensing using surface-enhanced Raman spectroscopy. Adv Funct Mater 18 (16):2348–2355
24. Synder P, Joshi A, Serna JD (2014) Modeling a nanocantilever-based biosensor using a stochastically perturbed harmonic oscillator. Int J Nanosci 13(2):1450011, 8 pp
25. Hsieh S, Hsieh S-L, Hsieh C-W et al (2013) Label-free glucose detection using cantilever sensor technology based on gravimetric detection principles. J Anal Methods Chem 2013, Article ID 687265, 5 pp

Chapter 11
Spintronics

Abstract The field of spintronics is introduced and differentiated from magneto-electronics. Augmentation of the capabilities of nanoelectronics by the addition of two spin degrees of freedom to the preexisting two charge degrees of freedom is explained. The spin degrees of freedom can also be used alone to create functional devices. The role of spintronics as a bridge between semiconductor ICs and magnetic storage is elucidated. The technologically recognized spintronic device working on giant magnetoresistance effect is compared with normal magnetoresistance. The operation of magnetic tunnel junction devices for providing high magnetoresistance ratios is described. Performance of MRAM is compared with SRAM, DRAM and flash memory devices. Besides fast access, the capability of spin transfer torque RAM to decrease the write current in comparison to MRAM is indicated. The main application areas of spintronics in computer hard disks and magnetic random access memory devices are highlighted.

11.1 Introduction

Information is stored in digital computers in two principal forms: in hard disks and as random access memory. Spintronic technology has contributed immensely to both types of memory by merging magnetism with electronics.

11.1.1 Defining Spintronics

Spintronics is the science and technology aimed at understanding and controlling a fundamental property of electrons known as the electron spin, in nanoscale structures and devices, and applying this knowledge to sensing, information processing, and communication circuits. Spintronics (abbreviated form of "spin transport electronics"), spin-based electronics, spin electronics, and flextronics are different names for the same specialization, viz., the utilization of intrinsic spin of the

V.K. Khanna, *Integrated Nanoelectronics*,
NanoScience and Technology, DOI 10.1007/978-81-322-3625-2_11

electron and related magnetic moment, along with its charge, in solid-state devices and circuits. Spintronics is sometimes called magnetoelectronics. However a deeper introspection reveals the difference between spintronics and magnetoelectronics because spins can be skillfully managed by electric as well as magnetic fields. So, magnetoelectronics is not spintronics in the strict sense of the term. Another similar term is "nanomagnetics", which is primarily concerned with magnetic interaction between nanomagnets. Spintronics deals with the use of spin polarized currents in memory and logic devices [1]. Besides integrated circuits, the extensive spectrum of spintronics covers several disciplines such as mathematics, physics, material science, and nanomedicine.

11.1.2 Spintronics and Semiconductor Nanoelectronics

The spin is a quantum mechanical property of the electron. This quantum property has not received the deserved attention in mainstream nanoelectronics. In fact, electron spin is one of the three inherent properties of an electron. Its other two deep-seated properties are the mass and charge. The spin represents the rotation of the electron about its axis. It is the intrinsic angular momentum of the electron. It is parameterized by the spin quantum number m_s. Two values of spin quantum number are permissible: $m_s = \pm 1/2$. The electron is compared with a spinning top. Like a top, it can either spin clockwise or anticlockwise. The spin value $m_s = +1/2$ is called spin up. It is symbolized as an arrow pointing upwards. The spin $m_s = -1/2$ is known as spin down. An arrow pointing in the downward direction indicates this value

In spintronics, both the spin and charge degrees of freedom of an electron are used. The intent is to realize hitherto inaccessible functions to enable enhanced functionality devices. Adding the electron spin degree of freedom to electronics working with only charge degree of freedom augments the performance of devices. Thus semiconductor nanoelectronics is based only on electronic charge. Spintronics is a further step ahead because it uses electron spin apart from its charge to extend the capability. New functionalities are therefore expected from semiconductor devices that make use of both charge and spin.

As we know, electron spin can have one of the two orientations, up or down. Currents or voltages can have one of the two values, high or low. Thus the two states of spin, up or down, combined with the two states of the current/voltage, high or low, sums up to four possible states. Therefore, instead of the binary states in digital electronics, one has quarternary states in spintronics. These states can be expressed as: down-low, down-high, up-low, and up-high. They are called quantum bits or qbits. The doubling of the number of states provides higher operating speed, and greater processing power. It increases the memory density and thereby the storage capacity. However, it is necessary that electron spin can be controlled efficiently much like the electronic charge in electronics, for practical applications.

Today, information technology is based on the charge degree of freedom of electrons for processing information in semiconductors (a part of electronics) and their spin degree of freedom for mass storage of information in magnetic materials (a component of spintronics). Thus, spintronics bridges two main disciplines underpinning information technology. These two disciplines are: (i) semiconductor devices and integrated circuits and (ii) magnetic information storage. By doing so, it brings nonvolatility to semiconductor-based circuits, an essential requirement for reducing power in many applications. It also introduces the concept of circuit to magnetic storage field. Other potential applications of spintronics are in quantum computation and in the development of the quantum computer.

Spintronics is primarily based on magnetism and magnetic materials such as iron, cobalt, and nickel. Such materials are not normally used in semiconductor electronics. Hence, problems concerned with etching and patterning thin films of these materials and their unification into the silicon process are of paramount significance for manufacturing spintronic devices.

11.1.3 Branches of Spintronics

Metallic spintronics has pervaded all the present-day reader heads of hard disks in computers. Metal-based spintronic devices can function as switches or valves. But they cannot amplify signals like semiconductor electronic devices. Therefore, efforts are needed to fabricate semiconductor spintronic devices which could function at par with or better than their semiconductor electronic counterparts. Further, spintronic devices need to be seamlessly assimilated with traditional semiconductor electronic components.

Semiconductor spintronics is still in infancy. By producing ferromagnetism in semiconductors, one can build devices such as light-emitting and laser diodes for light-wave communication systems, transistors, logic and memory chips. Remote sensor systems can be constructed employing magnetic detection with on-chip circuits for signal processing and optical communication done off-chip [2]. Spintronics utilizing semiconductors offers opportunities for integrating the developed devices directly with semiconductor nanoelectronics. Furthermore, the magnetic properties are controllable by electric field instead of the magnetic field leading to smaller devices than metallic spintronics.

Semiconductor spintronics includes three types of semiconductor materials. First category is the nonmagnetic semiconductors. Second class is magnetic semiconductors. Third group is a hybrid combination consisting of metal with semiconductor. A magnetic semiconductor is a material in which both electron charges and electron spins can be controlled. It is made by doping a semiconductor material with magnetic impurities. Not all semiconductors can be rendered magnetic by the doping method. Moreover, some of these semiconductors have a very low Curie temperature above which magnetic properties are lost. A few semiconductors which show Curie temperature above room temperature are: GaP:Mn, ZnO:Co, TiO_2:Co, etc. [3].

11.2 Giant Magnetoresistance (GMR) in Magnetic Nanostructures

A spintronic device which is already established industrially is the GMR sandwich structure. It is widely used in magnetic sensors. It is called the spin valve or GMR valve. Data storage technology has vastly benefitted from it. Spintronic technology has already revolutionized memory storage of our computers in the form of hard disks utilizing the giant magnetoresistance effect

GMR effect is a quantum mechanical magnetoresistance effect. By magnetoresistance (MR) effect is meant the change in resistance of a conductor subjected to an applied magnetic field. In magnetic materials like iron, cobalt and nickel, an increase in resistance is observed to the flow of current parallel to the lines of magnetization, and a decrease in resistance in the perpendicular direction; hence called anisotropic magnetoresistance (AMR). Usually, the MR effect is very small ~2–5%.

GMR is observed in magnetic sandwiches/multilayers formed by stacking alternating layers of ferromagnetic and nonmagnetic metals (Fig. 11.1). An example is single crystal Fe/Cr/Fe sandwich having (100) orientation. Another example is (100) oriented Fe/Cr multilayer. The thickness of the individual layers is only a few nm. Hence, they comprise only a small number of atomic layers. It is not noticed in thicker layers outside nano domain. The change in magnetoresistance in GMR effect is much higher than in normal MR effect. It is around 10–80%. In most modern GMR devices, it is 20–25%. The magnetoresistance is found to decrease drastically in the presence of a magnetic field.

When the external magnetic field is not applied on the nanostructure, the direction of magnetization of the adjoining ferromagnetic layers is antiparallel. This

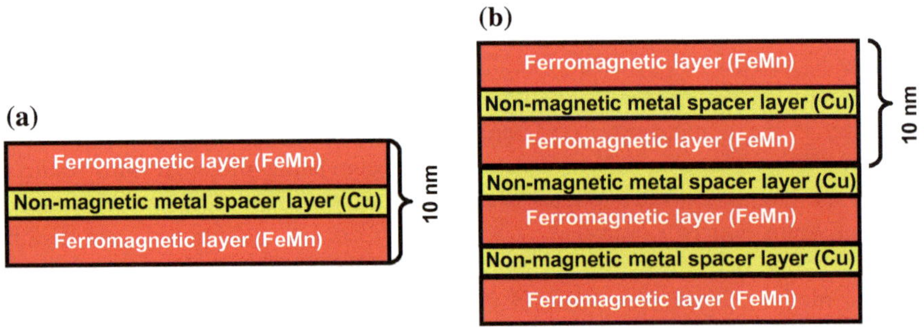

Fig. 11.1 Layered structures of GMR devices: **a** trilayer and **b** multilayer

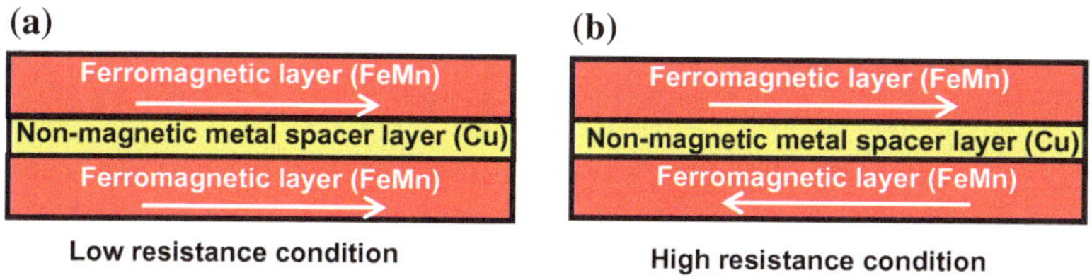

Fig. 11.2 Two possible magnetization states in a GMR structures: **a** parallel and **b** antiparallel

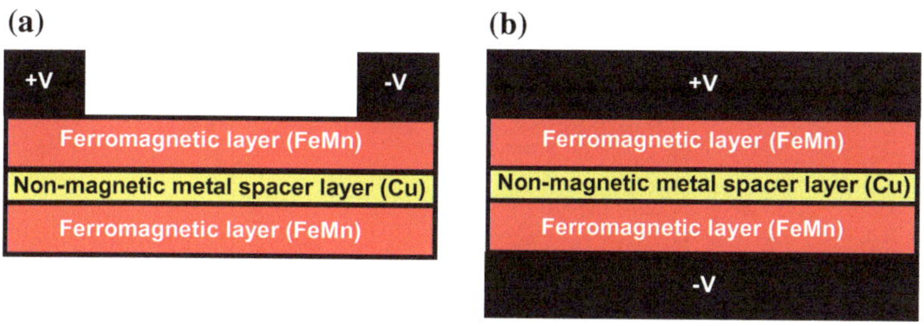

Fig. 11.3 Two current directions in a GMR structure: **a** current-in-plane (CIP) and **b** current-perpendicular-to-plane (CPP)

results in higher magnetic scattering and therefore higher electrical resistance. On placing the nanostructure in a magnetic field, the magnetization of the adjacent ferromagnetic layers becomes parallel amongst them. As a consequence, magnetic scattering is reduced and so also the electrical resistance.

Figure 11.2 shows the parallel and antiparallel magnetization states in GMR structures. Further, there are two possible current flow directions in these structures, in-plane and perpendicular to it, as shown in Fig. 11.3.

A spin valve is a multilayer GMR structure, which changes its resistance in accordance with the relative alignment of its constituent layers in the presence of an applied magnetic field. There are three kinds of spin valve structures (Fig. 11.4): top, bottom and symmetrical depending on which ferromagnetic layer in the structure is soft and can undergo orientation change in an external magnetic field. Figure 11.4a shows the top spin valve structure. In this structure, there are two ferromagnetic layers separated by a nonmagnetic spacer layer. The magnetic orientation of the upper ferromagnetic layer is fixed and unalterable. It is held in this condition by the nearby antiferromagnetic layer. Hence, the upper ferromagnetic layer is said to be magnetically hard. It is the pinned layer. However, the lower ferromagnetic layer is free to orient in the presence of a magnetic field. It s known as the free layer, and is magnetically soft. Due to the fact that the top ferromagnetic layer is magnetically hard, the structure is called top spin valve. In the bottom spin valve, Fig. 11.4b, the lower ferromagnetic layer is magnetically hard and the upper one is soft. In the symmetrical spin valve, Fig. 11.4c, both the upper and lower ferromagnetic layers are hard whereas the central ferromagnetic layer is soft.

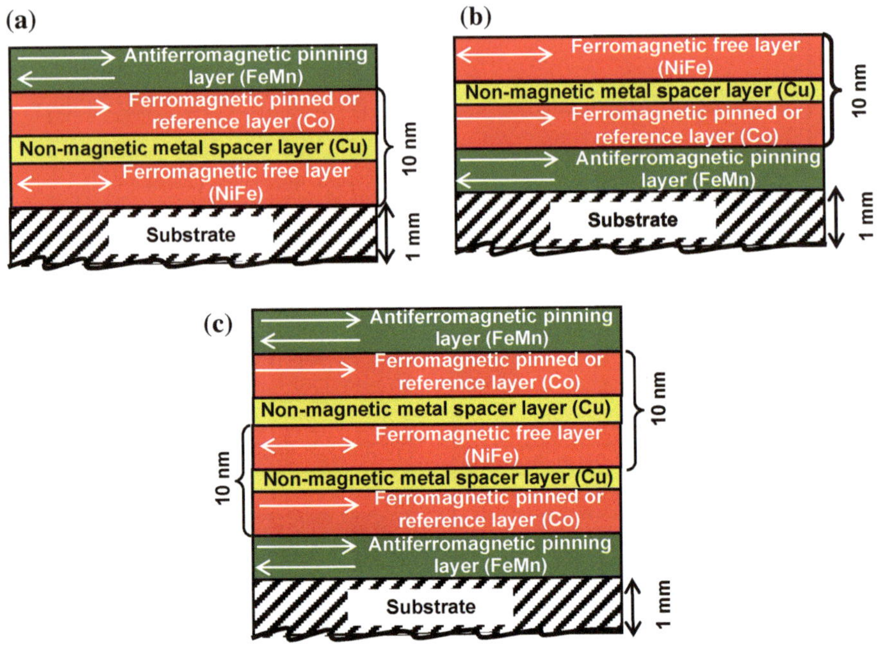

Fig. 11.4 Spin valve structures: **a** top, **b** bottom and **c** symmetrical

In the beginning stages, the films used for GMR studies were deposited by molecular beam epitaxy, which is a sophisticated and expensive technique. However, it was soon realized that GMR effect was also displayed by films obtained from simpler and inexpensive sputtering methods. Co/Cu layers were found to exhibit much pronounced GMR effect [4].

11.3 Magnetic Tunnel Junction (MTJ)

The MTJ is a thin film magnetoresistive device. It consists of two ferromagnetic electrodes made of, e.g., CoFeB, and separated by a thin insulating barrier layer (AlO_x or MgO) of thickness ~ 1 nm. Such a small thickness corresponds to barely 5–10 atomic monolayers [5], Fig. 11.5. The AlO_x layers are formed by depositing elemental aluminum and subjecting it to plasma oxidation. One of the two ferromagnetic electrodes has its magnetic orientation fixed by coupling with an antiferromagnetic pinning layer. This ferromagnetic electrode of fixed magnetic orientation is called the pinned or reference layer. The other ferromagnetic electrode can respond freely to external magnetic field. This electrode with reversible magnetization is therefore called the free or recording layer [6].

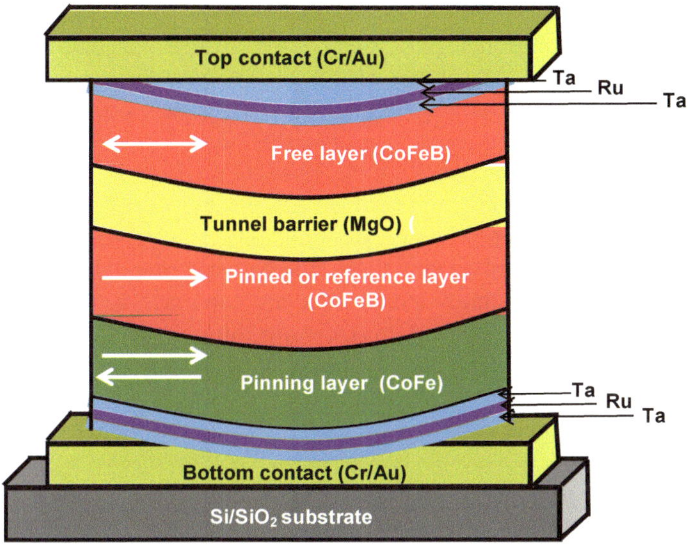

Fig. 11.5 Cross-sectional diagram of a magnetic tunnel junction showing the constituent layers

Under the influence of an applied magnetic field, the relative magnetic orientation between the two electrodes is altered. When the orientations of two electrodes are parallel, resistance of the structure is minimum and hence the tunneling current flowing through it is maximum. If they become antiparallel, the resistance increases, and therefore the current decreases. Magnetoresistance ratios are typically ~100–200% which is much higher than achieved in GMR. Magnetic tunnel junctions are used as read/write heads for disk drives.

MTJ sensors must be differentiated from the current perpendicular-to-plane GMR sensors. The main point of difference is that the GMR sensor works on the spin-dependent scattering effect. This effect occurs in the ferromagnetic layers as well as at the ferromagnetic/nonmagnetic interfacial regions. MTJ is based on the spin-dependent quantum mechanical tunneling. This tunneling takes place across a thin potential barrier.

11.4 Magnetic Random Access Memory (MRAM)

In Sect. 11.2, application of spintronic technology in computer hard disks was described. But digital data are stored in computers in different ways depending on the frequency of access, the speed of access, the period of storage necessary and volume of data to be accumulated. Data in hard disks need not be always accessed. Also, these data need not be instantly available to the computer. But they must be preserved for a long time. Further, it is essential that a high data density must be provided in the range of terabytes so that large quantities of data can be amassed. The need for long-term storage of data makes it mandatory that data is not lost when the power is switched off. Such type of memory is said to be nonvolatile. The hard disk memory is slow to access because the read/write magnetic head contains moving mechanical components. In a computer, these are the equivalents of the needle in an audio record player or the laser of a compact disc player. Access requires a time span of a few milliseconds. On the contrary, data that are required by the computer processor to execute its operations must be immediately available, i.e., fast access must be provided to these data on nanosecond time scale. But these data need not be everlastingly stored after the computer is switched off. So, a volatile memory can be used. These volatile memories are based on semiconductor devices. An externally applied electric field pushes electrons into or pulls them out of capacitors to create a charge pattern for encoding the data. This kind of memory is referred to as random access memory (RAM). A limitation of this memory is that it cannot be downscaled below a certain limit. The impediment to downscaling is that it is charge dependent. Below a limiting value, the signal-to-noise ratio increases to intolerable proportions. In opposition to data stored in charged/discharged states of capacitive devices, data stored in the form of relative orientation of magnetization of two magnetic layers is more appropriate for downscaling.

In a computer, SRAM (static RAM) is fast in operation but volatile. More memory density at cheaper cost is available from DRAM (dynamic RAM) but it is also volatile and needs periodical refreshing. However, one type of RAM, which is non-volatile is the flash memory. Flash memory is non-volatile but is still comparatively expensive. It is used in mobile phones. It suffers from slow speed and low persistence. A memory system, which combines the best features of the different types of aforementioned types of memory will be non-volatile like hard disks and will provide fast access like RAMs.

A major application area of spintronics is non-volatile memory devices such as magnetoresistive random access memory (MRAM), Fig. 11.6. Spintronic devices operate according to a simple scheme involving three stages: (i) writing the information as a particular orientation of spin, either up or down, (ii) transference of the written information by conduction of electrons along a circuit, and (iii) reading or recovery of the information. The spin orientation of electrons is preserved for a relatively long interval of time. It is not erased after the power source is removed. Therefore, it is useful for exploitation in memory storage. MRAM provides greater storage density, reduced power consumption and non-volatility. MRAM suffers from manufacturing reliability in the less-than-1 nm thick dielectric film separating the ferromagnetic electrodes of MTJ. So, the memory vanishes in less than a year.

In MRAM, the information is written by magnetic field produced by electric current. The information is stored in the direction of magnetization. The information is read through magnetoresistance changes caused by spin polarized currents.

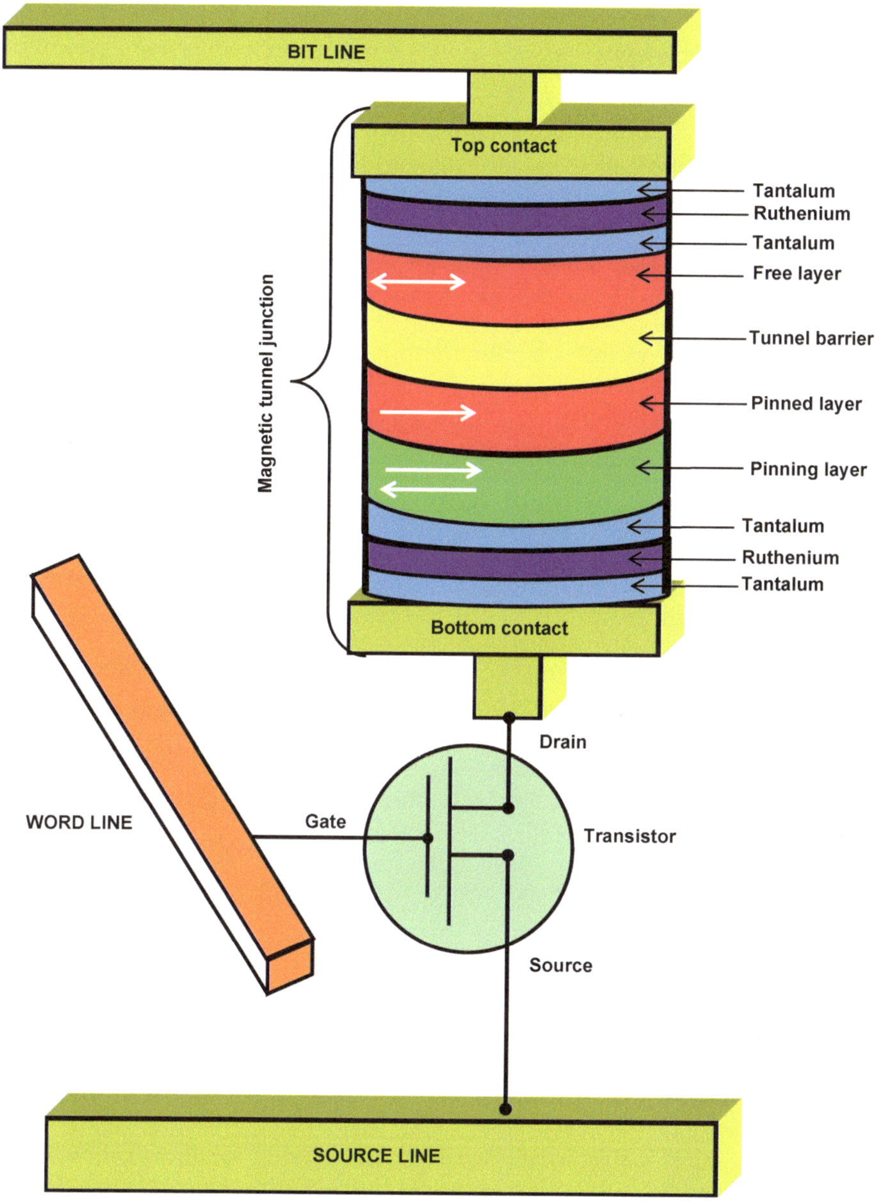

Fig. 11.6 MRAM cell using magnetic tunnel junction. Information is written by manipulating the magnetization of the elements through magnetic fields produced by the currents flowing in the bit and word lines. Information is read from voltage measurements across these lines. For bit selection, the word line and transistor are used

11.5 Spin Transfer Torque Random Access Memory (STT-RAM)

Generally an electric current is unpolarized consisting of half the total number of electrons in spin up position and the remaining half in spin down position. If by any means, the number of electrons in spin up or spin down position is increased to >1/2, the current is said to be spin polarized. In STT-RAM, a spin polarized current is used to change the magnetic orientation of the free layer in an MTJ. Instead of using a magnetic field as a read/write head, information is written electrically and also read electrically. By getting rid of the power-hungry read/write head and moving parts, the memory system becomes more rugged. Thus it provides fast-access, nonvolatile data storage. Power consumption may not appear to be a serious concern in a mains-operated desktop computer but is definitely not a trivial anxiety matter in a battery-operated laptop computer or mobile phone. Efforts have been constantly made to decrease the writing current of the MTJ storage element. At the same time, thermal stability for data retention and other functions must be retained. A writing current density of $1–2 \times 10^6$ A/cm^2 is reported for in-plane MTJ materials [1].

STT-RAM promises all the advantages of MRAM with scalability beyond 65 nm, thus providing a universal memory. This is possible by reducing write current and also through simpler memory architecture and manufacturing than MRAM. ST-RAM provides access times $\sim$10 ns along with high endurance.

MRAMs of today vastly differ from the magnetic memories of bygone years. Earlier memories detected the storage state by an inductive signal. Today's MRAM does the same utilizing either the AMR effect, GMR effect or is based on magnetic tunnel junctions. These modifications help in two ways. First, the big coils are eliminated from the memory. Second, the readout sensitivity is greatly enhanced.

11.6 Discussion and Conclusions

Presently DRAM is used as the main memory, SRAM as cache memory along with hard disk or solid-state drive. The speed gap between these memories limits the performance of computers. Using MRAM, a new memory hierarchy can be designed. The speed gap can be reduced due to the high speed and large capacity obtained from MRAM [7].

Three pertinent research areas of spintronics are: (i) Improvement in performance of present-day GMR and magnetic tunnel junction devices by exploring new materials.(iii) Designing and fabricating improved high-density, highly reliable, long life, low cost, and low power consumption MRAMs. Radiation-hard MRAMs are required for space applications. (iii) Paying attention to semiconductor spintronics so that active device functionalities can be built in spintronics.

Review Exercises

11.1 What is the full form of 'spintronics'? What are the other names by which 'spintronics' is called?

11.2 Define spintronics. What is the difference between spintronics and magnetoelectronics?

11.3 Name two intrinsic properties of electron besides its spin. What are the two allowed values of electron spin and how are they represented?

11.4 Explain how does spintronics extend the capability of semiconductor nanoelectronics by adding two spin degrees of freedom to the existing two charge degrees of freedom. What advantages accrue from the additional degrees of freedom?

11.5 How does spintronics act as a bridge between semiconductor integrated circuits and magnetic storage devices? Elaborate.

11.6 What are the problems faced in incorporating magnetic materials used in spintronics into semiconductor nanoelectronics?

11.7 What are the application areas of metallic spintronics and semiconductor spintronics? Which of the two branches of spintronics is still in infant stage?

11.8 Name a spintronic device, which is established industrially. Where is it used?

11.9 What is magnetoresistance effect? What is the meaning of 'anisotropic magnetoresistance (AMR)'? What is the extent of change in resistance?

11.10 How does giant magnetoresistance effect differ from normal magnetoresistance effect? How much does the degree of resistance change produced in GMR differ from that in AMR? Is GMR restricted only to films of thickness in the nanoscale?

11.11 Explain the origin of GMR effect in terms of the parallel/antiparallel alignment of adjacent ferromagnetic layers.

11.12 Does GMR effect only take place in nanostructures fabricated by molecular beam epitaxy? Name a less expensive method of producing such nanostructures.

11.13 A magnetoresistive device can provide a much higher magnetoresistance ratio than the GMR device. What is this device called? What value of the ratio is achieved with this device?

11.14 Describe how are the different layers of a magnetic tunnel junction arranged. How is the magnetization of one layer fixed? How does the free layer respond to an applied magnetic field? What is the effect on magnetoresistance of this device in a magnetic field?

11.15 What is the main application of spintronics? What are the three stages in the operational scheme of spintronic devices?

11.16 How is information written in MRAM? How is it read from MRAM? How does MRAM compare in performance with respect to SRAM, DRAM and flash memory?

11.17 What is meant by spin polarized current? Where is it used? In what ways is STT-RAM superior to MRAM?

11.18 What kind of tasks the metallic spintronic devices cannot perform? Why is it necessary to develop semiconductor spintronic devices?

References

1. Wolf SA, Lu J, Stan MR et al (2010) The promise of nanomagnetics and spintronics for future logic and universal memory. Proc IEEE 98(12):2155–2168
2. Pearton SJ, Norton DP, Frazier R et al. (2005) Spintronics device concepts. IEE Proc-Circuits Dev Syst 152(4):312–322
3. Wu Y (2003) Nano spintronics for data storage. In: Nalwa HS (ed.) Encyclopedia of nanoscience and nanotechnology, vol X, pp 1–50, ©2003 American Scientific Publishers
4. Parkin SSP (1995) Giant magnetoresistance in magnetic nanostructures. Annu Rev Mater Sci 25:357–388
5. Micro Magnetics: Magnetic tunnel junction sensor development for industrial applications. http://www.micromagnetics.com/pdfs/mtj.pdf. Accessed 1st Sept 2015
6. Choudhary P, Sharma K, Balecha S et al (2015) A review on magnetic tunnel junction technology. Int Res J Eng Technol (IRJET) 2(4):1635–1639
7. Fukami S, Sato H, Yamanouchi M et al (2014) Advances in spintronics devices for microelectronics–from spin-transfer torque to spin-orbit torque. In: 19th Asia and South Pacific Design Automation Conference (ASP-DAC), 20–23 Jan. 2014, Singapore, pp 684–691

Part IV
Beyond-CMOS Nanoelectronics

Chapter 12
Tunnel Diodes and Field-Effect Transistors

Abstract The concept of quantum mechanical tunneling is introduced. Degenerate and nondegenerate semiconductors are defined and distinguished. Possibility of carrier tunneling across extremely thin depletion regions is explained. Operation of a tunnel diode is described in terms of its energy band diagram. Current flow through the diode increases/decreases according to the availability/unavailability of vacant energy states in the valence band of the P-side that are aligned with respect to electron energy states on the N-side. Worthy of notice is the occurrence of negative resistance region in the current–voltage characteristic of a tunnel diode. The origin of such anomalous region is interpreted. The probability of resonant tunneling through a double barrier heterostructure is put in plain words on basis of the wave nature of electron. Acquisition of understanding of tunnel diode operation helps to bring out the dissimilarity between a tunnel diode and a resonant tunnel diode. Advantages, limitations and applications of resonant tunnel diodes in digital logic circuits and other areas are elaborated. The tunnel FET is proposed as an alternative to MOSFET. It is based on band-to-band tunneling for injection of carriers. It is a steep-slope switch offering the possibility of a subthreshold slope <60 mV/decade. This value is restricted in MOSFETs due to the tail of the Fermi distribution. Tunnel FETs cater to the very low power applications. They can operate at low voltages $V_{DS} < 0.5$ V. At such low voltages, CMOS performance is considerably worsened, which is a favorable aspect of tunnel FETs.

12.1 Introduction

The constant downscaling of MOS devices has brought about a drastic reduction in the thickness of insulating layers used in devices. The thickness of insulators has decreased to the level that the phenomenon of tunneling has acquired immense relevance.

V.K. Khanna, *Integrated Nanoelectronics*,
NanoScience and Technology, DOI 10.1007/978-81-322-3625-2_12

12.2 Quantum Mechanical Tunneling Across a P–N Junction

Tunneling is a consequence of quantum mechanics. It has no counterpart in classical physics. Quantum mechanical tunneling represents the passage of a carrier through an energy state, which may be disallowed by classical mechanics. In the MOS structure, the transition of an electron from the semiconductor body across a thin gate dielectric, to the gate metal is possible by tunneling mechanism. The gate dielectric constitutes an energy barrier. But the electron may cross it even if it possesses lesser energy than the energy barrier. Quantum-mechanically, there is a finite probability of this transition. The explanation lies in the quantum mechanical assertion that particles can behave like waves. In quantum or wave mechanics, the probability of finding an electron at a certain position in space is expressed in terms of a wave function. The wave function may penetrate the barrier and even extend to the other side. Quantum mechanics hypothesizes the existence of a nonzero probability for finding the electron on the other side. If the insulator thickness is very less, the probability may be significant.

Refer to Fig. 12.1a. For a thick barrier, both Newtonian and quantum mechanics unanimously declare that the electrons cannot cross the barrier. They can only pass through the barrier if they have more energy than the barrier height. In Fig. 12.1b, for a thin barrier, Newtonian mechanics still insists that electrons cannot cross the barrier. However, quantum mechanics differs here. According to quantum

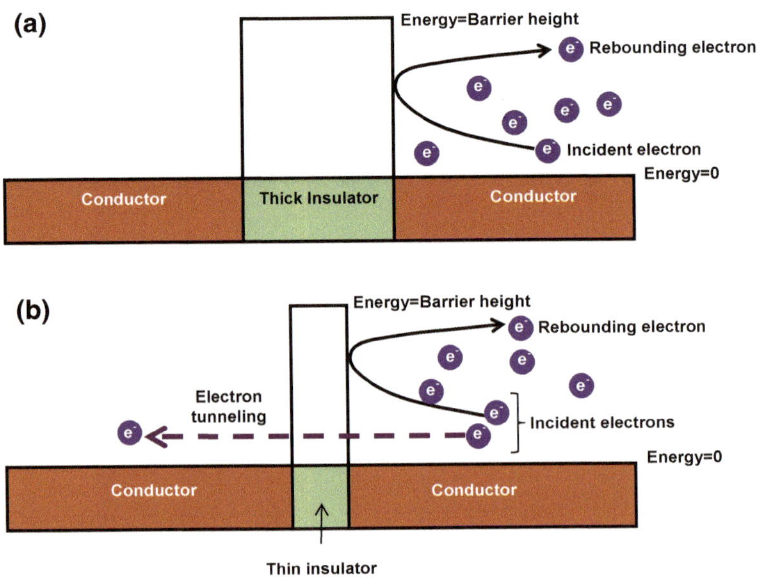

Fig. 12.1 Electrons impinging on **a** thick barrier and **b** thin barrier

mechanics, the wave nature of the electron allows it to tunnel through the barrier but the probability of tunneling decreases with barrier thickness. The thicker the barrier the lower the probability. Tunneling is outcome of the wave characteristics of electron. A tunnel is an artificial underground passage. It is built through a hill or under a building, road, or river.

Let us examine the possibility of tunneling across a P–N junction, Fig. 12.2. Consider the situation in which the P and N regions are highly doped. Then the depletion region becomes very thin (~ 10 nm). In such a case, there is a finite probability of tunneling of electrons across the depletion region. They can tunnel from the conduction band of N-region to the valence band of P-region. Note that during the tunneling, the energy of a particle does not change.

12.3 Nondegenerate and Degenerate Semiconductors

A nondegenerate semiconductor is one into which only a small concentration of impurity atoms, donor and/or acceptor, has been introduced in comparison to that of the host atoms. Hence, the impurity atoms are spaced far apart from each other. So, there is no interaction among donor electrons or acceptor holes.

A semiconductor is said to be degenerate if it has been doped to such a high concentration that the dopant atoms have become an appreciable fraction of the host atoms. In such a semiconductor, the impurity atoms are situated very near to each other. They are close enough to allow their donor electrons (or acceptor holes) to work together. Then single discrete donor/acceptor energy levels interact. Through this interaction, they split into bands of energies. This may cause overlapping of the donor/acceptor energy level with the bottom of the conduction band/top of valence band.

Overlapping occurs when the donor/acceptor impurity concentration becomes comparable with the effective densities of states N_C or N_V for the conduction and valence bands. For donors, the Fermi level moves upward. Its new position is inside the conduction band when the concentration of holes in the conduction band $> N_C$. For acceptors, the Fermi level shifts downwards. Its new position lies inside the valence band when the concentration of holes in the valence band $> N_V$. Overlapping in a degenerate semiconductor changes its electrical properties. It makes the semiconductor behave more like a conductor than a semiconductor.

In an intrinsic semiconductor, the Fermi level E_F is adjacent to the middle of energy gap E_g. In an extrinsic semiconductor, location of E_F depends on the doping level. Heavy N-type doping (N^+-doped) raises the Fermi level. This may happen to the extent that it may enter the conduction band. Heavy P-type doping (P^+-doped) causes a lowering of the Fermi level. At high acceptor doping concentrations, the Fermi level may move inside the valence band.

Likely positions of Fermi level in nondegenerate and degenerate semiconductors are indicated in Fig. 12.3.

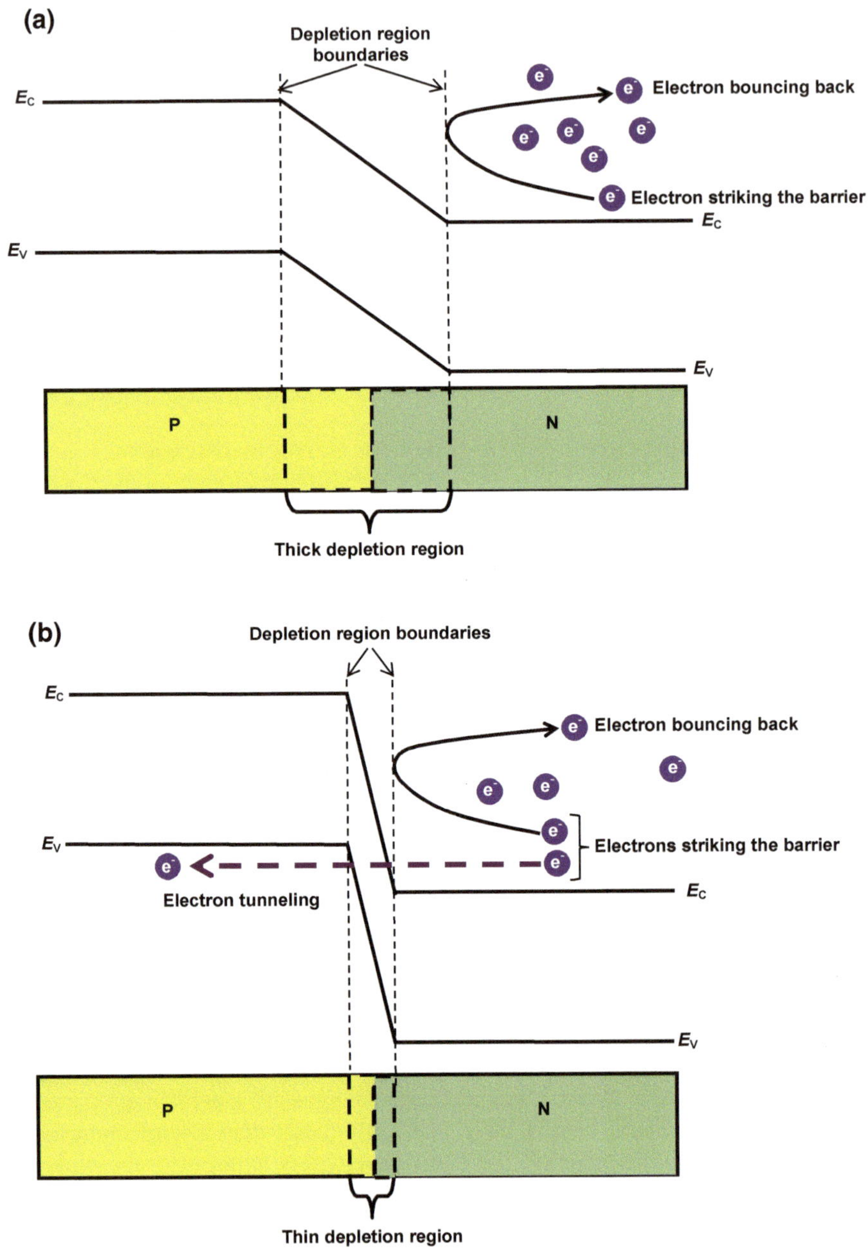

Fig. 12.2 Electron motion in a P–N junction across: **a** thick depletion region and **b** thin depletion region

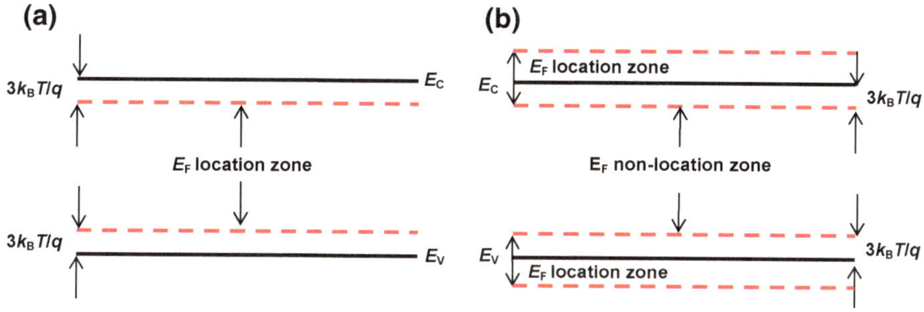

Fig. 12.3 Fermi level location in **a** nondegenerate semiconductor and **b** degenerate semiconductor

12.4 Negative Differential Resistance (NDR)

Generally, when the voltage V_{IN} across an ordinary ohmic resistor of constant resistance R is increased, the current I_{OUT} flowing out increases according to Ohm's law ($I_{OUT} = V_{IN}/R$). Also, if the current I_{IN} through the resistor is increased, the voltage V_{OUT} across it increases as well ($V_{OUT} = I_{IN} \times R$). These relationships show that the voltage and current are directly proportional. They change in the same directions. When one increases, the other also increases. The ratio of voltage/current = resistance remains unchanged. But there are some puzzling two-terminal electronic components. They are called negative differential resistors (NDRs). They display the opposite behavior in the middle part of their current–voltage (I–V) characteristics. The voltage across them and the current flowing through them change in opposite directions (Fig. 12.4).

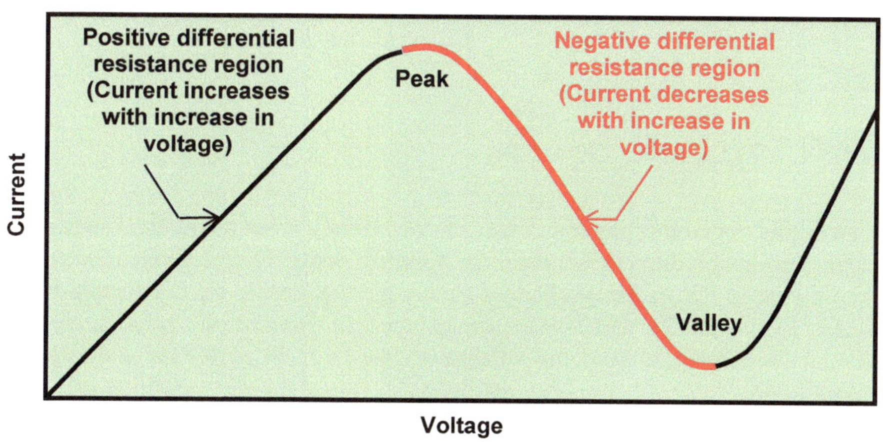

Fig. 12.4 Positive and negative resistance concepts

Negative resistance is a characteristic feature of specific electronic components. In these components, an increase in the applied voltage increases the resistance. A proportional decrease in current is thereby produced. Let us define the term differential resistance (r_{diff}). It is also called dynamic resistance or incremental resistance. It is obtained by finding the derivative of the voltage with respect to the current. It expresses the ratio of a small change in voltage to the corresponding change in current. It is the inverse slope of the I–V curve at a point. Differential resistance has relevance for currents whose values change with time. At certain points on the curve, the slope is negative meaning that an increase in voltage causes a decrease in current. These points have negative differential resistance: $r_{\text{diff}} < 0$. Like positive resistance, negative resistance is measured in ohms.

12.5 Tunnel Diode (TD)

The tunnel or Esaki diode was invented by Leo Esaki in 1958 [1, 2]. For experimental discovery of the electron tunneling effect in semiconductors, he was awarded the in Nobel Prize in Physics in 1973. The tunnel diode (Fig. 12.5) consists of a special kind of P–N junction. In this P–N junction, both the N- and P-regions are degenerately doped with impurities at concentrations $>10^{19}$ cm^{-3}. The tunnel diode exhibits a region in its voltage–current characteristic where the current decreases with increasing forward voltage. This region is known as negative resistance region. Three important aspects of the I–V characteristic curve of a tunnel diode must be highlighted: (i) The forward current increases with applied forward voltage and rises to a peak value (I_{P}) at a small applied forward bias. (ii) After attaining the peak value, the forward current decreases with increasing forward bias to a minimum or valley current (I_{V}). (iii) With further increase in the bias voltage, the forward current again begins to increase, and this increase in current with forward bias takes place like a normal diode. The portion of the characteristic curve between I_{P} and I_{V} is the region of negative resistance.

12.5.1 TD Under Zero Bias

In a highly doped P-type semiconductor, the Fermi level is below the valence band edge, i.e., inside the valence band. In a highly doped N-type semiconductor, the Fermi level is above the conduction band edge, i.e., inside the conduction band.

On joining the P- and N-type semiconductors together, the position of Fermi level adjusts itself to be same all through out from P- to N-side under thermal equilibrium. Hence, at zero bias, the Fermi level is at the same position on both P- and N-sides. Therefore, the valence band of the P-material overlaps with the conduction band of the N-material (Fig. 12.6a). The majority electrons and holes are at the same energy level. Thermal energy may cause movement of charge carriers

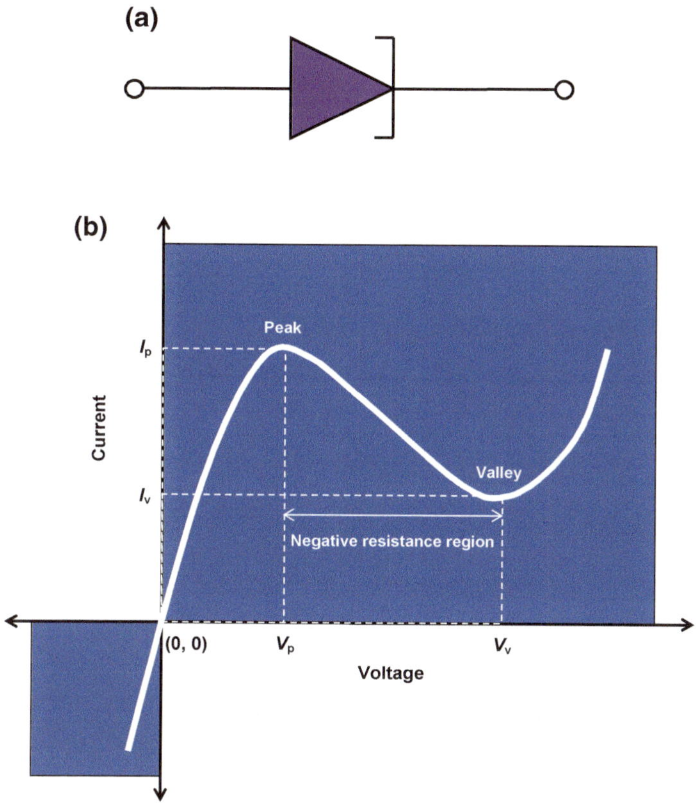

Fig. 12.5 **a** Circuit diagram symbol and **b** current–voltage characteristics of a tunnel or Esaki diode

across the depletion region. But at any given instant of time, the net current flow will be zero. This is because equal numbers of charge carriers will be flowing in the opposite directions.

12.5.2 TD Under Forward Bias

12.5.2.1 Step 1: Small Forward Bias

When a small forward bias (200 mV) is applied, the Fermi level positions change on both the P- and N-sides of the structure, relative to their zero-bias positions (Fig. 12.5b). On the P-side, the Fermi level descends to some extent. On the N-side, it ascends a little. The conduction-to-valence band overlapping becomes somewhat less. But the modifications in Fermi level positions affect the potential barrier at the junction, which is lowered by a small amount. This lowering is not large enough to enable the carriers to cross the forbidden gap in the normal fashion. However, the applied forward bias brings the energy levels of a few electrons in the conduction band of N-region in the same straight line as the vacant levels in P-region. Due to this alignment of energy levels, even though very little, some electrons in the conduction band of the N-region tunnel to the vacant states of the valence band in P-region. As a result, the charge carriers tunneling across the overlapping region of the bands create a small forward bias tunnel current.

12.5.2.2 Step 2: Slightly Larger Forward Bias

At a little larger voltage than in the previous step (Fig. 12.5c), an increasing number of electrons in the conduction band of the N-region gain energies equal to that of the holes in the valence band of P-region. In terms of energy band picture, a significant number of empty energy states of holes in the valence band of P-region become aligned with the electron energy states of N-region. As the degree of this alignment increases, the forward bias current continues to rise. In consequence, the stage of maximum tunneling current is reached.

12.5.2.3 Step 3: Still Larger Forward Bias

As the forward bias maintains its ascent (400 mV), the overlap between the two energy bands: conduction band of N-region and valence band of P-region, is diminished (Fig. 12.6d). Hence, in terms of energies, lesser number of electrons in the N-side is directly opposite to the empty states in the valence band. Alignment of hole energy levels on P-side with electron energy levels on N-side is upset. So, tunneling is not permissible for most electrons. As fewer and fewer electrons can tunnel across the junction, decrease in the tunneling current begins. The portion of the curve from peak point to valley point shows the decreasing current. The current decreases as the bias is increased, and the overlapping area becomes smaller. This portion of the curve is the fascinating, important and commonly exploited region of the tunnel diode characteristic. In this portion, the current decreases as the voltage increases. It is the negative resistance region of the tunnel diode. Very high frequency applications are possible using the tunnel diodes. The tunneling action occurs so rapidly that there is no transit time effect. The signal is therefore not distorted at all.

12.5.2.4 Step 4: Continued Increase in Forward Bias

As more forward voltage is applied, the valence and the conduction bands do not overlap any longer (Fig. 12.6e). The tunneling current ceases to flow. But the potential barrier is low. Hence, there is a change in the mechanism of conduction.

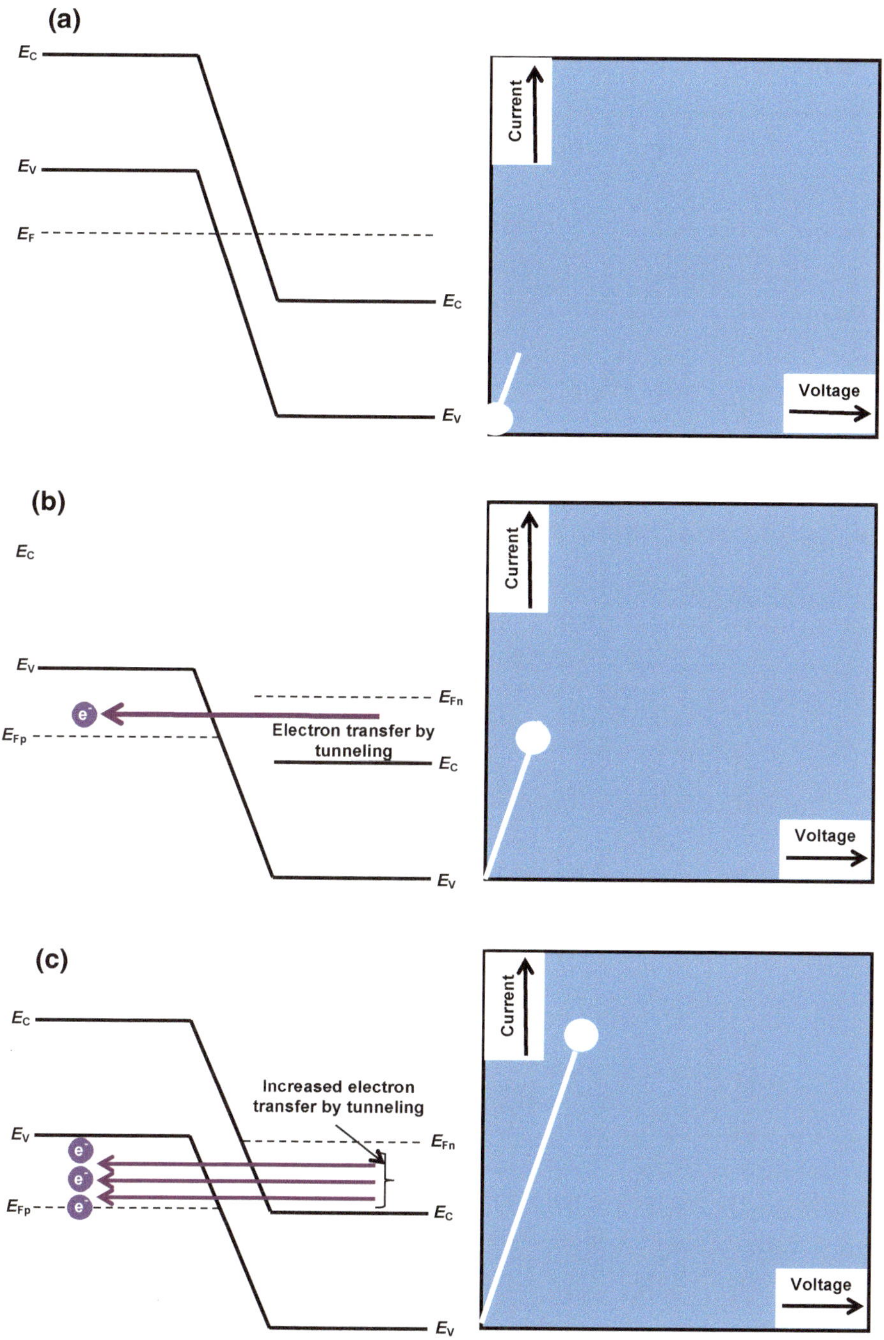

Fig. 12.6 Energy band diagrams and correlated current–voltage characteristics of a tunnel diode under **a** zero bias and rising forward bias: **b** small forward bias, **c** slightly larger forward bias, **d** still larger forward bias, **e** continued increase in forward bias and **f** further continuation of increase in forward bias

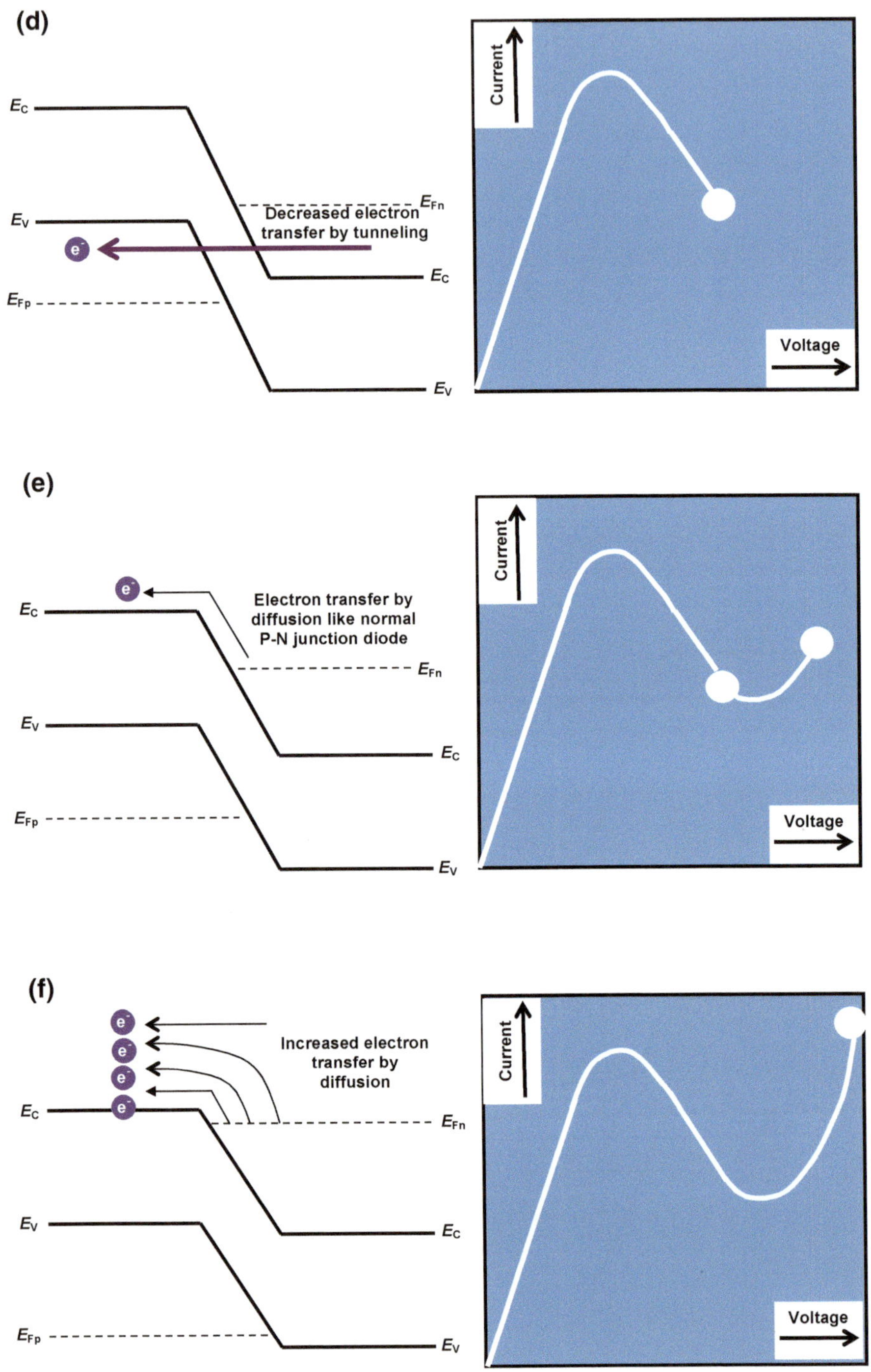

Fig. 12.6 (continued)

Regular diode forward current due to electron–hole injection increases. Thus tunneling current drops to zero whereas diffusion current starts rising.

12.5.2.5 Step 5: Further Continuation of Increase in Forward Bias

As the voltage increases, the tunnel diode *I–V* characteristic takes the shape of the forward characteristic of a standard P–N junction diode that we are accustomed with (Fig. 12.6f).

Summarizing, the forward operation of tunnel diode is as follows: Under forward bias operation, as voltage begins to increase, a larger number of electron states in the conduction band on the N-side become aligned with hole states in the valence band on the P-side. Electrons start moving by tunneling through the P–N junction barrier. In this regime, the current flows by tunneling mechanism. The forward current rises as the applied voltage increases because more and more number of electron states are getting aligned opposite to hole states. But after the voltage has reached a particular level, any further increase in voltage has the opposite effect. The electron and hole states on the opposite sides become increasingly misaligned. Due to this misalignment of states, the current falls with increasing voltage. The current flow mechanism is still tunneling, and the portion of decreasing current is called negative resistance because current decreases as voltage increases. As voltage continues to increase, the diode begins to work as a normal diode, in which the electrons travel by diffusion across the PN junction. The electrons no longer move by tunneling through the PN junction barrier. Diffusion mechanism dominates over the tunneling. The negative resistance region corresponding to decrease in forward current with forward voltage is the most interesting segment in the *I–V* characteristic of a tunnel diode.

12.5.3 TD Under Reverse Bias

Direct tunneling of electrons takes place (Fig. 12.7). This electron tunneling is from the valence band of the P-side towards the vacant states present in the conduction band of the N-side. A large tunneling current is thereby produced. The tunneling current increases with reverse voltage. The reverse *I–V* characteristic of a tunnel diode shows resemblance to that of the Zener diode. The reverse breakdown voltage of the tunnel diode is around zero.

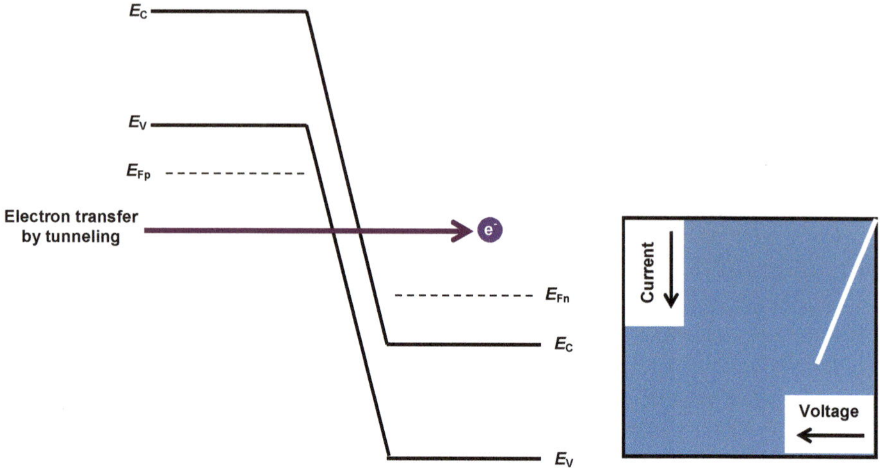

Fig. 12.7 Energy band diagram and current–voltage characteristic of a tunnel diode under reverse bias

12.6 Resonant Tunneling

It is known that the probability of tunneling or coefficient of transmission of an incident particle through a barrier is always <1. The higher the barrier, the lower is the probability. Similarly, the probability decreases for a wider barrier. Notwithstanding, it is possible that two barriers in a sequence can show complete transparency for certain energies of the incoming particles. The transmission coefficient ~ 1 for electrons having an energy = the resonant energy level of the quantum well. This implies that an electron possessing energy = the resonant energy, does not suffer any reflection from the double barrier. Instead, it penetrates the double barrier and is able to reach the opposite side. The occurrence of resonant tunneling is revealed by plotting the transmission coefficient of the electron through the structure against the energy of incident electrons. Pronounced and spiky maxima are noticed in this graph. These maxima correspond to particular values of energy. The peaks of the transmission coefficient are identified as the locations of resonant tunneling.

12.7 Resonant Tunneling Diode (RTD)

A major shortcoming of tunnel diode is its high reverse bias leakage current. This problem is overcome in a resonant tunneling diode. This will be explained while describing its operation in reverse bias mode.

12.7.1 RTD Heterostructure

Structurally, this diode consists of a quantum well enclosed by a tunnel barrier on each side (Fig. 12.8). Beyond each tunnel barrier is a Fermi sea of electrons formed by a heavily doped contact. Thus this diode has the structure: Contact (Region I)-Tunnel barrier (Region II)-Quantum well (Region III)-Tunnel barrier (Region IV)-Contact (Region V). The layers are numbered as Region I, Region II, ... The RTD structure is called double barrier quantum well (DBQW) structure.

Regions I and V are made of a small bandgap material such as GaAs (E_g = 1.42 eV). They are doped with a high carrier concentration $\sim 10^{18}\ \text{cm}^{-3}$, and are called emitter and collector. Regions II, III and IV are very thin $\sim$1–10 nm. Regions II and IV are made of a large bandgap material, e.g., $Al_xGa_{1-x}As$ ($E_g = 1.424 + 1.247x$ eV for $x < 0.45$ and $E_g = 1.9 + 0.125x + 0.143x^2$ eV for $x > 0.45$). They are the tunnel barriers. Region III is made of a small bandgap material (GaAs). It is the quantum well.

Composite stacks of GaAs/AlGaAs described above, comprising two or more semiconductors of different bandgaps but similar crystalline structures are referred to as heterostructures. These heterostructures consist of interfaces at the boundaries between different semiconductor materials. The interfaces are characterized by band offsets. The heterostructures are usually formed by molecular beam epitaxy (MBE). The MBE is a slow (1000 nm/h) epitaxial growth process.

12.7.2 Physical Phenomena in RTD

It is worthwhile to deliberate from the perspective of wave–particle duality and enquire about the magnitude of wavelength of an electron wave vis-a-vis the dimensions of DBQW structure. Let us calculate according to de Broglie relationship, the wavelength associated with an electron of mass $m = 9.1 \times 10^{-31}$ kg. We shall assume that the electron is traveling with a typical velocity $v = 10^6$ m/s. The electron wave has a wavelength given by the well-known de Broglie equation

$$\lambda = h/(mv) = 6.63 \times 10^{-34}/\left(9.1 \times 10^{-31} \times 10^{6}\right) = 7.286 \times 10^{-10}\,\text{m} = 0.73\,\text{nm} \tag{12.1}$$

where the symbol h denotes Planck's constant = 6.63×10^{-34} J s. It unequivocally transpires that the dimensions of the DBQW structure $\sim$1–10 nm are comparable with the wavelength of an electron $\sim$1 nm. A logical expectation is that wave phenomena are bound to take place in these structures. These phenomena include interference of waves. The multiplicity of phenomena that takes place in an RTD are illustrated in Fig. 12.9.

In the heavily doped contact regions I and V, the distribution of electrons with respect to energy obeys the Fermi–Dirac distribution function. The electrons are in

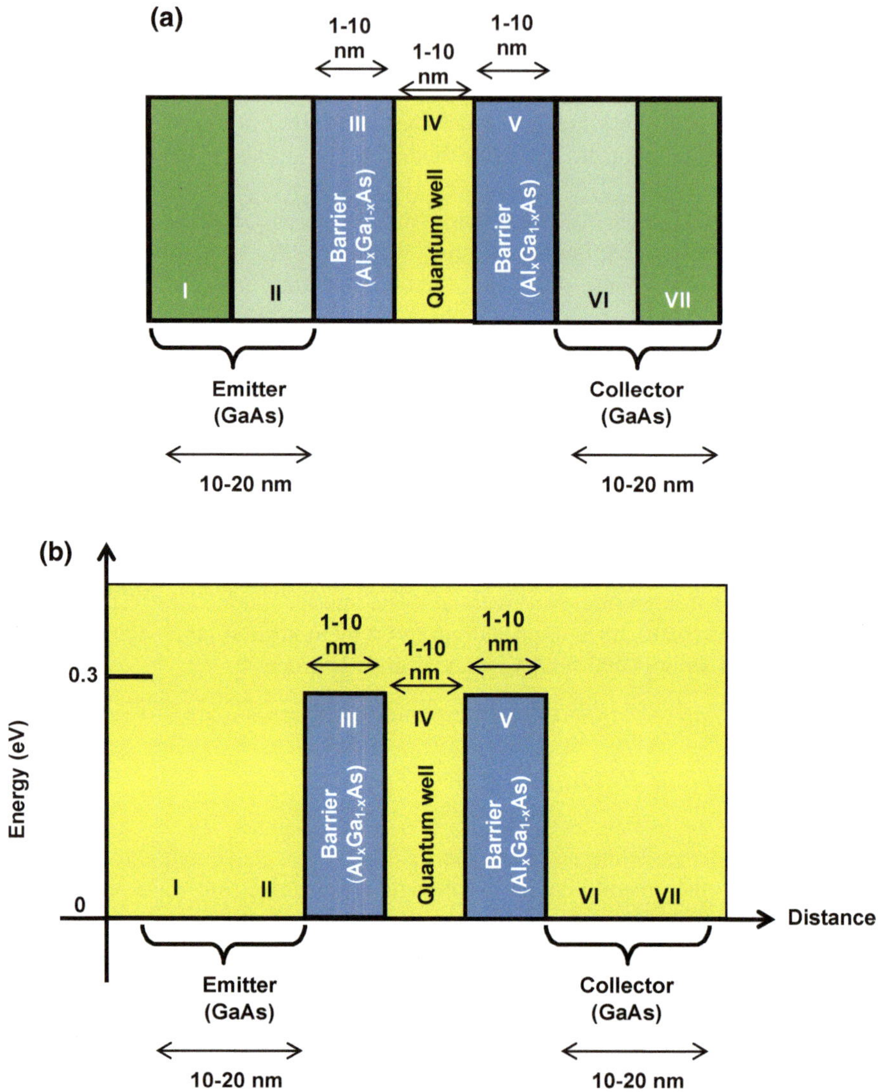

Fig. 12.8 Resonant tunneling diode: **a** layered structure showing typical thicknesses and **b** energy-band diagram showing the double barrier structure

thermal equilibrium with their surroundings. Consider a distribution of electrons under an applied forward bias. An electron wave traveling through DBQW structure experiences multiple reflections at the different interfaces. Depending on the wavelength of the electron, the interference may be either constructive or destructive. Wavelengths and therefore energies at which constructive interference occurs correspond to resonant tunneling. Due to broadening processes, these energy

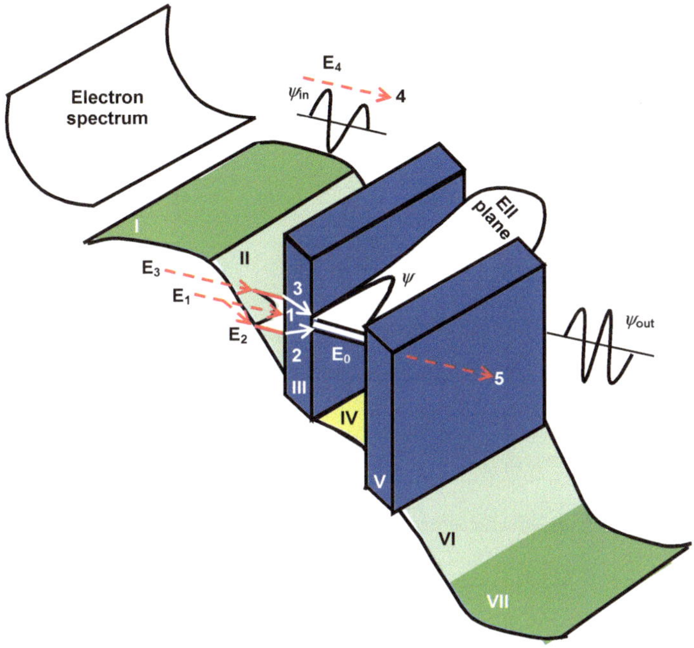

Fig. 12.9 Physical phenomena taking place inside a resonant tunneling diode

levels may show a finite spread and hence width. For these energy levels, the transmission probability through the barrier is nearly one [3].

Another important phenomenon that takes place when the electrons move into the quantum well is the dimensionality change. The emitter/collector region has three-dimensional (3-D) density of states while the quantum well has two-dimensional (2-D) density of states. Hence the electrons moving in a 3-D contact region strike the 2-D quantum well. The tunneling electrons do not see any potential change in the transverse direction. Hence, their transverse momentum is conserved. However, their longitudinal momentum changes with respect to distance.

Electron transport in an RTD takes place through several processes, a few of which are: (i) In resonant tunneling process, an electron having energy E_1 coincident with resonant energy level E_0 moves across the barrier. (ii) An electron is scattered into an energy level E_2 in the emitter. After absorbing a phonon, it tunnels through resonant level E_0. (iii) An electron having energy E_3 interacts with lattice vibration by phonon emission and then tunnels through E_0. (iv) An electron with high energy E_4 may surpass the thermionic emission barrier. (v) Apart from above processes, a few electrons may tunnel through nonresonant energy levels lying between the resonant levels.

A multitude of phenomena are involved in RTD operation. Particle and energy exchanges with the battery make the RTD open to surroundings. It is not an isolated quantum arrangement. Any impurities in the lattice may interfere with electron motion. Roughness of any interface or disorder of an alloy may affect the electron behavior. The electrons also undergo interactions amongst themselves. These scatterers produce scattering potentials in addition to the potential profile of the heterostructure.

12.7.3 Simplified Operation of RTD

The operation of RTD is explained with reference to its energy band diagrams under different biasing conditions (Fig. 12.10).

12.7.3.1 Low Forward Bias

The Fermi level E_{FE} of the emitter rises up a little but is not aligned with the resonant energy level E_0 (Fig. 12.10a). Hence, a small current flows by mechanisms such as nonresonant tunneling, scattering-assisted tunneling, thermionic emission over barriers and in the form of leakage current through surface states.

12.7.3.2 Larger Forward Bias

The Fermi level E_{FE} of the emitter moves upward. During its upward climb, it reaches nearer to the resonant energy level E_0 (Fig. 12.10b). As the Fermi level E_{FE} approaches closer to the resonant energy level E_0, the current steadily increases. When the Fermi level E_{FE} becomes collinear with the resonant energy level E_0, the current reaches the peak value I_{peak}. Steps (i) and (ii) comprise the first region of positive differential resistance of the RTD.

12.7.3.3 Still Larger Forward Bias

After the current has reached the peak value, further increase in forward voltage raises the Fermi level E_{FE} upwards so that once again it moves away from the resonant energy level E_0(Fig. 12.10c). Hence, the alignment between E_{FE} and E_0 is disturbed. This misalignment between the two energy levels leads to a decrease in forward current. Ultimately, the current attains the minimum value. The portion of current–voltage characteristic showing the decrease in forward current with applied voltage is the negative resistance segment. Step (iii) represents the region of negative differential resistance of the RTD.

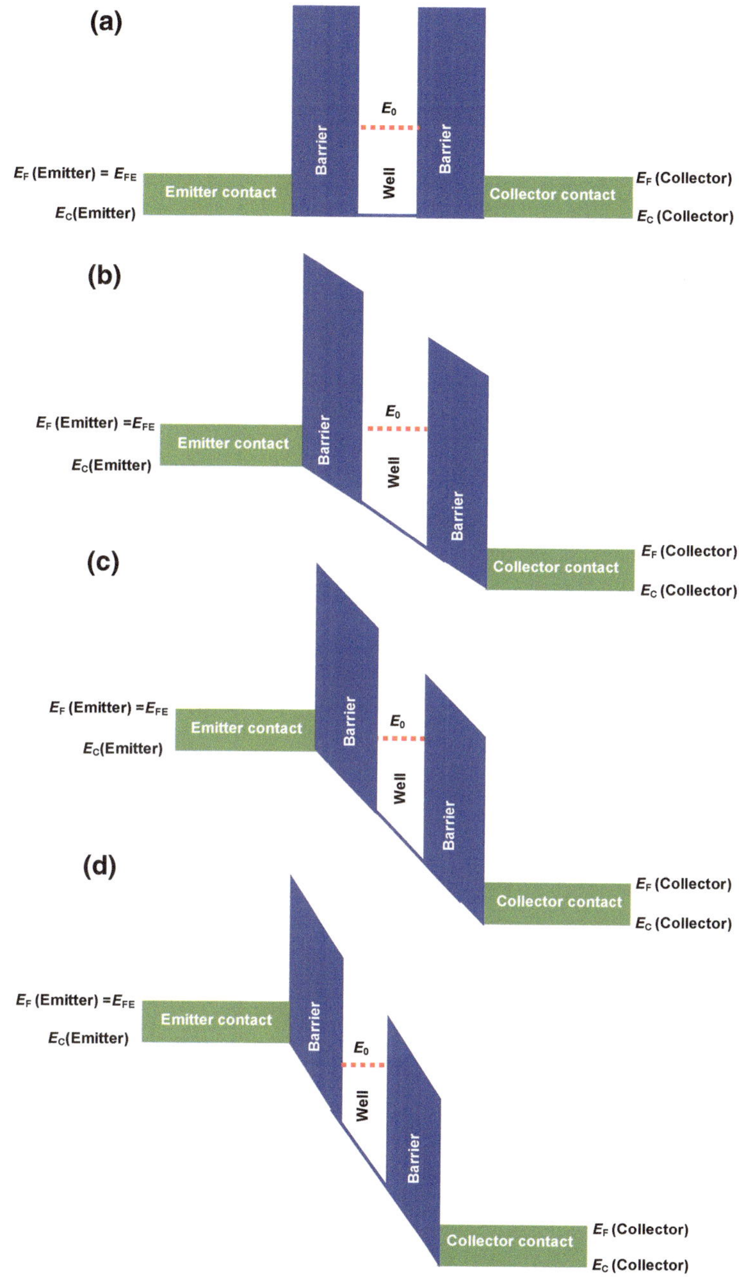

Fig. 12.10 Energy-band diagrams of a resonant tunneling diode under different forward bias conditions: **a** zero bias, **b** low bias, **c** increased bias, and **d** high bias

12.7.3.4 Continued Rise of Forward Bias

After the forward current has fallen to the I_{valley} level, if the forward voltage is continuously increased, the current begins to rise again (Fig. 12.10d). This rise of current results from the RTD following the behavior of a conventional diode. The rise of current after its fall to the minimum value results in a valley-shaped I–V characteristic. The I–V characteristic in the shape of a valley has a minimum valley current I_{valley}. Step (iv) is the second region of positive differential resistance of RTD.

12.7.3.5 Reverse Bias

RTDs have a symmetrical structure. The type of doping and concentration is the same on both the emitter and collector sides. So, the operation of the RTD under reverse bias is similar to that under forward bias. Their current–voltage characteristics are symmetrical in nature. The RTDs do not suffer from high leakage problem under reverse bias, as observed with TDs. This is a great advantage allowing the use of RTDs as rectifying elements.

12.8 Advantages of RTD

Amongst the several advantages offered by RTDs may be mentioned their capability to operate at room temperatures at extremely high switching speeds in the THz range with a very low power consumption. RTDs also reduce the complexity of circuits in terms of component counts.

12.9 Challenges of RTD

(i) Downscaling RTDs below certain limits increases the surface leakage current to undesirably high proportions. (ii) On integration of RTDs with other components, the operating speed of the resulting integrated device is lower than the intrinsic speed of RTDs alone. (iii) Precise control of layer thicknesses in RTDs and their properties requires the use of expensive MBE technique. (iv) RTDs provide a low I_{ON}/I_{OFF} ratio ~ 10 as compared to the ratio 10^5 required by CMOS. (v) Fabrication on a silicon platform and solving incompatibility issues is a formidable task.

12.10 Applications of RTD

RTDs have been combined with III–V compound semiconductor devices like heterojunction bipolar transistors (HBTs) and modulation-doped field-effect transistors (MODFETs) to fabricate bistable logic families providing NAND, NOR and inverter gate functions along with additional functionalities [4]. Using these elements, high-speed adder circuits have been designed. RTDs have also made entry into analog domain such as analog-to-analog converters and microwave circuits. RTDs hold enormous potential for futuristic high-performance, compact, close-packed memories [5].

12.11 Tunnel Field-Effect Transistor

12.11.1 Recalling MOSFET Principle

The MOSFET device operates by modulating the thermionic emission of charge carriers across a potential energy barrier through changes in gate-substrate voltage. It works by increasing or decreasing the height of the energy barrier. By such barrier raising or lowering, the charge carrier flow from one side of the barrier to the opposite side is either reduced or augmented resulting in off- and on-states of the MOSFET switch.

12.11.2 Tunnel FET Principle

The tunnel FET works in a distinctively different manner. It does not alter the barrier height. Instead, it always keeps the energy barrier high [6]. Here, the attention is focused on controlling the probability that a charge carrier, either electron or hole, on one side of the high energy barrier makes its appearance on the other side and vice versa. In a tunnel FET, the responsibility assigned to the gate potential is not alteration of barrier height but the thickness or width of the barrier because the barrier width determines the likelihood of carrier penetration across it through quantum mechanical tunneling mechanism.

12.11.3 Tunnel FET Structure

In so far as the structure is concerned, an N-channel MOSFET contains N^+ source and drain regions in a P-type substrate. In a tunnel FET, the source and the drain are oppositely doped, e.g., the source is P^+-doped while the drain is N^+-doped. Also, the substrate is intrinsic (I) in nature, meaning that in the substrate, the number of electrons = number of holes. Intrinsic substrate is a high-resistivity material. Thus the tunnel FET structure consists of P–I–N and N–I–P regions.

12.11.4 Tunnel FET Operation

Application of a gate voltage causes electron accumulation in the region underneath the gate. At a particular gate bias, the application of gate voltage causes alignment of the valence band of the source region with the conduction band of the channel. When the thickness of the energy barrier is <10 nm, there exists a finite probability for tunneling of electrons across the energy barrier from the source to the channel. Consequent upon the alignment of valence band of the source with conduction band of the channel, a window is created for tunneling of electrons from the source to the channel in the intrinsic region and into the drain, thus establishing a continuous current flow. This is the on-state of the tunnel FET. No sooner than the gate bias is withdrawn, the alignment of valence and conduction bands referred to above, is destroyed. Upon misalignment of bands, current flow stops. This is the off-state of tunnel FET. Energy band diagrams of a tunnel FET in the off- and on-states are drawn in Fig. 12.11.

12.11.5 Participation of Valence and Conduction Bands in Tunnel FET Operation

The main idea to be noted when understanding tunnel FET operation is that electrons in the valence band of the source are transferred into the conduction band of the channel. This behavior is strikingly different from that of a MOSFET in which the carriers move in one energy band only, either electrons in the conduction band or holes in the valence band, and not from one band to another.

The mechanism of tunneling across a barrier instead of climbing over a barrier requires less energy for switching from off-state to on-state of a tunnel FET, and conversely. Hence, the tunnel FET is more energy-efficient than a MOSFET. Tunnel FET configuration allows the realization of switches with subthreshold swing <60 mV/decade.

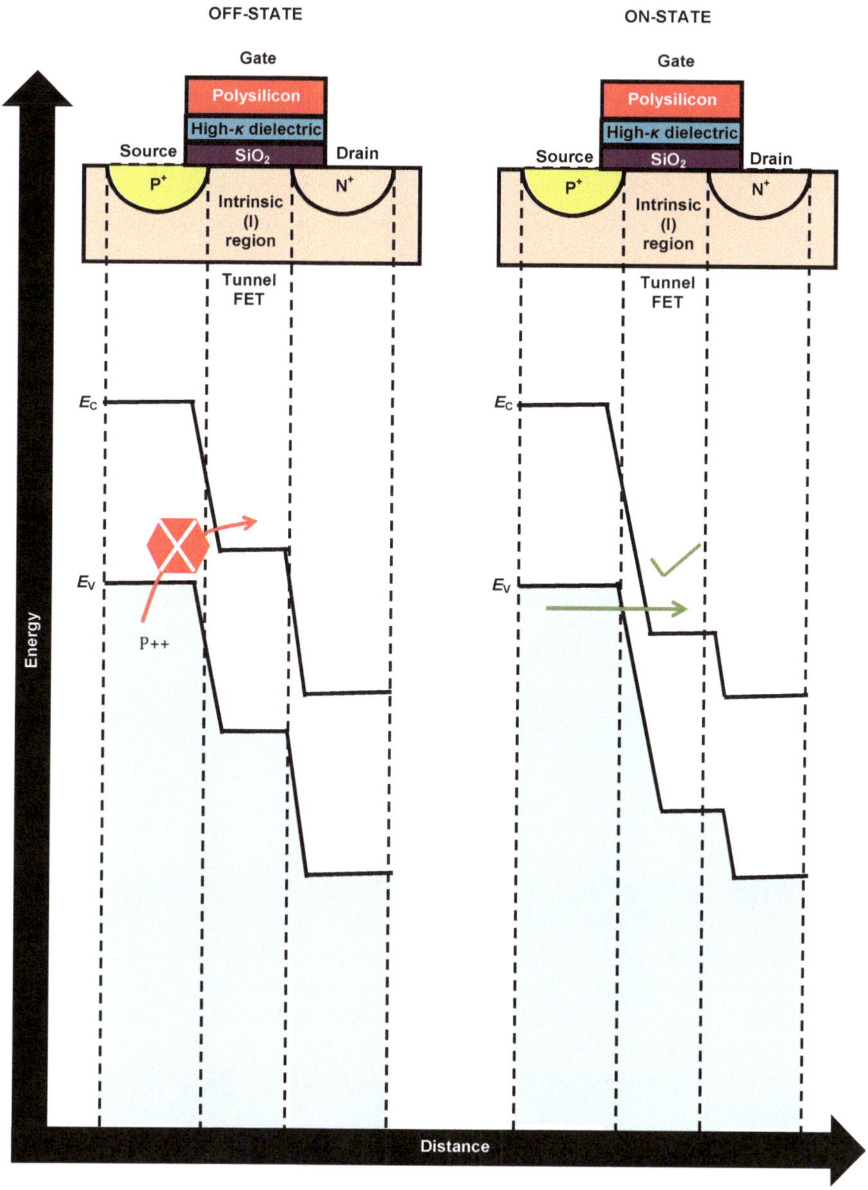

Fig. 12.11 Energy band diagrams of a tunnel FET in off- and on-states

12.12 Discussion and Conclusions

As flat hetero-interfaces can be easily realized with III–V semiconductors such as GaAs/AlGaAs and InGaAs/AlAs, they are much suited to RTD fabrication [7]. For digital logic circuit applications of RTDs, very large scale integrated circuits made from III-V compound semiconductors are attractive. Heterostructures, e.g., Si/CaF_2/CdF_2 and GaN/AlN have also been investigated. By developing suitable processes to fabricate RTDs using silicon, it will be possible to integrate them with monolithic silicon IC technology. Resonant tunneling has been reported across a few atomic layers thick boron nitride layer enclosed between two graphene electrodes, compared to tens of nm thick conventional RTD devices, promising ultrafast transit times for applications in high-frequency logic devices [8].

The tunnel FETs are sharper in turn-on characteristics than MOSFETs. Whereas MOSFETs work by diffusion over the barrier, the main phenomenon in TFETs is tunneling. To supply a large current, the barrier must be thin over a large effective area. Additionally, adequate density of states must be available on both sides. The TFET can be looked upon as a gated P–I–N diode working under reverse bias condition. Unlike the unipolar MOSFET device, the TFET is ambipolar with both electrons and holes acting as charge carriers. Ambipolar conduction can be suppressed by tailoring the doping profile or using heterostructure design [9].

Review Exercises

12.1 Explain the idea of tunneling in quantum mechanics. How is tunneling explained from the wave nature of electron?

12.2 What do quantum mechanics and classical mechanics declare about the probability of penetrating a thick potential barrier? Under what condition, tunneling is possible across a depletion region?

12.3 What is the effect of a very high doping concentration on: (i) the position of Fermi level in a semiconductor, and (ii) the electrical properties of the semiconductor.

12.4 A P-type semiconductor is heavily doped? What is the location of the Fermi level?

12.5 An N-type semiconductor is doped with a high concentration of impurity atoms? Where is the Fermi level of the material situated?

12.6 How are voltage and current related in an ohmic resistor? How does an electronic component showing negative resistance differ from an ohmic resistor?

12.7 Define differential resistance. At a certain point of the current–voltage characteristic of a device, the slope of the curve was found to be negative. What is the sign of its differential resistance?

12.8 How does a tunnel diode differ from a conventional P–N junction diode in construction? State three salient features of the current–voltage characteristics of a tunnel diode.

12.9 A tunnel diode is subjected to increasing values of forward bias. Explain its current–voltage characteristics from energy band picture. Draw neat and labeled diagrams supporting your explanation.

12.10 What is the reason for high reverse leakage current in a tunnel diode? What is the reverse breakdown voltage of a tunnel diode?

12.11 Explain the phenomenon of resonant tunneling across two potential barriers arranged in a row.

12.12 What is a heterostructure? What technique is commonly applied for fabricating heterostructures?

12.13 Show on a diagram the different structural layers in a resonant tunnel diode. What are the typical thicknesses of quantum barriers and the quantum well?

12.14. Explain the resonance phenomenon in a double barrier quantum well structure using the ideas of interference of electron waves.

12.15 Mention five carrier transport mechanisms contributing to current flow in a resonant tunneling diode.

12.16 A resonant tunneling diode is connected in forward bias mode. How do the different segments in its current–voltage characteristics originate?

12.17 How does the reverse bias operation of a resonant tunneling diode differ from that of a tunnel diode?

12.18 Mention some applications of resonant tunneling diodes in digital and analog circuits.

12.19 Differentiate between a tunnel FET and conventional MOSFET from the point of view of: (a) device structure, (b) operating principle, and (c) subthreshold slope.

12.20 Draw the energy band diagrams of a tunnel FET in off- and on-states. Point out the differences between the two cases and their effects on carrier flow.

References

1. Esaki L (1958) New phenomenon in narrow germanium p-n junctions. Phys Rev 109(2):603–604
2. Esaki L, Arakawa Y, Kitamura M (2010) Esaki diode is still a radio star, half a century on. Nature 464(7285):31. doi:10.1038/464031b
3. Sun JP, Haddad GI, Mazumder P et al (1998) Resonant tunneling diodes: models and properties. Proc IEEE 86(4):641–661
4. Mazumder P, Kulkarni S, Bhattacharya M et al (1998) Digital circuit applications of resonant tunneling devices. Proc IEEE 86(4):664–686
5. Uemura T, Mazumder P (1999) Design and analysis of resonant tunneling diode (RTD)-based high-performance memory system. IEICE Trans Electron E82-C(9):1630–1637
6. Seabaugh A, The tunneling transistor, IEEE Spectrum. http://spectrum.ieee.org/semiconductors/devices/the-tunneling-transistor. Accessed 7 April 2016

7. Nagase M, Tokizaki T (2014) Bistability characteristics of GaN/AlN resonant tunneling diodes caused by intersubband transition and electron accumulation in quantum well. IEEE Trans Electron Devices 61(5):1321–1326
8. Britnell L, Gorbachev RV, Geim AK (2013) Resonant tunneling and negative differential conductance in graphene transistors. Nat Commun 1–5. doi:10.1038/ncomms2817
9. Esfandyarpour R (2012) Tunneling field effect transistors http://large.stanford.edu/courses/2012/ph250/esfandyarpour1/. Accessed 7 April 2016

Chapter 13
Tunnel Junction, Coulomb Blockade, and Quantum Dot Circuit

Abstract Conceptual development regarding single electron transfer phenomena is presented. It is shown that energy necessary to place a single electronic charge on one plate of a capacitor with equal opposite charge on its opposite plate is not a clearly distinguishable event at micro- and milliscales. But it becomes a meaningful event at the nanoscale due to the significant amount of energy involved. Further, it is shown that the existence of a voltage requirement for tunneling to occur across the plates of a capacitor, the so-called Coulomb blockade effect, is a noticeable phenomenon exclusive to nanoscale. Moving further, it is found that the Coulomb blockade is observable at or near room temperatures only in the scale of nano dimensions. From this understanding, the notion of a tunnel junction is put forward as a barrier in the form of an electrical potential or thin dielectric film across which tunneling occurs. The tunnel junction is modeled as an ideal capacitor with a parallel-connected tunnel resistance whose value must be $\gg 4.2\ k\Omega$ for Coulomb blockade to become recognizable. The capacitor of the tunnel junction behaves in a different way from a normal capacitor. Upon excitation by a constant current source, the voltage across this capacitor oscillates between two values, which are referred to as single electron tunneling oscillations. Applying the tunnel junction model, the operation of a quantum dot circuit consisting of a quantum dot and two tunnel junctions is analyzed. In both cases, for electron tunneling into the quantum dot across one tunnel junction and for electron tunneling off the quantum dot across the other tunnel junction, Coulomb blockade occurs in a quantum dot circuit like a nanocapacitor. The energy band diagram for a small quantum dot circuit exhibits a discretized nature. The electron tunneling does not take place in a continuous fashion but in discrete voltage steps. The resulting current–voltage characteristic of the quantum dot circuit has the shape of a staircase which is called the Coulomb staircase.

V.K. Khanna, *Integrated Nanoelectronics*,
NanoScience and Technology, DOI 10.1007/978-81-322-3625-2_13

13.1 Introduction

In this chapter, we introduce Coulomb blockade effect, which can be witnessed at room temperature only in the nano regime. We model a tunnel junction and analyze a quantum dot circuit. Distinctive phenomena taking place in low-dimensionality structures are described [1].

13.2 Coulomb Blockade in a Nanocapacitor

13.2.1 Energy Required to Transfer a Single Electronic Charge

A generic tunnel junction [2] consists of two conducting terminals separated by a thin insulating space. This insulator behaves as a tunnel barrier across which electrons can pass by quantum-mechanical tunneling. Capacitance C of this capacitor is defined as the amount of charge stored on the plates ($+Q$ on one plate and $-Q$ on the other plate) by an applied potential difference V between its two plates:

$$C = Q/V \tag{13.1}$$

To calculate the electrostatic energy stored by the capacitor, we find the work done on transferring an elementary charge dQ from the one plate to its other plate against repulsive force due to pre-existing charge residing on that plate. This work is given by

$$\mathrm{d}W = V\mathrm{d}Q \tag{13.2}$$

The total work done to transfer the charge Q between the plates is obtained by integrating the above equation from $Q = 0$ to $Q = Q$ after substituting for V from (13.1)

$$W = \int \mathrm{d}W = \int_{Q=0}^{Q=Q} V\mathrm{d}Q = \int_{Q=0}^{Q=Q} (Q/C)\mathrm{d}Q = Q^2/2C \tag{13.3}$$

This is the work done by the power source to establish the charge configuration $+Q/-Q$ on the two plates of the capacitor.

The parallel-plate capacitor is a simple structure, which can be used for our discussion. It consists of two plane, parallel metallic plates of cross-sectional area A separated by a dielectric material of dielectric constant ε_r and thickness d. If ε_0 denotes the permittivity of free space = 8.854×10^{-12} F/m, the capacitance of the parallel-plate capacitor is expressed as

$$C = \frac{\varepsilon_0 \varepsilon_r A}{d} \tag{13.4}$$

Let us calculate the energy required to put one electronic charge $-q_e$ on one plate and equal positive charge $+q_e$ on the other plate for three capacitors: (i) Nanocapacitor, (ii) Microcapacitor and (iii) Millicapacitor, which are named in accordance with their areas. In each case, ε_r is taken as 1.

13.2.1.1 Nanocapacitor

$A = 5$ nm $\times$ 5 nm, $d = 1$ nm

$$\begin{aligned} C &= 8.854 \times 10^{-12} \times 1 \times 5 \times 10^{-9} \times 5 \times 10^{-9}/\left(1 \times 10^{-9}\right) = 2.2135 \times 10^{-19}\,\text{F} \\ W &= \left(1.6 \times 10^{-19}\,\text{Coulomb}\right)^2/\left(2 \times 2.2135 \times 10^{-19}\,\text{Coulomb/Volt}\right) \\ &= 5.78 \times 10^{-20}\left\{\text{Coulomb}^2 \times (\text{Volt/Coulomb})\right\} \\ &= 5.78 \times 10^{-20}(\text{Coulomb} \times \text{Volt}) = \left\{5.78 \times 10^{-20}/\left(1.6 \times 10^{-19}\right)\right\}\text{eV} = 0.36\,\text{eV} \end{aligned} \tag{13.5}$$

13.2.1.2 Microcapacitor

$A = 5$ μm $\times$ 5 μm, $d = 1$ nm

$$\begin{aligned} C &= 8.854 \times 10^{-12} \times 1 \times 5 \times 10^{-6} \times 5 \times 10^{-6}/\left(1 \times 10^{-9}\right) = 2.2135 \times 10^{-13}\,\text{F} \\ W &= \left(1.6 \times 10^{-19}\text{Coulomb}\right)^2/\left(2 \times 2.2135 \times 10^{-13}\text{Coulomb/Volt}\right) \\ &= 5.78 \times 10^{-26}\left\{\text{Coulomb}^2 \times (\text{Volt/Coulomb})\right\} \\ &= 5.78 \times 10^{-26}(\text{Coulomb} \times \text{Volt}) \\ &= \left\{5.78 \times 10^{-26}/\left(1.6 \times 10^{-19}\right)\right\}\text{eV} = 3.61 \times 10^{-7}\,\text{eV} \end{aligned} \tag{13.6}$$

13.2.1.3 Millicapacitor

$A = 5$ mm $\times$ 5 mm, $d = 1$ nm

$$\begin{aligned} C &= 8.854 \times 10^{-12} \times 1 \times 5 \times 10^{-3} \times 5 \times 10^{-3}/\left(1 \times 10^{-9}\right) = 2.2135 \times 10^{-7}\,\mathrm{F} \\ W &= \left(1.6 \times 10^{-19}\mathrm{Coulomb}\right)^2/\left(2 \times 2.2135 \times 10^{-7}\mathrm{Coulomb/Volt}\right) \\ &= 5.78 \times 10^{-32}\left\{\mathrm{Coulomb}^2 \times (\mathrm{Volt/Coulomb})\right\} \\ &= 5.78 \times 10^{-32}(\mathrm{Coulomb} \times \mathrm{Volt}) \\ &= \left\{5.78 \times 10^{-32}/\left(1.6 \times 10^{-19}\right)\right\}\mathrm{eV} = 3.61 \times 10^{-13}\,\mathrm{eV} \end{aligned} \tag{13.7}$$

These calculations lead to interesting observations: (i) The energy required to transfer a charge equivalent to a single electron decreases from the large value (0.361 eV) for a nanocapacitor to an appreciably small value (3.61×10^{-7} eV) for a microcapacitor and an almost negligible value (3.61×10^{-13} eV) for a millicapacitor. (ii) The transfer of a single electron is a noticeable energy event for a nanocapacitor while the same does not apply to micro and millicapacitors.

Thus, the size of the capacitor plays a very important role at nanoscale and the transfer of a single electronic charge is a clearly recognizable event at this scale. This makes possible the realization of a device, which has the precision of operation depending upon transfer of a single electron. It is therefore necessary to modify our line of thinking, which has hitherto developed from milli/microscale phenomena when nanoscale devices are being talked about.

13.2.2 Change in Energy Stored on Electron Tunneling

Suppose, a single electron of charge q_e tunnels across the dielectric of a nanocapacitor from one plate to another, so that the charge on this plate becomes $Q + q_e$. Then the energy W_e stored in the electric field of the capacitor is

$$W_e = (Q + q_e)^2/(2C) \tag{13.8}$$

so that the change in energy (ΔW) stored is

$$\begin{aligned} \Delta W &= W_e - W = (Q + q_e)^2/(2C) - Q^2/(2C) = \left(Q^2 + 2Qq_e + q_e^2 - Q^2\right)/(2C) \\ &= 2q_e(Q + q_e/2)/(2C) = q_e(Q + q_e/2)/C \end{aligned} \tag{13.9}$$

For tunneling to occur if $\Delta W < 0$, remembering that $q_e < 0$,

$$\begin{aligned} &(Q + q_e/2) > 0 \\ &\text{or, } Q > -q_e/2 \\ &\text{or, } CV > -q_e/2 \\ &\text{or, } V > -q_e/(2C) \end{aligned} \tag{13.10}$$

Similarly, for tunneling of charge of opposite polarity; hence

$$-q_e/(2C) < V < +q_e/(2C) \tag{13.11}$$

Thus, tunneling is only possible if a large voltage exists across the plates of the capacitor. The magnitude of this voltage is

$$|V| = |q_e|/(2C) \tag{13.12}$$

The existence of this voltage requirement manifested in the form of opposition of the electronic device to flow of current at small bias voltages is called the Coulomb blockade effect [3–5]. Physically, it originates from the repulsive force experienced by an incoming charge due to electric field of pre-existing charges of the same kind present on a conductor. The current–voltage characteristic of the nanocapacitor is shown in Fig. 13.1. Before a voltage $V = q_e/2C$ is reached, no electron is delivered to the conductor. So, current is zero. Then, when the voltage reaches $2q_e/2C$, the delivery of second electron takes place.

Non-observance of Coulomb blockade in micro and milli-sized capacitors arises from the fact that the energy necessary for electron transfer called the charging

Fig. 13.1 *I–V* characteristic of a nanocapacitor

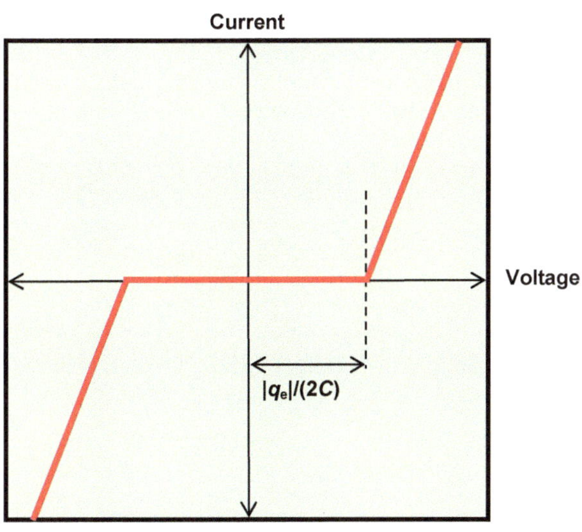

energy = $q_e^2/(2C)$ is very small. In such capacitors, when the energy is raised by a small amount, a large number of electrons are transferred. Therefore, the discreteness of electron transfer is not obvious. It appears as if the process is continuous. The discrete electron-by-electron transfer will be discernible only if a single electron is transferred by a large amount of energy. Since this happens at nano level, Coulomb blockade is a phenomenon, which is exclusive to nanoscale.

13.3 Effect of Temperature

The discussions so far are based on the assumption that the ambient temperature is $T = 0$ K. At any higher temperature, including the room temperature, thermal energy plays an important role and cannot be ignored. This energy is given by

$$E_{\text{Thermal}} = (1/2)k_B T/q \tag{13.13}$$

If the temperature is too high, the thermal energy will have a large value. So, the electron transfer by thermal energy will far exceed that taking place when the applied voltage is sufficient to enable tunneling. So, the thermally induced electron flow will dominate over that due to the applied voltage. Thus, random temperature-influenced electron transfers will be perceived and actual effects will not be visible. Therefore, the charging energy must be appreciably higher than the thermal energy at the required operating temperature, i.e.,

$$\begin{aligned} &q_e^2/(2C) \gg (1/2)k_B T \\ &\text{or, } q_e^2/(k_B C) \gg T \\ &\text{or, } T \ll q_e^2/(k_B C) \end{aligned} \tag{13.14}$$

$$\begin{aligned} &\ll \left(1.6 \times 10^{-19}\,\text{Coulomb}\right)^2 / \left\{\left(1.38 \times 10^{-23}\,\text{Joule/Kelvin}\right) \times C\,\text{Coulomb/Volt}\right\} \\ &= \left(1.855 \times 10^{-15}/C\right)\text{Coulomb}^2 \times \frac{\text{Kelvin}}{\text{Joule}} \times \frac{\text{Volt}}{\text{Coulomb}} \\ &= \left(1.855 \times 10^{-15}/C\right)\text{Coulomb} \times \frac{\text{Kelvin}}{\text{Joule}} \times \text{Volt} \\ &= \left(1.855 \times 10^{-15}/C\right)\frac{\text{Kelvin}}{\text{Joule/Coulomb}} \times \text{Volt} \\ &= \left(1.855 \times 10^{-15}/C\right)\frac{\text{Kelvin}}{\text{Volt}} \times \text{Volt} \\ &= \left(1.855 \times 10^{-15}/C\right)\text{Kelvin} \end{aligned} \tag{13.15}$$

For the nanocapacitor,

$$T \ll 1.855 \times 10^{-15}/C = \left(1.855 \times 10^{-15}\right)/\left(2.2135 \times 10^{-19}\right) = 8380.39\,\mathrm{K} \tag{13.16}$$

For the microcapacitor,

$$T \ll 1.855 \times 10^{-15}/C = \left(1.855 \times 10^{-15}\right)/\left(2.2135 \times 10^{-13}\right) = 0.00838\mathrm{K} \tag{13.17}$$

For the millicapacitor,

$$T \ll 1.855 \times 10^{-15}/C = 1.855 \times 10^{-15}/2.2135 \times 10^{-7} = 8.38 \times 10^{-9}\mathrm{K} \tag{13.18}$$

This exercise tells us that only for the nanocapacitor, Coulomb blockade effect can be seen at room temperature. Neither for the microcapacitor nor for the millicapacitor, there is any likelihood of observing Coulomb blockade at any temperature close to room temperature. Only when one goes much below subKelvin temperature, there is any probability to see this effect. Thus, Coulomb blockade is a phenomenon which is observable at a practical temperature only uniquely with nanotechnology.

13.4 Correlation of Uncertainty in the Number of Electrons with Capacitor Size

The inference drawn in Sect. 13.3 can be obtained through another line of reasoning. In this route, we presuppose that the device has to work at room temperature and then estimate the size of capacitor suitable to achieve this objective.

If instead of a single electron, suppose n electrons can tunnel as a consequence of thermal fluctuations. Then, the equation for charging energy is written as

$$W_c = (nq_\mathrm{e})^2/(2C) \tag{13.19}$$

Since the thermal energy is responsible for this transference of n electrons

$$\begin{aligned}
&(nq_\mathrm{e})^2/(2C) = (1/2)k_BT \\
&\text{or, } n = \sqrt{k_BTC}/q_\mathrm{e} = \sqrt{1.38 \times 10^{-23} \times 300\mathrm{K} \times C}/\left(1.6 \times 10^{-19}\right) \\
&= 4.02 \times 10^8\sqrt{C}
\end{aligned} \tag{13.20}$$

at a temperature $T = 300$ K, giving for $n = 1$, $C = 6.188 \times 10^{-18}$ F; for $n = 10$, $C = 6.188 \times 10^{-16}$ F; for $n = 100$, $C = 6.188 \times 10^{-14}$ F. Thus, the number of electrons increases as one moves towards larger capacitances. Single electron effect is observed at smaller capacitance values. Hence, the need of nanocapacitor arises if we insist upon operation at room temperature.

13.5 Modeling the Tunnel Junction

A tunnel junction (Fig. 13.2) can be modeled as a parallel combination of an ideal capacitance C and a resistance R_T called the tunnel resistance = V/I where V is the applied DC bias across the junction and I is the resultant current flowing through the junction by tunneling. The tunnel resistance is not an ohmic resistance, which arises from scattering of charge carriers by the atoms in a conductor in the classical picture. It is a result of quantum-mechanical phenomena. Therefore, it should be related to quantum mechanics.

13.5.1 Tunnel Resistance

Bearing the comments made in preceding subsection in mind, the number N of electrons localized on an island are described in terms of an average value $\langle N \rangle$. Coulomb blockade imposes the condition

$$|N - \langle N \rangle|^2 \ll 1 \tag{13.21}$$

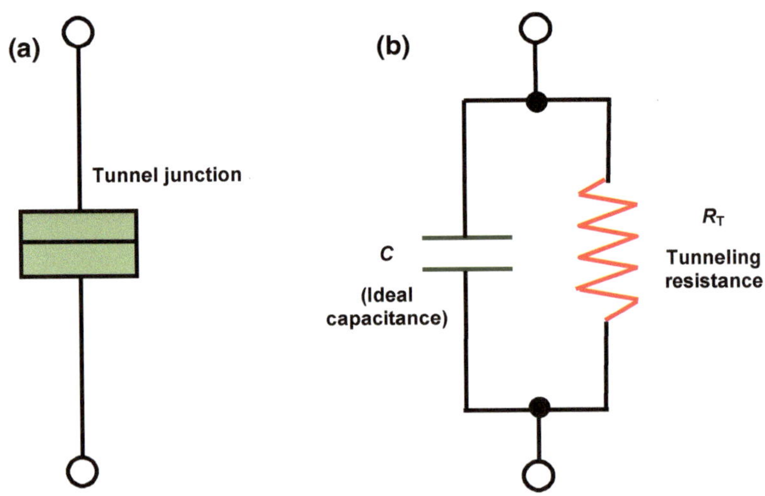

Fig. 13.2 Tunnel junction: **a** Circuit symbol and **b** equivalent circuit representation

In absence of tunnel barriers or in case of their inadequate opacity, one cannot think about charging an island or localizing an electron thereupon because it is the tunnel barrier, which exerts the necessary restraining force on the electron to confine it within a given volume.

According to the Uncertainty principle, the energy of a particle and time cannot be simultaneously measured accurately. The minimum value of the product of uncertainty of energy ΔE and that of time Δt is given by

$$\Delta E \Delta t \geq \hbar/2 \tag{13.22}$$

where $\hbar$ is Planck's constant = 1.05×10^{-34} J-s.

Recalling that the time constant of a parallel resistance–capacitance (*R–C*) circuit is given by

$$\tau = RC \tag{13.23}$$

the time Δt here is the time between two tunneling events written as

$$\Delta t = R_{\mathrm{T}} C \tag{13.24}$$

Since the uncertainty in energy is

$$\Delta E = q_{\mathrm{e}}^2/2C \tag{13.25}$$

we obtain

$$\Delta E \Delta t \geq \left(q_{\mathrm{e}}^2/2C\right)(R_{\mathrm{T}} C) = q_{\mathrm{e}}^2 R_{\mathrm{T}}/2 \tag{13.26}$$

Combining Eqs. (13.22) and (13.26), we have

$$\begin{aligned} & q_{\mathrm{e}}^2 R_{\mathrm{T}}/2 \gg \hbar/2 \\ & \text{or, } R_{\mathrm{T}} \gg \hbar/q_{\mathrm{e}}^2 \gg 1.05 \times 10^{-34}/\left(1.6 \times 10^{-19}\right)^2 = 4.1 \times 10^3 \Omega \end{aligned} \tag{13.27}$$

For observation of Coulomb blockade effect, the tunnel resistance R_{T} must be $\gg$4.1 kΩ.

13.5.2 A Constant Current Source Exciting a Tunnel Junction

An interesting phenomenon occurs when a constant current supply is placed across a tunnel junction (Fig. 13.3). Let us first look at a perfect capacitor. This capacitor behaves in accordance with our classical notion of capacitor in which the current is completely blocked by the dielectric. There is no tunneling of charge carriers across

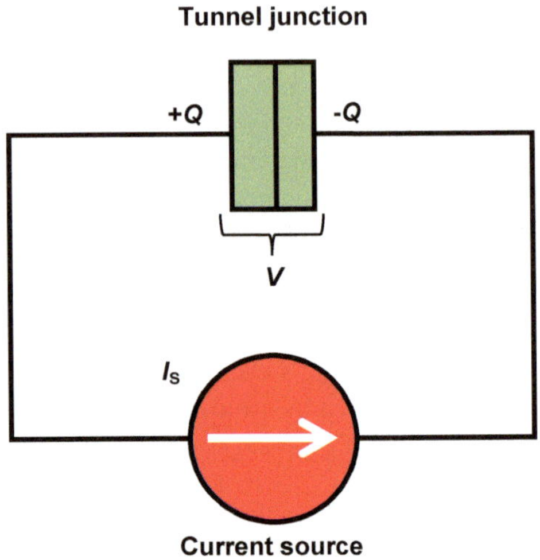

Fig. 13.3 A constant current source feeding a tunnel junction

the dielectric. For this capacitor, the relationship between voltage applied across the capacitor and current flowing through it is the classical formula

$$I_S = C\mathrm{d}V/\mathrm{d}t \tag{13.28}$$

so that the voltage across the capacitor plates is expressed as

$$V(t) = (1/C)\int_0^t I_S \mathrm{d}t = I_S t/C \tag{13.29}$$

and the charge is

$$Q(t) = I_S t \tag{13.30}$$

In reality, no current flows across the dielectric, but the current source propels electrons so that a charge density is built up on the capacitor plates.

Now let us focus our attention on the capacitor under consideration. It differs from the ideal capacitor because it allows tunneling of an electron no sooner than $|V| > |q_e|/(2C)$. Let us examine how allowance of tunneling makes this capacitor behave in a different way from the normal capacitor. Consequent upon the electron tunneling, the positive charge on the plate giving an electron increases by an amount q_e whereas the positive charge on the plate receiving an electron decreases by an amount q_e. The voltage across the capacitor plates changes from $-q_e/(2C)$ to $+q_e/(2C)$, and the net change in voltage is

$$\Delta V = (q_e/2C) - (-q_e/2C) = q_e/C \tag{13.31}$$

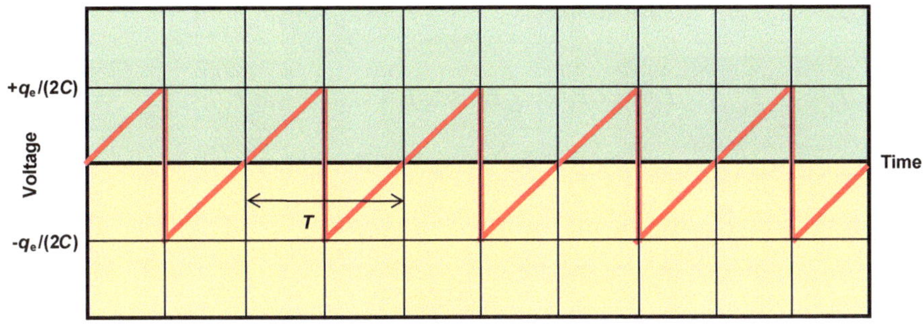

Fig. 13.4 Oscillations produced by single electron tunneling

After completion of tunneling when the voltage reaches the value $+q_e/2C$, the voltage immediately falls back to its previous value $-q_e/2C$. With passage of time, again when the voltage $|V| > |q_e|/(2C)$, the same incident is repeated. Thus, as time goes by, the voltage across the capacitor repetitively changes from $-q_e/(2C)$ to $+q_e/(2C)$. Overall, we can say that the voltage across the capacitor is oscillating between these two values. These oscillations are called single electron tunneling oscillations (Fig. 13.4).

The periodic time T of the above oscillations is obtained by finding $V(t)$ at $t = T/2$ and equating it to $q_e/(2C)$. Thus

$$V(t = T/2) = (1/C) \int_{t=0}^{t=T/2} I_S \mathrm{d}t = (1/C)I_S(T/2) = q_e/2C \tag{13.32}$$

from which

$$T = q_e/I_S \tag{13.33}$$

These oscillations are observed with difficulty due to the high lead capacitance of practical systems.

13.6 Basic Analysis of Quantum Dot Circuit

The circuit being studied consists of a metallic quantum dot QD, which is coupled to two external leads through tunnel junctions TJ_a and TJ_b (Fig. 13.5). Insulating regions separating the quantum dot from the lead-fixation metal contacts, form the tunnel junctions. These junctions have a metal/insulator/metal structure.

The equivalent circuit models of tunnel junctions can substitute the junctions, viz., an ideal capacitance C in parallel with a tunnel resistance R. For the tunnel junction TJ_a, the capacitance is C_a and tunnel resistance is R_a. For the tunnel junction TJ_b, the corresponding symbols are C_b and R_b. Let V_s be the supply voltage

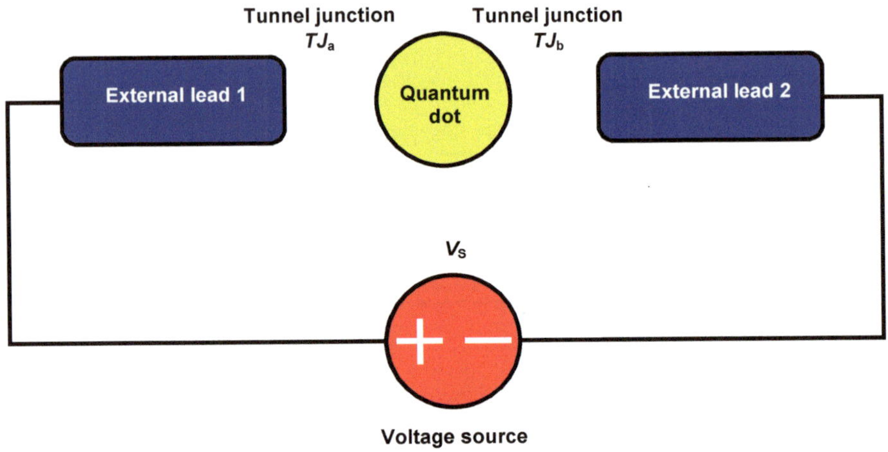

Fig. 13.5 A quantum dot circuit

of the circuit. Also, let V_a^i and V_b^i denote the voltages across the tunnel junctions TJ_a and TJ_b, respectively. If the charges associated with the tunnel junctions are Q_a^i, Q_{b}^i, we may write

$$Q_a^i = C_a V_a^i \tag{13.34}$$

$$Q_b^i = C_b V_b^i \tag{13.35}$$

Hence, the total charge on the QD island is

$$Q^i = Q_b^i - Q_a^i \tag{13.36}$$

By tunneling, electrons are transferred and stored on the QD island. Suppose when the experiment had begun, a discrete number n of electrons had already tunneled into the island. So, n is the initial number of electrons, and

$$Q^i = nq_{\mathrm{e}} \tag{13.37}$$

The total energy stored in the capacitors C_a, C_b is given by

$$\begin{aligned} E_{\mathrm{se}} &= \left(Q_a^i\right)^2/(2C_a) + \left(Q_b^i\right)^2/(2C_b) = (1/2)\left\{\left(C_a V_a^i\right)^2/C_a + \left(C_b V_b^i\right)^2/C_b\right\} \\ &= (1/2)\left\{C_a\left(V_a^i\right)^2 + C_b\left(V_b^i\right)^2\right\} \end{aligned} \tag{13.38}$$

Application of Kirchoff's law to the quantum dot circuit yields

$$\begin{aligned} V_s &= V_a^i + V_b^i = V_a^i + Q_b^i/C_b = V_a^i + \left(Q^i + Q_a^i\right)/C_b \\ &= V_a^i + \left(Q^i + C_a V_a^i\right)/C_b = V_a^i + \left(nq_e + C_a V_a^i\right)/C_b \end{aligned} \tag{13.39}$$

from Eq. (13.37).

The values of V_a^i and V_b^i are obtained as

$$\begin{aligned} &V_a^i = V_s - \left(nq_e + C_a V_a^i\right)/C_b = \left(V_s C_b - nq_e - C_a V_a^i\right)/C_b \\ &\text{or, } V_a^i C_b = V_s C_b - nq_e - C_a V_a^i \\ &\text{or, } V_a^i C_b + C_a V_a^i = V_s C_b - nq_e \\ &\text{or, } V_a^i = (V_s C_b - nq_e)/(C_b + C_a) = (V_s C_b - nq_e)/C_s \end{aligned} \tag{13.40}$$

since

$$C_b + C_a = C_s \tag{13.41}$$

Similarly,

$$V_b^i = (V_s C_a + nq_e)/C_s \tag{13.42}$$

Substituting for V_a^i and V_b^i, the equation for total energy stored, Eq. (13.38), becomes

$$\begin{aligned} E_{se} &= (1/2)\left\{C_a\left(V_a^i\right)^2 + C_b\left(V_b^i\right)^2\right\} \\ &= (1/2)\left[C_a\{(V_s C_b - nq_e)/C_s\}^2 + C_b\{(V_s C_a + nq_e)/C_s\}^2\right] \\ &= \{1/(2C_s^2)\}\left(C_a V_s^2 C_b^2 + C_a n^2 q_e^2 - 2nq_e C_a C_b V_s + C_b V_s^2 C_a^2 + C_b n^2 q_e^2 + 2nq_e C_a C_b V_s\right) \\ &= \{1/(2C_s^2)\}\left[C_a V_s^2 C_b^2 + C_b V_s^2 C_a^2 + C_a n^2 q_e^2 + C_b n^2 q_e^2\right] \\ &= \{1/(2C_s^2)\}\{V_s^2 C_a C_b (C_b + C_a) + n^2 q_e^2 (C_a + C_b)\} = \{1/(2C_s^2)\}\{V_s^2 C_a C_b (C_s) + n^2 q_e^2 (C_s)\} \\ &= \{1/(2C_s)\}\left\{V_s^2 C_a C_b + (nq_e)^2\right\} \end{aligned} \tag{13.43}$$

During tunneling of the electron across the junction, an amount of work W is done by the voltage source V_s in the circuit over a period of time Δt in which a quantity of charge Δq is transported across the junction. This work W is obtained as follows:

$$W = \int_0^{\Delta t} V_s I(t)\mathrm{d}t = \int_0^{\Delta t} V_s(\mathrm{d}q/\mathrm{d}t)\mathrm{d}t = \int_0^{\Delta q} V_s \mathrm{d}q = V_s \Delta q \tag{13.44}$$

13.6.1 Electron Tunneling into the Quantum Dot Island Through Tunnel Junction TJ_b

Suppose one single electron tunnels to the QD island through the tunnel junction TJ_b. By this transfer of electron, the voltage drops across both the junctions TJ_a and TJ_b change to V_a^f and V_b^f. The new voltage drops are given by

$$V_a^f = \{V_s C_b - (n+1)q_e\}/C_s \tag{13.45}$$

and,

$$V_b^f = \{V_s C_a + (n+1)q_e\}/C_s \tag{13.46}$$

These voltage drops can be expressed in terms of initial voltage drops V_a^i and V_b^i using Eqs. (13.40) and (13.42) as

$$V_a^f = \{V_s C_b - nq_e\}/C_s - q_e/C_s = V_a^i - q_e/C_s \tag{13.47}$$

$$V_b^f = \{V_s C_a + nq_e\}/C_s + q_e/C_s = V_b^i + q_e/C_s \tag{13.48}$$

So,

$$V_a^f = V_a^i - q_e/C_s \quad \text{or, } V_a^f - V_a^i = -q_e/C_s \quad \text{or, } \Delta V_a = -q_e/C_s \tag{13.49}$$

$$V_\mathrm{b}^\mathrm{f} = V_\mathrm{b}^i + q_e/C_s \quad \text{or, } V_\mathrm{b}^\mathrm{f} - V_\mathrm{b}^i = +q_e/C_s \quad \text{or, } \Delta V_\mathrm{b} = +q_e/C_s \tag{13.50}$$

These equations may be recast in terms of charges Q_a, Q_b as

$$\Delta V_a = \Delta Q_a/C_a = -q_e/C_s \quad \text{or, } \Delta Q_a = -C_a q_e/C_s \tag{13.51}$$

$$\Delta V_b = \Delta Q_b/C_b = +q_e/C_s \quad \text{or, } \Delta Q_b = +C_b q_e/C_s \tag{13.52}$$

The change in charge $\Delta Q_a = Q_a^f - Q_a^i$ is related with work W performed by the supply

$$W = V_s \Delta Q_a = V_s C_a q_e/C_s \tag{13.53}$$

whereas the change in charge $\Delta Q_b = Q_b^f - Q_b^i$ is related with the tunneling event.

The change in total energy caused by the tunneling of the electron into the quantum dot island through the tunnel junction TJ_b is found by subtracting the work done from the change in energy stored by the capacitors as

$$
\begin{aligned}
\Delta E_{\mathrm{t}} &= \Delta E_{\mathrm{se}} - W = \{1/(2C_{\mathrm{s}})\}\left\{V_{\mathrm{s}}^2 C_a C_b + (nq_{\mathrm{e}})^2\right\} \\
&\quad - \{1/(2C_{\mathrm{s}})\}\left[V_{\mathrm{s}}^2 C_a C_b + \{(n+1)q_{\mathrm{e}}\}^2\right] - V_s C_a q_{\mathrm{e}}/C_{\mathrm{s}} \\
&= \{1/(2C_{\mathrm{s}})\}\left\{V_{\mathrm{s}}^2 C_a C_b + (nq_{\mathrm{e}})^2 - V_{\mathrm{s}}^2 C_a C_b - (nq_{\mathrm{e}})^2 - 2nq_{\mathrm{e}}^2 - q_{\mathrm{e}}^2 - 2V_s C_a q_{\mathrm{e}}\right\} \\
&= -(q_{\mathrm{e}}/C_{\mathrm{s}})\{q_{\mathrm{e}}(n+1/2) + V_{\mathrm{s}} C_a\}
\end{aligned}
\tag{13.54}
$$

In order that the tunneling event is energetically favorable, this change in energy must be positive, so that $-q_{\mathrm{e}}/C_{\mathrm{s}}\{q_{\mathrm{e}}(n+1/2)+V_{\mathrm{s}}C_a\} > 0$

$$
\begin{aligned}
&\text{or, } q_{\mathrm{e}}(n+1/2) + V_{\mathrm{s}} C_{\mathrm{a}} > 0 \\
&\text{or, } V_{\mathrm{s}} C_{\mathrm{a}} > -q_{\mathrm{e}}(n+1/2) \\
&\text{or, } V_{\mathrm{s}} > -(q_{\mathrm{e}}/C_{\mathrm{a}})(n+1/2)
\end{aligned}
\tag{13.55}
$$

13.6.2 Electron Tunneling off the Quantum Dot Island Through Tunnel Junction TJ_a

After tunneling, the voltage drops across the tunnel junctions are

$$
\begin{aligned}
V_a^f &= \{V_{\mathrm{s}} C_b - (n-1)q_{\mathrm{e}}\}/C_{\mathrm{s}} = (V_{\mathrm{s}} C_b - nq_{\mathrm{e}} + q_{\mathrm{e}})/C_{\mathrm{s}} \\
&= (V_{\mathrm{s}} C_b - nq_{\mathrm{e}})/C_{\mathrm{s}} + q_{\mathrm{e}}/C_{\mathrm{s}} = V_a^i + q_{\mathrm{e}}/C_{\mathrm{s}} \\
\text{or, } & V_a^f - V_a^i = \Delta V_a = +q_{\mathrm{e}}/C_{\mathrm{s}}
\end{aligned}
\tag{13.56}
$$

and,

$$
\begin{aligned}
V_b^f &= \{V_{\mathrm{s}} C_a + (n-1)q_{\mathrm{e}}\}/C_{\mathrm{s}} = (V_{\mathrm{s}} C_b + nq_{\mathrm{e}} - q_{\mathrm{e}})/C_{\mathrm{s}} \\
&= (V_{\mathrm{s}} C_b + nq_{\mathrm{e}})/C_{\mathrm{s}} - q_{\mathrm{e}}/C_{\mathrm{s}} = V_b^i - q_{\mathrm{e}}/C_{\mathrm{s}} \\
\text{or, } & V_b^f - V_b^i = \Delta V_b = -q_{\mathrm{e}}/C_{\mathrm{s}}
\end{aligned}
\tag{13.57}
$$

The changes in charges are

$$\Delta Q_a = +C_a q_{\mathrm{e}}/C_{\mathrm{s}} \tag{13.58}$$

$$\Delta Q_b = -C_b q_{\mathrm{e}}/C_{\mathrm{s}} \tag{13.59}$$

The change in charge ΔQ_b is associated with work W performed by the supply

$$W = V_s \Delta Q_b = V_s C_b q_e / C_s \tag{13.60}$$

while the change in charge $\Delta Q_a = Q_a^f - Q_a^i$ is linked with the tunneling event.
The total energy changes by an amount

$$\begin{aligned}\Delta E_t = \Delta E_{se} - W &= \{1/(2C_s)\}\left\{V_s^2 C_a C_b + (nq_e)^2\right\} \\ &- \{1/(2C_s)\}\left[V_s^2 C_a C_b + \{(n-1)q_e\}^2\right] - V_s C_b q_e / C_s \\ &= \{1/(2C_s)\}\left\{V_s^2 C_a C_b + (nq_e)^2 - V_s^2 C_a C_b - (nq_e)^2 + 2nq_e^2 - q_e^2 - 2V_s C_b q_e\right\} \\ &= -(q_e/C_s)\{q_e(1/2 - n) + V_s C_b\}\end{aligned} \tag{13.61}$$

Since the system must transform from higher energy to lower energy state, this change in energy is positive; hence

$$-(q_e/C_s)\{q_e(1/2 - n) + V_s C_b\} > 0 \tag{13.62}$$

from which

$$V_s > -(q_e/C_b)(1/2 - n) \tag{13.63}$$

Assuming that $C_a = C_b = C$ and $n = 0$ (there were no previous electrons on the island), for the case of electron tunneling into the island through TJ_b, we have

$$V_s > -(q_e/C_a)(n + 1/2) = -(q_e/C)(0 + 1/2) = -(q_e/2C) \tag{13.64}$$

For the case of electron tunneling off the island through TJ_a, we get

$$V_s > -(q_e/C_b)(1/2 - n) = -(q_e/C)(1/2 - 0) = -(q_e/2C) \tag{13.65}$$

Thus for both cases, we find

$$V_s > -(q_e/2C) \tag{13.66}$$

13.6.3 *Electron Tunneling into the QD Island Through TJ_a and Tunneling off the QD Island Through TJ_b*

Repeating the sequence of steps laid down above and for the opposite situation of the electron first tunneling through tunnel junction TJ_a onto the QD island and then tunneling off the QD island through tunnel junction TJ_b, it can be shown that

$$V_s < +(q_e/2C) \tag{13.67}$$

Equations (13.66) and (13.67) may be combined together into a single equation

$$\begin{aligned} & -q_e/(2C) < V_s < +q_e/(2C) \\ & \text{or, } |V_s| > |q_e|/2C \end{aligned} \tag{13.68}$$

This is the same equation as Eq. (13.12) in Sect. 13.2.2 for the Coulomb blockade in a nanocapacitor. So, the quantum dot circuit shows the similar behavior to the nanocapacitor.

13.7 Energy Band Diagram of Tunnel Junction/Quantum Dot/Tunnel Junction Structure

Two disparate situations will be considered, namely, large and small quantum dots, distinguished by the fact that the former is of sufficient dimensions that it can be treated by classical mechanics whereas in the latter, quantum-mechanical effects cannot be ignored.

13.7.1 Large Quantum Dot

In the case of a large quantum dot, the energy levels in the quantum dot are spaced at a very small distance apart. Indeed, they appear to be coalesced into a band, as in a bulk material (Fig. 13.6). Then the Fermi level on the island is the same as in bulk material. For simplification, the capacitances C_a and C_b are postulated to be equal. Suppose, these equal-value capacitances are represented by the symbol C, i.e., $C_a = C_b = C$. Further, suppose the quantum dot island and the connection leads are made from the same material, e.g., aluminum. In Fig. 13.6a, the energy level diagram of the TJ_a/QD/TJ_b structure is sketched under the circumstance when no bias is applied to the circuit. Next, let us understand the effect of application of a potential difference V_0 across the structure. Then the positions of energy bands are changed with respect to their previous locations. The shifted positions are shown in Fig. 13.6b. Looking at these altered positions, it is obvious that tunneling of electrons is now favored across the tunnel junction TJ_b into the quantum dot island. The tunneling is also allowed further onward from the island across the tunnel junction TJ_a. Consequently, a net current flows through the circuit.

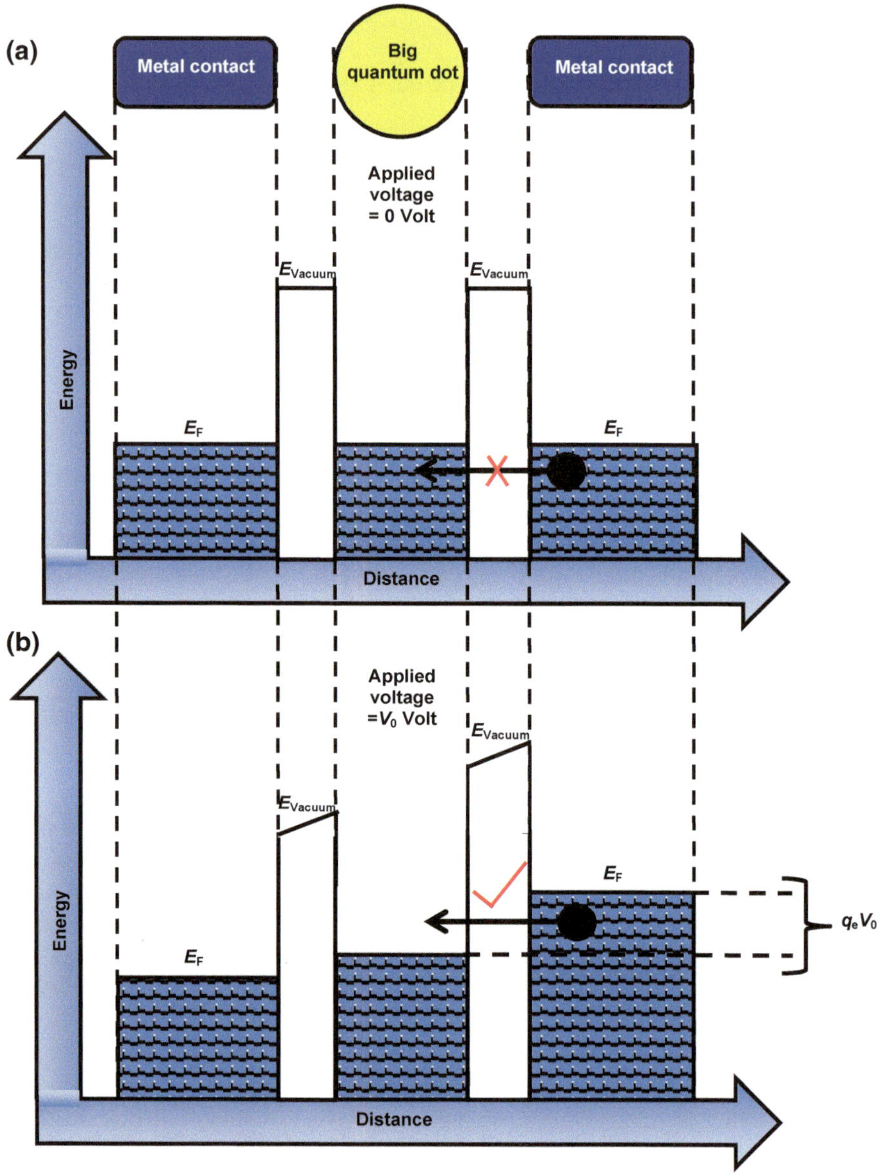

Fig. 13.6 Energy band diagram of a quantum dot circuit with *big quantum dot*: **a** without any applied bias and **b** after applying a bias V_0

13.7.2 Small Quantum Dot

In this quantum dot, the energy levels are discrete (Fig. 13.7). The charging energy, i.e., the energy required to add an electron to the quantum dot by surmounting the Coulomb blockade is represented by an energy gap, which is equally distributed about the Fermi level for symmetric junctions. In this energy gap, there are no states. Above the energy gap, there are empty discrete energy levels into which the electrons can tunnel. Figure 13.7a shows the energy band diagram of the structure without any bias and under thermal equilibrium. In Fig. 13.7b, the band diagram of the same structure is shown when a bias $V_0 >$ charging energy is applied. This bias raises the Fermi level on the side of junction TJ_b so that electrons can tunnel across TJ_b to the vacant states in the QD, and from there across TJ_a.

After one electron has tunneled from the tunnel junction TJ_b into the quantum dot, the energy on the quantum dot is increased by an amount $q_e^2/(2C)$ so that the total energy becomes $q_e^2/(2C) + q_e^2/(2C) = q_e^2/C$. For the second electron, the energy on the quantum dot $= q_e^2/(2C) + q_e^2/(2C) + q_e^2/(2C) = 3q_e^2/2C$. Physically, this can be comprehended by noting that the second electron faces opposition due to the repulsive force of the first electron. Therefore, more energy is required to place it on the quantum dot. The energy bands undergo re-adjustment correspondingly. However, if any electron leaves the quantum dot, the energy will decrease by an amount $= q_e^2/(2C)$.

If a generalized situation is contemplated, after one electron has tunneled to the quantum dot, the change in energy upon tunneling of the second electron is given by

$$\begin{aligned}\Delta E_t &= \Delta E_{se} - W = \{1/(2C_s)\}\left[V_s^2 C_a C_b + \{(n+1)q_e\}^2\right] \\ &\quad - \{1/(2C_s)\}\left[V_s^2 C_a C_b + \{(n+2)q_e\}^2\right] - V_s C_a q_e/C_s \\ &= \{1/(2C_s)\}\left\{V_s^2 C_a C_b + (nq_e)^2 + 2nq_e^2 + q_e^2 - V_s^2 C_a C_b - (nq_e)^2 - 4nq_e^2 - 4q_e^2 - 2V_s C_a q_e\right\} \\ &= \{1/(2C_s)\}\{-2nq_e^2 - 3q_e^2 - 2V_s C_a q_e\} \\ &= (1/C_s)\{-nq_e^2 - (3/2)q_e^2 - V_s C_a q_e\} \\ &= (q_e/C_s)[-\{n+(3/2)\}q_e - V_s C_a]\end{aligned} \tag{13.69}$$

For $\Delta E_t > 0$,

$$\begin{aligned}&(q_e/C_s)[-\{n+(3/2)\}q_e - V_s C_a] \\ &\text{or, } -\{n+(3/2)\}q_e - V_s C_a > 0 \\ &\text{or, } |V_s| > 3|q_e|/(2C_a)\end{aligned} \tag{13.70}$$

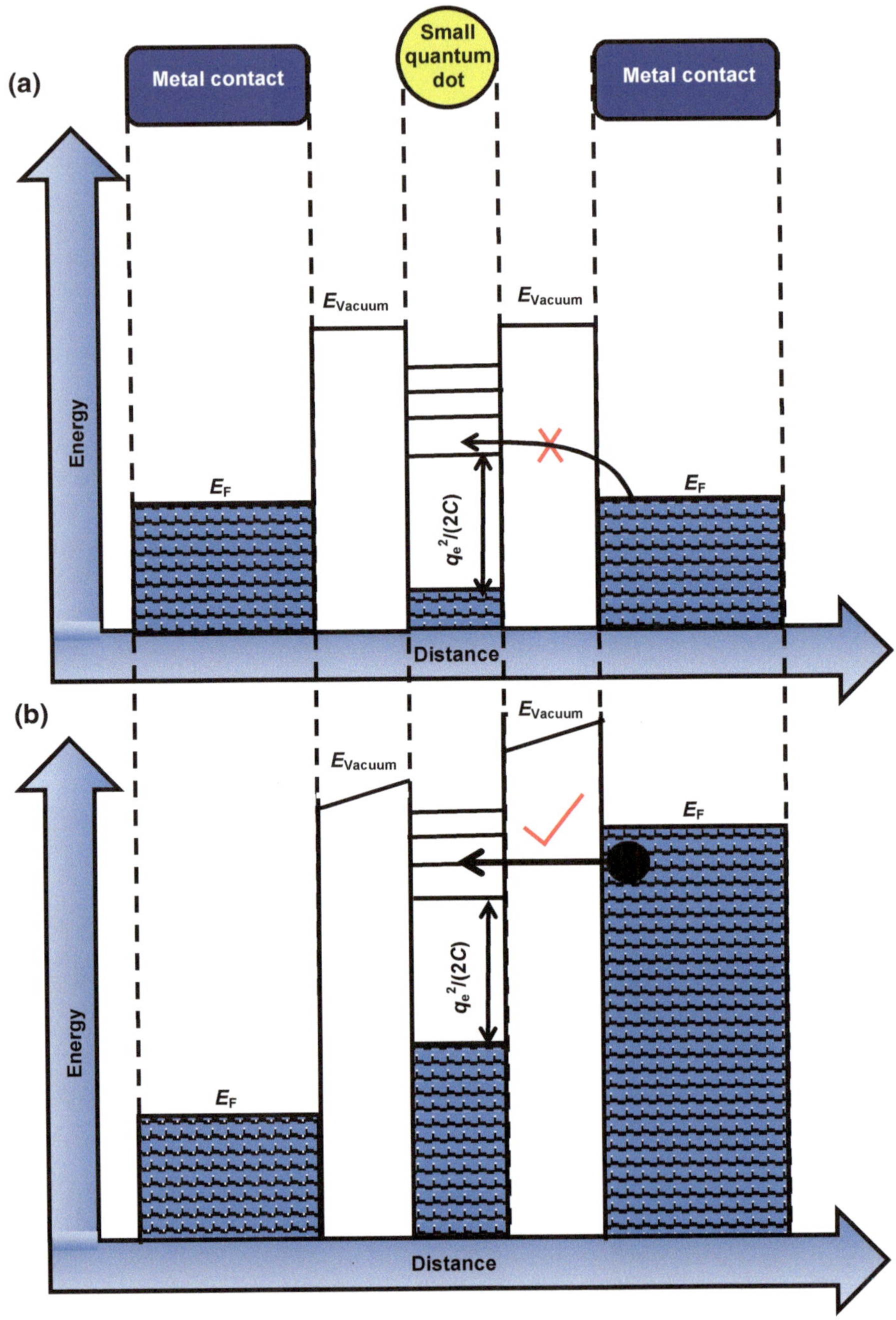

Fig. 13.7 Energy band diagram of a quantum dot circuit with *small quantum dot*: **a** without any applied bias $V_0 = 0$ Volt and **b** after applying a bias $V_0 > q_e^2/(2C)$

Similarly, for the third electron,

$$|V_s| > 5|q_e|/(2C_a) \tag{13.71}$$

For the fourth electron,

$$|V_s| > 7|q_e|/(2C_a) \tag{13.72}$$

Thus electron tunneling events and consequent electric current flow occurs at discrete steps of voltage

$$|V_s| > m|q_e|/(2C_a) \tag{13.73}$$

where m is an integer = 1, 3, 5, … for the tunneling of 1st, 2nd, 3rd, … electrons. The current–voltage characteristic generated by consecutive electron tunneling events has the contour of a staircase. This staircase is known as Coulomb staircase (Fig. 13.8).

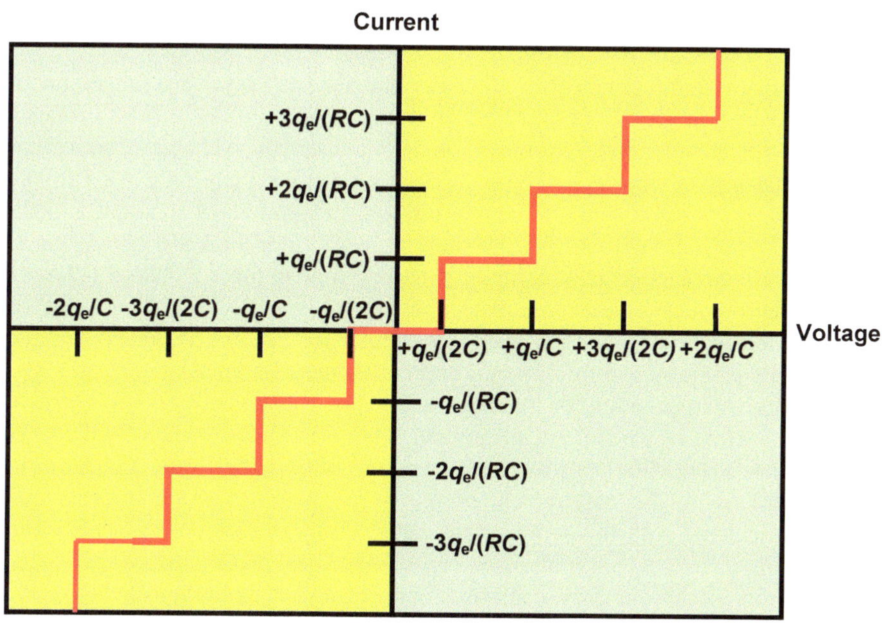

Fig. 13.8 Staircase-like current–voltage characteristic of a quantum dot circuit

13.8 Discussion and Conclusions

The large difference between the energies necessary to place an electronic charge on one plate of a capacitor and equivalent positive charge on its opposite plate reveals the dissimilarity between nanoscopic world and microscopic and milli-size worlds. Coulomb blockade effect is manifested as a voltage requirement for tunneling of electronic charge across a capacitor dielectric. It is only perceivable at room temperature at the nanoscale. In order that this effect is observed in a tunnel junction, the tunnel resistance must exceed 4.1 kΩ. When connected across a constant current source, a tunnel junction circuit produces single electron oscillations. In resemblance to a nanocapacitor, a quantum dot circuit shows Coulomb blockade phenomena. For small size quantum dots, electron tunneling events take place at disconnected voltage stages. This discreteness imparts the current–voltage characteristics the shape of a stairway. This ladder is called the Coulomb staircase. Thus, a multitude of phenomena were explained which happen exclusively at nanometer dimensions.

Review Exercises

13.1 What is a tunnel junction? How does the dielectric layer in a tunnel junction behave differently from a thick bulk dielectric?

13.2 Write the equation for the work done in transferring a charge Q between the plates of a capacitor. Three capacitors have the plate area A as follows: (i) $A = 5\text{ nm} \times 5\text{ nm}$; (ii) $A = 5\ \mu\text{m} \times 5\ \mu\text{m}$; and (iii) $A = 5\text{ mm} \times 5\text{ mm}$. All these capacitors have the same dielectric film thickness $d = 1$ nm. For these three capacitors, compare the work done to place a charge $+Q$ on one plate and $-Q$ on the other plate where $Q = 1.6 \times 10^{-19}$ C. What significant conclusion do you elicit from this comparison? Hence, differentiate nanoscale phenomenon from micro- and milliscale phenomena.

13.3 Derive an equation for the magnitude $|V|$ of voltage necessary to enable the tunneling of a charge of magnitude = the electronic charge = $|q_e|$ from one plate to the other plate of a nanocapacitor. What is the physical phenomenon underlying the requirement of this voltage? What is the relevant effect associated with this requirement called? Why is such an effect not noticeable in micro- and milli-sized capacitors?

13.4 From thermal energy considerations, show that Coulomb blockade effect will be observed for the nanocapacitor of Ex. 13.2 at room temperature but not for the micro- and milli-sized capacitors.

13.5 Reverse the line of thinking of Ex. 13.4. Supposing that the transference of electrons is activated by thermal energy, show that the number of electrons transferred at room temperature (300 K) decreases as the capacitor is made smaller in size.

13.6 How is a tunnel junction modeled? Is tunnel resistance an ohmic resistance? How does it originate? Show that Coulomb blockade will be observable only for tunnel resistance values exceeding 4.1 kΩ.

13.7 How does the allowance of tunneling across the dielectric of a tunnel junction make it behave differently from a normal capacitor? What happens when a tunnel junction is connected across a constant current supply?

13.8 Derive an equation for the energy stored in a quantum dot circuit in terms of voltages V_a and V_b across the tunnel capacitors C_a and C_b, the capacitance $C_s = C_a + C_b$, and the supply voltage V_s. Hence, show that for electron tunneling across the tunnel junction TJ_b,

$$V_s > -(q_e/C_a)(n + 1/2)$$

while for electron tunneling across the tunnel junction TJ_a

$$V_s > -(q_e/C_b)(1/2 - n)$$

13.9 Draw and explain the energy band diagram of a quantum dot circuit for a large quantum dot. Draw the same for a small size quantum dot and point out the difference between the energy band diagrams for the two cases.

13.10 What is Coulomb staircase? Explain the origin of Coulomb staircase from the discrete nature of energy band diagram of a quantum dot circuit with a small size quantum dot.

References

1. Wasshuber C, 1 Introduction: what is single electronics? http://www.iue.tuwien.ac.at/phd/wasshuber/node10.html. Accessed 1 April 2016
2. EEE5425 Introduction to nanotechnology, 2/17/2010, © Nezih Pala npala@fiu.edu, Coulomb Blockade and Single Electron Transistors. http://web.eng.fiu.edu/npala/EEE5425/EEE5425_Ch6_Coulomb%20Blockade%20and%20SETs_Handouts.pdf, Also, https://www.google.co.in/url?sa=t&rct=j&q=&esrc=s&source=web&cd=2&cad=rja&uact=8&ved=0ahUKEwi18eeLyuLKAhXBWI4KHbS2DpMQFgghMAE&url=http%3A%2F%2Fweb.eng.fiu.edu%2Fnpala%2FEEE5425%2520PPTs%2FEEE5425_6_Coulomb_Blockade_SETs_v1.pptx&usg=AFQjCNEioCePJIEUn5daa0Wt2SLXIbiOBA&bvm=bv.113370389,d.c2E. Accessed 6 Feb 2016
3. Beenakker CWJ (1991-II) Theory of Coulomb-blockade oscillations in the conductance of a quantum dot. Phys Rev B 44(4):1646–1656. doi:http://dx.doi.org/10.1103/PhysRevB.44.1646
4. Leobandung E, Guo L, Wang Y et al (1995) Observation of quantum effects and Coulomb blockade in silicon quantum-dot transistors at temperatures over 100 K. Appl Phys Lett 67(7):938–940
5. Park J, Pasupathy AN, Goldsmith JI et al (2002) Coulomb blockade and the Kondo effect in single-atom transistors. Nature 417:722–725. doi:10.1038/nature00791

Chapter 14
Single Electronics

Abstract Operational principle of single electron transistor is outlined. Based on energy band diagram, the influence of the gate voltage on the drain-source voltage for tunneling from source to drain is expounded. Derivation of the equation for total energy stored in the capacitors comprising a single electron transistor is presented. From energy viewpoint, necessary conditions favoring electron tunneling are deduced. These electron tunneling processes take place from one tunnel junction into the quantum dot island, and thereafter from this island across the other tunnel junction. Considering opposite sequences of operation of electron tunneling across tunnel junctions into and away from the quantum dot island, symmetrically placed triangular regions are sketched and combined to show Coulomb diamonds. Logic circuit operation based on single electron transistors is introduced. The reader is familiarized with voltage-based logic and charge-based logic used with SETs. Restrictions on increasing the low voltage gain of SETs are discussed. Elimination of the requirement of separately fabricating complementary SETs is both an advantage and a disadvantage. Difficulties faced in straightway adoption of CMOS logic circuits for SET logic are indicated. Operation of voltage-logic-based SET AND, NOT, and OR gates is described. Other applications of SETs as a supersensitive electrometer, as a standard of direct current and for IR detection are briefly touched upon.

14.1 Introduction

Single electronics deals with controlling the transport of a single/small number of electrons. The most important member of the family of single electronic devices is the single electron transistor [1]. A single electron transistor is a transistor, which turns on and off every time that a single electron is added to it. Construction wise, it consists of a central island, usually made of a metallic film, serving as a quantum dot. This island is coupled to the source and drain leads through two tunnel junctions [2]. A tunnel junction consists of a thin insulating film sandwiched between two conducting layers. To form the two tunnel junctions on the two sides

V.K. Khanna, *Integrated Nanoelectronics*,
NanoScience and Technology, DOI 10.1007/978-81-322-3625-2_14

of the central metal island, this island is separated from the source and drain leads through thin insulating films.

The central island is also coupled to the gate electrode via capacitive action. For this action, it is covered with a dielectric layer on which metal film is deposited. Thus, a single electron transistor is a three-terminal, nanoelectronic device in which a capacitively coupled input voltage signal applied at the gate terminal controls the current flowing between source and drain terminals [3]. To thoroughly understand its operation, let us begin with the phenomenon of Coulomb blockade in this structure and proceed further step-by-step.

14.2 Single Electron Transistor Action

The circuit being studied consists of a metallic quantum dot QD. This QD is coupled to two external leads through tunnel junctions TJ_a and TJ_b, as shown in Fig. 14.1a. Compare it with a common MOSFET shown in Fig. 14.1b. The equivalent circuit of the single electron transistor is shown in Fig. 14.1c. The tunnel junctions are formed by insulating regions separating the quantum dot from the metal contacts on which the leads are fixed. These junctions have a metal/insulator/metal structure. To this double tunnel junction, a gate terminal is added for additional control. This gate terminal is isolated from the QD island by an ideal non-tunneling capacitance. Thus, the single electron transistor is a Coulomb blockade structure in which gate action is provided. It differs from the quantum dot circuit discussed in previous chapter with regard to the provision of this gate terminal.

Figure 14.2 shows the application of biasing voltages on a SET device along with its equivalent circuit in the biased condition.

Refer to the energy band diagrams of SET shown in Fig. 14.3. To the gate terminal of the SET, a voltage V_g is applied. Application of a positive gate voltage $V_g > 0$, brings down the Fermi level E_F on the quantum dot island. A negative voltage $V_g < 0$, raises E_F. On the energy band diagram, the charging energy is represented by an energy gap or Coulomb blockade gap of magnitude $q_e^2/2C$, which is located equally on the opposite sides of the Fermi level. By choosing the proper value of gate bias, the upper limit of the energy gap can be shifted above, below or made to align with the Fermi levels on the gate. This means that the drain-source voltage necessary to push a current through the single electron transistor is dependent on the gate bias V_g.

The tunnel junctions may be substituted by their equivalent circuit models, viz., an ideal capacitance C in parallel with a tunnel resistance R. For the tunnel junction TJ_a, the capacitance is C_a and tunnel resistance is R_a. For the tunnel junction TJ_b, the corresponding symbols are C_b and R_b. Let V_s be the supply voltage of the circuit. Also, let V_a^i and V_b^i denote the voltages across the tunnel junctions TJ_a and

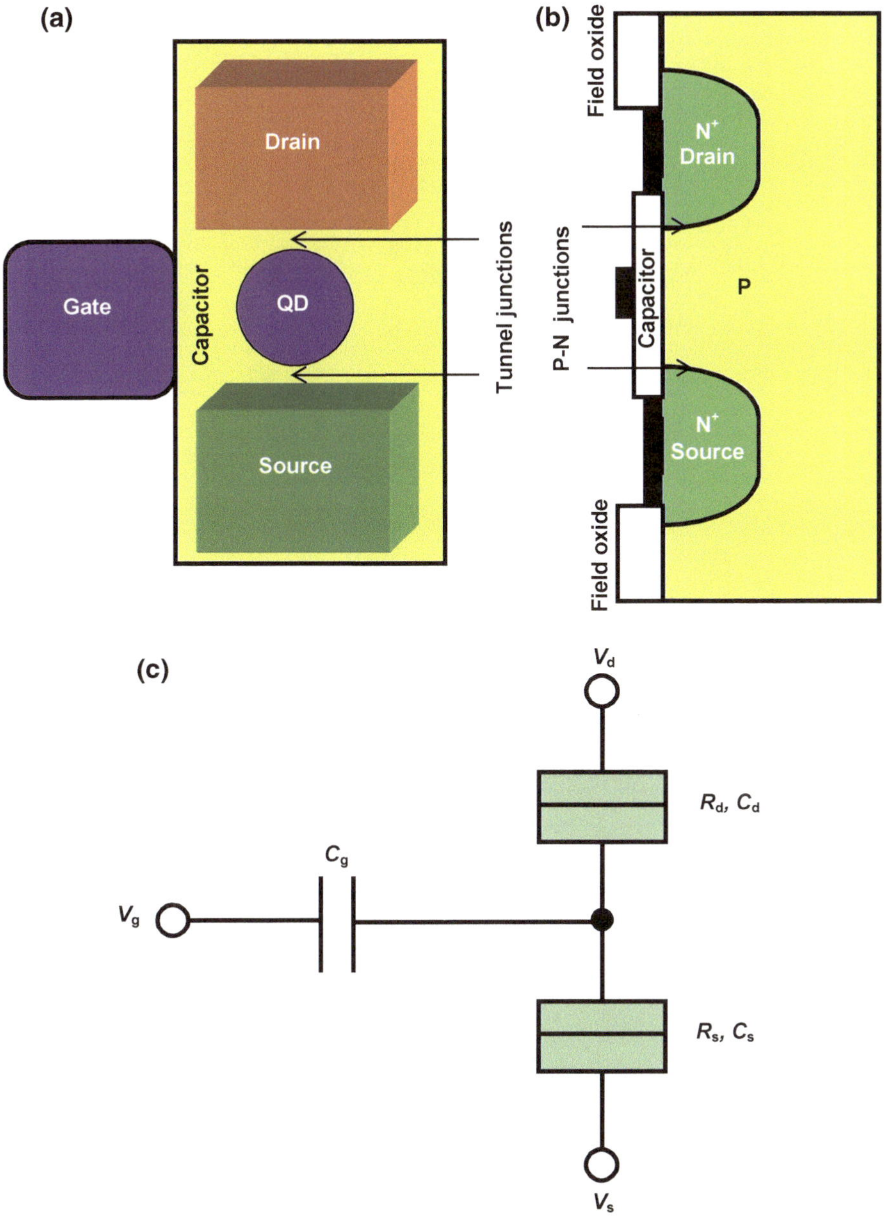

Fig. 14.1 **a** Single electron transistor, **b** MOSFET and **c** equivalent circuit of SET

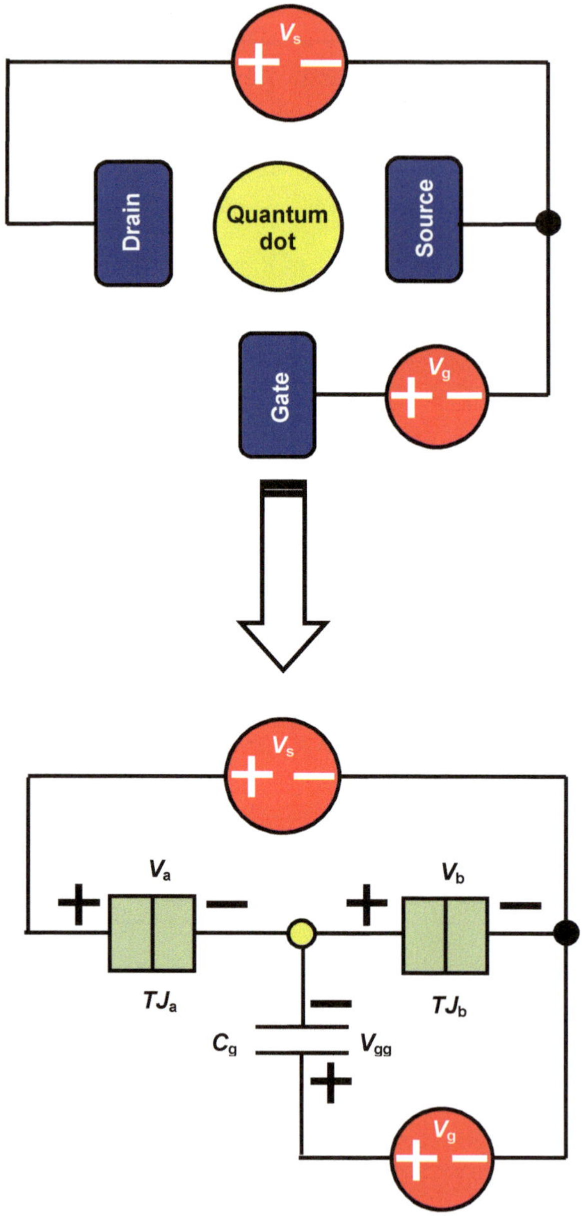

Fig. 14.2 A single electron transistor circuit with biasing voltages (*top*), and its equivalent circuit diagram (*bottom*) in biased condition

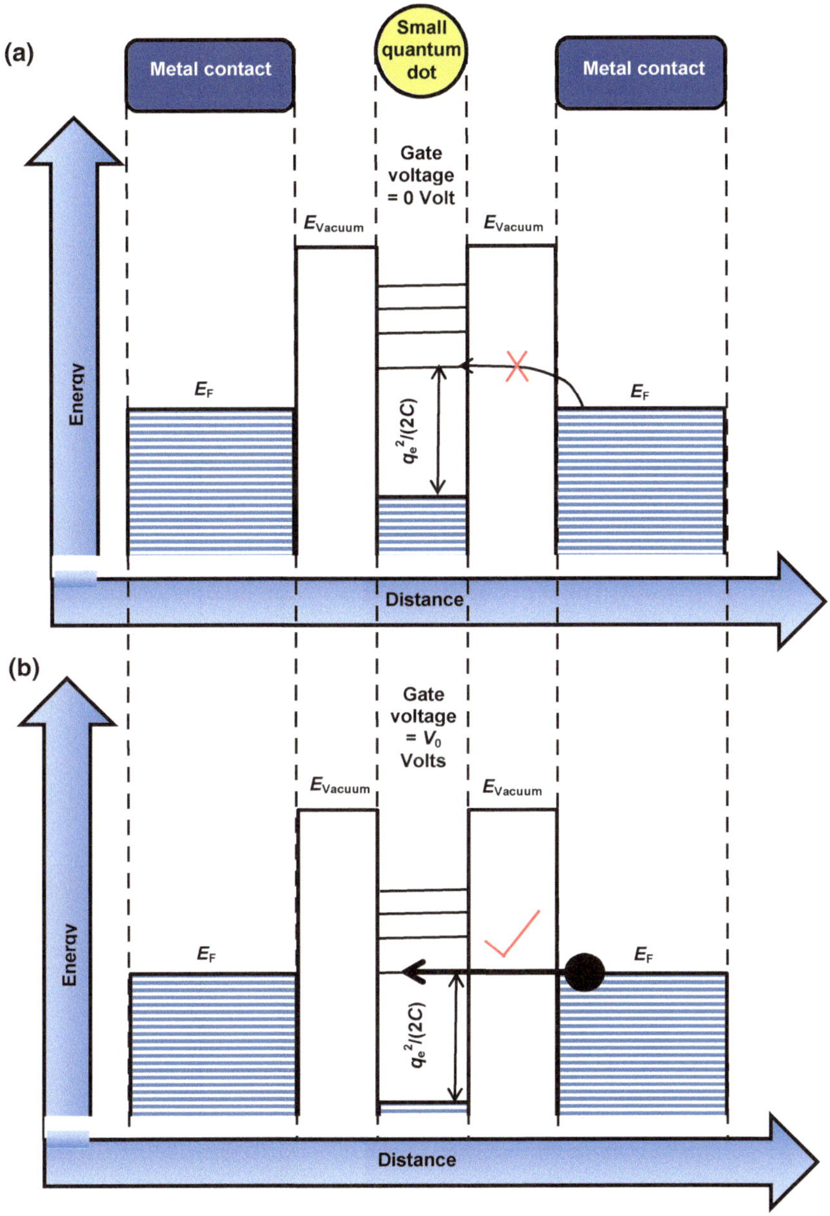

Fig. 14.3 Energy band diagram of a single electron transistor when: **a** no gate voltage is applied, and **b** a positive gate voltage = V_0 Volts is applied; the voltage V_0 depresses the Coulomb blockade gap

TJ_b, respectively. If the charges associated with the tunnel junctions are Q_a^i, Q_b^i, we may write

$$Q_a^i = C_a V_a^i \tag{14.1}$$

$$Q_b^i = C_b V_b^i \tag{14.2}$$

If Q_g^i is the charge on the gate terminal, C_g is the gate capacitance and V_g^i is the gate voltage,

$$Q_g^i = C_g\left(V_g^i - V_b^i\right) \tag{14.3}$$

Hence, the total charge on the QD island is

$$Q^i = Q_b^i - Q_a^i - Q_g^i \tag{14.4}$$

By tunneling, electrons are transferred and stored on the QD island. Suppose when the experiment had begun, a discrete number n of electrons had already tunneled into the island. So, n is the initial number of electrons, and

$$Q^i = nq_e \tag{14.5}$$

The total energy stored in the capacitors C_a, C_b, C_g is given by

$$\begin{aligned} E_{se} &= \left(Q_a^i\right)^2/(2C_a) + \left(Q_b^i\right)^2/(2C_b) + \left(Q_g^i\right)^2/(2C_g) \\ &= (1/2)\left\{\left(Q_a^i\right)^2/C_a + \left(Q_b^i\right)^2/C_b + \left(Q_g^i\right)^2/C_g\right\} \end{aligned} \tag{14.6}$$

Application of Kirchoff's law to the quantum dot circuit yields

$$V_g^i = V_{gg}^i + V_b^i = V_{gg}^i + Q_b^i/C_b = Q_g^i/C_g + Q_b^i/C_b \tag{14.7}$$

$$V_s = V_a^i + V_b^i = Q_a^i/C_a + Q_b^i/C_b \tag{14.8}$$

The values of Q_a^i, Q_b^i and Q_g^i are obtained as

$$\begin{aligned} Q_a^i &= C_a\left(V_s - Q_b^i/C_b\right) = C_a\left\{\left(V_s C_b - Q_b^i\right)/C_b\right\} \\ &= C_a\left[\left\{V_s C_b - \left(Q^i + Q_a^i + Q_g^i\right)\right\}/C_b\right] \end{aligned} \tag{14.9}$$

or,

$$\begin{aligned}Q_a^i C_b &= C_a V_s C_b - C_a Q^i - C_a Q_a^i - C_a Q_g^i \\ &= C_a V_s C_b - C_a Q^i - C_a Q_a^i - C_a C_g \left(V_g^i - V_b^i\right) \\ &= C_a V_s C_b - C_a Q^i - C_a Q_a^i - C_a C_g V_g^i + C_a C_g V_b^i \\ &= C_a V_s C_b - C_a Q^i - C_a Q_a^i - C_a C_g V_g^i + C_a C_g \left(V_s - Q_a^i / C_a\right)\end{aligned} \tag{14.10}$$

or,

$$Q_a^i C_b C_a = C_a^2 V_s C_b - C_a^2 Q^i - C_a^2 Q_a^i - C_a^2 C_g V_g^i + C_a^2 C_g V_s - C_a C_g Q_a^i$$

or,

$$Q_a^i C_b = C_a V_s C_b - C_a Q^i - C_a Q_a^i - C_a C_g V_g^i + C_a C_g V_s - C_g Q_a^i$$

or,

$$\begin{aligned}Q_a^i \left(C_b + C_a + C_g\right) &= C_a V_s C_b - C_a Q^i - C_a C_g V_g^i + C_a C_g V_s \\ &= C_a \left\{V_s \left(C_b + C_g\right) - C_g V_g^i - Q^i\right\}\end{aligned} \tag{14.11}$$

or,

$$Q_a^i C_s = C_a \left\{V_s \left(C_b + C_g\right) - C_g V_g^i - Q^i\right\} \tag{14.12}$$

where

$$C_s = C_a + C_b + C_g \tag{14.13}$$

or,

$$Q_a^i = C_a \left\{V_s \left(C_b + C_g\right) - C_g V_g^i - Q^i\right\} / C_s \tag{14.14}$$

Similarly,

$$\begin{aligned}Q_b^i &= C_b \left(V_s - Q_a^i / C_a\right) = C_b \left\{\left(V_s C_a - Q_a^i\right) / C_a\right\} \\ &= C_b \left[\left\{V_s C_a - \left(Q_b^i - Q_g^i - Q^i\right)\right\} / C_a\right]\end{aligned} \tag{14.15}$$

or,

$$
\begin{aligned}
Q_b^i C_a &= C_b V_s C_a - C_b Q_b^i + C_b Q_g^i + C_b Q^i \\
&= C_b V_s C_a - C_b Q_b^i + C_b C_g \left(V_g^i - V_b^i\right) + C_b Q^i \\
&= C_b V_s C_a - C_b Q_b^i + C_b C_g V_g^i - C_b C_g V_b^i + C_b Q^i \\
&= C_b V_s C_a - C_b Q_b^i + C_b C_g V_g^i - Q_b^i C_g + C_b Q^i
\end{aligned}
\tag{14.16}
$$

Since

$$C_b V_b^i = Q_b^i \tag{14.17}$$

or,

$$Q_b^i C_a + C_b Q_b^i + Q_b^i C_g = C_b V_s C_a + C_b C_g V_g^i + C_b Q^i$$

or,

$$Q_b^i (C_a + C_b + C_g) = C_b \left(V_s C_a + C_g V_g^i + Q^i\right)$$

or,

$$Q_b^i C_s = C_b \left(V_s C_a + C_g V_g^i + Q^i\right)$$

or,

$$Q_b^i = C_b \left(V_s C_a + C_g V_g^i + Q^i\right) / C_s \tag{14.18}$$

Likewise,

$$Q_g^i = C_g \left(V_g^i - V_b^i\right) = C_g \left[V_g^i - \left\{\left(V_s C_a - Q_a^i\right)/C_a\right\}\right] \tag{14.19}$$

or,

$$
\begin{aligned}
Q_g^i C_a &= C_g V_g^i C_a - C_g V_s C_a + C_g Q_a^i \\
&= C_g V_g^i C_a - C_g V_s C_a + C_g C_a \left\{V_s (C_b + C_g) - C_g V_g^i - Q^i\right\} / C_s
\end{aligned}
\tag{14.20}
$$

by putting Q_a^i from Eq. (14.14).

$$
\begin{aligned}
\therefore Q_g^i C_a C_s &= C_g V_g^i C_a C_s - C_g V_s C_a C_s + C_g C_a \left\{V_s (C_b + C_g) - C_g V_g^i - Q^i\right\} \\
&= C_g V_g^i C_a C_s - C_g V_s C_a C_s + C_g C_a \left(V_s C_b + V_s C_g - C_g V_g^i - Q^i\right) \\
&= C_g V_g^i C_a C_s - C_g V_s C_a C_s + C_g C_a V_s C_b + C_g^2 C_a V_s - C_g^2 C_a V_g^i - C_g C_a Q^i
\end{aligned}
\tag{14.21}
$$

or,

$$\begin{aligned} Q_{\mathrm{g}}^{i} C_{\mathrm{s}} &= C_{\mathrm{g}} V_{\mathrm{g}}^{i} C_{\mathrm{s}} - C_{\mathrm{g}} V_{\mathrm{s}} C_{\mathrm{s}} + C_{\mathrm{g}} V_{\mathrm{s}} C_{b} + C_{\mathrm{g}}^{2} V_{\mathrm{s}} - C_{\mathrm{g}}^{2} V_{\mathrm{g}}^{i} - C_{\mathrm{g}} Q^{i} \\ &= C_{\mathrm{g}} \left(V_{\mathrm{g}}^{i} C_{\mathrm{s}} - V_{\mathrm{s}} C_{\mathrm{s}} + C_{\mathrm{g}} V_{\mathrm{s}} + V_{\mathrm{s}} C_{b} - C_{\mathrm{g}} V_{\mathrm{g}}^{i} - Q^{i} \right) \\ &= C_{\mathrm{g}} \left\{ V_{\mathrm{g}}^{i} (C_{\mathrm{s}} - C_{\mathrm{g}}) - V_{\mathrm{s}} (C_{\mathrm{s}} - C_{\mathrm{g}} - C_{b}) - Q^{i} \right\} \\ &= C_{\mathrm{g}} \left\{ V_{\mathrm{g}}^{i} (C_{a} + C_{b}) - V_{\mathrm{s}} (C_{a}) - Q^{i} \right\} \end{aligned} \tag{14.22}$$

by putting

$$C_{\mathrm{s}} - C_{\mathrm{g}} = C_{a} + C_{b} \tag{14.23}$$

and

$$C_{\mathrm{s}} - C_{\mathrm{g}} - C_{b} = C_{a} \tag{14.24}$$

because

$$C_{\mathrm{s}} = C_{a} + C_{b} + C_{\mathrm{g}} \tag{14.25}$$

$$\therefore Q_{\mathrm{g}}^{i} = C_{\mathrm{g}} \left\{ V_{\mathrm{g}}^{i} (C_{a} + C_{b}) - V_{\mathrm{s}} (C_{a}) - Q^{i} \right\} / C_{\mathrm{s}} \tag{14.26}$$

Substituting the values of $Q_{a}^{i}, Q_{b}^{i}, Q_{\mathrm{g}}^{i}$, from Eqs. (14.14), (14.15), and (14.19), the equation for energy stored by the capacitors C_{a}, C_{b}, C_{g} becomes

$$\begin{aligned} E_{\mathrm{se}} = & \left[C_{a} \left\{ V_{\mathrm{s}} (C_{b} + C_{\mathrm{g}}) - C_{\mathrm{g}} V_{\mathrm{g}}^{i} - Q^{i} \right\} / C_{\mathrm{s}} \right]^{2} / (2C_{a}) \\ & + \left[C_{b} \left(V_{\mathrm{s}} C_{a} + C_{\mathrm{g}} V_{\mathrm{g}}^{i} + Q^{i} \right) / C_{\mathrm{s}} \right]^{2} / (2C_{b}) \\ & + \left[C_{\mathrm{g}} \left\{ V_{\mathrm{g}}^{i} (C_{a} + C_{b}) - V_{\mathrm{s}} (C_{a}) - Q^{i} \right\} / C_{\mathrm{s}} \right]^{2} / (2C_{\mathrm{g}}) \\ = & \ C_{a} \left\{ V_{\mathrm{s}} (C_{b} + C_{\mathrm{g}}) - C_{\mathrm{g}} V_{\mathrm{g}}^{i} - Q^{i} \right\}^{2} / (2C_{\mathrm{s}}^{2}) \\ & + C_{b} \left(V_{\mathrm{s}} C_{a} + C_{\mathrm{g}} V_{\mathrm{g}}^{i} + Q^{i} \right)^{2} / (2C_{\mathrm{s}}^{2}) + C_{\mathrm{g}} \left\{ V_{\mathrm{g}}^{i} (C_{a} + C_{b}) - V_{\mathrm{s}} (C_{a}) - Q^{i} \right\}^{2} / (2C_{\mathrm{s}}^{2}) \\ = & \ C_{a} \left\{ V_{\mathrm{s}} C_{b} + V_{\mathrm{s}} C_{\mathrm{g}} - C_{\mathrm{g}} V_{\mathrm{g}}^{i} - Q^{i} \right\}^{2} / (2C_{\mathrm{s}}^{2}) + C_{b} \left(V_{\mathrm{s}} C_{a} + C_{\mathrm{g}} V_{\mathrm{g}}^{i} + Q^{i} \right)^{2} / (2C_{\mathrm{s}}^{2}) \\ & + C_{\mathrm{g}} \left\{ V_{\mathrm{g}}^{i} C_{a} + V_{\mathrm{g}}^{i} C_{b} - V_{\mathrm{s}} C_{a} - Q^{i} \right\}^{2} / (2C_{\mathrm{s}}^{2}) \\ = & \left\{ C_{a} \left\{ V_{\mathrm{s}} C_{b} + V_{\mathrm{s}} C_{\mathrm{g}} - C_{\mathrm{g}} V_{\mathrm{g}}^{i} - Q^{i} \right\}^{2} + C_{b} \left(V_{\mathrm{s}} C_{a} + C_{\mathrm{g}} V_{\mathrm{g}}^{i} + Q^{i} \right)^{2} + C_{\mathrm{g}} \left\{ V_{\mathrm{g}}^{i} C_{a} + V_{\mathrm{g}}^{i} C_{b} - V_{\mathrm{s}} C_{a} - Q^{i} \right\}^{2} \right\} / (2C_{\mathrm{s}}^{2}) \end{aligned} \tag{14.27}$$

The numerator of this equation is

$$
\begin{aligned}
&C_aV_s^2C_b^2 + C_aV_s^2C_g^2 + C_aC_g^2V_g^{i^2} + C_aQ^{i^2} + 2V_s^2C_aC_bC_g \\
&- 2C_aV_sC_bC_gV_g^i - 2V_sC_aC_bQ^i - 2V_sC_aC_g^2V_g^i - 2C_aC_gV_sQ^i \\
&+ 2C_aC_gV_g^iQ^i + V_s^2C_a^2C_b + C_bC_g^2V_g^{i^2} + C_bQ^{i^2} + 2C_aC_bV_sC_gV_g^i \\
&+ 2C_bC_gV_g^iQ^i + 2C_bV_sC_aQ^i + C_gC_a^2V_g^{i^2} + C_gV_g^{i^2}C_b^2 + C_gV_s^2C_a^2 \\
&+ C_gQ^{i^2} + 2C_gV_g^{i^2}C_aC_b - 2C_gV_g^iV_sC_a^2 - 2C_gV_g^iC_aQ^i \\
&- 2C_gV_g^iC_bV_sC_a - 2C_gV_g^iC_bQ^i + 2C_gV_sC_aQ^i \\
&= C_aC_gV_s^2(C_g + 2C_b + C_a) + C_aC_gV_g^{i^2}(C_g + C_a + 2C_b) \\
&- 2C_aC_gV_sV_g^i(C_g + C_a + C_b) + C_aV_s^2C_b(C_b + C_a) \\
&+ C_bC_gV_g^{i^2}(C_b + C_g) + Q^{i^2}(C_a + C_b + C_g) \\
&= C_aC_gV_s^2(C_g + C_b + C_a) + C_aC_gV_g^{i^2}(C_g + C_a + C_b) \\
&- 2C_aC_gV_sV_g^i(C_g + C_a + C_b) + C_aV_s^2C_b(C_b + C_a) \\
&+ C_bC_gV_g^{i^2}(C_b + C_g) + Q^{i^2}(C_a + C_b + C_g) \\
&+ C_aC_gV_s^2C_b + C_aC_gV_g^{i^2}C_b \\
&= C_aC_gV_s^2(C_g + C_b + C_a) + C_aC_gV_g^{i^2}(C_g + C_a + C_b) \\
&- 2C_aC_gV_sV_g^i(C_g + C_a + C_b) + C_aV_s^2C_b(C_b + C_a) \\
&+ C_bC_gV_g^{i^2}(C_b + C_g) + Q^{i^2}(C_a + C_b + C_g) \\
&+ C_aC_gV_s^2C_b + C_aC_gV_g^{i^2}C_b \\
&= C_aC_gV_s^2(C_g + C_b + C_a) + C_aC_gV_g^{i^2}(C_g + C_a + C_b) \\
&- 2C_aC_gV_sV_g^i(C_g + C_a + C_b) + C_aV_s^2C_b(C_b + C_a + C_g) \\
&+ C_bC_gV_g^{i^2}(C_b + C_g + C_a) + Q^{i^2}(C_a + C_b + C_g) \\
&= C_aC_gV_s^2C_s + C_aC_gV_g^{i^2}C_s - 2C_aC_gV_sV_g^iC_s \\
&+ C_aV_s^2C_bC_s + C_bC_gV_g^{i^2}C_s + Q^{i^2}C_s \\
&= C_s\left\{C_aC_gV_s^2 + C_aC_gV_g^{i^2} - 2C_aC_gV_sV_g^i + C_aV_s^2C_b + C_bC_gV_g^{i^2} + Q^{i^2}\right\} \\
&= C_s\left\{C_aC_g\left(V_s^2 + V_g^{i^2} - 2V_sV_g^i\right) + C_aC_bV_s^2 + C_bC_gV_g^{i^2} + Q^{i^2}\right\} \\
&= C_s\left\{C_aC_g\left(V_s - V_g^i\right)^2 + C_aC_bV_s^2 + C_bC_gV_g^{i^2} + Q^{i^2}\right\}
\end{aligned}
\tag{14.28}
$$

$$
\begin{aligned}
\therefore E_{se} &= C_s\left\{C_aC_g\left(V_s - V_g^i\right)^2 + C_aC_bV_s^2 + C_bC_gV_g^{i^2} + Q^{i^2}\right\}/(2C_s^2) \\
&= \left\{C_aC_g\left(V_s - V_g^i\right)^2 + C_aC_bV_s^2 + C_bC_gV_g^{i^2} + Q^{i^2}\right\}/(2C_s)
\end{aligned}
\tag{14.29}
$$

If an electron tunnels into the quantum dot through TJ_b, the stored energy changes by an amount

$$\begin{aligned}\Delta E_{\mathrm{se}} &= \left[(nq_{\mathrm{e}})^2 - \{(n+1)q_{\mathrm{e}}\}^2\right]/(2C_{\mathrm{s}}) \\ &= \left(n^2q_{\mathrm{e}}^2 - n^2q_{\mathrm{e}}^2 - 2nq_{\mathrm{e}}^2 - q_{\mathrm{e}}^2\right)/(2C_{\mathrm{s}}) = -q_{\mathrm{e}}^2(2n+1)/(2C_{\mathrm{s}})\end{aligned} \tag{14.30}$$

As a consequence, the charges on TJ_a and the gate undergo changes. The charge on TJ_a changes by ΔQ_a given by the expression

$$\Delta Q_a = C_a q_{\mathrm{e}}/C_{\mathrm{s}} \tag{14.31}$$

and the charge on the gate changes by

$$\Delta Q_{\mathrm{g}} = C_{\mathrm{g}} q_{\mathrm{e}}/C_{\mathrm{s}} \tag{14.32}$$

The amount of work accompanying the change in charge ΔQ_a is

$$W_{\mathrm{a}} = \Delta Q_{\mathrm{a}} V_{\mathrm{s}} = C_{\mathrm{a}} q_{\mathrm{e}} V_{\mathrm{s}}/C_{\mathrm{s}} \tag{14.33}$$

and the work associated with the change in charge ΔQ_{g} is

$$W_{\mathrm{g}} = \Delta Q_{\mathrm{g}} V_{\mathrm{g}} = C_{\mathrm{g}} q_{\mathrm{e}} V_{\mathrm{g}}/C_{\mathrm{s}} \tag{14.34}$$

In order that the tunneling is favored from energy viewpoint

$$\Delta E_{\mathrm{se}} - \left(W_a + W_{\mathrm{g}}\right) > 0 \tag{14.35}$$

or,

$$-q_{\mathrm{e}}^2(2n+1)/(2C_{\mathrm{s}}) - C_a q_{\mathrm{e}} V_{\mathrm{s}}/C_{\mathrm{s}} - C_{\mathrm{g}} q_{\mathrm{e}} V_{\mathrm{g}}/C_{\mathrm{s}} > 0$$

or,

$$q_{\mathrm{e}}(2n+1)/(2C_{\mathrm{s}}) + C_a V_{\mathrm{s}}/C_{\mathrm{s}} + C_{\mathrm{g}} V_{\mathrm{g}}/C_{\mathrm{s}} > 0$$

or,

$$q_{\mathrm{e}}(2n+1)/2 + C_a V_{\mathrm{s}} + C_{\mathrm{g}} V_{\mathrm{g}} > 0$$

or,

$$q_{\mathrm{e}}(n+1/2) + C_a V_{\mathrm{s}} + C_{\mathrm{g}} V_{\mathrm{g}} > 0 \tag{14.36}$$

For the tunneling of an electron off the quantum dot across TJ_a, the change in energy is

$$\begin{aligned}\Delta E_{se} &= \left[(nq_e)^2 - \{(n-1)q_e\}^2\right]/(2C_s) \\ &= \left(n^2q_e^2 - n^2q_e^2 + 2nq_e^2 - q_e^2\right)/(2C_s) \\ &= -q_e^2(-2n+1)/(2C_s) = -q_e^2(-n+1/2)/C_s\end{aligned} \tag{14.37}$$

The charge on TJ_b changes by

$$\Delta Q_b = \left(C_b + C_g\right)q_e/C_s \tag{14.38}$$

As before, the charge on the gate changes by

$$\Delta Q_g = C_g q_e/C_s \tag{14.39}$$

The quantity of work for the change in charge ΔQ_b is

$$W_b = \Delta Q_b V_s = \left(C_b + C_g\right)q_e V_s/C_s \tag{14.40}$$

Moreover, as in previous case, the quantity of work for the change in charge ΔQ_g is

$$W_g = \Delta Q_g V_g = C_g q_e V_g/C_s \tag{14.41}$$

The tunneling of an electron off the quantum dot across TJ_a is energetically favored provided

$$\Delta E_{se} - \left(W_b - W_g\right) > 0 \tag{14.42}$$

or,

$$-q_e^2(-n+1/2)/C_s - \left(C_b + C_g\right)q_e V_s/C_s + C_g q_e V_g/C_s > 0$$

or,

$$q_e(-n+1/2) + \left(C_b + C_g\right)V_s - C_g V_g > 0 \tag{14.43}$$

Let us assume that in the beginning, i.e., for $n = 0$, the quantum dot is electrically neutral, and an electron tunnels across the junction TJ_b into the quantum dot. Then from Eq. (14.36), the potential required for tunneling to take place is

$$q_e \times (0 + 1/2) + C_a V_s + C_g V_g > 0 \tag{14.44}$$

or,

$$q_e/2 + C_a V_s + C_g V_g > 0$$

or,

$$\begin{aligned} C_a V_s &> -q_e/2 - C_g V_g \\ V_s &> -(1/C_a)(q_e/2 + C_g V_g) \end{aligned} \tag{14.45}$$

Region of V_s-V_g plane defined by this equation is drawn in Fig. 14.4a.

When $n = 1$, the electron will off the quantum dot across the junction TJ_a. From Eq. (14.43), the potential required for this transition is

$$q_e(-1 + 1/2) + (C_b + C_g)V_s - C_g V_g > 0 \tag{14.46}$$

or,

$$(C_b + C_g)V_s - q_e/2 - C_g V_g > 0$$

or,

$$(C_b + C_g)V_s > (q_e/2 + C_g V_g) \tag{14.47}$$

$$\therefore V_s > \{1/(C_b + C_g)\}(q_e/2 + C_g V_g) \tag{14.48}$$

Figure 14.4b shows the region of the V_s-V_g plane represented by this equation. It is easy to see that for observation of a current flow from TJ_b through the quantum dot to TJ_a, both Eqs. (14.45) and (14.48) must be satisfied. Figure 14.4c marks the triangular region of V_s-V_g plane conforming to the two Eqs. (14.45) and (14.48).

Now, in the initial condition, let there be one electron stationed on the QD island. Suppose, a second electron tunnels across the junction TJ_b on to the quantum dot island. Then the voltage required must be

$$V_s > -(1/C_a)(q_e + q_e/2 + C_g V_g) = -(1/C_a)(3q_e/2 + C_g V_g) \tag{14.49}$$

Also, for the electron on QD island to tunnel off the island across the junction TJ_a, the required voltage

$$V_s > \{1/(C_b + C_g)\}(q_e + q_e/2 + C_g V_g) = \{1/(C_b + C_g)\}(3q_e/2 + C_g V_g) \tag{14.50}$$

In this way, we can ponder about tunneling of 3rd, 4th, …, etc., electrons across TJ_b into QD and off QD across TJ_a. In each case, a triangular region in V_s-V_g plane is obtained conforming to the corresponding pair of equations for V_s.

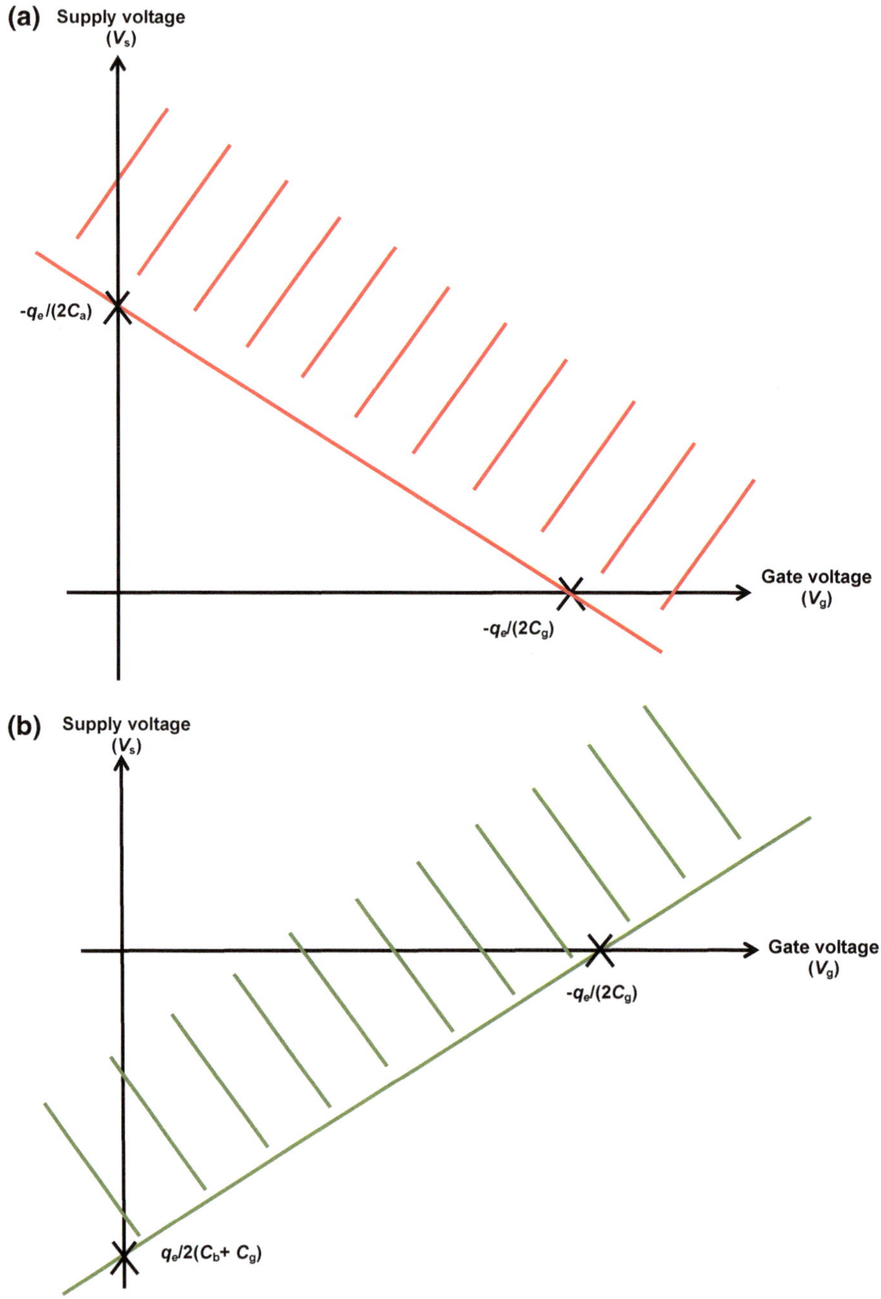

Fig. 14.4 Regions of the V_s-V_g plane representing the condition in which the island is charge neutral and: **a** an electron tunnels into the island through TJ_b, **b** the electron tunnels off the island into TJ_a, and **c** for both tunneling events

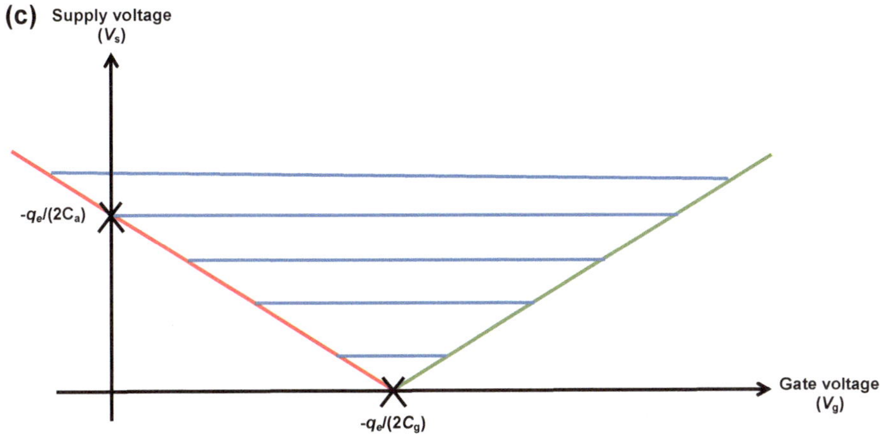

Fig. 14.4 (continued)

Next, consider the opposite sequence of tunneling events. In this sequence, an electron tunnels across the junction TJ_a on to the QD island. Subsequently, it tunnels off the QD island across the junction TJ_b. For this sequence of tunneling events, identical triangular regions are obtained symmetrically situated on opposite sides of the V_g axis. These regions can be combined with their symmetrical counterparts to obtain diamond-shaped regions, shown in Fig. 14.5 as forbidden areas for tunneling. The chief property of these grey shaded regions is that within the boundaries of these regions, electron tunneling is forbidden, the current $I = 0$ and charge stability is maintained. The diamond-shaped regions are termed Coulomb diamonds from their semblance to diamond shapes (Fig. 14.5). The zero conduction Coulomb diamonds represent the Coulomb blockade zones.

14.3 Types of Single Electron Transistor Logic

14.3.1 Voltage-Based Logic

The simplest strategy is to evolve an SET logic on similar lines to the common CMOS logic, with which we are familiar from the beginning [3]. It appears that considerable simplification in digital logic design will be possible by adopting this strategy. An existing CMOS library could be translated into the equivalent SET library. It seems that the designer need not bother about the physics of single electron transistor action and only remembers the dependence of drain-source current on gate-source and drain-source voltages. However, such a pathway is not easy to follow because of differences between SET and CMOSFET, which should be compulsorily taken into account:

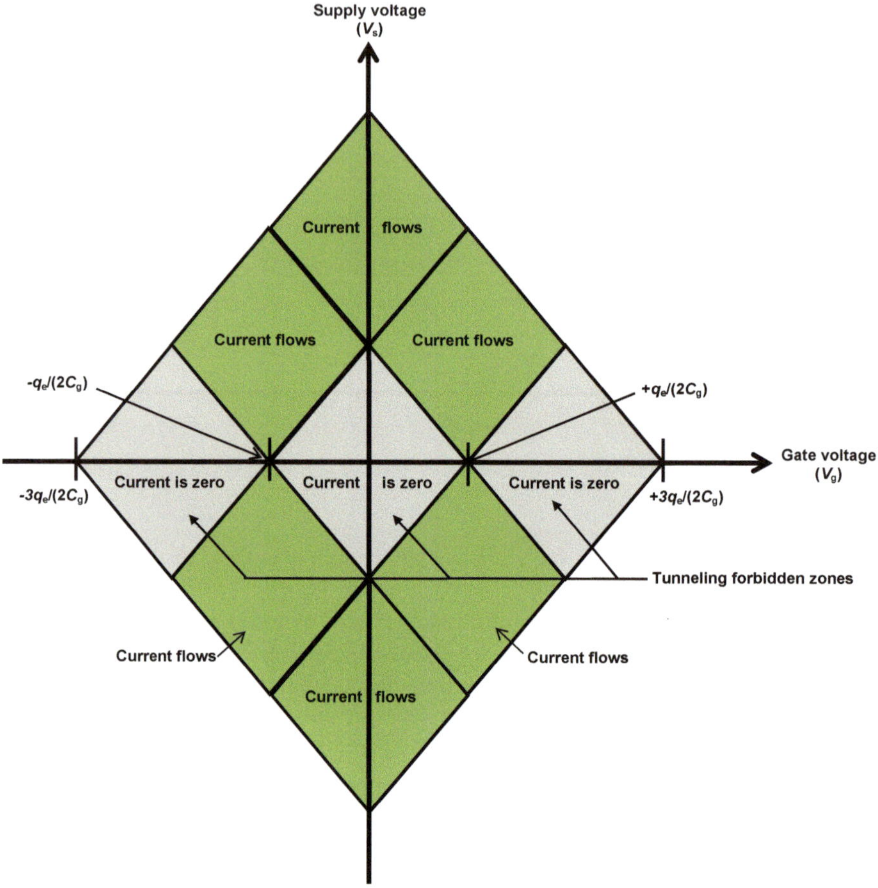

Fig. 14.5 *Diamond-shaped* structures called Coulomb diamonds for a single electron transistor

(i) Low voltage gain of SET One major deviation is that the voltage gain of the SET device is very low and cannot be increased much. This is because the gain is the ratio of the gate capacitance to tunnel junction capacitance. To increase the gain, the gate capacitance has to be increased with respect to the tunnel junction capacitance, keeping the total capacitance small. Gate capacitor is a non-tunnel capacitor. To increase its capacitance, its dielectric thickness cannot be decreased beyond a certain value. Tunnel junction is a tunnel capacitor with a thin dielectric film and hence a large capacitance value. On attempting to increase the gate capacitance, e.g., by increasing the area, the total capacitance of the central island increases. As a result, the device will require a low operating temperature, possibly in the cryogenic range. So, increasing the ratio of capacitances is restricted by strict boundaries. Operating temperature requirements constrain the increase of SET gain beyond a value of 5. A gain value around 2 is generally used. Further, it is found that the voltage gain decreases very rapidly as temperature increases. When the gain becomes <1, the circuit operation is adversely affected.

(ii) Physical similarity of complementary SET transistors In CMOSFET logic, the N-channel MOSFET is structurally a different device than P-channel MOSFET. Both devices are separately fabricated on the circuit. But in case of SETs, complementary action is achievable in identical SET devices. It depends on the biasing condition. The operating point of one transistor of the complementary pair is selected on the ascending branch of I_{DS}-V_{GS} characteristic while that of the other transistor is chosen on the descending branch of the above curve. Using extra capacitors allows us to invoke this behavior. However, no special effort may be necessary in case of symmetrical transistors having zero background charges.

The availability of alternating transconductance in SET devices is both a boon and a curse. It is a boon from the viewpoint that transistors of only one variety are needed to establish complementarity. It is a curse because CMOS logic cannot be straightway mimicked to develop SET logic. The simplest logic circuits call for redesigning to achieve appreciable margins of parameters. The ranges of operation of the logic circuits that have been redesigned and optimized narrow down under thermal fluctuations. A serious shortcoming of the designed logic circuits is that neither of the two transistors in the complementary pair is completely switched off leading to an increase in leakage current and thereby in static power dissipation. When high density SET logic circuits capable of threatening the CMOS logic are considered, the static power dissipation density becomes inadmissibly enormous >10 kW cm^{-2}.

14.3.2 *Charge-Based Logic*

To resolve the drawback of excessive power consumption encountered in voltage-based logic, another type of logic has been propositioned. Herein an electron is propagated alongside a shift register- like path. The format used for representing information is different. It is based on the presence or absence of an electron in a conducting island. Any static state does not involve the flow of a current. However, the signal is in a decomposed format, and the binary operations take place through resistively coupled transistors. In as much as the required circuits contain resistors, inadequate margins are available. Hence, capacitively coupled circuits are deemed more suitable, and had to be designed.

14.4 Digital Logic Gates

Only gates of voltage-based logic will be examined here.

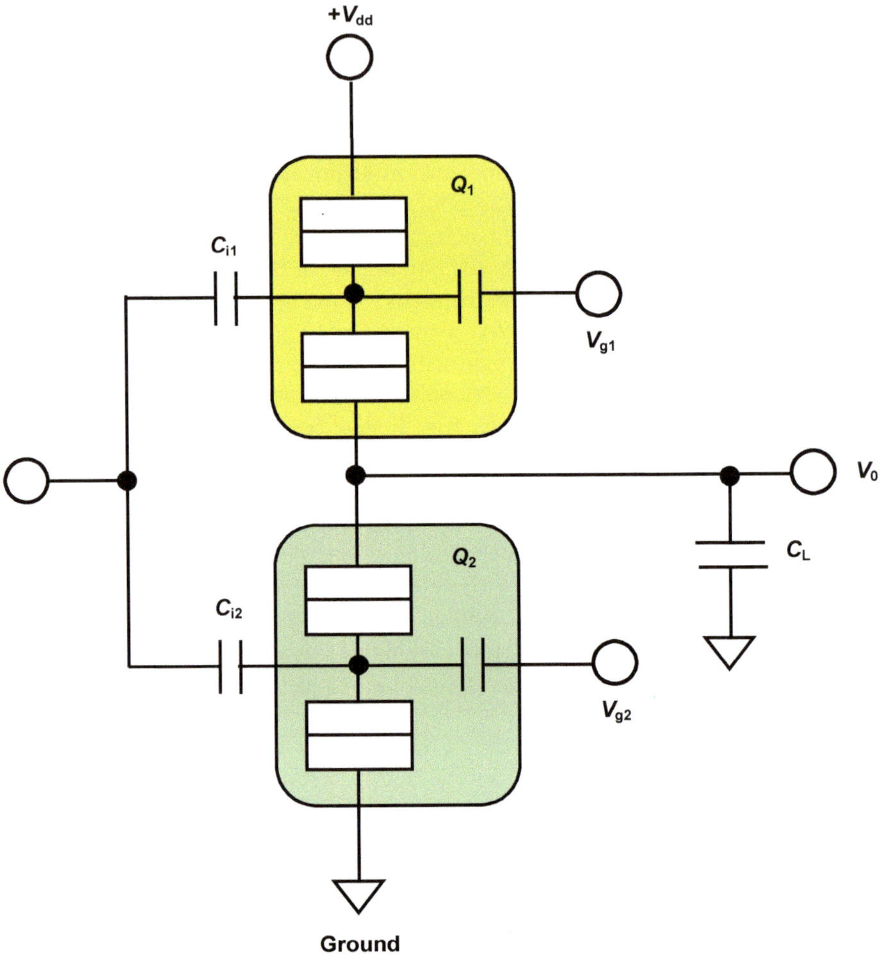

Fig. 14.6 Not gate made from SETs

14.4.1 SET NOT Gate

The NOT gate (Fig. 14.6) consists of two SET transistors Q_1, Q_2. These transistors are connected in series. The transistor Q_1 works as a P-channel MOSFET. The transistor Q_2 works as an N-channel MOSFET. The gates of the two transistors are knotted together. The input signal is applied at this common gate terminal. The two input capacitors C_{i1} and C_{i2} are included to provide a voltage gain ~ 2. The gates V_{g1} and V_{g2} are tuning gates. They are used to control the charges on the islands of the two SETs. The output signal is taken from the meeting point of the two SETs.

The capacitor C_L is a load capacitor. Its purpose is to prevent charging effects at the output terminal.

The operation of the circuit closely follows the CMOSFET logic NOT gate: The supply voltage $+V_{dd}$ is applied. The two tuning gates V_{g1}, V_{g2} are adjusted such that when the input signal V_i is low, the transistor Q_1 is turned on. At the same time, the transistor Q_2 being in Coulomb blockade state remains off. Thus, the two transistors operate in such a way that the output terminal is connected to the $+V_{dd}$ terminal through the short-circuited SET Q_1 and has a high resistance connection to ground through the open-circuited SET Q_2. Thus, the output signal is in logic high or 1 state when the input signal is in logic low or 0 state.

On application of a high voltage signal at the input, the induced charge on the SETs is altered by a fraction of an electron. This alteration brings the transistor Q_1 into Coulomb blockade state while the transistor Q_2 is rendered conducting. The $+V_{dd}$ terminal is cut off from the output terminal by the high resistance path through non-conducting transistor Q_1. Also, the output terminal is connected to ground through the low-resistance path across the conducting transistor Q_2. Thus, the output signal is in logic low or 0 state when the input signal is in logic high or 1 state.

Thus in both cases of low and high states of input signal, the output signal acquires the reverse polarity. Clearly, this circuit works as a NOT logic gate or inverter circuit.

14.4.2 SET AND Gate

The circuit (Fig. 14.7) contains six SETs: Q_1, Q_2, Q_3, Q_4, Q_5, Q_6. The SETs Q_1 and Q_3 work as a complementary pair. In CMOS logic analogy, Q_1 may be looked upon as a P-channel MOSFET and Q_3 as an N-channel MOSFET. Likewise, the SETs Q_2, Q_4; and Q_5, Q_6 behave as complementary pairs. Q_2, Q_4 behave like P-channel and N-channel MOSFETs. Similarly Q_5, Q_6 act as P-channel and N-channel devices, respectively. On these similarities between SET and CMOS circuits, the realization of NAND gate function by the SET circuit is easily understood. It may be noted that all the complementary pairs (Q_1, Q_3), (Q_2, Q_4) and (Q_5, Q_6) are driven by single input signals, A, B and C. The SETs (Q_1, Q_3), and (Q_2, Q_4) comprise the AND circuit. The SETs (Q_5, Q_6) are easily seen to resemble the SET NOT gate. Their inverter function need not be elaborated further. Let us focus attention on SETs (Q_1, Q_3), and (Q_2, Q_4).

Recalling discussions regarding the NOT circuit, when the input signal is in logic low or 0 state, the PMOS device is turned on and NMOS device is turned off. Hence, when $A = 0$, $B = 0$, in the pairs (Q_1, Q_3), and (Q_2, Q_4), the SETs (Q_1, Q_2) will be on and Q_3, Q_4 will be off. So, $C = 1$ because $+V_{dd}$ supply will be shorted through (Q_1, Q_2) to the point P. The output signal is the inverted form of C, i.e., $V_{out} = 0$.

If $A = 1$, $B = 1$, the SETs (Q_3, Q_4) will be on and (Q_1, Q_2) will be off. Therefore, $+V_{dd}$ will be cut off from point P. At the point P, $-V_{dd}$ will be applied though (Q_3, Q_4) combination. So, $C = 0$ and $V_{out} = 1$.

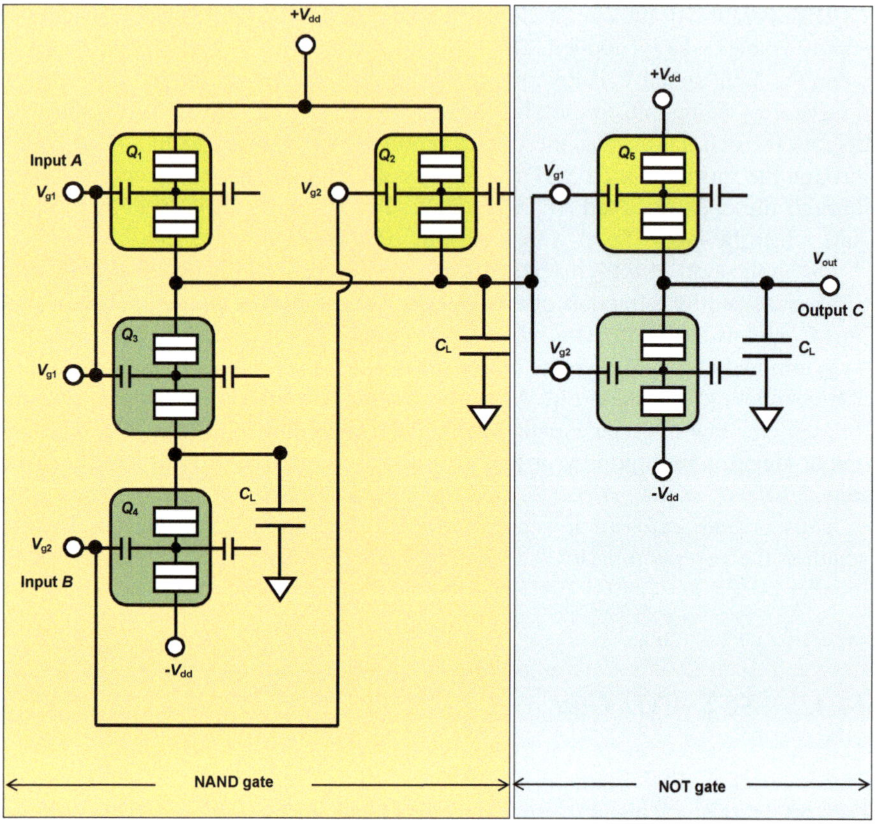

Fig. 14.7 AND gate made from single electron transistors

For $A = 1$, $B = 0$, the SETs (Q_2, Q_3) will be on and (Q_1, Q_4) will be off. Hence, $+V_{dd}$ will be connected through Q_2 to point P. So, $+V_{dd}$ will be applied at point P. Connection between $-V_{dd}$ and point P will be broken through Q_4. Then, $C = 1$ and $V_{out} = 0$.

Finally, when $A = 0$, $B = 1$, the SETs (Q_1, Q_4) will be on and (Q_2, Q_3) will be off. In this situation, $+V_{dd}$ will be connected to the point P due to conducting Q_1. Consequently, voltage at point P is $+V_{dd}$. Linkage between $-V_{dd}$ and point P will be interrupted through Q_3. Thus, $C = 1$ and $V_{out} = 0$.

Based on the above analysis, the conditions of the four SETs, and the truth table of the given logic circuit are compiled in Table 14.1.

Table 14.1 Truth table of the SET AND gate in relation to (Q_1, Q_2, Q_3, Q_4) transistor conditions

A	B	Q_1	Q_2	Q_3	Q_4	C	V_{out}
0	0	On	On	Off	Off	1	0
1	1	Off	Off	On	On	0	1
1	0	Off	On	On	Off	1	0
0	1	On	Off	Off	On	1	0

14.4.3 SET OR Gate

As for the AND gate, the OR gate (Fig. 14.8) also contains six transistors: Q_1, Q_2, Q_3, Q_4, Q_5, Q_6. But the circuit layout of OR gate differs from that of the AND gate. The transistors are arranged in a different configuration. The AND gate had two sourcing transistors Q_1, Q_2 in parallel connection, joined to $+V_{dd}$; and two sinking transistors Q_3, Q_4 in series connection, joined to $-V_{dd}$. In the OR gate, the sourcing transistors Q_1, Q_2 are series-connected while the sinking transistors Q_3, Q_4 are parallely arranged. The SETs (Q_1, Q_3) constitute one complementary pair of devices. The SETs (Q_2, Q_4) form another complementary pair. The transistors (Q_1, Q_3) are fed by one input signal A. Similarly, the transistors (Q_2, Q_4) are supplied a single input signal B. The resultant of signals A and B is signal C which is inverted by the NOT gate comprising SETs (Q_5, Q_6) as before. Further, as for the AND gate,

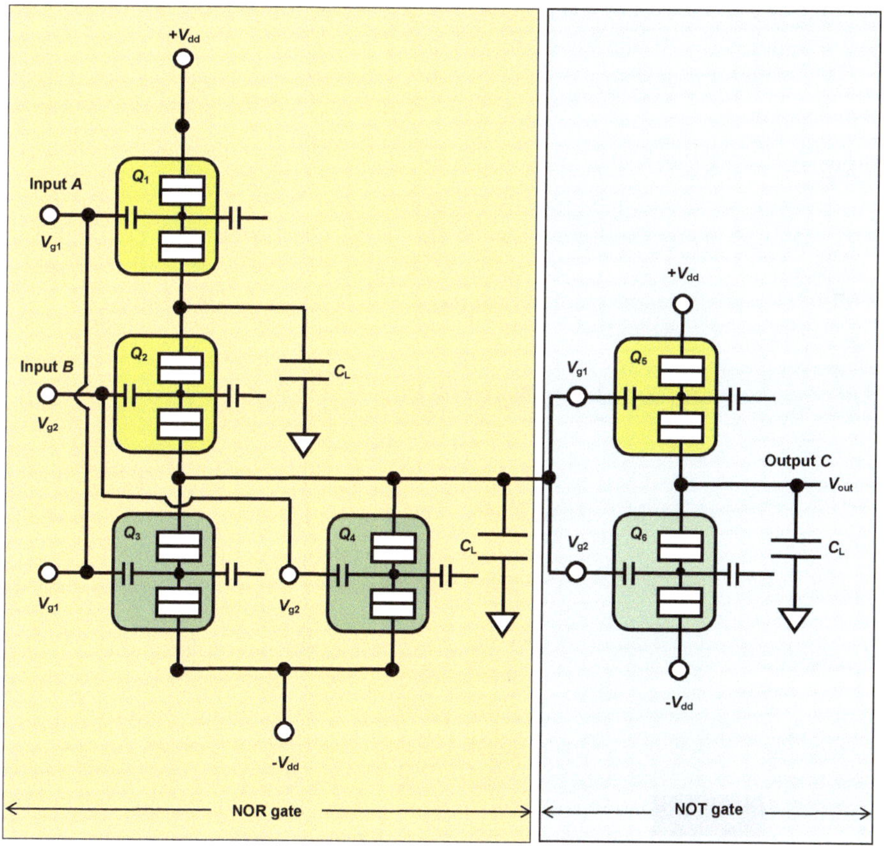

Fig. 14.8 OR gate using single electron transistors

Table 14.2 Conditions of the four SETs and truth table of the OR gate

A	B	Q_1	Q_2	Q_3	Q_4	C	V_{out}
1	0	Off	On	On	Off	0	1
0	1	On	Off	Off	On	0	1
0	0	On	On	Off	Off	1	0
1	1	Off	Off	On	On	0	1

the SETs (Q_1, Q_3) may be regarded as P- type and N-type devices in CMOS logic. The SETs (Q_2, Q_4) are also P- and N-type devices in CMOS logic.

If input A is high ($A = 1$, $B = 0$), the SET Q_3 will conduct, Q_1 will be non-conducting. If input B is high ($B = 1$, $A = 0$), SET Q_4 will be conducting, Q_2 will be non-conducting. So, if either input A is high or input B is high ($A = 1$, $B = 0$ or $B = 1$, $A = 0$), one of the two SETs Q_3 or Q_4 will be conducting. Also, one of the two SETs, Q_1 or Q_2 will be non-conducting. Conduction by either Q_3 or Q_4 implies that the point P is connected to $-V_{dd}$ through Q_3 or Q_4. Non-conduction by either Q_1 or Q_2 means the disconnection of point P from $+V_{dd}$ in both cases. Thus, in both the situations, $C = 0$ and hence $V_{out} = 1$.

When both the inputs A and B are low ($A = 0$, $B = 0$), the SETs (Q_1, Q_2) will conduct and (Q_3, Q_4) will not conduct. Then $+V_{dd}$ will be connected to point P through (Q_1, Q_2) and $-V_{dd}$ will be disconnected from it through (Q_3, Q_4). In this situation, $C = 1$ and $V_{out} = 0$.

For the case of both the inputs A and B being high ($A = 1$, $B = 1$), the SETs (Q_1, Q_2) will be non-conducting and the SETs (Q_3, Q_4) will be conducting. Under this condition, the connection between $+V_{dd}$ and point P will be broken (through Q_1, $Q2$) while that between point P and $-V_{dd}$ will be made through (Q_3, Q_4). In this circumstance, $C = 0$ and $V_{out} = 1$.

Thus, the on/off states of the four SETs Q_1, Q_2, Q_3, Q_4 and the truth table of the OR logic gate are presented in Table 14.2.

14.5 Other Applications

Besides digital logic, single electron transistors have found applications in several other fields, a common example being supersensitive electrometry. It is found that the drain-source current of a SET becomes an extremely sensitive function of the gate voltage when its drain-source voltage is fixed slightly above the threshold voltage for Coulomb blockade. Then the drain-source current variation is appreciable even when the external charge changes in sub-multipliers of elementary electronic charge. This behavior leads to the application of SET as an ultrasensitive electrometer to measure excessively small charge variations. The capability of single electron electrometry has been exploited in determining electron addition energies such as for finding the distribution of energy levels in quantum dots, and other nanosize entities.

Another suggested use of single electron transistor is in the form of a standard of direct current by phase locking the SET oscillations in an oscillator circuit with the help of an externally applied RF source whose frequency has been thoroughly characterized. During each period of the external RF signal, a certain number of electrons will be transferred, resulting in the generation of a direct current. This direct current will be related to the frequency of the RF signal.

Arrays of single electron devices can be used as detectors of infrared radiation due to their lower shot noise along with trouble-free regulation of threshold voltage.

14.6 Discussion and Conclusions

Differing from a normal MOSFET, which requires 1000–10,000 electrons to switch from on-state to off-state or vice versa, a single electron transistor can undergo switching with just one electron, which in itself is a dumbfounding phenomenon happening at ultra small dimensions. Therefore, faster logic circuits consuming very low power are expected by mastering the SET technology [4, 5].

Review Exercises

14.1 What is single electronics? What does a single electron transistor do? How many terminals does this device have? What are the functions of these terminals? Is the gate terminal connected to the quantum dot by a tunneling capacitance? How does the single electron transistor differ from a quantum dot circuit?

14.2 From the energy band diagram of a single electron transistor, explain how the gate voltage determines the drain-source voltage required for electron tunneling from source to drain?

14.3 Derive the equation for the total energy stored in the capacitors C_a, C_b, and C_g of a single electron transistor:

$$E_{se} = \left\{ C_a C_g \left(V_s - V_g^i \right)^2 + C_a C_b V_s^2 + C_b C_g V_g^{i^2} + Q^{i^2} \right\} / (2C_s)$$

where Q^i is the total charge on the quantum dot island, V_g^i is the gate voltage, V_s is the supply voltage and $C_s = C_a + C_b + C_g$.

14.4 Show that an electron will tunnel into the quantum dot through TJ_b if

$$q_e(n + 1/2) + C_a V_s + C_g V_g > 0$$

where q_e is the electronic charge and n is the initial number of electrons on the quantum dot island, C_a is the capacitance of tunnel junction TJ_a, V_s is the supply voltage, C_g is the gate capacitance and V_g is the gate voltage.

14.5 Prove that an electron will tunnel off the quantum dot through TJ_a if

$$q_e(-n+1/2)+(C_b+C_g)V_s-C_gV_g>0$$

where q_e is the electronic charge and n is the initial number of electrons on the quantum dot island, C_b is the capacitance of tunnel junction TJ_b, V_s is the supply voltage, C_g is the gate capacitance and V_g is the gate voltage.

14.6 Show that for $n = 0$, an electron will tunnel across the junction TJ_b into the quantum dot if

$$V_s > -(1/C_a)(q_e/2+C_gV_g)$$

Draw the region of V_s-V_g plane defined by this equation.
For $n = 1$, show that the electron will tunnel off the quantum dot across the junction TJ_a if

$$V_s > \{1/(C_b+C_g)\}(q_e/2+C_gV_g)$$

Draw the corresponding region of V_s-V_g plane. Mark the resultant triangular region of the V_s-V_g plane satisfying both the above equations.
Considering the opposite sequence of tunneling operations in which an electron tunnels across the junction TJ_a into the quantum dot and then tunnels off the quantum dot across the junction TJ_b, similar triangular regions are defined which are located symmetrically opposite on the V_g axis to the previously obtained triangular regions. What are the regions drawn by shading the symmetrically placed triangles called? What do they indicate and what do they represent?

14.7 Bring out the differences between SET and CMOS devices, which do not allow straightway adoption of CMOS logic circuits for implementation by SET devices.

14.8 What restrictions forbid raising the voltage gain of a single electron transistor? What value of voltage gain is generally used?

14.9 Is it necessary to fabricate complementary SETs separately? How is complementary behavior realized in SETs?

14.10 Explain the statement, "The availability of identical complementary SET transistors is both a blessing and a curse."

14.11 Why is static power consumption high in SET logic circuits? How does charge-based logic decrease power consumption? How does this logic work?

14.12 Draw and explain the operation of a voltage-logic based SET NOT gate.

14.13 Draw the circuit diagram of a SET AND gate. Identify the SETs identical to P-channel and N-channel transistors, and hence explain the operation of this circuit in analogy to CMOS AND gate.

14.14 Draw the circuit diagram of a SET OR gate. Point out differences in connections of transistors from the SET AND gate, and explain the working of the gate.

14.15 Explain how a SET acts an ultrasensitive electrometer? Cite some applications of this capability of SET?

14.16 Explain the use of SET as a standard of direct current. Why are SET arrays used as IR detectors?

References

1. Likharev KK (1999) Single electron devices and their applications. Proc IEEE 87(4):606–632
2. Kastner MA (2000) The single electron transistor and artificial atoms. Ann Phys (Leipzig) 9(11–12):885–894
3. Korotkov AN (1999) Single electron logic and memory devices. Int J Electron 86(5):511–547
4. Uchida K, Koga J, Ohba R et al (2003) Programmable single electron transistor logic for future low-power intelligent LSI: proposal and room-temperature operation. IEEE Trans Electron Devices 50(7):1623–1630
5. Amakawa S, Tsukagoshi K, Nakazato K et al (2004) Single electron logic based on multiple-tunnel junctions. In: Nakashima Hiroshi, Signpost Research (eds) Mesoscopic tunneling devices. Kerala, India, pp 1–34

Chapter 15
Semiconductor Nanowire as a Nanoelectronics Platform

Abstract Bottom-up approach to nanowire synthesis using vapour-liquid-solid technique is outlined. Pros and cons of this approach with top-down paradigms are highlighted. Using silicon nanowires, the fabrication of P-N junction diodes, bipolar and field-effect transistors as well as complementary inverters is described. Fabrication and operation of P-channel Ge/Si heterostructure and N-channel GaN/AlN/AlGaN heterostructure nanowire transistors is discussed. Placement of nanowires at desired locations and their interconnections to form logic circuits is addressed. Cross-bar architecture is an accepted structure, which has rendered possible the gainful utilization of the unique properties of nanowires. The nanowires can act as versatile building blocks for the assembly of nanoelectronic circuits.

15.1 Introduction

Semiconductor nanowires are 1-D silicon nanostructures with diameter <100 nm [1]. Lengths range from several microns to a few cm with aspect ratios $>10^3$. As they are radially restricted to sub-100 nm scale, their propertries are controlled by size confinement effects.

15.2 Nanowire Growth by Bottom-up and Top-Down Paradigms

Bottom–up approach through chemical synthesis provides a cost-effective methodology to grow high-quality nanowires but ordered positioning of nanowires and their integration with CMOS is difficult. Fabrication of intricate integrated circuits by accurately placing the wires at required ocations has so far been frustrating. Top-down approach using electron beam lithography, focused ion beam lithography or advanced optical lithography produces nanowires reproducibly. It is

V.K. Khanna, *Integrated Nanoelectronics*,
NanoScience and Technology, DOI 10.1007/978-81-322-3625-2_15

CMOS compatible involving no integration issues but highly capital intensive. As both bottom-up and top-down paradigms have relative merits and demerits, a combination of the two paradigms is practiced.

15.3 Metal-Catalyst-Assisted Vapor–Liquid–Solid (VLS) Method of Nanowire Growth

In this method [2], a metallic nanoparticle (Au, Ag, Al or Cu) adsorbs vapors of the element desired in the nanowire from gas-phase reactants at a high temperature (Fig. 15.1). As a result, catalytic liquid alloy droplets are formed. These droplets adsorb more vapors to reach supersaturation level. At this level, the concentration of adsorbed component in the alloy exceeds its equilibrium value. Therefore, the alloy system tries to attain the minimum free energy stage. Minimization of free energy is achieved by precipitation of the component at the liquid-solid interface. Such a nucleation site at the liquid-solid interface acts as a starting point for crystal growth. This crystal growth continues as long as the supply of vapor is maintained. Thus a nanowire of required length is formed. Evidently, the method requires the transportation of the element of the wire material as a vapor. Further, the method

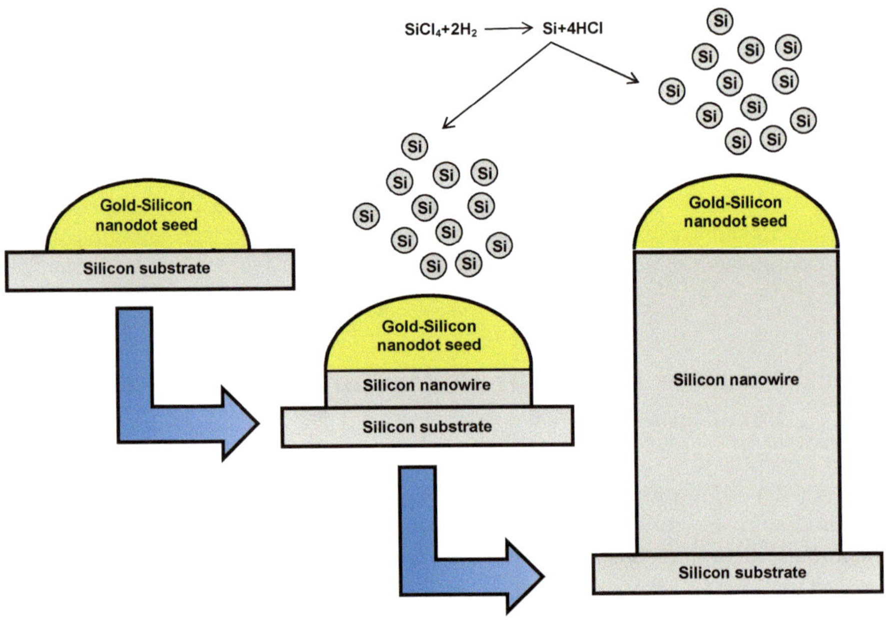

Fig. 15.1 Mechanism of VLS growth of Si nanowires from liquid gold-silicon seeds

requires the formation of a liquid alloy droplet on the substrate. Still further, the method involves solidification of the liquid droplet at the liquid-solid interface to form the nanowire crystal. In this way, all the three phases, viz., vapor, liquid and solid, are the partakers of this method; hence it is called vapor-liquid-solid (VLS) mechanism. The position of nanowire is the location of the metal catalyst nanoparticle. The diameter of nanowire depends on the size of this nanoparticle. Since the nanowire growth requires a high temperature, techniques such as laser ablation (LA), chemical vapor deposition (CVD), etc., are used to provide the required temperature.

15.4 Synthesis of Single Crystal Si Nanowires of Required Diameters

0.1% poly-L–lysine $(C_6H_{12}N_2O)_n$ is deposited on oxidized silicon wafers used as substrates [3], Fig. 15.2. Au nanoclusters of dimensions 5, 10, 20, and 30 nm are spread on the substrate. As the clusters are negatively charged, they cling and adhere to the positively charged lysine. The substrates are cleaned in O_2 plasma and loaded into a quartz reactor, evacuated to <100 mTorr and kept at a temperature of 440 °C in Ar atmosphere. Si nanowires are grown by pyrolysis of silane: $SiH_4 = Si + 2H_2$. The 5 nm nanocluster gave a nanowire of diameter 6 nm. The 10 nm nancluster yielded a nanowire of diameter 12 nm. The nanowire diameters for 20 and 30 nm nanoclusters were 20 and 31 nm respectively. This experiment shows that the diameter/position of a nanowire is determined by the diameter/position of the Au nanocluster. The nanowire diameters were only 1–2 nm larger than those of the respective nanocluster catalysts, testifying to their suitability for assembling nano-electronic devices using nanowires as structural units.

15.5 Laser-Assisted Catalytic Growth and Doping of Si Nanowires

Ablation of a gold target by an Nd:YAG laser produces Au nanoclusters. These Au nanoclusters are used as catalysts for nanowire growth in SiH_4 vapor phase reactant. For P-type doping of nanowires, B_2H_6 is introduced in the reactant gas stream. N-type doping is carried out with phosphorous dopant employing an Au-P target. Extra red phosphorous is incorporated in the reactant gas flow [4].

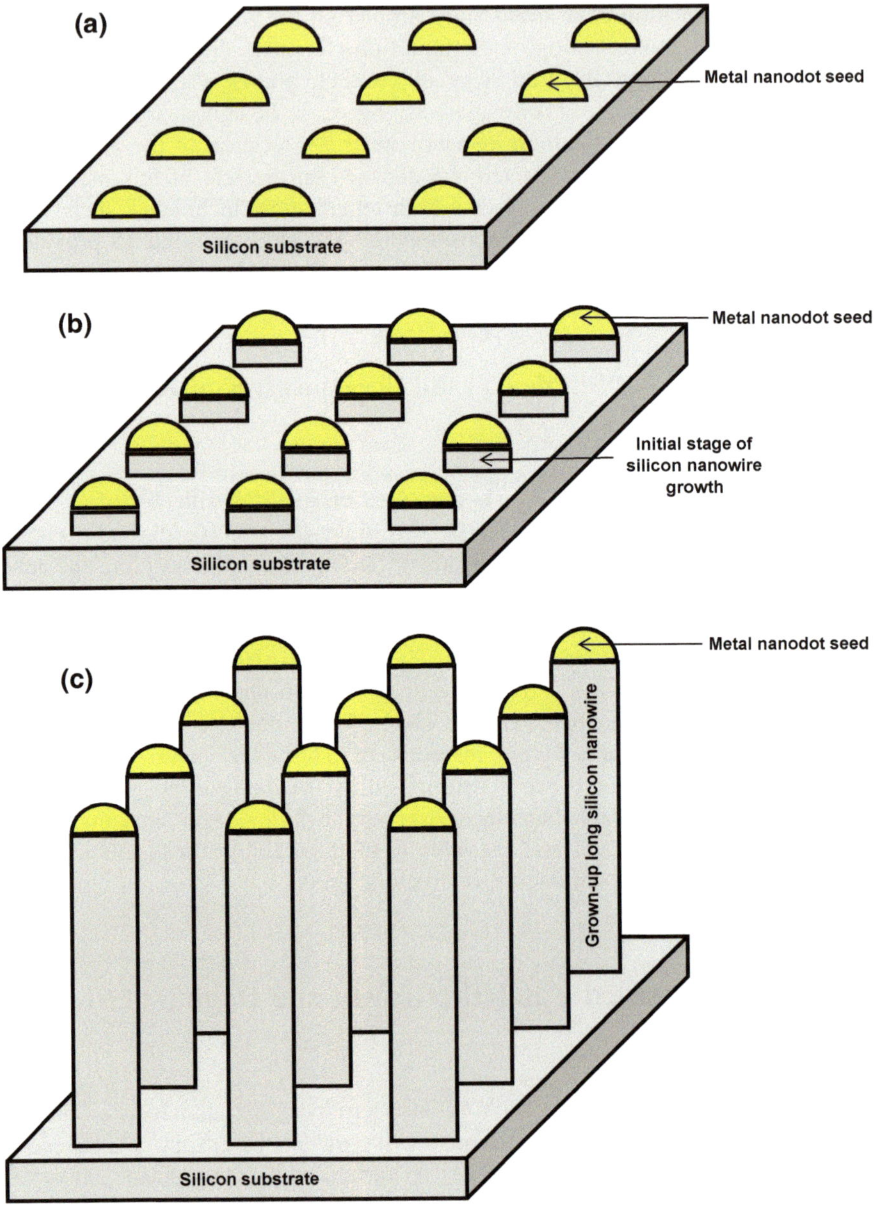

Fig. 15.2 Simultaneous growth of multiple silicon nanowires at precisely defined locations on a substrate **a** Spreading Au nanoclusters on the substrate, **b** initiation of nanowire growth, and **c** final long nanowires

15.6 Ohmic Contacts to Si Nanowires

After dispersing in acetone, the 20–50 nm diameter, P-type and N-type silicon nanowires are placed on a SiO_2/Si substrate with 600 nm thick SiO_2 layer. By thermal evaporation, Al (50 nm)/Au (150 nm) contacts are made to the nanowires and used in un-annealed state. Linear current-voltage characteristics are recorded for individual P-type and N-type nanowires. Hence, the contacts are ohmic in nature [5].

15.7 P-N Junction Diodes Made from Crossed Si Nanowires

P-N junction nanostructures are formed using crossed P- and N-type nanowires. These P-N junctions show rectifying diode behavior.

15.8 Bipolar Transistor Made from Crossed Si Nanowires

The N^+-P-N bipolar transistor is fabricated by putting together N^+ emitter and N-type collector nanowires across a P-type base nanowire [5]. A P-type nanowire is placed from solution onto the substrate at the intended location followed by attachment of N^+ and N-type nanowires to the metal pads and assembly across the P-type nanowire. Viewing is done through an optical microscope to perform this nanowire assembly. The nanowires are handled by micromanipulators. Thus a Si nanowire-based bipolar transistor is constructed from crossed Si nanowires. The assembled transistor worked like a standard planar transistor. It gave a current gain of 0.94 in the common-base configuration. The common-emitter current gain was 16 indicating high electron injection efficiency as well as high carrier mobility.

15.9 Field-Effect Transistors Using Si Nanowires

10 nm diameter P- and N-type Si nanowires are used as P- and N-channel field effect transistors by applying a gate voltage through heavily-doped Si substrate, used as a back gate [6]. At an applied gate voltage of 5 V, the current on-off ratio is $>10^4$.

In traditional field-effect transistors, degenerately doped source and drain regions act as the contacts of the channel region. In opposition, in the nanowire FET, the contacts are established through metal films. A Schottky barrier exists at the metal/semiconductor interface. Hence the nanowire FETs are effectively Schottky barrier devices. The contact resistances of source/drain terminals critically influence

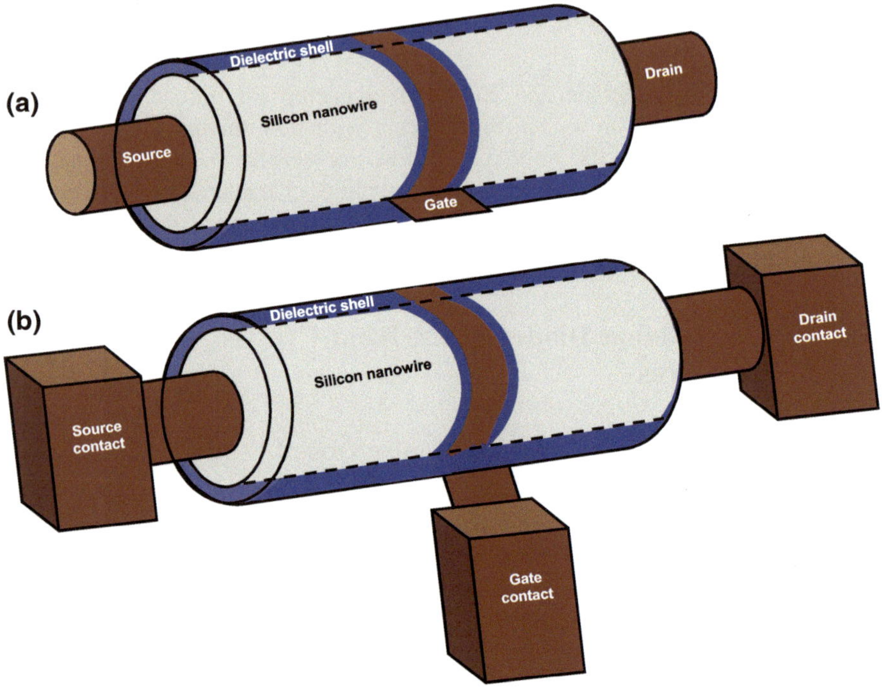

Fig. 15.3 Silicon nanowire structures: **a** nanowire in core/shell structure and **b** a nanowire FET

their electrical performance. The contacts can be made ohmic to a certain degree by annealing, thereby increasing the on-state current. Ohmic contacts with low contact resistance are obtained for heavily-doped nanowires but for lightly-doped nanowires, the contact resistance effects still dominate.

Figure 15.3a shows how a silicon nanowire in core/shell structure can be used as the channel of a FET device. In Fig. 15.3b, the complete nanowire FET is formed with the contact electrodes.

As displayed in Fig. 15.4, silicon nanowires can be stacked in horizontal (Fig. 15.4a) and vertical (Fig. 15.4b) configurations to fabricate multiple FET devices.

15.10 P-Channel, Ge/Si Core/Shell Nanowire Heterostructure Transistor

In the Ge/Si core/shell heterostructure, a valence band offset of 440 meV exists between germanium core of band gap 0.67 eV and silicon shell of band gap 1.11 eV. Therefore, the Fermi level pinned inside the band gap of silicon is located at a lower level than the edge of the valence band in the germanium core. Hence,

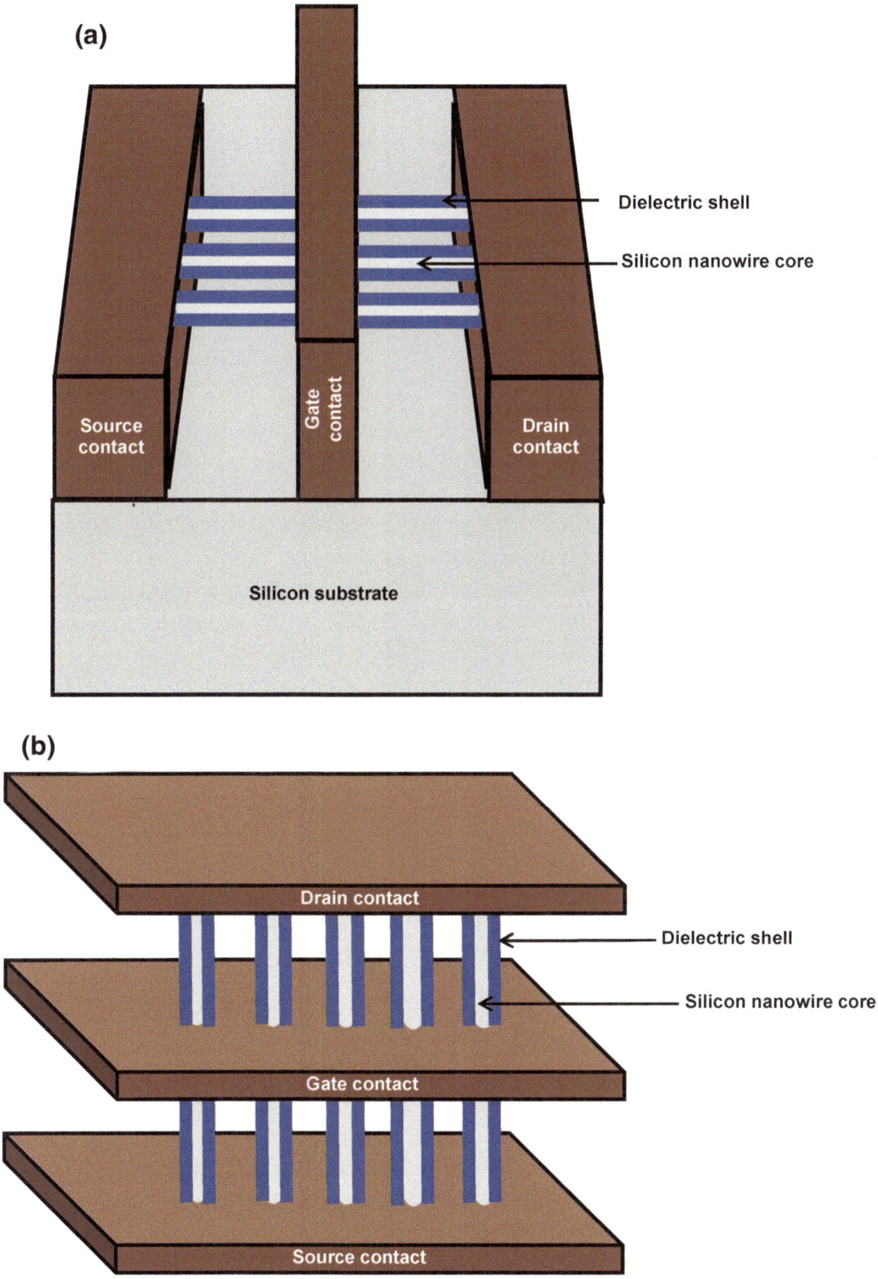

Fig. 15.4 Silicon nanowire FETs built with: **a** Horizontal stack and **b** vertical stack of core/shell nanowires

free holes accumulate in the channel formed in the Ge core. Consequently, a 1-D hole gas is formed. This hole gas is confined within the Ge core even when the Ge core and the Si shell materials are intrinsic.

The Ge/Si transistors are made of an epitaxial nanowire containing a Ge core (diameter 14.6 nm) surrounded by a silicon shell (diameter 1.7 nm). Germanium nanowire cores are synthesized by VLS method catalyzed by Au nanolusters [7]. The Au nanoclusters, deposited on SiO_2/Si substrates, are placed in a quartz furnace. Germanium core growth initiation is done by 1-min nucleation in 10 % GeH_4 (germane) in H_2 at 315 °C. The core is axially elongated at 280 °C. In the same reactor, changing the reactant from germane to silane produces the silicon shell at 450 °C.

FET devices are fabricated using electron beam lithography [8]. $NiGe_xSi_y$ is used for source/drain contact electrodes. It is formed by a solid-state reaction converting Si to silicide and NiSi-Si-NiSi heterostructures from thermally evaporated Ni. At the beginning, the channel length is 4 μm. By annealing, the $NiGe_xSi_y$ contact regions are lengthened and the Ge/Si nanowire channel is shortened to sub-100 nm. The top gate dielectric is 4 nm thick HfO_2 formed by atomic layer deposition using Tetrakis(ethylmethylamino)hafnium $[Hf(N(CH_3)(C_2H_5))_4]$ as precursor. Cr/Au (5 nm/50 nm) contact is formed by thermal evaporation for the gate electrode. Following this process, FET devices with channel lengths of 70 and 40 nm are fabricated.

For the 70 nm nanowire FET, the maximum drain current $I_{D(max)}$ is 121 μA and the peak transconductance gm is 78 μS with $V_{DS} = 0.5$ V. For the 40 nm FET device, these parameters are 152 μA and 91 μS, respectively. The scaled on-current for the 40 nm FET is 2.1 mA/μm, and the scaled transconductance is 6.2 mS/μm at $V_{DS} = 0.5$ V. Comparing with corresponding values for 35 nm P-channel Si MOSFET, the nanowire FETs emerge to be superior.

15.11 N-Channel, GaN/AlN/AlGaN Heterostructure Nanowire Transistor

In this heterostructure, the conduction band of GaN lies at a lower level than that of AlGaN [9]. A large internal electric field prevails across the heterojunction between the GaN core and AlN/AlGaN shells. This field arises from the intense spontaneous and piezoelectric polarization. Hence, a 1-D electron gas is formed inside the GaN/AlN/AlGaN nanowire heterostructure in undoped condition.

The heterostructure consists of an intrinsic GaN core. Upon this GaN core, AlN and AlGaN shells are deposited in a sequential manner. For synthesis of nanowires, nickel nanocluster precursor is used. The substrate is made of Al_2O_3. MOCVD reactor is used. Trimethylgallium is the source for Ga. Trimethylaluminum is the Al source. Ammonia is the source of nitrogen. Temperature of growth of GaN core is 775 °C. AlN and AlGaN shells are successively formed at 1040 °C in hydrogen.

The intrinsic electron mobility in the GaN/AlN/AlGaN heterostructure is 3100 cm^2/V-s at room temperature. At 5 K, it increases to 21000 cm^2/V-s. Top-gated field-effect transistors are fabricated with ZrO_2 high–κ dielectric. In these FETs, an on/off current ratio of 10^7 is achieved. The subthreshold slope is 68 mV/decade. The scaled on-current is 500 mA/mm. The scaled transconductance is 420 mS/mm.

15.12 Complementary Inverters Using P-Type and N-Type Si Nanowire Transistors

Like Si CMOS circuits, Si nanowires are used to make complementary inverter circuits. These inverter circuits are made from lightly doped silicon nanowires. The nanowires are very sensitive to gating action enabling full depletion. The circuits need low static power dissipation $\sim$0.5–5 nW in either high or low logic level conditions. In comparison, a single nanowire device requires 10^3–10^4 times larger power dissipation.

15.13 Nanowire Integration Methods for Building Nanowire Circuits

Following methods are applied to organize nanowires into arrays with defined position and orientation [10]: (i) Fluid-assisted organization: The duration of flow of fluid determines the surface coverage by nanowires. The flow rate controls the alignment of nanowires. The size of the fluidic channel restricts the size of the nanowire array. (ii) Langmuir–Blodgett (LB) method: A monolayer of a nanowire-surfactant is compressed along one axis on an aqueous subphase. As a result, the nanowires are aligned. The inter-wire spacing is also controlled. By transferring the compressed layer to a substrate, parallel nanowires are arranged on the substrate surface. (iii) Blown bubble film (BBF) method: The nanowires are dispersed in a homogeneous solution. Expansion of a bubble from the nanowire solution in a controlled direction and rate, and thereafter transfer of the bubble to the substrate helps to form nanowire films of high density on large size wafers.

Using the above techniques, complex nanowire structures are obtained by replication of the alignment and transfer processes of nanowires. A dry deposition scheme entails transfer of synthesized nanowires from the growth substrate to the device substrate by contact printing.

The “grow-in-place” method involves the synthesis of nanowires at the pre-chosen sites. The nanowires are grown either vertically or horizontally. The vertical growth is used for fabrication of vertical nanowire field-effect transistors. The horizontal growth has been applied to NEMS devices.

Crossbar structure is a popular pattern for fabricating active devices at the point of intersection of two groups of nanowires. Logic circuits are fabricated using crossed nanowires to form P-N junctions [11]. For realization of a 2-input NOR gate, a crossed P-N junction array is constructed from two P-type Si nanowires as inputs and one N-type GaN nanowire as output. An AND gate is formed from one P-type Si nanowire and three N-type GaN nanowires. A NOR gate is also assembled from a 1×3 crossed nanowire junction array. Interconnection of AND with NOR gates yields XOR gate. Thus, all vital logic gate and computation functions are explored.

15.14 Discussion and Conclusions

Semiconductor nanowires offer several advantages over carbon nanotubes. They can remain semiconducting in nature irrespective of their diameter unlike the carbon nanotubes which can be semiconducting or metallic depending on their diameters. Moreover, controlled doping of nanowires is possible by applying the vast knowledge from semiconductor industry. Boron or phosphorous dopants can be incorporated during laser-assisted catalytic growth of nanowires. By heavy doping, metallic regime is reached. Nanowires can therefore act both as functional devices and interconnections. On the opposite side, as-grown carbon nanotube samples comprise a mixture of semiconducting and metallic tubes, which need to be differentiated and separated.

Review Exercises

15.1 Why are nanowires said to be one-dimensional materials although they have both length and diameter dimensions.

15.2 Bring out the pros and cons of bottom-up and top-down paradigms for synthesizing semiconductor nanowires. Why is a hybrid approach using both paradigms advisable for practical use?

15.3 Explain the vapor–liquid–solid method of nanowire growth and justify the term “VLS.” On what factors do the diameter and location of the nanowire depend? Why is a high-temperature environment such as provided by laser ablation necessary?

15.4 How are silicon nanowires formed by laser-assisted catalytic growth? How are they doped P- and N-type?

15.5 How are P-N junction diodes and bipolar transistors made from crossed nanowires? How are the nanowires handled during assembly? How are they viewed?

15.6 How do nanowire FETs differ from conventional MOSFETs regarding the contact regions? Are complementary inverters implementable using P- and N-channel nanowire transistors?

15.7 How is a 1-D hole gas formed in a Ge/Si core/shell nanowire heterostructure? How is such a transistor fabricated?

15.8 How is a 1-D electron gas formed in a GaN/AlN/AlGaN heterostructure transistor? Explain the fabrication of this transistor.

15.9 Describe three methods of organizing nanowires into arrays. Name a pattern commonly used to make active devices with nanowires.

15.10 Point out a few advantages of semiconductor nanowires as opposed to carbon nanotubes as a nanoelectronic building block.

References

1. Mikolajick T, Weber WM (2015) Silicon nanowires: fabrication and applications. In: Li Q (ed) Anisotropic nanomaterials: preparation, properties and applications, nanoscience and technology. Springer International Publishing, Switzerland, pp 1–25
2. Wagner RS, Ellis WC (1964) Vapor-liquid-solid mechanism of single crystal growth. Appl Phys Lett 4(5):89–90
3. Cui Y, Lauhon LJ, Gudiksen MS et al (2001) Diameter-controlled synthesis of single-crystal silicon nanowires. Appl Phys Lett 78(15):2214–2216
4. Cui Y, Duan X, Hu J et al (2000) Doping and electrical transport in silicon nanowires. J Phys Chem 104(22):5213–5216
5. Cui Y, Lieber CM (2001) Functional nanoscale electronic devices assembled using silicon nanowire building blocks. Science 291:851–853
6. Cui Y, Zhong Z, Wang D et al (2003) High performance silicon nanowire field effect transistors. Nano Lett 3(2):149–152
7. Lu W, Xiang J, Timko BP et al (2005) One-dimensional hole gas in germanium/silicon nanowire heterostructures. PNAS 102(29):10046–10051
8. Hu Y, Xiang J, Liang G et al (2008) Sub-100 Nanometer channel length Ge/Si nanowire transistors with potential for 2 THz switching speed. Nano Lett 8(3):925–930
9. Li Y, Xiang J, Qian F et al (2006) Dopant-free GaN/AlN/AlGaN radial nanowire heterostructures as high electron mobility transistors. Nano Lett 6(7):1468–1473
10. Lu W, Xie P, Lieber CM (2008) Nanowire transistor performance limits and applications. IEEE Trans Electron Devices 55(11):2839–2876
11. Huang Y, Duan X, Cui Y et al (2001) Logic gates and computation from assembled nanowire building blocks. Science 294:1313–1317

Chapter 16
Carbon Nanotube-Based Nanoelectronics

Abstract Carbon nanotubes serve as ideal one-dimensional materials for nanoscale electronic circuitry, not only because of their small size but also due to their overall exceptional properties, providing the necessary mechanical and chemical stability to the devices. Amongst the three main processes developed for CNT growth, namely arc discharge, laser ablation and chemical vapor deposition, the last one stands out prominently for its adaptability to nanoelectronics manufacturing. A noteworthy feature of fabrication of CNT devices is that the process is doping-free. Instead of doping, the polarity of the FETs is determined by the metals used as contacting electrodes. By appropriate choice of metals, P-channel, N-channel and complementary symmetry CNT FETs are realized. Elimination of the doping requirement for fabrication of CNT devices makes them invulnerable to dopant-related fluctuations. Semiconducting CNTs form the basis of transistor circuits whereas metallic CNTs are used as interconnects. Self-aligned process for large-scale fabrication of P-channel, N-channel and complementary CNT configurations paves the way towards adoption of CNT technology for bulk production.

16.1 Introduction

Carbon nanotubes are long, hollow cylinders made up of a network of carbon atoms. A CNT cylinder can be imagined as formed by wrapping around a single sheet of graphite of one atomic layer thickness called graphene to form a continuous cylinder. The CNTs constitute one type of fullerenes, namely the cylindrical fullerenes. A fullerene is a carbon molecule. It is found as a hollow sphere, a tube and in other shapes as well. A spherical fullerene is called buckminsterfullerene (buckyball). The cylindrical fullerene is the carbon nanotube or bucky tube. Fullerene is one allotrope of carbon. The other allotropes are graphite and diamond. Allotropes represent various structural forms of an element differing in the arrangement of atoms in a molecule.

V.K. Khanna, *Integrated Nanoelectronics*,
NanoScience and Technology, DOI 10.1007/978-81-322-3625-2_16

16.2 Types of Carbon Nanotubes

CNTs either consist of a single cylinder or multiple concentric cylinders (Fig. 16.1). Depending on the number of cylinders and hence walls found in a CNT structure, the nanotubes are classified as either single-walled and multi-walled carbon nanotubes with acronyms SWCNT and MWCNT respectively. Naturally, they have thicknesses from 1 to 50 nm. On the average, diameter of an SWCNT is 1.2–1.4 nm. The length of CNTs is several microns. Their aspect ratio (length/diameter is ~ 1000).

16.3 Geometrical Structure and Chirality of a Carbon Nanotube

The graphene sheet can be rolled in several ways to construct a CNT cylinder. The resulting CNTs differing in the manner of rolling the graphene sheet display different physical properties. These CNTs have different atomic structures. The atomic structure of a CNT is described by the chirality or helicity of the tube. The chirality of the tube is represented by its chiral vector $\vec{\mathrm{r}}$ and the chiral angle θ (Fig. 16.2). The chiral vector $\vec{\mathrm{r}}$ is given by [1]

$$\vec{\mathrm{r}} = n\vec{\mathrm{a}} + m\vec{\mathrm{b}} \tag{16.1}$$

where n, m, the lattice translational indices are integers and $\vec{\mathrm{a}}$, $\vec{\mathrm{b}}$ are unit vectors shown in the diagram. The integral values of n, m are decided by the number of steps executed across the carbon bonds of the hexagonal lattice to calculate the magnitudes of $\vec{\mathrm{a}}$, $\vec{\mathrm{b}}$. The chiral angle θ is defined as the angle between the vectors $\vec{\mathrm{r}}$ and $\vec{\mathrm{a}}$. It determines the degree of twisting of the nanotube, $0 \leq |\theta| \leq 30°$. A CNT for which $\theta = 0$ is called a zigzag nanotube. A CNT for which $\theta = 30°$ is known as an armchair nanotube. When $0 \leq |\theta| \leq 30°$, the nanotube is said to be chiral. Figure 16.3 depicts the three kinds of CNTs.

16.4 Electrical Properties of Carbon Nanotubes

Electrical properties of a SWCNT strongly depend on the values of (n, m) indices. In other words, they are controlled by the wrapping or chirality of the nanotube [2]. Another vital parameter is the tube diameter. Diameter of SWCNT ranges from 0.4 to 2 nm. These dependences are non-monotonic. An SWCNT may show metallic or semiconducting properties with changes in (n, m) indices or diameter as the bandgap alters from 0 to 2 eV. On the other hand, the MWCNT is generally a metal

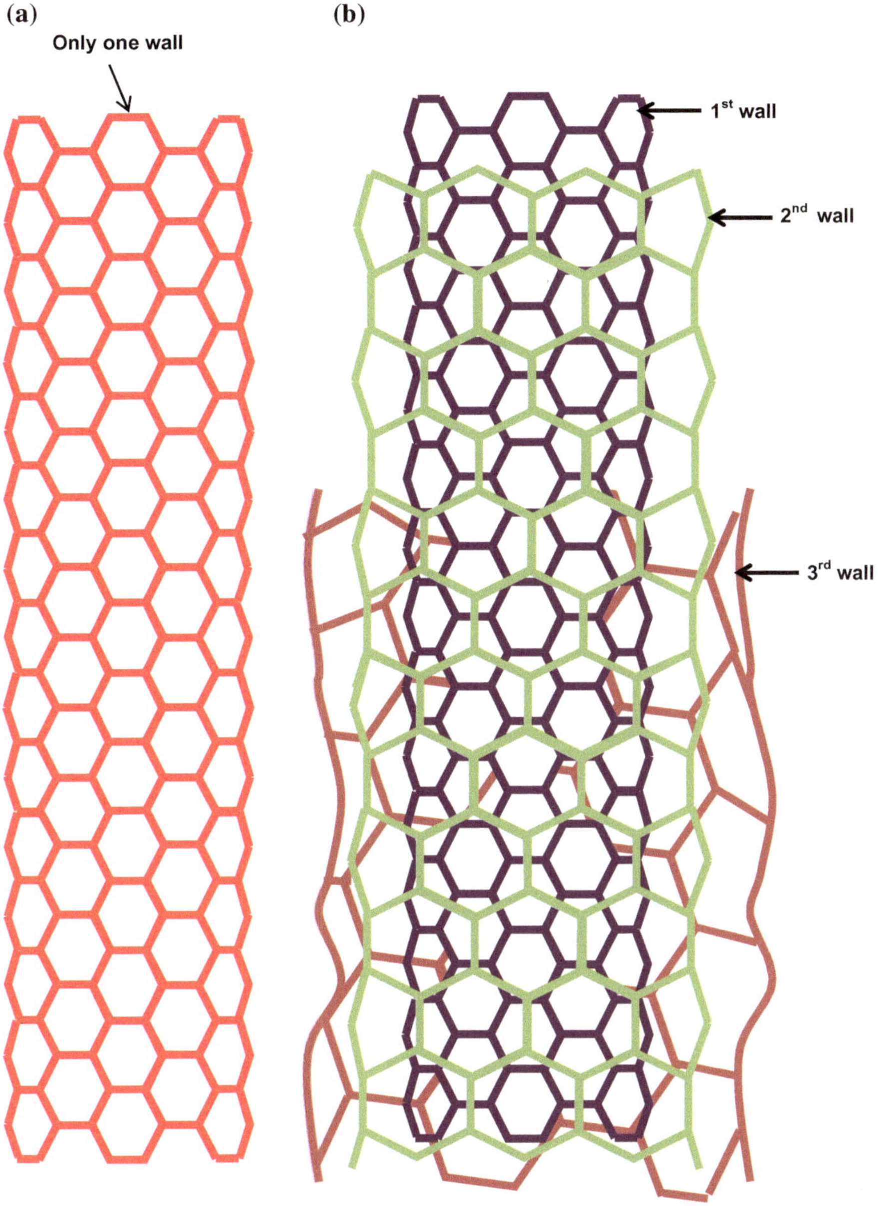

Fig. 16.1 Carbon nanotubes: **a** single-walled CNT (SWCNT) and **b** multi-walled CNT (MWCNT)

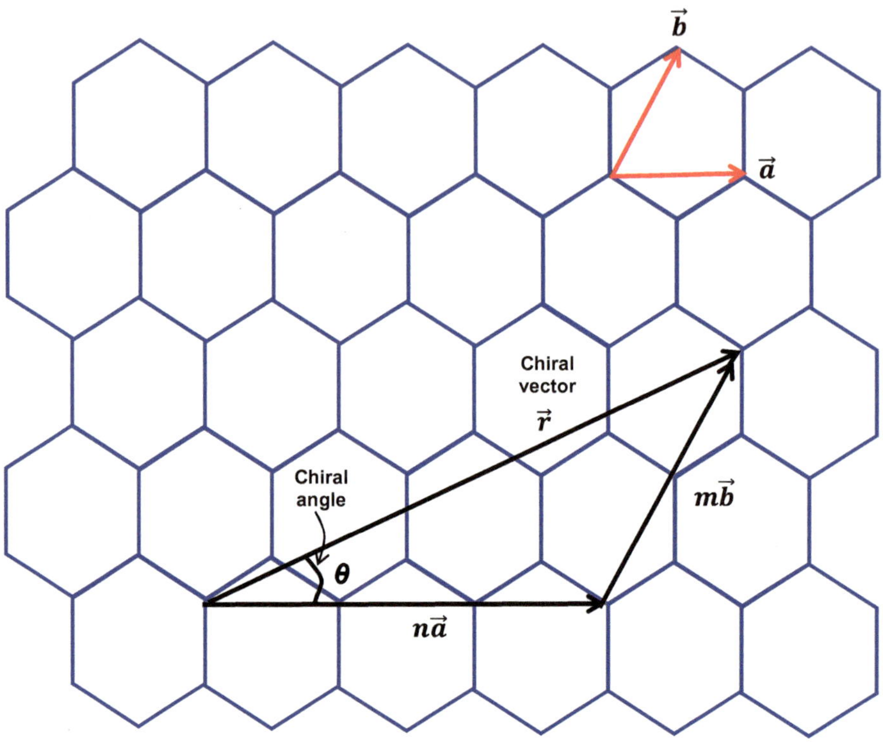

Fig. 16.2 Diagrammatic representation of chirality of a carbon nanotube

with zero bandgap. This happens because of statistical variations in the properties of its constituent SWCNTs due to which one individual SWCNT, and thus the entire whole MWCNT becomes conducting.

An armchair (n, n) SWCNT is always a good conductor like a metal and is usable as interconnect in IC. Resistivity of ropes of SWCNTs is 10^{-4} Ω-cm at 300 K [3]; resistivity of copper is 1.68×10^{-6} Ω-cm. Maximum sustainable current density is 10^7 A/cm^2. The zigzag and chiral SWCNTs are semiconducting and useful for fabricating transistors. A zigzag (n, 0) SWCNT is metallic only under the condition that n is a multiple of 3 [4]. In general, when $(n - m)$ is a multiple of 3, CNT is metallic; otherwise, it is a semiconductor with bandgap ~0.5 eV.

A semiconducting SWCNT has symmetric energy band structure between its conduction and valence bands. An obvious implication is the equality of electron and hole rest masses and also electron and hole mobilities in the SWCNT. Carrier mobilities in SWCNTs are very high ~3000–4000 cm^2/V-s, both for electrons and holes, owing to less scattering in 1D-nanostructure Additionally, ballistic transport is achievable in FET devices having channel length <200 nm.

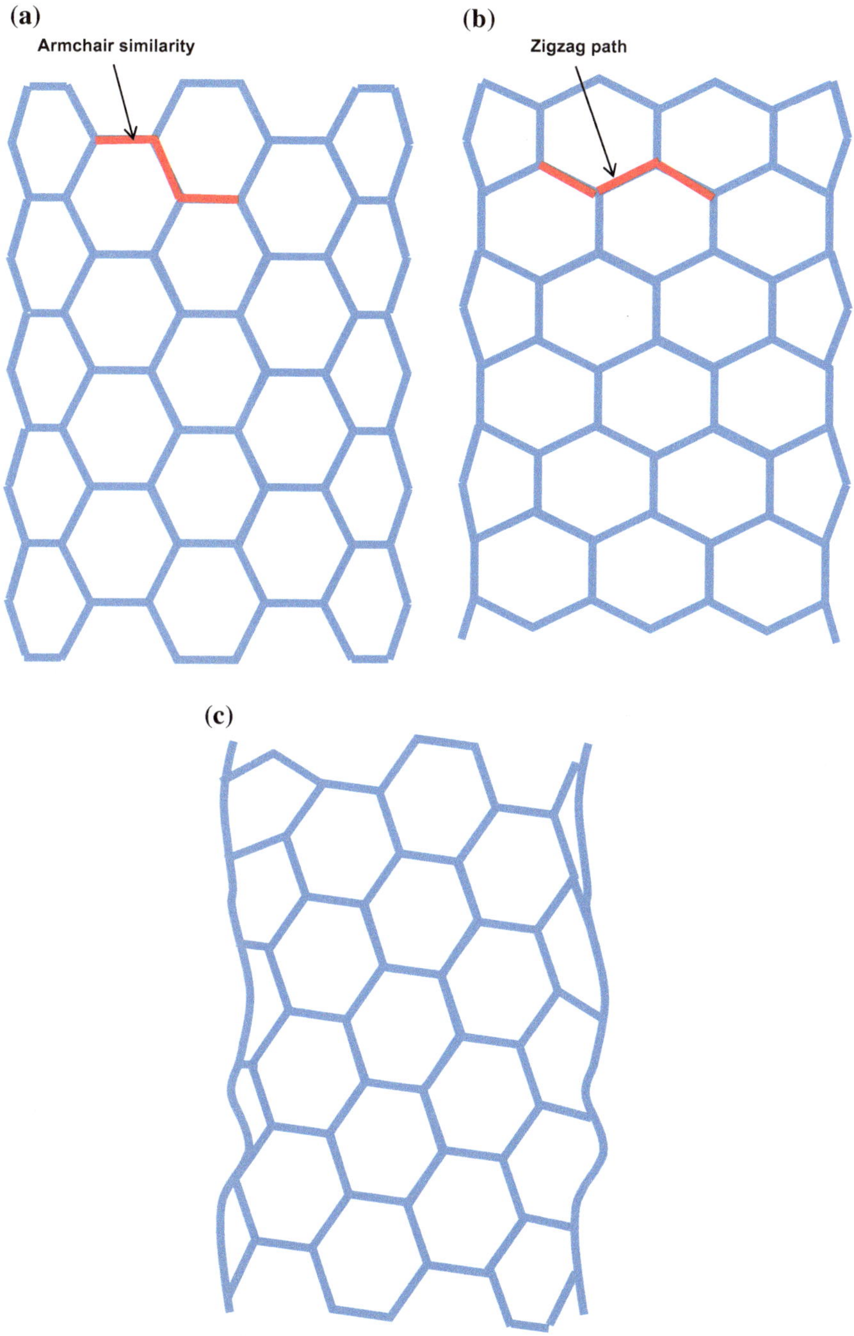

Fig. 16.3 Three types of CNTs: **a** arm chair, **b** zigzag and **c** chiral

Due to their extremely high aspect ratios, CNTs represent one of the sharpest inorganic objects serving as excellent field emitters. Field emission is the tunneling of electrons from a metallic tip into vacuum under a high electrostatic field.

16.5 Mechanical Properties of Carbon Nanotubes

Due to the strong chemical bonds between neighboring carbon atoms, the ultimate tensile strength of SWCNT is 13–52 GPa [5]; that of steel is 500 MPa. Young's modulus of SWCNT is 2.8–3.6 TPa; that of MWCNT is 1.7–2.4 TPa [6]; the value for steel is 760 MPa.

16.6 Thermal Properties of Carbon Nanotubes

Thermal conductivity of CNTs is temperature-dependent. It is 2000–6000 W/m-K at 300 K; that of steel is 50 W/m-K and copper is 385 W/m-K.

16.7 Synthesis of Carbon Nanotubes

A variety of methods for synthesis of carbon nanotubes are available. Amongst these, three methods have gained immense popularity. These methods will be discussed in the ensuing subsections.

16.7.1 Arc Discharge

It is the oldest technique for synthesizing high-quality CNTs [7]. In this technique (Fig. 16.4), an arc is struck between two graphite electrodes. These electrodes are put in an enclosure filled with an inert gas such as He or Ar. The inert gas filling is done at a low pressure of 50–700 mbar. The arc discharge process occurs at a high temperature ~4000 °C. The high temperature causes evaporation of 28% of the anode. The carbon vapor crystallizes on the extremity of the cathode. After the chamber has been de-pressurized and cooled, the CNTs and the by-products of the process are collected. Arc discharge with pure carbon gives MWCNTs. The MWCNT diameters range from 4 to 30 nm. Catalyzing the synthesis by Fe, Y, S, Ni and Mo leads to SWCNT production. The process yield expressing the carbon conversion into CNTs is 60%. The yield is enhanced to 70–90% by filling the anode with 1% Y and 4.2% Ni, and keeping the voltage drop at 30 V at 100 A [8].

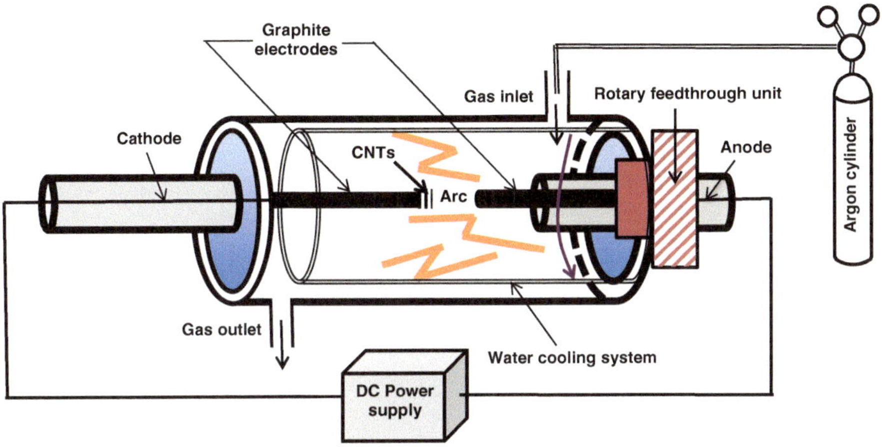

Fig. 16.4 Arc discharge device for CNT synthesis

16.7.2 Laser Ablation

Strong pulses of laser beam melt, vaporize and erode a graphite target containing small quantity of Ni and Co powder in a tubular furnace maintained at 1200 °C, and flushed by He or Ar flow (Fig. 16.5). The CNTs growing in the gas phase are carried away downstream to the cold copper collector where they condense as a spongy black deposit [9]. The nanotubes as well as the by-products are gathered. MWCNTs and SWCNTs are obtained in the same way as in arc discharge method. Both arc discharge and laser ablation methods require excessively high temperatures. At these high temperatures, the carbon source is unnecessarily evaporated. Moreover, the nanotubes produced are highly disordered and entangled ropes of CNTs, and need intense purification. The nanotubes produced in the small time intervals over which these processes run are short in length.

16.7.3 Chemical Vapor Deposition (CVD)

This method overcomes the disadvantages of arc discharge and laser ablation. It is based on thermal catalytic decomposition of hydrocarbon [10]. The substrate covered with metal catalyst nanoparticles (Ni, Co, or Fe or a combination) is heated to 700 °C in a reaction chamber (Fig. 16.6). The size of nanoparticles determines the diameters of resulting nanotubes. The nanoparticles are deposited, either by patterned or masked deposition of metal or by plasma etching of a pre-deposited metal layer. A mixture of two gases is fed to the chamber. One of these gases is the carrier gas such as H_2, N_2 or Ar. The other gas is a hydrocarbon gas, e.g., carbon monoxide (CO), acetylene (C_2H_2), methane (CH_4), ethylene (C_2H_4), ethanol

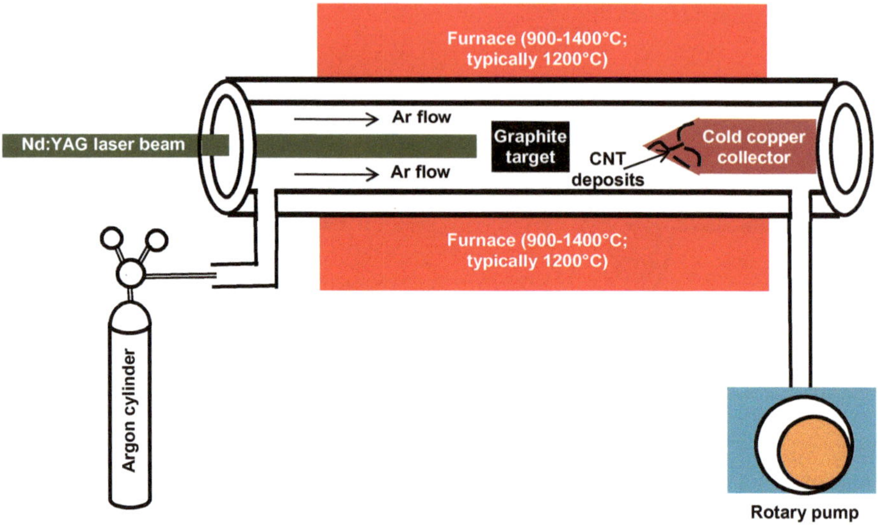

Fig. 16.5 Pulsed laser ablation method for synthesizing carbon nanotubes

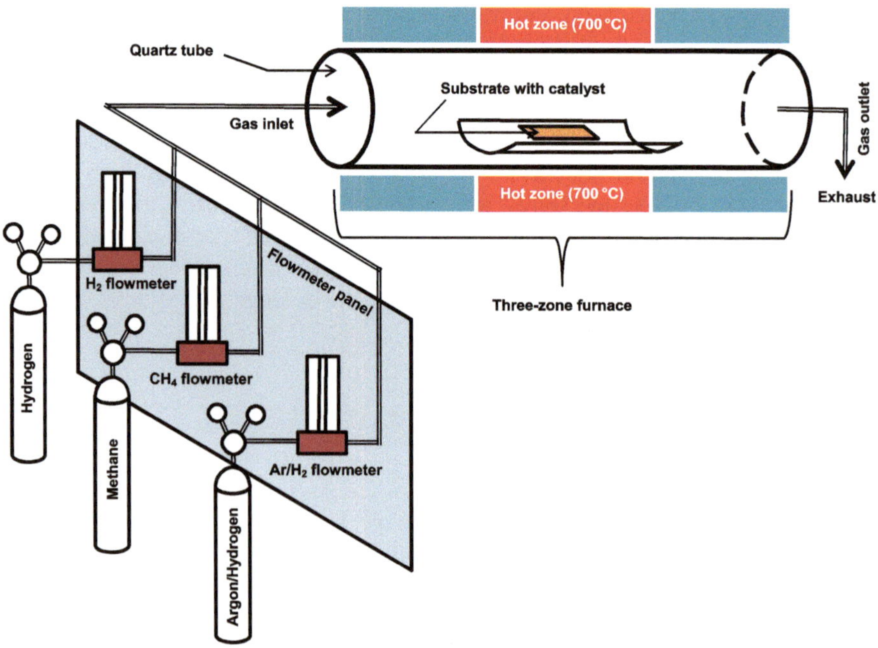

Fig. 16.6 Chemical vapor deposition apparatus for CNT growth

(C_2H_6O), or benzene (C_6H_6). The hydrocarbon works as carbon precursor, which is decomposed on the surface of the catalyst nanoparticle. The carbon is carried to the edges of the nanoparticle where the CNT grows. CNT production yield is high ~90%. Using the CVD method, by in situ growth of SWCNTs at pre-chosen positions on a substrate, nanochips can be fabricated so that the method is amenable to production of nanocircuitry. Besides offering controllability of CNT growth, the CVD method requires a low growth temperature, and is also economical.

16.8 Chirality-Controlled Synthesis of Carbon Nanotubes

A major bottleneck of all the aforementioned methods of CNT synthesis is that an inhomogeneous mixture of nanotubes of different types is deposited. These nanotubes have different chiralties, bandgaps and other properties, which make isolation of nanotubes of given properties, difficult.

To separate nanotubes of a specific chirality from a synthesized mixture, it is necessary to categorize nanotubes of particular chiralities. This is implemented using a DNA-based recognition procedure [11]. By this method, several important single-chirality semiconducting species can be purified from a mixture. A large number of short DNA sequences were identified. Each sequence recognizes and helps to purify a particular type of nanotube from a mixture. These sequences can hydrogen bond to form 2-D sheets. They can selectively fold on CNTs to form 3-D drums.

The DNA-wrapped seeds are spread on a substrate. They are annealed in air, water and hydrogen. Introduction of methane or ethanol commences vapor phase epitaxial growth of CNTs. The growth temperature is 900 °C. AFM measurements showed a significant increase in length of SWCNTs of a specific chirality after vapor phase epitaxy. In Raman spectroscopy radial breathing mode (RBM) measurement of a CNT agglomerates is a method of demarcating chiralities. Raman spectroscopy disclosed that VPE-grown SWCNTs had the same (RBM) positions. Original nanotube seeds were used for examining the different chiralities.

16.9 Doping-Free Fabrication of CNT FET

Fabrication of a CNT FET is a doping-free process [12]. In place of doping profile tailoring, the polarity of the FET is contact-controlled. It is based on the fact that element palladium (Pd) injects holes barrier-free into the CNT. Element scandium (Sc) injects electrons barrier-free into it. Hence, a P-channel FET can be fabricated by making ohmic contacts to the CNT using palladium. The element Pd makes a perfect ohmic contact with the valence band of the CNT, resulting in the P-channel device. Similarly, the N-channel FET is realized using as the contact metal. The element Sc makes an ideal ohmic contact with the conduction band of CNT.

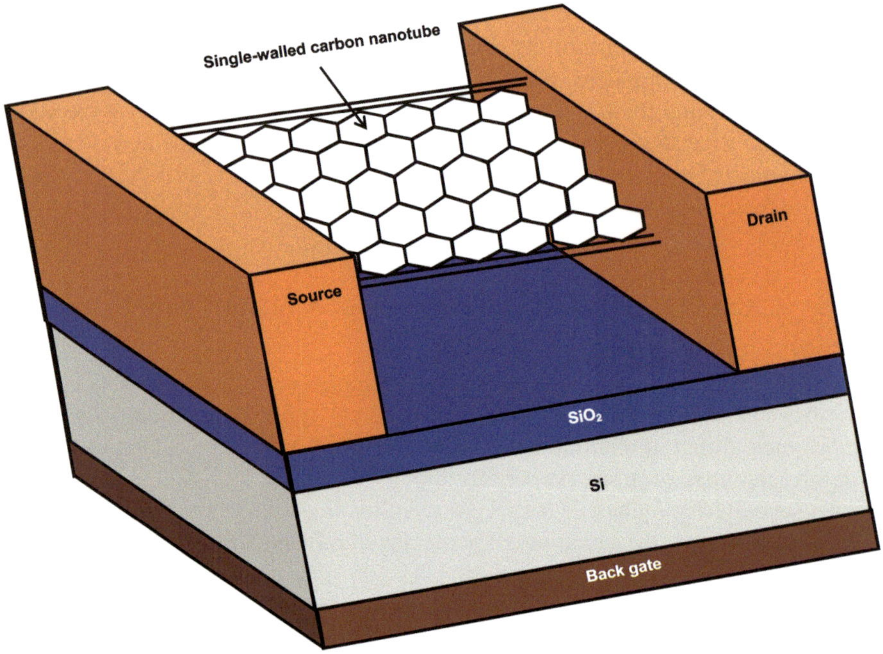

Fig. 16.7 Back-gated SWCNT FET

The main advantage of this approach of independence from impurity addition is that the CNT need not be doped P- or N-type. Besides process simplification, the damage inflicted on the CNT lattice by doping is avoided. Therefore, the lattice remains intact and the carrier mobility is preserved. Defects produced in the lattice during doping act as scattering centers decreasing the mobility. In both cases of P- and N-channel devices, the back gate configuration is used. The doped silicon substrate underlying the silicon dioxide layer is the back gate of the FET (Fig. 16.7).

16.10 Self-aligned Processes for Fabrication of CNT FET

For large-scale manufacturing of CNT FETs as components of integrated circuits, it is desired that the source, gate and drain electrodes must be positioned accurately. Suppose, the gate length in the mask is equal to the channel length. If a slight misalignment occurs during photolithography, it is likely that the channel may be broken on one side, either on the source or drain side. Then the source will not be connected to drain and current will not flow between them. To prevent this mishap, the gate length in the mask can be made slightly larger to overlap the source and drain regions. But the overlaps result in parasitic capacitances which deteriorate the high frequency characteristics by lowering the circuit speed. By a self-aligned process is meant a process which automatically guarantees the correct placement of these electrodes. The placement of the electrodes is correct when: (i) There are no gaps between the edges of source and gate as well as drain and gate to prevent channel discontinuity between source and drain. (ii) There are no overlapping regions at the above locations to avoid the unwanted parasitic effects. Self-aligned processes for fabrication of P-channel and N-channel CNT FETs are discussed in Sects. 16.11 and 16.12 respectively.

16.11 Fabrication of P-Channel CNT FET

A self-aligned process for the fabrication of P-channel CNT FETs using Pd source and drain contacts and HfO_2 gate dielectric (relative permittivity $\kappa = 15$) is developed [13]. It is applied to fabrication of CNT FET of short channel length 50 nm. It is based on the development of a lift-off process using low-temperature atomic layer deposition of HfO_2 at 90 °C [14]. The low temperature of this process allows deposition of HfO_2 and dielectrics such as aluminum oxide and zirconium oxide without impairing the function of the photoresist. Neither is there any hard baking of the photoresist nor is there any substantial outgassing from it. On P^{++}silicon substrates with 10 nm thick SiO_2 layer, SWCNT is grown by CVD. The gate region is defined in the polymer photoresist PMMA. ALD of HfO_2 is carried out with tetrakis(diethylamido)hafnium ($Hf[NEt_2]_4$) as precursor. A total of 80 ALD cycles are followed. Each cycle deposits 0.1 nm thick HfO_2 resulting in 8 nm thick HfO_2 layer. Following the DI H_2O dose, the purging time is 350 s. After the $Hf[NEt_2]_4$ dose, it is 150 s. The HfO_2 layer is covered with 50 nm thick Al layer. Spontaneously 4–8 nm thick native Al_2O_3 grows on the Al film. So, the gate is covered with a three-layer stack $HfO_2/Al/Al_2O_3$. Next, 7 nm thick Pd is deposited by highly directional electron beam evaporation. The three-layer stack at the gate subdivides the deposited palladium into two regions. These two regions are the source and drain, which are formed in perfect alignment with the gate stack as a natural consequence of this process. The insulating native Al_2O_3 layer grown over the Al metal together with the directionally deposited Pd film ensure that the gate, source and drain regions are not electrically shorted. Nonetheless, a series resistance $\sim$1.7 kΩ exists at each source/drain electrode because of the thin Pd metallization. The SWCNT lies underneath the Pd contacts; the contacts are annealed at 175 °C in argon ambient for 5 min. The SWCNT is also buried under the $HfO_2/Al/Al_2O_3$ gate stack. The reported CNT FET showed a peak transconductance $\sim$30 μS per CNT. The maximum linear on-state conductance was $0.5 \times (4q^2/h)$ where q is the electronic charge and h is Planck's constant. The saturation current was $\sim$25 μA. The ratio of currents in on-state to off-state was $>10^3$ at $V_{DS} = 0.3$ V. The subthreshold swing was $\sim$110 mV/decade. The bottom gate, i.e., the silicon substrate was grounded during measurements. In Si MOSFETs, surface roughness, Coulomb and phonon scattering effects at the

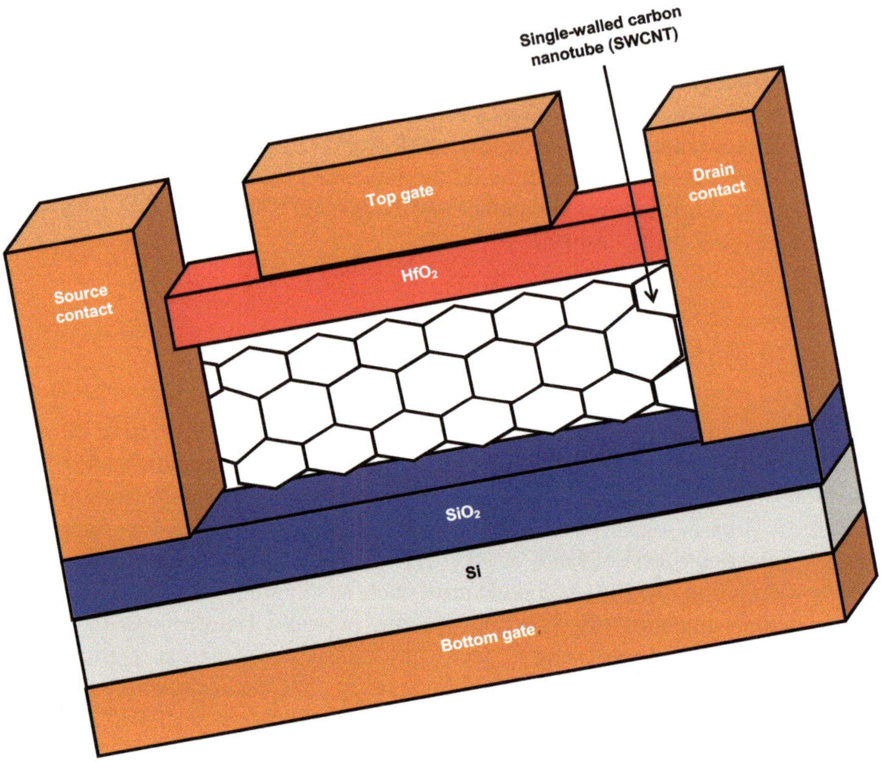

Fig. 16.8 SWCNT FET with *top* (*front*) and *bottom* (*back*) gates

interface between Si and high-κ dielectric seriously hamper carrier transport, degrading mobility. These effects are less troublesome at the van der Waals interface with HfO_2 film. Therefore CNT FETs promise superior performance.

Figure 16.8 shows the schematic diagram of a dual-gated (top and bottom) CNT FET.

16.12 Fabrication of N-Channel CNT FET

In the self-aligned process for fabrication of a P-channel CNT FET, the growth of native oxide on deposited Al metal was beneficially utilized for isolation between source, gate and drain terminals. But aluminum is a reactive metal. In a P-channel device, low work function reactive metals like aluminum (4.08 eV) are usable. But the same does not apply to N-channel FETs where high work function metals such as Pd (5.12 eV), Au (5.1 eV), Pt (6.35 eV) are required. These metals are non-reactive or inert. Therefore, the self-aligned process for P-channel CNT FET is not applicable to an N-channel CNT FET.

The process starts with a silicon substrate with a thin SiO_2 layer [15]. Over the SiO_2 layer lays the SWCNT. The photoresist PMMA is coated on the wafer. In the photoresist, the windows for the source and drain regions are defined by electron beam lithography. Scandium film of thickness 60 nm is deposited over the patterned photoresist layer. Lift-off process is used to form source and drain electrodes. These electrodes have high vertical sidewalls. At their locations, they cover the underlying SWCNT while the remaining SWCNT is exposed. The next step is delineation of the gate window by electron beam lithography. Then 15 nm thick layer of HfO_2 is formed by atomic layer deposition at 90 °C. The HfO_2 layer is followed by 5 nm thick titanium film deposited by electron beam evaporation. After Ti evaporation, the completed HfO_2/Ti gate stack is defined by lift-off process.

Mechanisms of growth of HfO_2 and Ti films are different. It is the difference in growth mechanisms of these films, which is responsible for success of this process. The HfO_2 film covers the whole surface. It also covers the sidewalls of the scandium layer over source and drain. Thus it covers both the horizontal surfaces and vertical sidewalls uniformly. It provides a good coverage of the step at the scandium layer. This is essential to ensure insulation between source and gate, and drain and gate. On the opposite side, the titanium film covers only the horizontal surfaces of HFO_2 film. It is not formed on the vertical sidewalls of the HfO_2 film. Its step coverage is not good. Discontinuity of titanium film at the sidewalls of the HfO_2 layer over source and drain electrodes isolates the source, gate and drain electrodes adequately to enable satisfactory device operation. In absence of this discontinuity, these electrodes will be shorted together leading to device malfunction. The final fabricated N-channel CNT FET showed that the boundaries of the source, gate and drain regions were well placed at prescribed positions with great precision.

The process was applied to fabricate both a short channel (120 nm) CNT FET as well as a long channel CNT FET (2 μm). For the 120 nm CNT FET, the peak transconductance was 25 μS, on-state conductance was $0.32 \times (4q^2/\mathrm{h})$, the saturation current was 25 μA, the ratio of on- sate and off-state currents was 10^4 at $V_{DS} = 0.3$ V, and the subthreshold swing was 100 mV/decade. These experimental results for N-channel CNT FET are comparable with aforementioned data for P-channel CNT FET. The comparability of performance metrics shows that Pd-contacted P-channel CNT FETs can be fabricated alongside Sc-contacted N-channel CNT FETs to realize complementary structures like the coveted CMOS configuration of Si ICs.

The intrinsic delay time of 120 nm CNT FET was 0.86 ps and energy delay product was 6×10^{-28} Js/μm, which are comparable with that for a 40-nm channel length Si MOSFET. By quantitatively fitting the experimental data, the mean free path of electrons in the 120 nm channel length CNT FET was found to be 190 nm. Since the electron mean free path is much longer than the channel length, this transistor is a ballistic device. From the 2 μm channel length device, the peak mobility of electrons was extracted to be 4650 cm^2/V-s, which shows that HfO_2 deposition over SWCNT does not adversely influence its charge transport properties.

The above process has the advantage that it can be extended to other suitable gate metals. This provides a tool to adjust the threshold voltage of the N-channel CNT FET, e.g. the threshold voltage with Ti gate metal was −0.67 V. For the Pd gate CNT FET, it was displaced by 0.51 V along positive V_{GS} direction.

16.13 Complementary Symmetry SWCNT FET Devices

Following in the footsteps of CMOS technology, pairs of opposite polarity CNT transistors are fabricated using Pd and Sc contacts for P- and N-channel devices [16]. Cu-catalyzed CVD method was used for directional growth of SWCNTs on a silicon wafer with 500 nm thick SiO_2 [17]. For catalyzing the growth of SWCNTs, monodispersed Cu nanoparticles are put on the substrates. The growth is done in methanol (or ethanol or isopropanol) with Ar and/or H_2 as carrier gas at a temperature of 825–850 °C for 15 min. Horizontally aligned arrays of SWCNTs with length of about a few hundred microns are synthesized. Using the silicon substrate as a back gate, semiconducting SWCNTs are recognized.

P-channel and N-channel CNT FETs are fabricated. SWCNT diameter is 2 nm and channel length is 1 μm. First the P-channel CNT FET is fabricated using the following sequence of process steps: Lithography for source, drain electrodes-Pd film deposition-Lift-off process-Lithography for gate window definition-ALD of HfO_2-Pd film deposition-Lift-off process. Then N-channel CNT FET is fabricated by the process sequence: Source, drain electrode lithography-Sc film deposition-Lift-off process-Gate window opening lithography-HfO_2 film ALD-Sc film deposition-Lift-off process. Note that this process differs from that for P-channel CNT FET only in replacement of palladium by scandium.

For the P-channel CNT FET, the maximum ratio of on-and off-state currents is $I_{on}/I_{off} = 10^5$ at $V_{DS} = 0.1$ V, subthreshold swing (*SS*) is 90 mV/decade, on-state conductance is $0.16 \times (4q^2/h)$, transconductance is 17 μS and peak mobility is 3300 cm^2/V-s. For the N-channel CNT FET, the ratio I_{on}/I_{off} is 10^5, SS is 100 mV/decade, on-state conductance is $0.13 \times (4q^2/h)$, transconductance is 14.5 μS and peak mobility is 3000 cm^2/V-s. The intrinsic gate delay is ~10 ps for both P- and N-channel CNT FETs. Thus the electrical performance parameters of P- and N-channel CNT FETs are quite matched with each other. The symmetric energy gap of CNT with nearly identical electron and hole effective masses results in almost equal electron and hole mobilities in CNT. Both static and dynamic characteristics of the two polarities of CNT FETs are equivalent.

Resemblance of characteristic parameters of P- and N-channel CNT FETs places the CNT devices in a more favorable circumstance as compared to Si CMOSFETs. In Si, the electrons have a smaller effective mass than holes. Consequently the electron mobility is about 1350/480 = 2.8 times higher than hole mobility in silicon. So, in silicon, the PMOS device is inferior to the NMOS device. Of course, the DC characteristics of PMOS device can be brought at par with the NMOS device by increasing the channel width of the PMOS device by a factor of 2.8. Although the

DC underperformance is compensated, the dynamic performance of PMOS is not changed. It is still slower than NMOS and lags behind it.

Besides the above advantages of CNT FETs, it is easy to see that the fabrication process of a complementary pair of P-channel and N-channel devices on the same SWCNT is much simpler than that for fabrication of a complementary pair of PMOS and NMOS in standard Si CMOS technology. Firstly, it requires no doping. Secondly, it is isolation-free.

16.14 Pass Transistor Logic (PTL)

In CMOS logic, the primary input signals are fed to the gate terminals. But in pass transistor logic, the primary input signals may be used to drive any of the three terminals: gate, source and drain. PTL eliminates redundant transistors. For executing the same logic function, PTL requires less power consumption than normal CMOS logic. Also, it provides faster operational speed. But for PTL, a mandatory requirement is zero threshold voltage. This is arduous to accomplish with Si technology. In Si technology, doping is the controlling parameter. The main disconcerting issues are plummeting of threshold voltage and decrease in gain. *Tout au contraire*, threshold voltage control is a relatively straightforward process in CNT technology. Since the threshold voltage of a CNT FET is proportional to the work function of the gate metal, the work function of this metal can be altered in desired direction as per demand. The threshold voltages of P-channel and N-channel CNT FETs were centered by threshold voltage engineering towards approximately 0 V [18].

16.15 Discussion and Conclusions

CNT nanoelectronics shows striking dissimilarities with Si nanoelectronics. Doping via ion implantation and thermal diffusion is a very important process in Si technology. But its non-requirement in CNT technology constitutes a major difference between the two technologies. Doping is associated with creation of lattice defects. Absence of doping is a significant advantage in favor of CNT technology. Moreover, at the nanoscale, random doping fluctuations are a primary cause of concern because they are responsible for unpredictable variability amongst devices, which is particularly haunting at low dimensions. A second primary difference between CNT nanoelectronics and Si nanoelectronics arises from the inequality of performance of P-channel and N-channel Si CMOS devices, which makes it essential to make amends for this shortfall during device design. On the other hand, P-channel and N-channel CNT are quite matched with each other so that complementary nanostructures are much easier to implement.

A major challenge is that the as-grown CNTs contain both semiconducting and metallic CNTs, from which the semiconducting variety needs to be sorted out. Techniques for selectively growing CNTs with enriched semiconducting component need to be developed.

Review Exercises

16.1 What is meant by a fullerene? What is a spherical fullerene called? What is the name of a cylindrical fullerene?
16.2 Is fullerene an allotrope of carbon? Name the other two allotropes of carbon.
16.3 What is the average diameter of a single-walled carbon nanotube?
16.4 What is meant by chirality of a carbon nanotube? Name and define the parameters used to represent chirality of a CNT.
16.5 A CNT has a chiral angle of 0°. What is this CNT called?
16.6 The chiral angle of a CNT is 30°. Identify the type of CNT.
16.7 How do the electrical properties of a CNT depend on the lattice translational vectors and tube diameter?
16.8 Why does an MWCNT usually show metallic character while an SWCNT can be metallic or semiconducting?
16.9 Does an armchair SWCNT shows metallic or semiconducting properties? What is the condition under which a zigzag (n, 0) SWCNT shows metallic character?
16.10 What is the effect of a symmetric energy band structure of SWCNT between conduction and valence bands on the rest masses and mobilities of charge carriers in the SWCNT?
16.11 Why are CNTs excellent field emitters?
16.12 How do the tensile strength and Young's modulus of CNTs compare with the same metrics for steel? Which is a better thermal conductor: CNT or copper?
16.13 Describe the arc discharge method of synthesizing CNTs. How can SWCNTs be synthesized by this method?
16.14 Describe the laser ablation method of producing CNTs? What are the common shortcomings shared by the laser ablation and arc discharge methods?
16.15 What is the role of catalyst nanoparticles in the CVD method of CNT synthesis? What are the carrier gases used? What gas is used as a carbon precursor in this method? How can this method be applied to nanochip manufacturing?
16.16 Does CVD method produce CNTs of specified chiralities? How can chiralities of the nanotubes be controlled during production?
16.17 What are the advantages of fabricating a CNT FET without doping? Name the contact metals used for fabricating: (i) P-channel FET, (ii) N-Channel FET.

16.18 What is a self-aligned process for fabricating a FET device? What are the benefits of using such a process?

16.19 How does low-temperature atomic layer deposition of hafnium oxide help in self-aligned fabrication of a P-channel CNT FET? Describe the main steps of this process.

16.20 Why is the self-aligned process for fabrication of P-channel FET not applicable to an N-channel device? Describe the self-aligned process by which an N-channel CNT FET can be fabricated. How do different growth mechanisms of HfO_2 and Ti contribute to the successful implementation of this process?

16.21 For qualification of a FET as a ballistic device, the mean free path is longer than the channel length. True or false?

16.22 Describe the fabrication of a complementary symmetry SWCNT device. What is the principal advantage in favor of SWCNT FETs regarding matching of transistors as compared to silicon CMOS? In what respects is the fabrication of opposite polarity CNT transistors on the same SWCNT simpler than Si CMOS technology?

16.23 How does pass transistor logic differ from CMOS logic? What is the necessary condition for its implementation? Why can it be easily realized with CNT FETs than with Si MOSFETs?

16.24 Point out two major dissimilarities between silicon nanoelectronics and CNT nanoelectronics. Mention one challenge facing CNT nanoelectronics.

References

1. Structure of CNTs (2016). https://sites.google.com/site/cntcomposites/structure-of-cnts. Accessed 16 Jan 2016
2. Bandaru PR (2007) Electrical properties and applications of carbon nanotube structures. J Nanosci Nanotechnol 7:1–29
3. Electrical Transport (2016). http://www.pa.msu.edu/cmp/csc/ntproperties/electricaltransport.html. Accessed 16 Jan 2016
4. Matsuda Y, Tahir-Kheli J, Goddard WA (2010) Definitive band gaps for single-wall carbon nanotubes. J Phys Chem Lett 1:2946–2950
5. Ruoff RS, Qian D, Liu WK (2003) Mechanical properties of carbon nanotubes: theoretical predictions and experimental measurements. C R Physique 4:993–1008
6. Lourie O, Wagner HD (1998) Evaluation of Young's modulus of carbon nanotubes by micro-Raman spectroscopy. J Mater Res 13(9):2418–2422
7. Arora N, Sharma NN (2014) Arc discharge synthesis of carbon nanotubes: comprehensive review. Diam Relat Mater 50:135–150
8. Journet C, Maser WK, Bernier P et al (1997) Large-scale production of single-walled carbon nanotubes by the electric-arc technique. Nature 388:756–758
9. Hornbostel B, Haluska M, Cech J et al (2006) Arc discharge and laser ablation synthesis of single-walled carbon nanotubes. In: Popov VN, Lambin P (eds) Carbon Nanotubes, Springer, pp 1–18
10. Kumar M, Ando Y (2010) Chemical vapor deposition of carbon nanotubes: a review on growth mechanism and mass production. J Nanosci Nanotechnol 10(6):3739–3758

11. Tu X, Manohar S, Jagota A et al (2009) DNA sequence motifs for structure-specific recognition and separation of carbon nanotubes. Nature 460(7252):250–253
12. Peng L-M, Zhang Z, Wang S (2014) Carbon nanotube electronics: recent advances. Mater Today 17(9):433–442
13. Javey A, Guo J, Farmer DB et al (2004) Self-aligned ballistic molecular transistors and electrically parallel nanotube arrays. Nano Lett 4(7):1319–1322
14. Biercuk MJ, Monsma DJ, Marcus CM et al (2003) Low-temperature atomic-layer-deposition lift-off method for microelectronic and nanoelectronic applications. Appl Phys Lett 83 (12):2405–2407
15. Zhang Z, Wang S, Ding L et al (2008) Self-aligned ballistic n-type single-walled carbon nanotube field-effect transistors with adjustable threshold voltage. Nano Lett 8(11):3696–3701
16. Zhang Z, Wang S, Wang Z et al (2009) Almost perfectly symmetric SWCNT-based CMOS devices and scaling. ACS Nano 3(11):3781–3787
17. Zhou W, Han Z, Wang J et al (2006) Copper catalyzing growth of single-walled carbon nanotubes on substrates. Nano Lett 6(12):2987–2990
18. Ding L, Zhang Z, Pei T et al (2012) Carbon nanotube field-effect transistors for use as pass transistors in integrated logic gates and full subtractor circuits. ACS Nano 6(5):4013–4019

Chapter 17
Graphene-Based Nanoelectronics

Abstract The ultrahigh room-temperature carrier mobility in graphene makes it very useful for microwave and high-frequency devices. Additionally, high current-carrying capability >10^8 A/cm^2 together with high thermal conductivity $\sim$2000–4000 Wm^{-1}K^{-1} for freely suspended graphene and $\sim$600 Wm^{-1}K^{-1} for SiO_2-supported graphene establish its superiority among nanoelectronic materials. Graphene flakes are easily isolated from graphite by mechanical exfoliation. Graphene can be grown on metal films and transferred to desired substrates. It can be grown epitaxially on silicon carbide. Graphene sheets can also be synthesized by a substrate-free process in the gas phase. Planarity of graphene makes widely practiced planar processes of semiconductor industry applicable to graphene. On the downside, the bandgap of graphene is zero. Hence, graphene transistors cannot be switched off effectively. However, single-layer graphene transistors show excellent performance in GHz analog circuits. By quantum confinement, a bandgap is opened in graphene when cut into nanoribbons. Bandgap is also created by applying a perpendicular electric field to bilayer graphene. However, carrier mobility in nanoribbons is lower than in large-area graphene. Present status of graphene nanoribbon and bilayer transistors is described. Although they display higher on-off current ratios than transistors fabricated on original graphene, intensive efforts are required to realize the full potentiality of graphene for nanoelectronics.

17.1 Introduction

Graphene is the name ascribed to a single atomic layer of graphite (Fig. 17.1). It is actually two-dimensional graphite. It extends along two dimensions only: length and breadth. The third dimension (height) is taken as zero because of its small monoatomic thickness. It is a sp^2-carbon allotrope. Graphene is a honeycomb-type, hexagonal lattice. This lattice is remindful of benzene ring-like structure. At each corner of the hexagonal lattice is placed a carbon atom. This lattice is made of tightly bound carbon atoms through 0.142 nm long chemical bonds. Its lattice constant is 0.246 nm.

V.K. Khanna, *Integrated Nanoelectronics*,
NanoScience and Technology, DOI 10.1007/978-81-322-3625-2_17

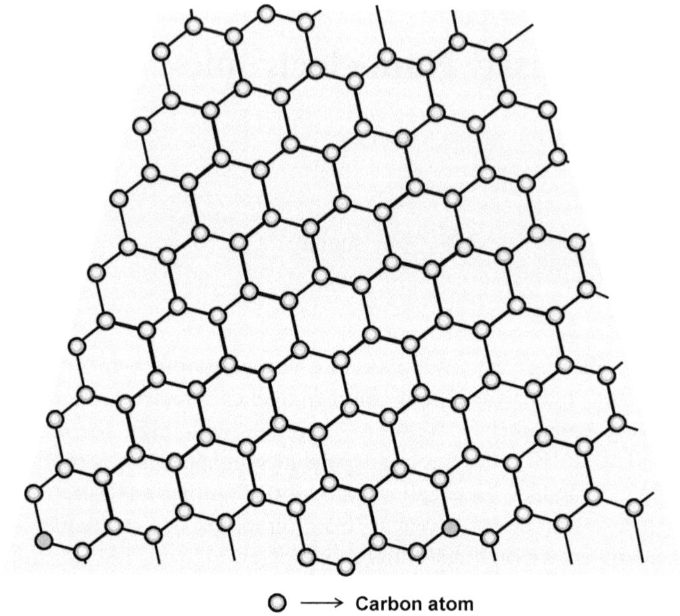

Fig. 17.1 The molecular structure of graphene

17.2 Electrical Properties of Graphene

By wrapping around graphene into spherical shape, zero-dimensional bucky balls are made. By rolling it into sheets, one-dimensional carbon nanotubes are constructed. By stacking graphene layers, three-dimensional graphite is built. It is the mother of buckminsterfullerene, CNT and graphite. The properties of graphene are critically controlled by its quality, determined by the amount of defects and impurities present.

Carbon atom contains 4 electrons in its outermost shell. In graphene, each carbon atom is bound to three neighboring carbon atoms. Three valence electrons of carbon atoms participate in chemical bonding on the 2-D plane. The fourth electron is left free in the 3rd dimension. This free electron contributes to electrical conduction. It can be located above or below the grapheme sheet. It is called a π-electron.

Graphene is a semimetal. In its energy band diagram, there is zero overlap between conduction and valence bands. It can also be looked upon as a semiconductor whose bandgap is zero. Therefore, in graphene, both electrons and holes act as charge carriers. The effective mass of these electrons and holes is zero. The carriers behave as a 2D gas comprising massless Dirac fermions. As these electrons lack mass, their behavior is similar to photons. Carriers can travel distances up to 300 nm without any scattering. Carrier mobility in graphene is extra-ordinarily high

[1]. For epitaxial graphene, it is 27000 cm^2/V-s [2]. Suspended single layer of graphene shows a mobility >200,000 cm^2/V-s at electron density $\sim 2 \times 10^{11}$ cm^{-2} [3]. This high value is achievable by reducing impurities because temperature dependence of mobility is weak. Graphene stands at the pinnacle of all known materials regarding mobility in intrinsic state. It has the highest carrier mobility. For comparison, the electron mobility in silicon is 1400 cm^2/V-s and in GaAs, it is 8500 cm^2/V-s.

Discouragingly, the semimetallic behavior of graphene allows a low ratio <10 between on-state and off-state currents of a graphene FET. This ratio is very low than the desired value $>10^3$ required for logic circuit operation. Nonetheless, the high mobility makes graphene very useful for ultra-fast circuits.

The 2-D planar nature of graphene allows easy upscaling of drive current of graphene FETs by increasing the channel width.

17.3 Mechanical Properties of Graphene

The eventual tensile strength of graphene is 130 GPa. This is much higher as compared with 13–52 GPa of SWCNT. Graphene is excessively light ~ 0.77 mg/m^2. This turns out to be 1/1000th fraction of that of paper. Young's modulus of graphene is 0.5 TPa.

17.4 Optical Properties of Graphene

Graphene is optically transparent. A solo layer of graphene absorbs 2.3% of white light. A second layer absorbs another 2.3% of white light [4].

17.5 Preparation of Graphene

17.5.1 Micromechanical Exfoliation

Graphene is peeled off by Scotch adhesive tape from highly oriented pyrolytic graphite (HOPG). In graphite, graphene sheets are loosely fastened by van der Waals forces. Repeated steps of peeling may be necessary from the multi-layer graphene obtained. The detached graphene is rubbed against the surface of silicon dioxide-on-silicon substrate giving random graphene flakes of different sizes and thicknesses.

17.5.2 Growth on Metals Followed by Transfer to Insulating Substrates

In this low-cost, scalable method, CVD is carried out by exposing Ni thin films to diluted hydrocarbon flow at ambient pressure at 900–1000 °C [5]. The Ni thin films are polycrystalline. They are formed on SiO_2/Si substrates by e-beam evaporation. They are annealed to form crystalline grains resembling those of the single-crystal substrates used for epitaxial graphene growth. Thus grown gaphene films, 1–12 layers thick and up to 20 μm in lateral size, are transferred to non-specific substrates by wet etching of Ni after coating a support material such as PMMA on the Ni/graphene surface.

17.5.3 Thermal Decomposition of Silicon Carbide

Single-crystal graphene is epitaxially grown on a 4H-SiC or 6H-SiC substrate by its thermal decomposition during high-temperature annealing at 1250–1450 °C under ultrahigh vacuum. In this bottom-up growth mode, the number of atomic layers is controllable by changing silicon desorption rate through annealing temperature and time. At the high annealing temperature, Si atoms leave the SiC surface by sublimation. The surface therefore becomes rich in carbon. Hence, the process may be considered as controlled graphitization of SiC surface. The growth depends on the polarity of the SiC surface, i.e., Si or C-face. On a (0001) silicon-terminated face, growth rate is low. The growth self-terminates after a short time producing a monolayer. On a (000-1) carbon-terminated face, the growth rate is high, not self-terminating and gives multiple layers up to 5–100 layers. Because of co-existence of different layers, the graphene formed is inhomogeneous.

Graphene derived from vacuum decomposed SiC contains small grains of size 30–200 nm. Ex situ graphitization of SiC in a dense atmosphere of inert gas Ar at a pressure of 1 bar can control the sublimation of silicon to provide sizeable, homogeneous monolayers of graphite with large domain size [6].

17.5.4 Substrate-Free Deposition

A substrate–free method to deposit graphene nanosheets uses atmospheric pressure microwave plasma reactor [7]. An aerosol containing argon gas and ethanol drops is sent into argon plasma through an alumina tube. The products of the reaction deposit on nylon membrane filters. The graphene sheets are sonicated in methanol. The sheets showed good stability in ambient storage. The atmospheric pressure reactor for graphene production is simple in operation.

17.6 First Graphene Top-Gated Transistor-like Field-Effect Device

P-type boron-doped silicon wafers with acceptor concentration = 1×10^{15} cm^{-3} are thermally oxidized (thickness of 300 nm) [8]. Mechanically Oexfoliated graphene films are deposited on SiO_2. Ti/Au contacts are deposited and gate region is defined lithographically. A gate stack of SiO_2 (20 nm)/ Ti (10 nm)/Au (100 nm) is evaporated followed by lift-off process. The device has a gate length = 500 nm.

It is found that when there is no top gate, the bottom gate bias modulates the drain-source current by almost an order of magnitude [8]. Without top gate, the extracted electron mobility is 4780 cm^2/V-s and the hole mobility is 4790 cm^2/V-s at an effective electric field = 0.4 MV/cm. After deposition of the top gate, electron mobility is 530 cm^2/V-s and hole mobility is 710 cm^2/V-s at the same effective field. The decrease in mobilities takes place due to involvement of the π-orbitals on the top surface in van der Waals bonds with SiO_2. Nevertheless, the electron and hole mobilities in graphene remain comparatively higher than universal mobility values for Si MOSFETs in which electron mobility is 490 cm^2/V-s and hole mobility is 95 cm^2/V-s at 0.4 MV/cm.

17.7 High-Frequency Graphene Transistor

Top-gated graphene FETs with gate length 240 nm are fabricated by epitaxially forming graphene on the Si face of a SiC wafer annealed at 1450 °C [9]. Hall-effect mobility is 1000–1500 cm^2/V-s. Following the spin coating of an interfacial polymer layer, 10 nm thick HfO_2 is formed by ALD, maintaining the mobility between 900 and 1520 cm^2/V-s. The transition frequency f_T is 100 GHz, which is far above 40 GHz value for Si FET of identical gate length. As radiofrequency FETs are not required to turn off, such FETs are usable for RF circuits. However, the low on-off ratio, in the range of 2–20, renders them inadequate for logic circuit applications. Another drawback of graphene FETs is their linear output characteristic, either without saturation or with weak saturation, or a saturation-like nature with a second linear region, which adversely affects RF performance as well.

17.8 Opening a Bandgap in Graphene

A significant and definite energy bandgap can be created in graphene in three ways: (i) By constraining graphene in one direction to produce narrow strips <10 nm wide, either by cleaving graphene lithographically, by synthesizing from polycyclic molecules or by unzipping MWCNTs by intercalation of K/Na alloy between their walls [10]. For widths <20 nm, the graphene nanoribbons (GNRs) have a bandgap >200 meV. The bandgap of armchair as well as zigzag GNRs is inversely proportional to the width of the nanoribbon. As the bandgap increases, the shape of conduction and valence bands changes to parabolic from conical shape, resulting in increase of effective mass and decrease of mobility of carriers. (ii) By applying an electric field on bilayer graphene in a direction orthogonal to the bilayer. Bilayer graphene is gapless. It has parabolic bands. A bandgap ~200–250 meV opens up in bilayer graphene in a perpendicular field of $1–3 \times 10^7$ V/cm. The size of the bandgap can be adjusted by varying the strength of the applied field. (iii) By application of a global uniaxial strain on graphene.

17.9 GNR Transistor

A top-gated GNR transistor is fabricated by integrating ultrathin high-permittivity dielectric layer with graphene [11]. The dielectric is in the form of Si/HfO_2 core/shell nanowires. Si nanowire is highly doped. Its diameter is 30 nm and thickness of HfO_2 film is 1–2 nm. Gate length is 500 nm and gate width is 10 nm. HfO_2 is formed by ALD at 250 °C. Tetrakis(dimethylamido) hafnium is used as the precursor and water as oxidant. The substrate with resistivity <0.001 Ω-cm is the back gate electrode while the 300 nm thermal SiO_2 layer is the back gate dielectric. The source, gate and drain electrodes are made of Ti (50 nm)/Au(50 nm).

For fabricating the FET structure, the Si/HfO_2 nanowires are aligned on the graphene flakes mounted on Si substrate. This is done by a physical dry transfer process. In subsequent lithographic and metallization processes, source and drain electrodes are formed. Exposed graphene is removed by oxygen plasma etch. Graphene below the nanowire is left behind. It is connected to two graphene blocks underneath the source and drain electrodes. The core of the nanowire is connected to the external electrode by etching away the top half of the dielectric shell in oxygen plasma, and depositing the top gate electrode metal.

The on-state current is 27 μA at $V_{DS} = 1$ V and $V_{GS} = -1$ V. The ratio I_{on}/I_{off} is ~70 at $V_{DS} = 0.1$ V. The maximum transconductance g_m is 32 μS at $V_{DS} = 1$ V. The scaled value of I_{on} is ~2.7 mA/μm, and that of g_m is 3.2/μm, which are much higher than those of sub-100 nm Si PMOSFET and NMOSFET using high-κ dielectric.

Two types of graphene FET structures, one single and another dual-gated, are shown in Fig. 17.2.

17.10 Graphene Bilayer Transistor

A graphene bilayer transistor is fabricated as follows [12]: Graphene bilayer flake identification—Lithography and Ti (0.5 nm)/Pd(20 nm)/Au(20 nm)/Ti(5 nm) source/drain metallization—Lithography for defining graphene bilayer channel—Spin

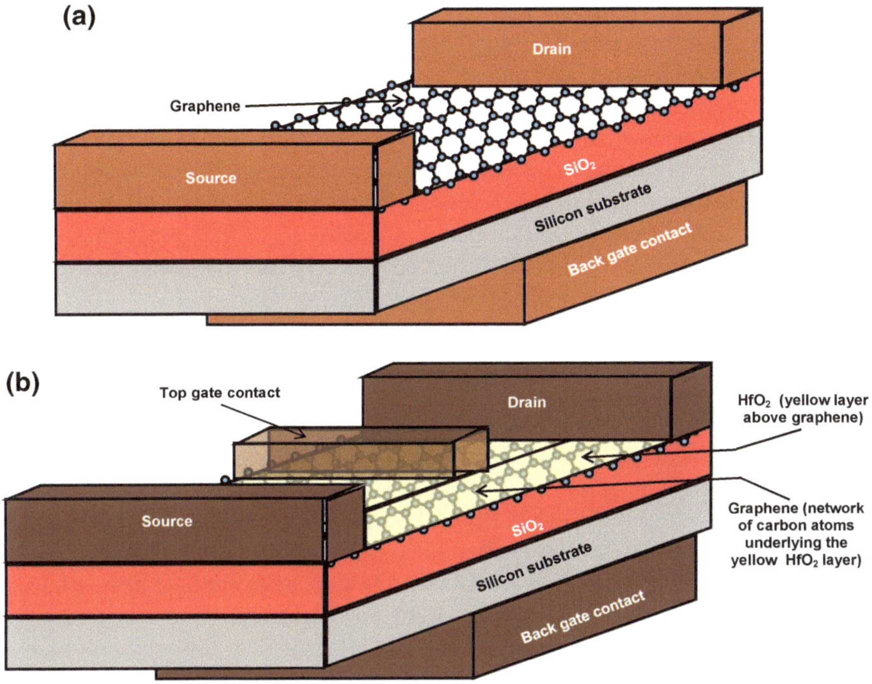

Fig. 17.2 Graphene FETs: **a** Back-gated, and **b** top and bottom gated

coating of organic seed layer for ALD- HfO_2 ALD at <200 °C-Lithography and top Ti (5 nm)/Au (25 nm) gate metallization. The top gate consists of organic seed layer and 10 nm thick HfO_2 film. The bottom gate is 300 nm thick SiO_2. The channel is 3 μm long and 1.6 μm wide. The ratio I_{on}/I_{off} is ~100 at room-temperature. The bandgap is >130 meV.

17.11 Hexagonal Boron Nitride (h-BN)-Graphene-Hexagonal Boron Nitride FET

Graphene FETs have been fabricated using thermal silicon dioxide or silicon carbide as a substrate. Both these substrates have a derogatory influence on the carrier mobility in graphene. In SiO_2, the high surface roughness combined with a high density of charge trapping centres and defects leads to mobility degradation while the terraced rough substrate of SiC increases the scattering of charge carriers causing mobility reduction. Due to elimination of substrate effects, very high carrier mobilities have been observed in suspended graphene. So, the substrate effects must be minimized if fast devices are needed. Keeping this in view, hexagonal boron nitride has been used as both the substrate and top gate dielectric in FETs. Devices have been fabricated with BN-graphene-BN structure [13]. The motivation for using h-BN is its similarity of atomic structure to graphene. From this resemblance, it is often known as white graphene. But unlike graphene, it is not semimetallic in character. It is an insulator with energy gap of 5.97 eV. The performance of graphene FETs in BN sandwich structure has been compared with that of control FETs in which the bottom substrate dielectric was silicon dioxide and the top gate dielectric was aluminum oxide. This study revealed that the BN-graphene-BN FETs had much higher current gain cutoff frequency $\sim$33 GHz in opposition to 18 GHz for the Al_2O_3-graphene-SiO_2 FETs.

17.12 Discussion and Conclusions

The main charisma of graphene is the high mobility of carriers flowing through it. This benefit may not be available always because of the requisite bandgap creation, during which mobility invariably decreases. Nonetheless, the advantage of making a very thin circuit of either a monolayer or a few monolayers thickness is no small achievement. A major challenge of graphene nanoelectronics is to create a bandgap in a controlled fashion for large-scale manufacturing of transistors reproducibly. This would lead the way for fabricating transistors with switching off capability for use in logic circuits. This would also enable the avoidance of the second linear regime in RF transistors.

Review Exercises

17.1 Why is graphene called the mother of buckminsterfullerene, CNT and graphite? Justify.

17.2 What is the bandgap of graphene? Is graphene optically opaque?

17.3 Comment on the following electrical parameters for graphene: (i) carrier mobility and (ii) bandgap. How do the values of these parameters for graphene affect the performance of electronic devices fabricated from it?

17.4 Can graphene be obtained mechanically by exfoliation? How is it grown on metallic thin films? How is it transferred to the required substrate?

17.5 Explain the mechanism by which graphene is epitaxially grown on a silicon carbide substrate? How does the growth depend on the polarity of silicon carbide face? How can large size grains be obtained from the small size grains deposited by thermal decomposition of SiC?

17.6 Is it possible to deposit graphene nanosheets without any substrate? If yes, how is it carried out?

17.7 Describe the fabrication and operation of a high-frequency graphene transistor? Discuss its relative merits and demerits for high-frequency and logic circuit applications.

17.8 Is it possible to create a bandgap in graphene artificially? How is its carrier mobility affected by bandgap creation?

17.9 How is a graphene nanoribbon formed from graphene? Describe the fabrication and working of a graphene nanoribbon transistor? How does its on-off current ratio differ from that of a pristine graphene transistor?

17.10 What is the typical on-off current ratio achieved in a bilayer graphene transistor? What is the typical bandgap value of the bilayer?

17.11 "The primary advantage offered by graphene is not its mobility, but the ability to provide extremely thin electronic devices, either a monolayer or a few atomic layers thick." Express your views on this statement.

References

1. Novoselov KS (2011) Graphene: the magic of flat carbon. In: The electrochemical society interface spring, pp 45–46
2. Berger C, Song Z, Li X et al (2006) Electronic confinement and coherence in patterned epitaxial graphene. Science 312(5777):1191–1196
3. Bolotin KI, Sikes KJ, Jiang Z et al (2008) Ultrahigh electron mobility in suspended graphene. Solid State Commun 146(9–10):351–355
4. de La Fuente J, Graphenea, http://www.graphenea.com/pages/graphene-properties#.VpphkhZvbdk. Accessed on 16th January 2016
5. Reina A, Jia X, Ho J et al (2009) Large area, few-layer graphene films on arbitrary substrates by chemical vapor deposition. Nano Lett 9(1):30–35
6. Emtsev KV, Bostwick A, Horn K et al (2009) Towards wafer-size graphene layers by atmospheric pressure graphitization of silicon carbide. Nat Mater 8:203–207
7. Dato A, Radmilovic V, Lee Z et al (2008) Substrate-free gas-phase synthesis of graphene sheets. Nano Lett 8(7):2012–2016
8. Lemme MC, Echtermeyer TJ, Baus M et al (2007) A graphene field-effect device. IEEE Electron Device Lett 28(4):282–284
9. Lin Y-M, Dimitrakopoulos C, Jenkins KA et al (2010) 100-GHz transistors from wafer-scale epitaxial graphene. Science 327(5966):662. doi:10.1126/science.1184289
10. Dimiev AM, Tour JM Graphene nanoribbons: production and applications. Sigam-Aldrich, http://www.sigmaaldrich.com/technical-documents/articles/materials-science/graphene-nanoribbons-production-and-applications.html. Accessed on 20th January 2016
11. Liao L, Bai J, Cheng R et al (2010) Top-gated graphene nanoribbon transistors with ultra-thin high-k dielectrics. Nano Lett 10(5):1917–1921
12. Xia F, Farmer DB, Lin YM et al (2010) Graphene field-effect transistors with high on/off current ratio and large transport band gap at room temperature. Nano Lett 10(2):715–718
13. Wang H, Taychatanapat T, Hsu A et al (2011) BN/Graphene/BN transistors for RF applications. IEEE Electron Device Lett 32(9):1209–1211

Chapter 18
Transition Metal Dichalcogenides-Based Nanoelectronics

Abstract Transition metal dichalcogenides (TMDs) serve as a two-dimensional, layered-structure, semiconducting material option to the gapless graphene in which carrier mobility is degraded by present bandgap engineering methods, predominantly by edge scattering. The TMD family includes compounds made of a transition metal, commonly Mo, W, Nb, Ta, Ti with a chalcogen atom, e.g., S, Se, Te. Preparation methods of atomically thin films of transition metal dichalcogenides, their properties and applications in nanoelectronics are described. Mechanical exfoliation has been widely used for laboratory studies of these thin films. The CVD approach is capable of producing homogenous films over large surface areas. It offers a facile route for nanoelectronic device manufacturing on wafers. Its compatibility with existing semiconductor fabrication facilities is additionally favorable. Single layer, bilayer, and multiple layer FET devices have shown the dependence of carrier mobility on TMD layer thickness and quality of contacts. Present status of these devices is presented. Their promising electrical characteristics call for more efforts for integrating them with silicon electronics.

18.1 Introduction

Pristine graphene has the shortcoming that it is not a semiconductor. The resulting low current on/off ratio makes it ineligible for transistor fabrication. For its application as a transistor, bandgap engineering is necessary. It is a complex process during which mobility is degraded to the level observed in strained Si films. TMDs are the semiconducting analogs of graphene with definitive bandgaps. They are intrinsically semiconductor materials with electron mobility as good as silicon. They are two-dimensional, layered materials comprising a few atomic strata.

V.K. Khanna, *Integrated Nanoelectronics*,
NanoScience and Technology, DOI 10.1007/978-81-322-3625-2_18

18.2 Composition and Mechanical Properties of TMDs

They are inorganic materials represented by the formula MX_2. Here, *M* denotes an atom of a transition metal, and *X* is a chalcogen atom. *M* = Mo, W, Nb, V, Ta, etc., whereas *X* = S, Te, Se, etc. [1]. Stacking of X-M-X layers results in the formation of single crystals of TMD [2]. Thickness of each layer is ~0.6–0.7 nm. Within each layer, the atoms are fastened together by mixed bonds of covalent-ionic type. These bonds are fairly strong. The M-X bond length varies from 0.315 to 0.403 nm. Contrary to strong bonding between atoms of a given layer, the layer-to-layer bonding is weak because it is through van der Waals forces of attraction. This

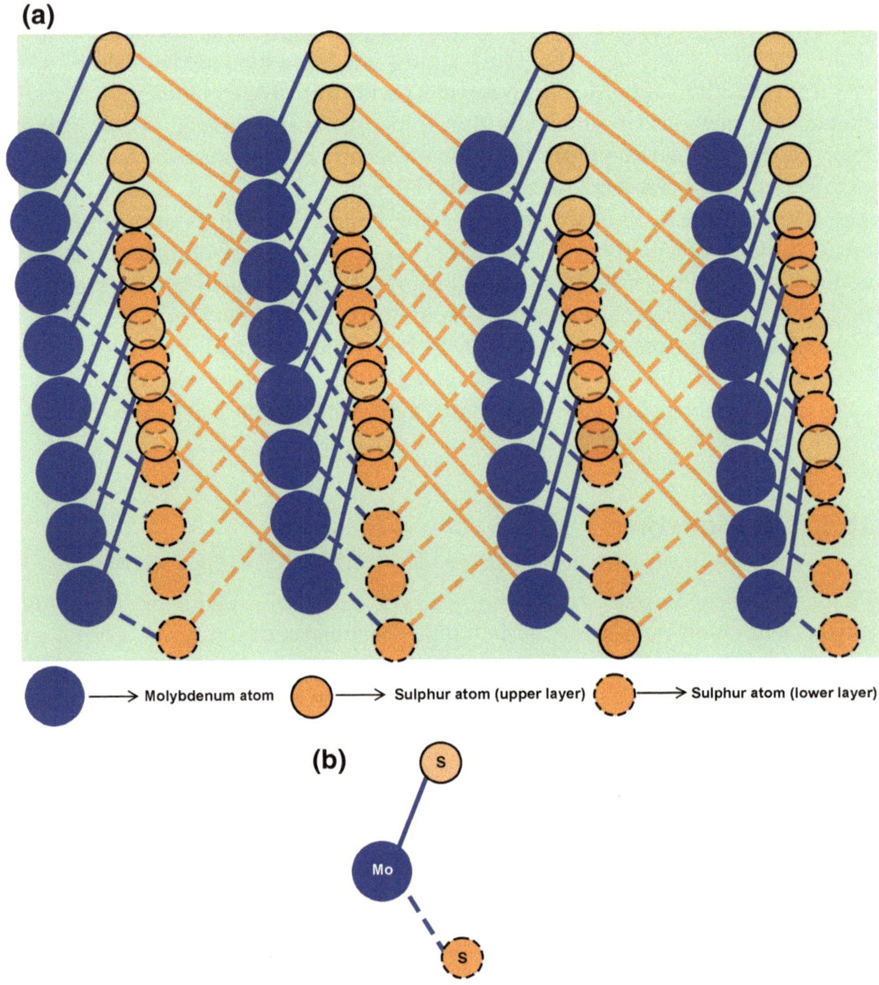

Fig. 18.1 Molybdenum disulphide: **a** the atomic arrangement in the crystal lattice and **b** isolated MoS_2 molecule

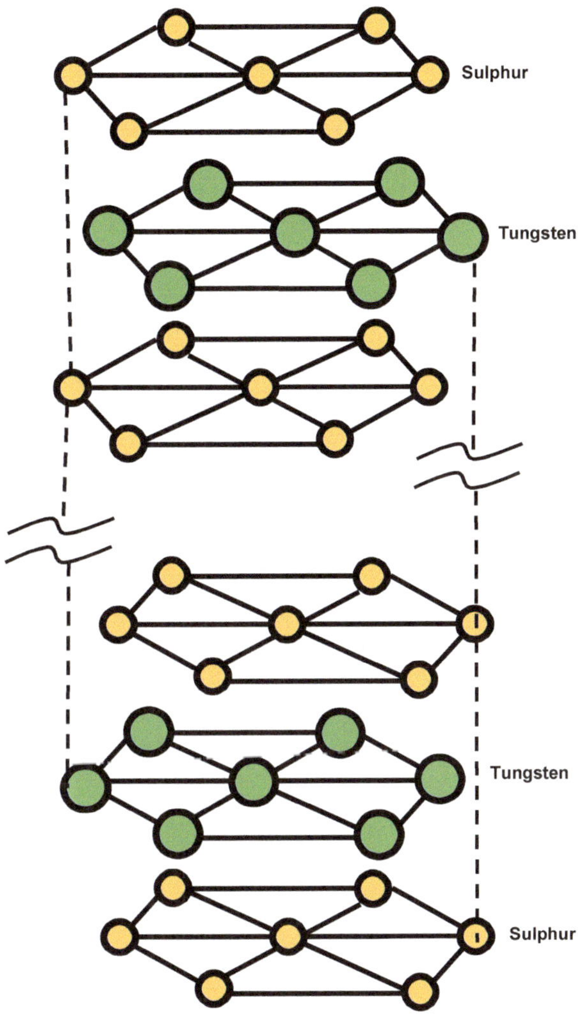

Fig. 18.2 Sulfur and tungsten atomic layers in the crystal structure of WS_2

anisotropy in bonding makes the thin, small, plate-like crystals, flexible, and easy-to-cleave structures. The crystal structure of MoS_2 is shown in Fig. 18.1. Similar to graphene, a monolayer of MoS_2 is mechanically strainable by 25%. Young's modulus of a single layer of MoS_2 is 270 ± 100 GPa [3]. Its in-plane stiffness is 180 ± 60 N/m. It has an average breaking strength = 15 ± 3 N/m. Figure 18.2 illustrates the structure of WS_2 crystals.

18.3 Electrical Properties of TMDs

Electrical properties of TMDs originate from the filling of the *d* bands from group 4 to group 10 elements. These *d* bands are nonbonding. A partial occupation of the orbitals leads to metallic behavior whereas full occupancy results in semiconducting properties. The electronic structure is less affected by the chalcogen atoms compared to the metal atoms. Nonetheless, widening of the *d* bands causes bandgap reduction by raising the atomic number of the chalcogen. Electrical properties of TMDs depend on several variables offering numerous opportunities for tuning the properties at discretion: (i) Their properties transform with composition. MoS_2 and WS_2 are semiconducting. WTe_2 and $TiSe_2$ display semimetallic character. NbS_2 and VSe_2 exhibit metallic nature. $NbSe_2$ and TaS_2 are superconductors. (ii) Their properties change with the structure of the crystalline phase, e.g., hexagonal phase 2H-MoS_2 is a semiconductor while tetragonal phase 1T-MoS_2 is a metal. (iii) Their properties are determined by the manner of superimposing of layers in their bulk crystal and in thin film forms, e.g., hexagonal 2H-MoS_2 phase is an indirect bandgap semiconductor in bulk form but becomes a direct bandgap semiconductor upon exfoliation into monolayers. Bilayer MoS_2 is an indirect bandgap semiconductor. It has a bandgap of 1.3 eV. Single-layer MoS_2 is a direct bandgap semiconductor with a bandgap of 1.8 eV.

18.4 Optical Properties of TMDs

Direct bandgap makes the monolayers of a TMD suitable for optoelectronic devices such as phototransistors and photodetectors. Also, TMDs exhibit a high optical absorption capability. A monolayer of TMD material absorbs 5–10% of incident light.

18.5 Preparation of TMDs

18.5.1 Micromechanical Exfoliation

This method is similar to the exfoliation of graphene from graphite. Single-layer MoS_2 is a stack of three hexagonally packed layers. These layers are S-Mo-S. These are held together by strong covalent bonds. In bulk MoS_2, the single-layer stacks are bound by feeble van der Waals forces. Thus intralayer interaction is strong whereas interlayer interaction is weak. Hence, the different layers are separable by mechanical force. Single crystal two-dimensional materials produced from bulk crystals by Scotch tape-based exfoliation are of good quality. But they are quite small in size, around a few tens of μm.

18.5.2 Liquid Exfoliation

In an all-inclusive, mixed-solvent scheme for TMDs like WS_2, MoS_2, BN, etc. [4] solvents are selected with suitable compositions. These solvent compositions comprising mixtures of volatile solvents produce highly stable suspensions of TMDs, e.g., pure water and ethanol are incapable of dispersing any TMDs. But when the correct ethanol/water ratio is chosen, the above TMDs can be dispersed in nanostructured form despite the fact that both ethanol and water are poor solvents for these TMDs. Dispersions of WS_2 and MoS_2 are dark green while those of BN are milky. The dispersions did not precipitate even after storing in ambient conditions for a week. The mixed-solvent scheme can be scaled up easily. Besides cost-effectiveness, the scheme affords freedom to decide solvents according to application.

18.5.3 Low-Temperature Decomposition of Precursors

Free-standing nanosheet crystals of single or few layers of MoS_2 and WS_2 covered by protective oleylamine organic coating are produced by decomposition of thio salts as single-source precursors in oleylamine at a low temperature ~360 °C [5]. For two-dimensional MoS_2, $(NH4)_2MoS_4$ is stirred in oleylamine under flowing nitrogen at 100 °C for 15 min. The temperature is maintained at 360 °C for half an hour. After cooling to room temperature, the required product is collected. For two-dimensional WS_2, the above steps are repeated with $(NH_4)_2WS_4$ in oleylamine. As before, upon cooling from 360 °C, the desired product is obtained. Excess surfactant is removed by washing with ethanol and centrifuging. In both cases, the oleylamine coating stabilizes the suspension. It prevents aggregation and avoids oxidation. The method offers a simple one-pot synthetic route.

18.5.4 Chemical Vapor Deposition

Aqueous reduced graphene oxide (rGO) is spin coated on the Si/SiO_2 substrate and dried at 50 °C [6]. The Si/SiO_2 substrate is suspended above the ceramic boat containing the MoO_3 powder with the SiO_2 surface facing downwards towards the MoO_3 powder. Sulfur powder is kept in another boat alongside the MoO_3 powder boat. Nitrogen environment is maintained inside the reaction chamber. The chamber is heated up to 650 °C. At this high temperature, sulfur vapor reduces the MoO_3 powder forming volatile suboxide $MoO_{3\text{-}x}$, which diffuses to the substrate and undergoes further reaction with sulfur vapor to produce MoS_2 film. Aqueous rGO molecules are spin coated to promote the layered growth of MoS_2.

Sulphurization of a preformed metallic film provides a simple technique to deposit atomic thickness disulphide layers on insulating substrates [7]. MoS_2 is deposited by direct sulphurization of Mo thin film preformed on a SiO_2/Si substrate. This sulphurization is carried out by thermally annealing the Mo film in sulfur atmosphere. The thickness and size of the MoS_2 film are determined by the respective values of the Mo film.

18.6 Single-Layer Dual-Gate MoS_2 FET

Single-layer FETs or those made of a few atomic layers are immune to short channel effects that plague Si devices. This transistor is made of a monolayer of MoS_2 of thickness 0.65 nm, which is obtained by micromechanical exfoliation of a single layer from crystals of molybdenite [8]. This single layer of MOS_2 is transferred upon 270 nm thick SiO_2 film grown by thermal oxidation on degeneratively doped silicon substrate serving as a back gate. Following a lithographic step, gold film of 50 nm thickness is deposited to form source and drain electrodes. Annealing is done at 200 °C for decreasing the contact resistance. The carrier mobility at this stage of the process is very low $\sim$0.1–10 cm^2/V-s. The next process step is a mobility booster stage in which 30 nm thick HfO_2 film is formed by atomic layer deposition. A probable mechanism for mobility upgradation is inhibition of Coulomb scattering in the ambience of high-κ dielectric material. It is a type of dielectric screening. The extracted low-field mobility now increases to 200 cm^2/V-s. This thin device has a conducting channel of 0.65 nm thickness. The top gate length is 500 nm and gate width is 4μm. It exhibited a current on/off ratio $>10^8$. The subthreshold swing is 74 mV/decade. The bottom gate length is 1.5 μm. With back gate bias of 10 V and V_{DS} = 10 mV, the on-resistance is 27 kΩ. For any proposed substitution of Si in CMOS logic type circuits, a current on/off ratio in the range 1×10^4–1×10^7 is desirable along with a bandgap > 400 meV. Both these requirements are fully complied by the single-layer MoS_2 transistor.

Figure 18.3 depicts the constructional features of a dual-gated MoS_2 FET.

18.7 Bilayer Back-Gated MoS_2 FET

Bilayer MoS_2 FETs sparked interest because of smaller bandgap of the bilayer MoS_2 than single-layer MoS_2, lower sensitivity to ambience due to smaller surface area and possibility of bandgap modification by applying a vertical perpendicular electric field, as done with graphene [9]. These devices are fabricated on heavily doped silicon substrates acting as the back gate. On the substrate, 300 nm thick SiO_2 is deposited. Bilayer MoS_2 is formed by mechanical exfoliation on the SiO_2 film.

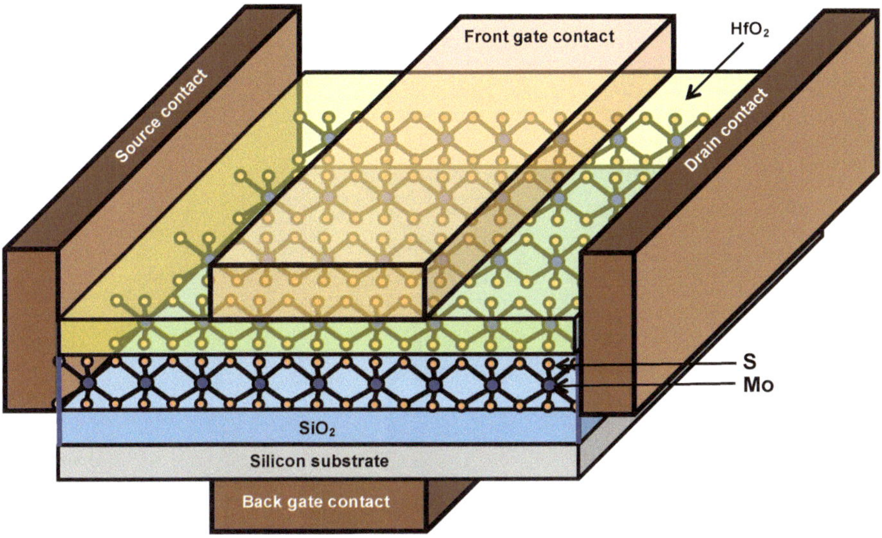

Fig. 18.3 Dual-gated molydenite (MoS_2) FET

40 nm thick Ti is deposited after source/drain lithography followed by lift-off. The source/drain contacts are annealed in ($Ar + H_2$) mixture at 400 °C. The channel length as well as channel width is 800 nm. The on-state conductance is 13 nS and extrinsic mobility is 0.12 cm^2/V-s. The low values are ascribed to adsorption of oxygen and water from the ambient environment. After high vacuum annealing $\sim 10^{-6}$ Torr, the adsorbates are expunged so that on-conductance increases to 756 nS. The on/off ratio is 10^7 and the extrinsic mobility is $\sim$2.4 cm^2/V-s. The devices exhibit ohmic contacts with Ti and also current saturation.

Figure 18.4 shows a back-gated MoS_2 field-effect transistor.

18.8 Multilayer Dual-Gate MOS_2 Transistor

This FET is fabricated using 23 layers of MoS_2 (14 nm thick) [10]. The MoS_2 flakes are mechanically exfoliated on SiO_2/Si substrate (with SiO_2 thickness = 300 nm). The Si layer of the substrate is the back gate electrode and the SiO_2 film is the back gate dielectric. The top gate is made of 16 nm thick Al_2O_3 film formed by atomic layer deposition using trimethylaluminum (TMA) and water as precursors at 200 °C. Source/drain metallization is Ni/Au. Gate contact is Ti/Au. The gate length is $\sim 3 \mu m$. Using back gate modulation, the maximum drain current density is 7.07 mA/mm. For top gate, it is 6.42 mA/mm. The on/off current ratio is 10^8. The highest field-effect mobility of electrons is 517 cm^2/V-s from back gate control.

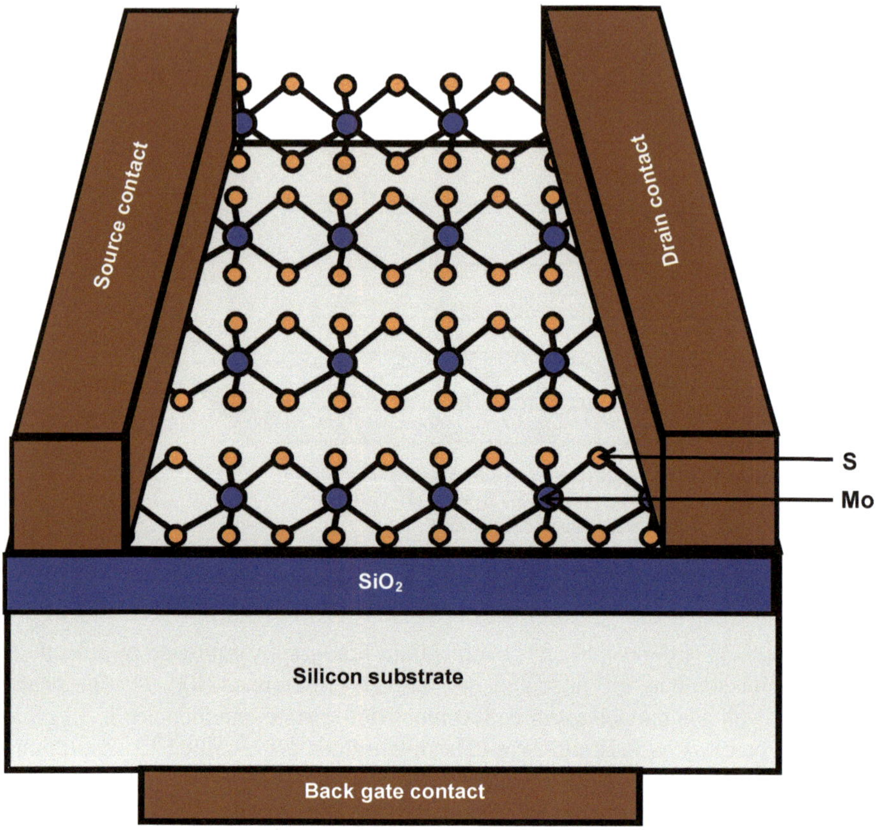

Fig. 18.4 Back-gated molydenite (MoS_2) FET

18.9 Mobility Dependence on MoS_2 Layer Thickness and Contact Quality

Systematic investigational studies of carrier mobility dependence on MoS_2 dielectric layer thickness are carried out [11]. For performing these studies, back-gated FETs are fabricated and characterized. In these FETs, widely varying thickness of MOS_2 layers is taken. It is varied from 2 to 70 nm. Each MoS_2 layer is 0.65 nm thick. The corresponding number of layers in the range of thicknesses examined increases from as low as 5 to 107. For the source and drain contacts, the metal used is scandium. The work function of scandium is 3.5 eV.

Interestingly, the variation of mobility with the number of MoS_2 layers does not follow a monotonic trend. For 3 nm thick MoS_2 layer, the mobility is $\sim$25 cm^2/V-s and increases to 184 cm^2/V-s for 10 nm thick layer. Thereafter, further increase in the number of MoS_2 layers is accompanied by a decline of mobility, which falls to $\sim$60 cm^2/V-s for 50 nm thickness. So, 10 nm thick MoS_2 containing 15 layers gives the highest mobility. This shows that the thickness of MoS_2 must be kept between 6 and 12 nm. Further, when 15 nm thick Al_2O_3 is deposited on the top surface of the back-gated FET with 10 nm thick MoS_2 film, the mobility increases from 184 cm^2/V-s to 700 cm^2/V-s by a factor of 3.8. These FETs show a maximum current density of 240 μA/μm and a transconductance of 4.7 μS/μm at V_{DS} = 1 μm for 5 μm channel length.

For the above FETs, mobility also changes when different contact metals are used. For the 10 nm FET without top Al_2O_3 covering, the effective mobility in 10 nm FET is 21 cm^2/V-s. For Pt source/drain contacts, it is 125 cm^2/V-s. For Ti contacts on source and drain, it is 184 cm^2/V-s for Sc. These data show the effect of contact resistance on mobility.

18.10 Discussion and Conclusions

Transition metal dichalcogenides have generated significant research interest because of their matchless electrical, optical, mechanical, chemical, and other properties. They have bandgaps ~1–1.2 eV. So, logic device operations can be implemented with TMD-based devices. Ability to isolate single and a few atomic layer thick films have allowed the fabrication of ultrathin devices, which outperform Si devices at the nanoscale limit.

Review Exercises

18.1 In what ways transition metal dichalcogenides overcome the zero bandgap disadvantage of graphene?

18.2 Write the general chemical formula of transition metal dichalcogenides and explain the symbols used. Why is it easy to clean single or small number of layers of TMDs from bulk material?

18.3 How do the electrical properties of transition metal dichalcogenides vary with composition, structure of the crystalline phase and the style of placement of layers in a bulk crystal?

18.4 Bilayer MoS_2 is an indirect bandgap semiconductor whereas a monolayer of the same material is a direct bandgap semiconductor. True or false?

18.5 In MOS_2, which interaction is strong: interlayer or intralayer? How are these interaction strengths related to separating single layers from bulk material by exfoliation?

18.6 Describe the liquid exfoliation scheme for preparing nanostructured dispersions of transition metal dichalcogenides. Are these dispersions stable over time or they precipitate out?

18.7 Describe a one-pot synthetic process to prepare free-standing nanosheet crystals of MoS_2 covered by oleylamine.

18.8 How is MOS_2 formed on an oxidized silicon substrate by chemical vapor deposition? What is the role of reduced graphene oxide coating on the substrate?

18.9 Describe the operation and fabrication of a single-layer MoS_2 FET. Does it fulfill the criteria for replacing a MOSFET in logic circuits?

18.10 What are the attractive features for fabricating a bilayer MoS_2 FET? How is carrier mobility affected by adsorption of water and oxygen from the environment? How are the adsorbates driven out? Do the electrical characteristics improve after removing the adsorbates?

18.11 What is the magnitude of carrier mobility achieved in MoS_2 FET containing 23 layers of MoS_2? Does mobility depend on the number of layers? What is the optimum number of layers to achieve a high mobility?

References

1. Lv R, Robinson JA, Schaak RE et al (2015) Transition metal dichalcogenides and beyond: synthesis, properties, and applications of single- and few-layer nanosheets. Acc Chem Res 48 (1):56–64
2. Podzorov V, Gershenson ME, Kloc Ch et al (2004) High-mobility field-effect transistors based on transition metal dichalcogenides. Appl Phys Lett 84(17):3301–3303
3. Han SA, Bhatia R, Kim S-W (2015) Synthesis, properties and potential applications of two-dimensional transition metal dichalcogenides. Nano Convergence 2:17, 14pp. doi:10.1186/s40580-015-0048-4
4. Zhou K-G, Mao N-N, Wang H-X et al (2011) A mixed-solvent strategy for efficient exfoliation of inorganic graphene analogues. Angew Chem Int Ed 50:10839–10842
5. Altavilla C, Sarno M, Ciambelli P (2011) A novel wet chemistry approach for the synthesis of hybrid 2D free-floating single or multilayer nanosheets of MS_2@oleylamine (M = Mo, W). Chem Mater 23(17):3879–3885
6. Lee Y-H, Zhang X-Q, Zhang W et al (2012) Synthesis of large-area MoS_2 atomic layers with chemical vapor deposition. Adv Mater 24:2320–2325
7. Khan Y, Liu Z, Najmaei S et al (2012) Large-area vapor-phase growth and characterization of MoS_2 atomic layers on a SiO_2 substrate. Small 8(7):966–971
8. Radisavljevic B, Radenovic A, Brivio J et al (2011) Single-layer MoS_2 transistors. Nat Nanotechnol 6:147–150
9. Qiu H, Pan L, Yao Z, et al. (2012) Electrical characterization of back-gated bi-layer MoS_2 field-effect transistors and the effect of ambient on their performances. Appl Phys Lett 100:123104-1 to 123104-3 (2012)
10. Liu H, Ye PD (2012) Dual-gate MOSFET with atomic-layer-deposited Al_2O_3 as top-gate dielectric. IEEE Electron Device Lett 33(4):546–548. doi:10.1109/LED.2012.2184520
11. Das S, Chen H-Y, Penumatcha AV et al (2013) High performance multilayer MoS_2 transistors with scandium contacts. Nano Lett 13:100–105

Chapter 19
Quantum Dot Cellular Automata (QDCA)

Abstract For representation of binary information and performing computations on them, cells containing quantum dots at defined locations are used. Tunnel barriers separate the neighboring dots. Under the control of a back plane voltage, electrons can tunnel between dots. But intercell barriers strictly prevent tunneling of electrons across cells. Information is encoded in the form of positions of electrons in the cell. Electrons in each cell interact Coulombically. The cells are also coupled through Coulomb forces between electrons. The utilization of QDCA as a wire, as a majority voter, and for performing logic AND/OR operations is explained. Salient features and applications of the QDCA approach are described. The quantum dot-based architecture is experimentally proven to work in the mKelvin temperature range. This field-coupled nanocomputing model is likely to challenge and succeed the CMOS for room-temperature operation when technological capability develops to the level of easily fabricating quantum dots of molecular size.

19.1 Introduction: Moving Towards Transistorless Computing Paradigms

Although FETs today are vastly improved versions of FETs of 1970s, still they are ultimately switches like mechanical relays. A possible course for nanoelectronics could be to break away from FET-based paradigms to those using nanostructures. In these paradigms, the basic nanoelectronic element is no longer a switching device. Neither currents nor voltages are used for encoding information. Instead, one works with different arrangements of individual electrons.

19.2 Tougaw-Lent Proposition of a Quantum Device

Tougaw and Lent [1], researchers at the University of Notre Dame, USA, noted that quantum structures should not be used in traditional architectures because their output current, typically in the nanoampere range, will be able to drive other similar

V.K. Khanna, *Integrated Nanoelectronics*,
NanoScience and Technology, DOI 10.1007/978-81-322-3625-2_19

devices only by a large increase in their input voltage. Further, the operation of the device will be overshadowed by the stray capacitance of interconnecting wires. They therefore proposed a scheme called Quantum Dot Cellular Automata (QDCA). It contains an array of quantum dot cells. These cells are interconnected through Coulomb electrostatic force acting between electrons contained in the cells. In this scheme, both the problems, one arising from insignificant output currents and that due to parasitic capacitances of the wiring are eliminated. This happens because there is no current flow between cells. No connecting wires are required to connect the cells together. The individual cells are Coulomb coupled. No wire is used to supply power to cells.

19.3 Role of Quantum Dots in the Scheme

In Tougaw–Lent scheme, the information is encoded in the form of position of electron in a vessel. Therefore, a vessel is the basic element required to trap the electron so that it can be seen as present or absent in its assigned place in the vessel. The quantum dot, a nanometer size structure, serves the purpose of this vessel. It is made by creating an island of conducting material surrounded by an insulating material. It consists of a central region of low potential enclosed by a ring of high potential. Such a ring is called a potential well. The potential well is a region in a field of force where the potential is appreciably lower that at points in its immediate neighborhood. An electron inside the quantum dot requires a high energy to escape. So, it remains confined inside unless sufficient energy is available.

19.4 The Standard QDCA Cell

19.4.1 Four Quantum Dot, Two-Electron Arrangement

The QDCA cell is square-shaped (Fig. 19.1). It is a four quantum dot cell. Four quantum dots representing four potential wells are located at the four corners of the square. Between any two quantum dots or potential wells, there is a tunnel barrier. Besides the four quantum dots, the QDCA cell contains two mobile electrons. These electrons can tunnel between adjoining sites on the cell. The compensating positive charges remain stationary. Potential barriers between cells do not allow any tunneling of electrons between cells, i.e., from one cell to another. Hence, the electrons are localized on their respective quantum dots within a cell. The electrons of a cell cannot move out of it. Two possible orientations of QDCA cells at 90° and 45° are shown in Fig. 19.2.

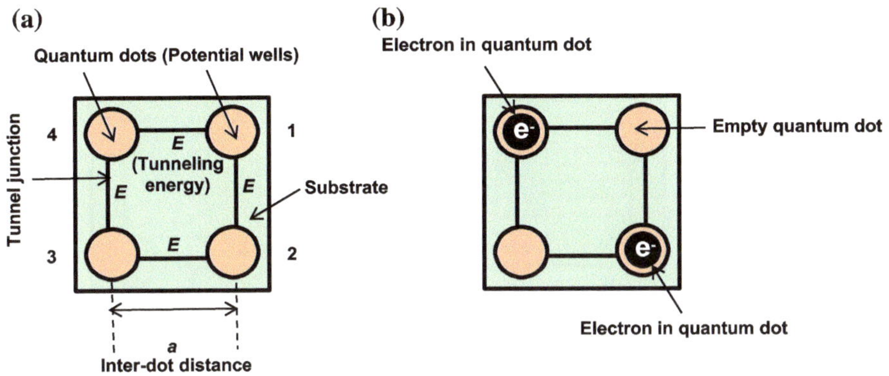

Fig. 19.1 Basic four-site quantum dot cellular automata (QDCA) cell: **a** cell anatomy and **b** electrons in QDCA cell. *E* is the tunneling energy between adjacent quantum dots and *a* is the inter-dot spacing

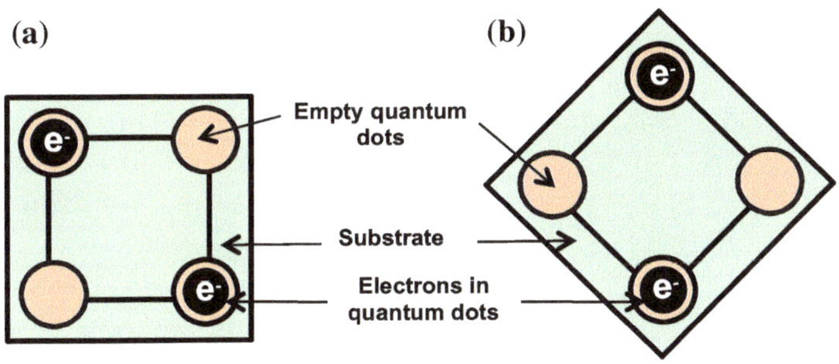

Fig. 19.2 Orientations of QDCA cells: **a** 90° cell and **b** 45° cell

19.4.2 Null and Polarization States of the QDCA Cell

A null state is defined in which the tunnel barriers between quantum dots are lowered so that electrons are free to hop from one quantum dot to another. However, potential barriers compelling the electrons to remain within each cell are not relaxed. In this null state, the repulsive force between the negatively charged electrons pushes them as far apart as possible but confined within the cell. The locations of maximum separation between electrons in a square cell can be on the extremities of a diagonal. There are two diagonals in a square. So, the pair of electrons can be supposed to lie on the ends of one of the two diagonals. Such diametrically opposite sides are referred to as antipodal sites. Hence, for any isolated cell there are two energetically equivalent states of the cell corresponding to the localization of electrons on the ends of the two diagonals. These will be called polarization states (Fig. 19.3). Polarization here does not mean that the cell possesses a dipole moment. It only implies two possible arrangements of electrons in

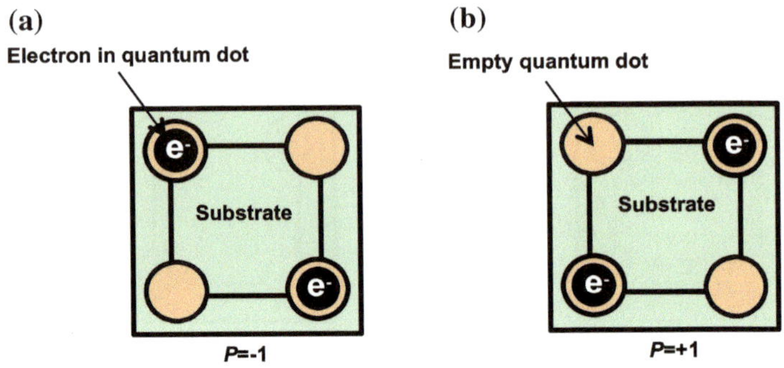

Fig. 19.3 QDCA cell polarizations: **a** $P = -1$ or 0 and **b** $P = +1$ or 1

the cell. A polarization $P = +1$ will be used to denote binary 1. A polarization $P = -1$ will stand for binary 0. Once a cell has attained a particular polarization state, it will be assumed that the tunnel barriers are once again raised so that electrons can change their positions only after the barriers are lowered.

Briefly, in the QDCA paradigm, (i) the structural unit of the logic circuit is the quantum dot cell. It is not a switch in the classical sense, although it is said to switch between different states. (ii) The logic states 0 and 1 of this structural unit are encoded in the positions of electrons in the quantum dot cell. (iii) These logic states are represented by polarization $P = +1$ or logic 1 and $P = -1$ or logic 0.

19.4.3 Changing the Polarization States of a QDCA Cell and Reading These States

The polarization state of a cell can be changed by applying a clock signal. By applying this signal, the tunnel barriers between any two quantum dots in a cell are either raised or lowered, as desired. This signal acts through a capacitively coupled gate on the QDCA cell.

To read the positions of electrons in a QDCA cell, i.e., electron configuration, and hence the polarization state of a cell, capacitively coupled electrometers are used.

19.5 QDCA Cell Fabrication

The quantum dots can be made of aluminum deposited by thermal evaporation. The tunnel barrier is made of insulating aluminum oxide (AlO_x). Optical lithography cannot be used to make small quantum dots because the size of dots is less than the wavelength of light used. Electron beam lithography is used to define the pattern of quantum dots. Unlike optical lithography, this technique does not use a mask containing the pattern to be formed. Instead, the pattern is directly written by an electron beam on the substrate. The technique is capable of producing dots with small variability in sizes and positions. As electron beam lithography is a fairly expensive process, a suitable simpler technique is being explored for large-scale manufacturing of QDCA cells at a low cost. A suitable process is self-organization technique. In self-organization, the deposited film has a different lattice structure than the layer on which the film is deposited. Stresses are developed in the interfacial region of the two crystalline materials which cause the deposited film to clump into small-sized dots in the same way as oil droplets are formed on the surface of water.

19.6 Advantages of QDCA Cell

The main benefit of QDCA cells is the high packing density provided by them. If quantum dots of diameter 20 nm are used, a full adder circuit is accommodated in an area $\sim 1\ \mu m \times 1\ \mu m$ [2]. The simplification of interconnections between any two cells is another advantage in their favor. No current flow is involved. The absence of current flow drastically reduces the power consumption and the associated losses.

19.7 Binary Wire

A wire can be woven from standard QDCA cells in two ways: (i) By placing standard cells end-to-end in a linear arrangement. (ii) By rotating the standard cell by an angle of 45°, and placing these cells rotated by 45° along a straight line. Accordingly, two types of binary wires are obtained: the 90° wire and 45° wire.

19.8 The 90° Wire

Figures 19.4 and 19.5 show how this wire is used to transfer information and data. First consider one standard cell. Another standard cell is placed close to it. If the tunnel barriers are raised on both the cells, the two cells cannot influence each other

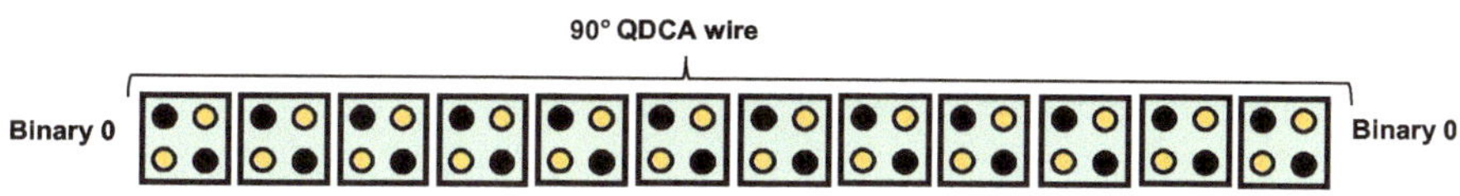

Fig. 19.4 Transmission of binary 0 on a 90° QDCA wire

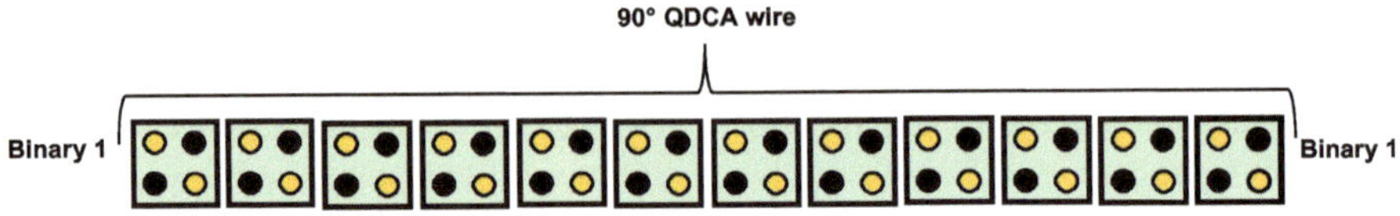

Fig. 19.5 Transmission of binary 1 on a 90° QDCA wire

and will remain in their original states. Now, suppose, the left cell is called the driver cell, then the tunnel barriers on this cell will remain raised, as before. The right cell is the driven cell on which the tunnel barriers will be lowered to enable driving. On lowering of tunnel barriers, the electrons in the driven cell will acquire their new positions due to the repulsive forces exerted on them by the electrons of the driver cell. Thus electrons of the driven cell will be in the matching positions as those of the driver cell. Hence, the driver and driven cells will have the same polarization state.

Similarly, by adding a third cell adjacent to the second cell and making the third cell as the driven cell by decreasing its tunnel barriers and the second cell as the driving cell by increasing its tunnel barriers, the third cell takes on the same polarization state as the second cell. Effectively, the polarization state of first and second cells has now been transferred to the third cell. Thus all the cells: the first, second, and third are in the same state of polarization.

In the above manner, addition of any number of cells and repetition of above operations places all the cells in the same polarization state. It can be said that the polarization state of the first cell has been transmitted through the intervening cells to the last cell. Thus this positioning of standard cells along a straight line serves as a binary wire communicating the data fed in the first cell to the last cell.

The above string of cells is sometimes called a "pseudo-wire" because in contradistinction to a real wire, there is no flow of current through the wire.

19.9 The 45° Wire

Let us follow the procedure analogous to the 90° wire. Let us place the first cell in an orientation rotated by 45° with respect to the standard cell configuration. Let us call it the driving cell (Fig. 19.6). The tunnel barriers of the driving cell are

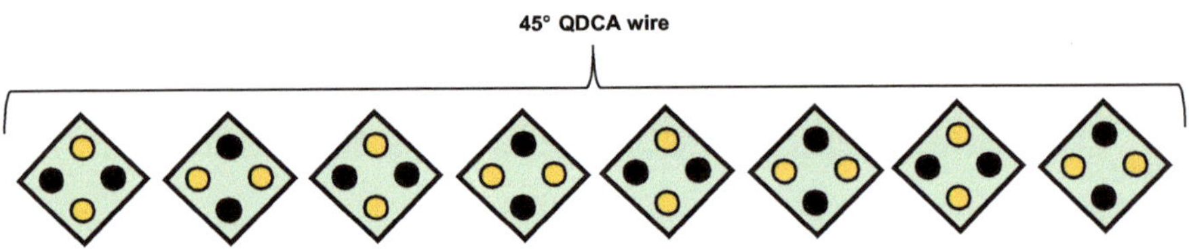

Fig. 19.6 Binary 0 and binary 1 signals available on the same 45° QDCA wire

increased. The second cell rotated by 45° is the driven cell. The tunnel barriers of the driven cell are decreased. It is evident that the electrons in the second cell will be shifted to positions which are opposite to those of electrons in the first cell due to mutual repulsion. In other words, the driven cell will be in opposite state of polarization with respect to the driving cell. Now, suppose a third cell is placed adjoining the second cell. Now, this third cell becomes the driven cell in which tunnel barriers are decreased, and the second cell is the driving cell in which tunnel barriers are increased. Perceptibly, the third cell will reconcile down by readjustment of electron positions. The polarization state of the third cell in this condition is reverse of the second cell. In this fashion, a long arrangement of cells along a straight line can be organized. The long linear arrangement can be configured by placement of any number of cells rotated by 45°. Each succeeding cell of this arrangement has the opposite polarization relative to its predecessor cell. It is worth noting that this arrangement of cells differs from that of 90° wire. In the 45° wire, each cell has inverted polarization with respect to its neighbor, both the left neighbor and the right neighbor. The polarization state alternates from each cell to the next cell. Hence, the normal and inverted polarization states are available on the same wire whereas in the 90° wire only one polarization state, either normal or inverted, was obtained. If an odd number of cells is used in the 45° wire, the same polarization is found on the last cell as the starting cell. For an even number of cells, the polarization of the last cell is the inverted version of the first cell. Thus knowledge of length of wire helps us to know the signal that was sent. Looking at the anti-alignment of the consecutive elements in this wire, it is usual to call this wire as an "inverter chain".

19.10 QDCA Inverter or NOT Gate

In the QDCA inverter (Fig. 19.7), the input signal is fed to a 90° wire. Then all the cells in this wire will have the same polarization. At the opposite extremity, the above wire is subdivided into two parallel branching segments, both of which are 90° wires. At the point of subdivision, one extra cell extends from the wire receiving the input signal and overlaps with both the branching segments on its sides. Due to this overlapping cell, the branching segments have the same polarization as the wire into which the input signal was applied. After traveling a short distance, the branching segments are again united in the manner shown such that the branches touch the uniting wire only at the corners and there is no cell protruding in the space between the branches. Then remembering that electrons will occupy positions such that no two electrons are placed in vicinity, it is evident that the electron positions in the wire uniting the branches will be opposite to that in the branching segments. Since cells of the branching segments have the same polarization as the input signal, the cells in the uniting wire will have opposite polarization to that of the input signal. Thus inversion or NOT function has been implemented.

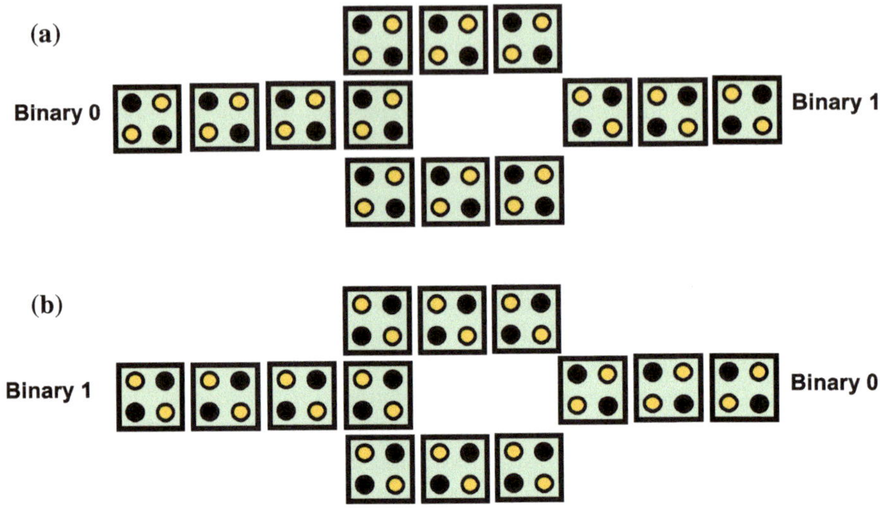

Fig. 19.7 QDCA inverters for converting **a** binary 0 to binary 1 and **b** binary 1 to binary 0

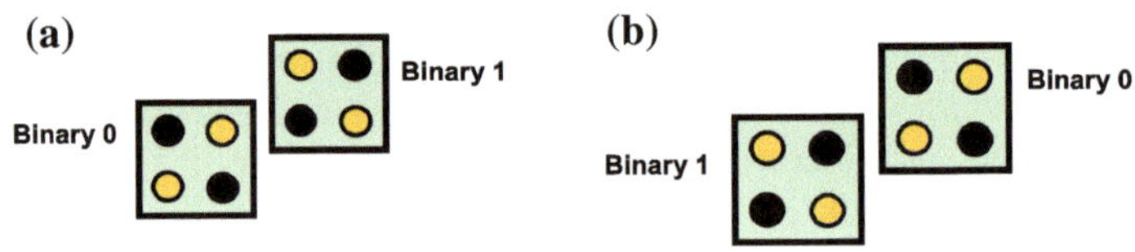

Fig. 19.8 QDCA inverters using off-center cells for converting **a** binary 0 to binary 1 and **b** binary 1 to binary 0

Thus in a NOT gate, the input signal splits up and diverges into two parallel wires. It is made to converge at the output wire at which its inverted version appears. Due to geometrical symmetry of the design, if $P = +1$ or logic 1 at the input, the output is $P = -1$ or logic 0. Conversely, for input $P = -1$ or logic 0, the output is $P = +1$ or logic 1.

NOT function implementation can also be done using off-center cells as shown in Fig. 19.8.

19.11 QDCA Majority Voter

The QDCA majority voter consists of five cells placed in the positions shown in Fig. 19.9. One cell placed in the center performs the calculation. It is called the "device cell". The three cells A, B, C at the left, on the top and on the bottom of the device cell are the input cells. The fifth cell on right of the device cell is the output cell. The majority voter works on the principle that the electrostatic forces exerted

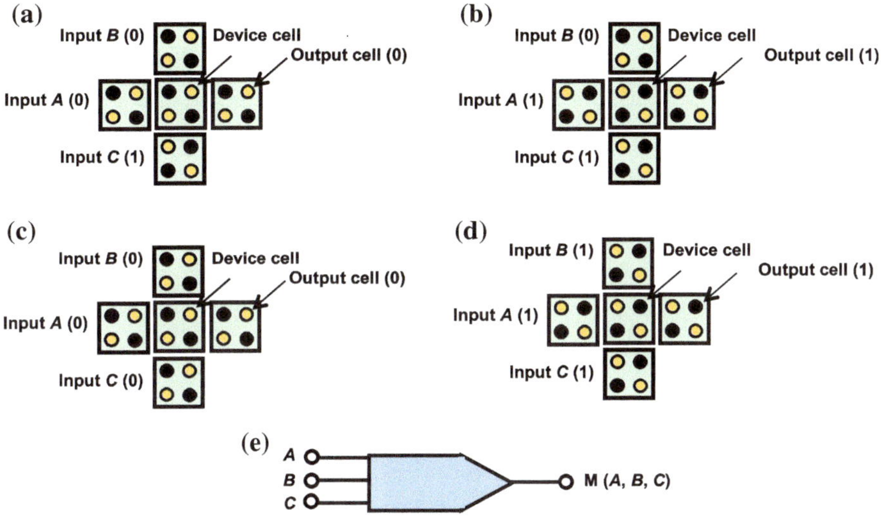

Fig. 19.9 QDCA majority gates **a** $A = 0$, $B = 0$ and $C = 1$ so that output = 0; **b** $A = 1$, $B = 0$ and $C = 1$ so that output = 1; **c** $A = 0$, $B = 0$ and $C = 0$ so that output = 0; and **d** $A = 1$, $B = 1$ and $C = 1$ so that output = 1; in all the cases, the device cell undergoes adjustment conforming to the resultant of the Coulomb forces from surrounding cells, as determined by influence of the majority. **e** Symbol of majority gate

by the electrons in the cells sum up to a resultant force. Therefore, the device cell acquires the polarization state that majority of the three neighboring input cells *A*, *B*, and *C* have. Once the device cell has adjusted to its polarization state in accordance with this majority rule, the output cell immediately takes up the same polarization as the device cell. Considering any case in Fig. 19.9a–d, as an example, suppose in Fig. 19.9b, two of the QDCA cells have $P = +1$ or logic 1 state. Naturally, the device cell and therefore the output cell will show $P = +1$ or logic 1 state. Similar procedure applies to other input combinations.

19.12 QDCA OR Gate

Figure 19.10 shows one revised version of the majority voter. In this version, the left QDCA cell is named as the program cell and its polarization is fixed at $P = +1$ or logic 1. The top QDCA cell is used for feeding the input signal *A*. The bottom QDCA cell is used for applying the input signal *B*. The right QDCA cell is the output cell. The choice of which cell will be the program cell is flexible. Any one of the three cells: the left one, the top one, or the bottom one could have been chosen as the program cell. The left cell has no special characteristic which makes it suitable to be selected as a program cell. The point to remember is that the polarization of the program cell remains constant.

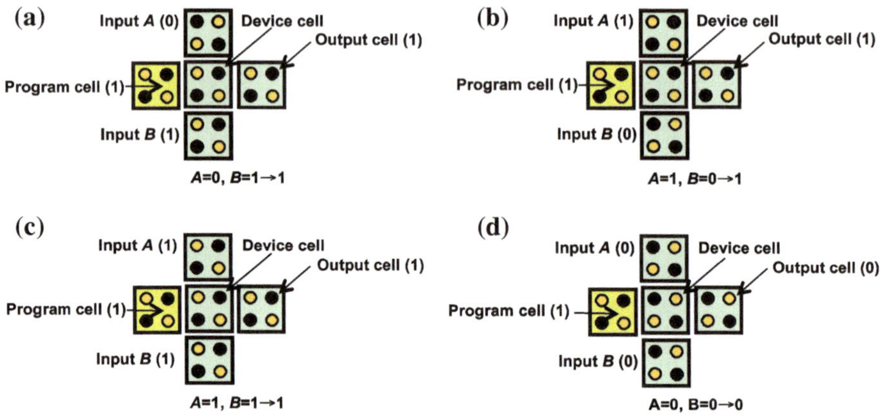

Fig. 19.10 Four possible combinations of values of input signals A and B, and corresponding device cell and output cell configurations for a QDCA OR gate. **a** $A = 0$, $B = 1 \rightarrow 1$. **b** $A = 1$, $B = 0 \rightarrow 1$. **c** $A = 1$, $B = 1 \rightarrow 1$. **d** $A = 0$, $B = 0 \rightarrow 0$

When input signal A = logic 1, but input signal B = logic 0, since program cell is at logic 1, the majority is for logic 1 and the output signal is logic 1. Same remarks apply to the situation in which input signal A = logic 0, but input signal B = logic 1. Since program cell is at logic 1, the majority is for logic 1 and the output signal is again logic 1. Thus in both situations when either of the two input signals is in logic 1 state and the other in logic 0 state, the output signal is logic 1.

Further, if input signal A = logic 1 as well as input signal B = logic 1, the majority is for logic 1 because the program cell is at logic 1. Then the output signal is logic 1. If input signal A = logic 0, and input signal B = logic 0, the majority is for logic 0 by virtue of the program cell being at logic 1. Hence, the output signal = logic 0. Hence, when both inputs are at logic 1, the output is logic 1. When both inputs are at logic 0, the output is logic 0.

The above combinations of input signals with the resulting output signals are compiled in the truth table (Table 19.1). This is easily identified as the OR gate truth table. So, the arrangement of quantum dot cells behaves as an OR gate.

In Fig. 19.11, the easy method for transforming an OR gate into a NOR gate is illustrated.

Table 19.1 Input and output signal values for the two-input OR gate

Input A	Input B	Output = $A + B$
0	1	1
1	0	1
1	1	1
0	0	0

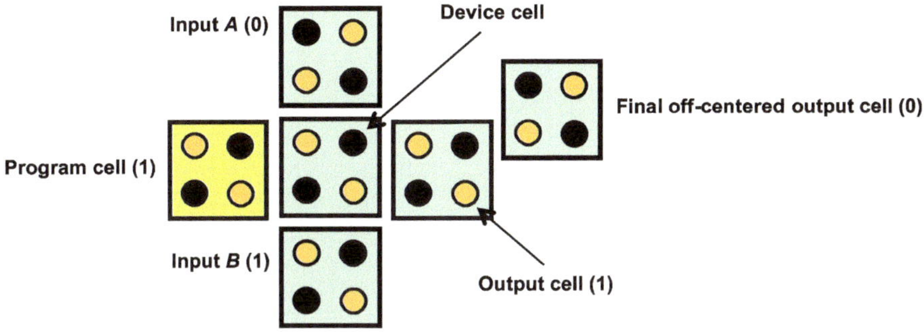

Fig. 19.11 Conversion of an OR gate into a NOR gate by placing an off-centered cell near the output cell of OR gate: (A = 0, B = 1) → 0

19.13 QDCA AND Gate

Figure 19.12 is a modification of the OR gate of Fig. 19.10. In Fig. 19.12, the program cell is taken as logic 0. This is the main demarcating feature with regard to the OR gate wherein the program cell is taken at logic 1. Let us go through the sequence of steps in the previous section and find out the output signal in each case. When A = logic 1, B = logic 0, program cell = logic 0, the majority is for logic 0 and output = logic 0. For A = logic 0, B = logic 1, program cell = logic 0, the majority is for logic 0 and output signal = logic 0. If A = logic 1, B = logic 1, program cell = logic 0, the majority is for logic 1 and output signal = logic 1. Finally, when A = logic 0, B = logic 0, program cell = logic 0, majority is for logic

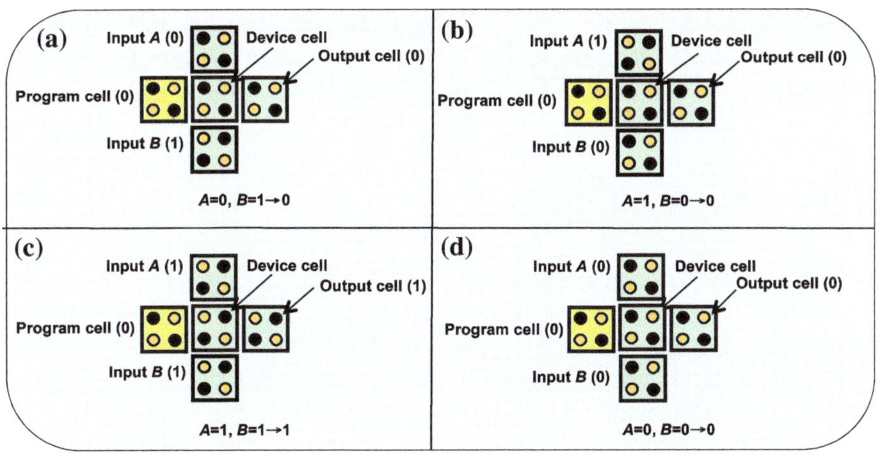

Fig. 19.12 Different possible combinations of values of input signals A and B, and corresponding device cell and output cell configurations for a QDCA AND gate. **a** A = 0, B = 1 → 0. **b** A = 1, B = 0 → 0. **c** A = 1, B = 1 → 1. **d** A = 0, B = 0 → 0

Table 19.2 Input and output signal values for the two-input AND gate

Input *A*	Input *B*	Output = *A* + *B*
0	1	0
1	0	0
1	1	1
0	0	0

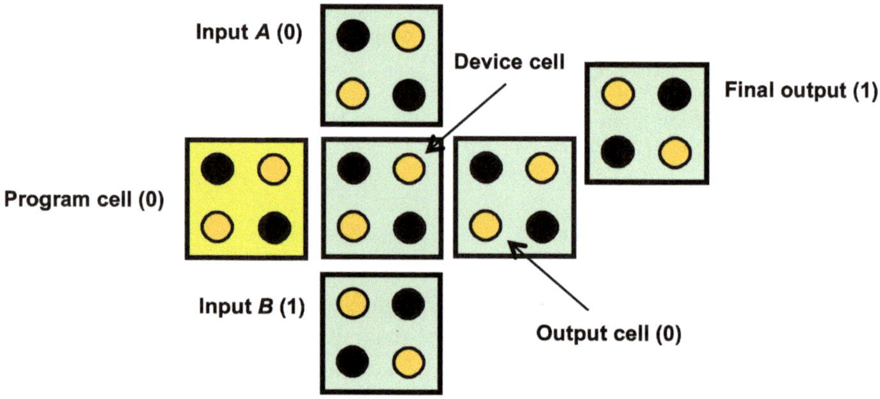

Fig. 19.13 Conversion of an AND gate into a NAND gate by placing an off-centered cell near the output cell of AND gate: $(A = 0, B = 1) \rightarrow 1$

0 and the output signal = logic 0. The above combinations of signals lead to the truth table (Table 19.2), which is the truth table of an AND gate. Hence AND gate function is verified. Thus an AND gate is obtained from the OR gate simply by changing the polarization of the program cell from logic 1 to logic 0.

As done previously for NOR gate realization, a NAND gate is obtained by the introduction of an off-centered cell at the output terminal of an AND gate. This scheme of execution of NAND gate function is shown in Fig. 19.13.

19.14 Clocking of QDCA

In order to operate a QDCA logic circuit, it must be supplied the required clock signals [3]. Judicious timing of clock signals is necessary to guide the synchronized flow of data avoiding random adjustments of QDCA cells. The clock signals control the tunnel barriers between adjacent quantum dots in each cell. They provide the power to run the circuit.

There are four stages in the modulation of tunnel barriers. In the first stage, the tunnel barriers are increasing in magnitude so that a restriction on electron tunneling is being applied. This is the switching stage. In the second stage, the tunnel barriers have been increased to the final maximum value so that tunneling is completely

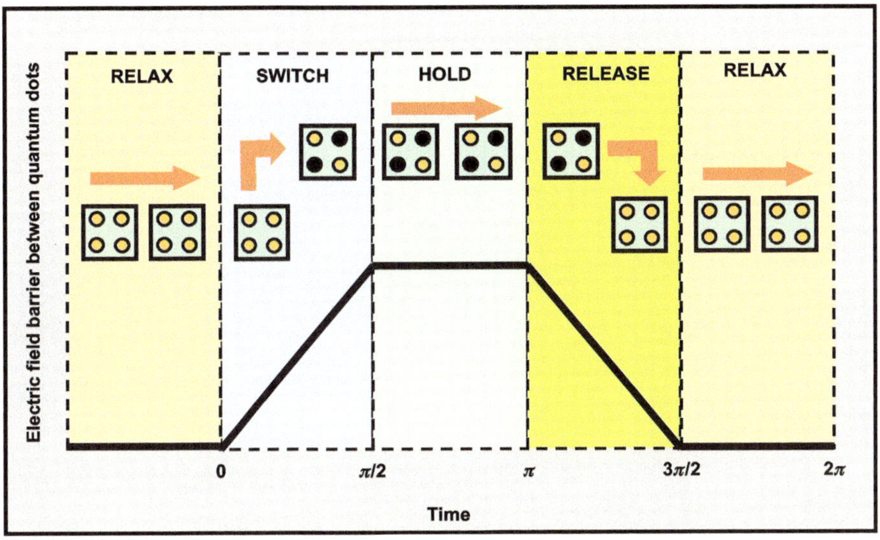

Fig. 19.14 Four stages in a QDCA clock

forbidden. This is the hold stage. In the third stage, the tunnel barriers are decreasing so that electrons are becoming free. This is the release stage. In the fourth stage, the tunnel barriers have reached the ultimate minimum value so that the electrons are totally free to tunnel between quantum dots. This is the relaxed stage.

According to the above four stages, one clock cycle consists of four clock signals (Fig. 19.14). The four clock signals are shifted in phase by a quarter of the complete clock cycle, i.e., by 90° among each other. Hence, the clock signal is referred to as a four-phase clock signal. The four phases are called switch, hold, release, and relax. A group of QDCA cells being controlled by the clock comprise a clocking zone.

The four-phase clock signal for operating a QDCA cell is shown in Fig. 19.15. The clock signal changes phase when the potential barriers affecting a clocking zone are increased/decreased or remain increased/decreased [4]. During the switch phase of clock signal, the tunnel barrier is slowly increased. The QDCA cell stabilizes to one of the two polarization states $P = +1$ or $P = -1$ under the action of contiguous cells. In the hold phase, the tunnel barrier is fixed at the highest value. The consequence is repression of electron tunneling between dots and maintenance of the polarization state of cell as such. During the release phase, the tunnel barrier is gradually decreased. The electrons are becoming free and gaining mobility. The QDCA cell is unpolarized. In the last relaxed phase, the tunnel barrier has reached its lowest value. The QDCA cell continues in its unpolarized state. The polarization of a QDCA cell is determined during its passage through switch phase by the polarizations of nearby cells that are either in switch or hold phases. Cells in release and relax phases are unable to affect the polarization of a cell.

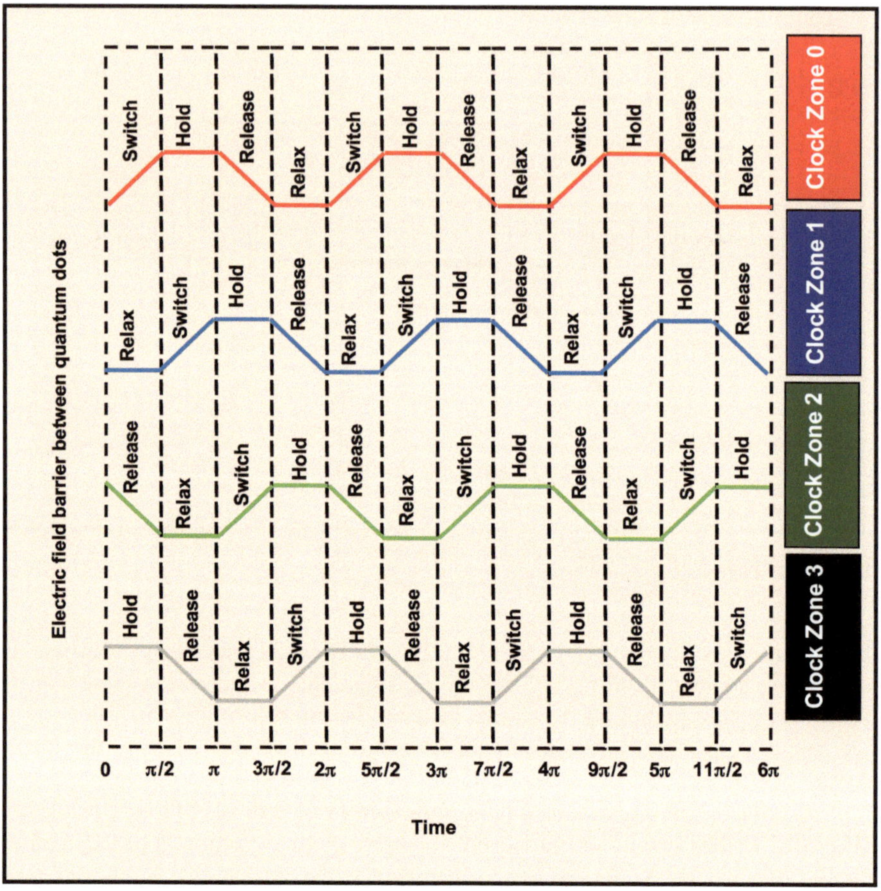

Fig. 19.15 The four-phase QDCA clock signal

Figure 19.16 shows the propagation of data through four adjoining clocking zones. Due to the phase delay, when the cells in a given zone are in the hold stage, those in the succeeding zone are entering the switch stage. By maintaining this sequence, the data is transferred from one region to the next.

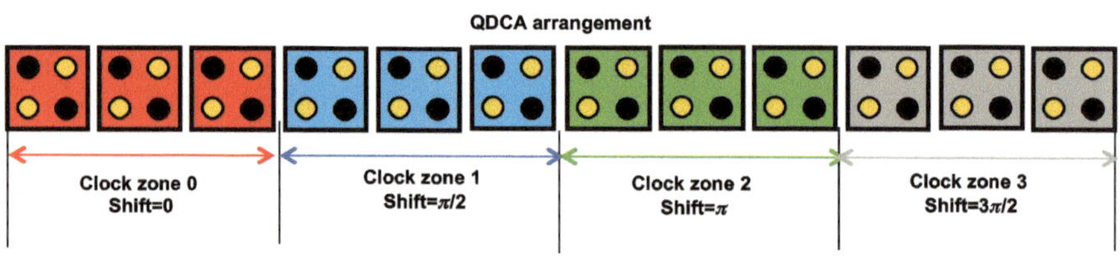

Fig. 19.16 Managing the flow of data along a QCA wire

The inquisitive reader may be tempted to ask about the method of applying the clock signal because the cells are not connected by any wires. The clock signals are applied through an induced electric field. This electric field is produced by CMOS circuitry. The circuitry is embedded under the QDCA plane.

19.15 Experimental Validation of QDCA Cell and QDCA Logic Functionality

From the preceding discussion, it appears as if the QDCA concept is purely hypothetical. But this is not true. The operation of QDCA cell was experimentally demonstrated by Orlov et al. [5] using a 4-dot QDCA cell (Fig. 19.17). This cell contained the dots D_1, D_2, D_3, and D_4, coupled in a circle through tunnel barriers. The area enclosed by the tunnel barrier (60 nm × 60 nm) determines the charging energy of the quantum dot and thereby the operating temperature of QDCA cell. Because the Coulomb force for this size is very weak $\sim k_B \times 1$ K, the demonstration required exceedingly low temperature in the milliKelivin range. For ensuring a cold environment, the cell was placed on the finger of a dilution refrigerator. The dilution refrigerator is a cryogenic device providing temperatures in mK range by a method called 3He/4He dilution; 3He and 4He are helium isotopes. A temperature ~ 10 mK was maintained. The dots D_1 and D_2 are gated by voltage sources V_1 and V_2 ($V_1 = -V_2$). A differential voltage is applied between

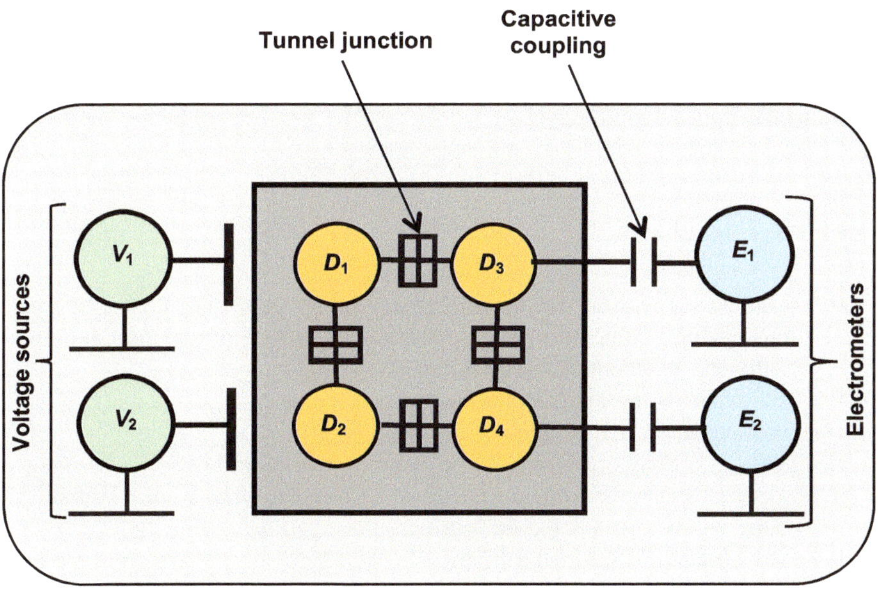

Fig. 19.17 Experimental set up comprising the QDCA cell with voltage sources and electrometers

these gates. A voltage sweep is performed from negative to positive voltage. Gates of the remaining dots are held at a constant voltage. Dots D_3 and D_4 are capacitively coupled to electrometers E_1, E_2 to measure the potential difference between these two dots. On application of the voltage sweep from negative to positive, the electron on the dot D_2 begins to move. It hops to dot D_1. Consequent upon the electron jumping from D_2 to D_1, there is an unbalance of forces acting between electrons, whereby electron in dot D_3 leaps to dot D_4. The potential difference $V_{D3} - V_{D4}$ between dots D_3 and D_4 is recorded by the electrometers E_1, E_2. From this potential difference, the charges on the dots D_3 and D_4 are evaluated. On the basis of this calculation, it is found that the electron has changed its position from dot D_3 to dot D_4. Thus the experiment shows that an electron switches its location on the side of output dots when an electron does so, on the side of input dots. Hence the QDCA cell considered is proved to be working.

Further onwards, Amlani et al. [6] operated logic AND and OR gates by applying input signals to them. They could verify that the gates were actually operating and giving output signals in compliance with the truth tables of the respective gates.

19.16 Discussion and Conclusions

The QDCA paradigm described in this chapter offers new opportunities as a computational architecture. CMOS topological layouts cannot be directly translated to work in this model. New designs have to be formulated. For harnessing the full potential of QDCA, the line of thinking must deviate from Boolean logic circuits. Then many features outside CMOS are likely to be availed. The QDCA offers incredible increase in speed and packaging density with downscaling, which can be done to the molecular/atomic level.

Review Exercises

19.1 Outline the reasons due to which quantum structures should not be used in customary architectures. How does the proposition of quantum dot cellular automata overcome these problems?

19.2 Elucidate the part played by quantum dot in the cellular automata. What are the special properties of quantum dot, which led to its selection for this role? Why is a tunnel barrier required between quantum dots?

19.3 How is binary information encoded in four-dot QDCA cell? Without any current flow, how is the information conveyed between cells?

19.4 How does a QDCA cell differ from a MOSFET switch? Which of the two is a better structural unit for logic circuits? Why?

19.5 Explain the two polarization states of a QDCA cell? How is the cell switched between the two states? How is the information read from a QDCA cell? What is the null state of the cell?

19.6 How does electron beam lithography differ from optical lithography? Why optical lithography cannot be used to delineate the pattern of quantum dots?

19.7 Explain with the help of a diagram the propagation of a binary value along the length of a 90° QDCA wire? How does the binary value traverse the length of a 45° QDCA wire? What is the difference between the two cases?

19.8 Draw a QDCA NOT gate. Explain with the help of this diagram how does this gate invert the polarization state of an input signal?

19.9 Why is the 3-input majority gate a fundamental QDCA logic circuit? Describe its operation with a labeled diagram.

19.10 How is the QDCA majority gate converted into an AND gate by fixing one of its inputs in the 0 state?

19.11 How is the QDCA majority gate reduced to an OR gate by fixing one of its inputs in the 1 state?

19.12 Describe the four stages in the modulation of tunnel barriers in a QDCA cell.

19.13 What is meant by clocking of QDCA cells? Why is a 4-phase clock signal necessary for QDCA? What happens during each phase?

19.14 How was the functioning of a QDCA cell experimentally demonstrated? Why was a low temperature necessary for performing this experiment?

References

1. Tougaw PD, Lent CS (1994) Logical devices implemented using quantum cellular automata. J Appl Phys 75(3):1818–1825
2. Snider GL, Orlov AO, Amlani I et al (1999) Quantum-dot cellular automata. J Vac Sci Technol A 17(4):1394–1398
3. Tougaw D (2014) A clocking strategy for scalable and fault-tolerant QDCA signal distribution in combinational and sequential devices. In: Anderson NG, Bhanja S (eds) Field-coupled nanocomputing. Springer, Berlin, Heidelberg, pp 61–72
4. Mustafa M, Beigh MR (2013) Design and implementation of quantum cellular automata based novel parity generator and checker circuits with minimum complexity and cell count. Indian J Pure Appl Phys 51:60–66
5. Orlov AO, Amlani I, Bernstein HG et al (1997) Realization of a functional cell for quantum-dot cellular automata. Science 277:928–930
6. Amlani I, Orlov AO, Toth G (1999) Digital logic gate using quantum-dot cellular automata. Science 284:289–291

Chapter 20
Nanomagnetic Logic

Abstract Diverting away from the monotony of charge-based paradigms, nanoelectronics looks toward nanomagnetics for help. CMOS logic is likely to become unacceptably energy inefficient below 10 nm gate length. QDCA-based logic too is confronted with the problem of allowing operation only at temperatures close to 0 K at the present technological competence. Nanomagnetics comes to rescue with a technology offering orders of magnitude lower heat dissipation than CMOS and capable of providing room-temperature operation. Nanomagnetic logic in the form of magnetic quantum cellular automata could serve as the holy grail of IC industry after CMOS has reached the end of the roadmap. This chapter highlights the limitations of CMOS and QDCA paradigms. Single-spin logic is introduced as a probable option. But the necessity of very cold environments for its deployment discourages us to tread this path. Then ferromagnetic dot-based logic is discussed and the actualization of magnetic quantum cellular automata (MQCA) through reconfigurable array of magnetic automata (RAMA) is treated.

20.1 Introduction

In this chapter, the two charge-based nanoelectronics approaches: MOSFET and quantum dot-based, are briefly discussed. Then a new approach established on electron spin is introduced.

20.2 Departing from Charge-Based Nanoelectronics

20.2.1 Charge-Based MOSFET Nanoelectronics

At the present juncture of nanoelectronics, a need is being felt for withdrawal from the prevailing practice relying on the quantity of charge present in the MOSFET channel as the basis for designing logic circuits, and to explore alternative avenues.

V.K. Khanna, *Integrated Nanoelectronics*,
NanoScience and Technology, DOI 10.1007/978-81-322-3625-2_20

As we know, MOSFET nanoelectronics is based primarily on the magnitude of electric charges. When the MOSFET is in ON-state, the channel contains a large number of electrons. This state of plentiful electronic charges is represented by the code '0'. In the OFF-state of MOSFET, the channel is exhausted of electrons. This state of scarcity of electronic charges is encoded by the symbol '1'. The MOSFET switch changes states between 0 and 1, i.e., from a state of presence of abundant charges to one having a negligible proportion of charges. In order to change the states of a MOSFET switch, the channel has to be either flooded with charges or charges have to be removed from a filled channel. Flooding of the channel or removal of charges from it is brought about by the current flowing through the MOSFET switch circuit. This means that a current flow is indispensable to alter the switch states. The flow of current is invariably associated with production of heat. According to Joule's law of heating effects of electric current, the quantity of heat Q generated by a current of magnitude I flowing through a resistor of resistance R for an interval of time t is given by the familiar equation

$$Q = I^2Rt/J, \tag{20.1}$$

where J denotes the mechanical equivalent of heat = 4.2 J/Calorie.

The wastage of energy in the form of heat is a serious disadvantage of the existing charge-based MOSFET nanoelectronics.

20.2.2 Charge-Based QDCA Nanoelectronics

To deviate from charge-based MOSFET nanoelectronics, the QDCA nanoelectronics was proposed. This nanoelectronics is based on encoding of binary information in the location of positions of electrons in quantum dot cells. Computation is performed through Coulombic interactions between cells without any flow of current and without supplying any power to particularized cells. So, this method removes the power dissipation issue of MOSFET-based nanolectronics. However, it is also charge-based. Furthermore, it mandates the use of very small quantum dots < 2 nm if the system is required to work at room temperature. Otherwise, cryogenic operation is essential with its attendant cooling assembly and exorbitant expenditure. At the present state of human technological knowledge, the manufacturability of such quantum dots is not within reach.

20.3 Single-Spin Logic

Fine structure of hydrogen spectral lines and Stern–Gerlach experiment revealed that the electron possessed an intrinsic angular momentum and magnetic moment. In the classical picture this was possible only if the electron were a charged spinning ball. Spin is a pseudovector or axial vector, whose sign is invariant on reversal of coordinate axes. It is distinguished from a polar vector or true vector or simply vector which changes sign upon such reversal of axes, e.g., velocity, momentum, force.

From classical analogy, the intrinsic angular momentum of electron was associated with spin. In line with the quantization of orbital angular momentum and assuming that electron spin behaved similarly, an angular momentum quantum number = 1/2 was required to give two possible states for this angular momentum that were predicted by experiments.

Spin is fixed in magnitude but variable in direction. It is governed by Pauli's exclusion principle. This principle asserts that an orbital is occupied by a maximum of two electrons with opposing spins, one of which is represented by an upward arrow, spin quantum number $m_s = +1/2$ and the other by a downward arrow, $m_s = -1/2$.

For a particle placed in a magnetic field, two polarizations are allowed quantum-mechanically, one parallel to the field and other antiparallel to the field. These two conditions can be used to represent logic 0 and logic 1 states.

To switch from one state to another state, the electron need not undergo any displacement from its position. It is not physically moved in space. Consequently, the switching between states can take place without flow of current. It is merely required to reverse the spin of the electron. Due to this reason, the energy dissipation accompanying the current flow is significantly reduced. This is a major advantage favoring single-spin logic.

The energy consumed during spin reversal = Difference of energies between the two polarization states = Zeeman energy ΔE in a magnetic field B given by

$$\Delta E = g\mu_{\mathrm{B}}B, \tag{20.2}$$

where g is the electron spin g-factor with value 2, and μ_{B} is the Bohr magneton = 5.79×10^{-5} eV/T. It is acknowledged that spin couples feebly with phonons. This fact can be put to advantage. The energy ΔE can be reduced to a value < thermal energy $k_{\mathrm{B}}T$. The constant k_{B} is the Boltzmann constant and T is the absolute temperature. By doing so, unwanted random flips of spins will not create any disturbance to the logic. At room temperature, $k_{\mathrm{B}}T \sim 25$ meV. The feeble coupling of spin with phonons leads to another favorable circumstance that spin can be maintained in a position out of equilibrium for a longer interval of time. Thus single-spin logic is extremely efficient in saving energy. But considering the state-of-the-art semiconductor facilities and taking cognizance of technological incapability of producing spin-carrying particles of small dimensions, these devices must be operated at very low temperatures ~ 1 K. This is because the spins interact through Coulomb's law of magnetostatics and the magnetostatic force between opposite spins varies inversely as the square of the distance between the spin-carrying particles. Larger the size of the particles, greater is the distance between spins and smaller is the force of attraction or repulsion. Therefore, it is presently not a lucrative proposal.

20.4 The Notion of Room-Temperature Nanomagnetic Logic

For avoidance of low-temperature environments, a new direction for logic has been pursued, namely through dipole–dipole interactions between magnetic dots. This direction makes us follow the footsteps of quantum dots. Thinking about magnetic dots, one immediately turns attention toward ferromagnetic materials. These materials include iron, cobalt and nickel, rare earths, lodestone, etc. The atoms of these materials exhibit magnetic moments as their unpaired spins line up parallel to each other in regions called domains. In the unmagnetized condition, the domains are randomly oriented but in the presence of an external magnetic field, they are aligned with each other so that the material displays intense magnetism.

In a ferromagnetic dot, a large number of electron spins $\sim 10^4$ act together in a united manner. Therefore, the dot acts as a material with colossal spin. If a grain of ferromagnetic material has anisotropic shape, the dot shows bistable behavior. Its bistability property can be utilized to represent logic 1 and logic 0 for realization of digital logic. Magnetic quantum cellular automata (MQCA) works in the same way as quantum dot cellular automata (QDCA). The dots are addressable electrically. They can be gated and clocked by electric fields. But MQCA does not impose the requirement of extremely low temperatures; rather it works at room temperature. *Prima facie*, it may appear that the energy necessary to switch 10^4 spins contained in a ferromagnetic dot will be 10^4 times the energy dissipated for a single spin so that this logic will be immensely wasteful of energy. Fortunately, the situation is vastly different because the spins interact amongst themselves. This interaction decreases the number of degrees of freedom appreciably. So, the switching energy for 10^4 spins is only 35 times that required to switch a single spin. From the viewpoint of power dissipation, it is quite encouraging.

Not only from power dissipation standpoint, the above value of energy is also optimal for ferromagnetic dots of sizes reachable by present technology without any disturbance due to thermal energy at room temperature. Energy required for flipping ferromagnetic dots of submicron size containing such large number of spins from spin up to spin down state or vice versa at room temperature ($\sim$several eV) is >> than the thermal energy k_BT (25 meV) at this temperature. So, the flipping of spins occurring during circuit operation is not perturbed by their random flipping due to thermal agitation. Contemporary technological expertise in photolithography easily allows fabrication of ferromagnetic dots of submicron size. Thus, for reasonably sized ferromagnetic dots within current technological capabilities, the flipping energy is neither very large to be worrisome for power dissipation nor it is too small to be vexatious for being upset by thermal energy of molecules. Therefore, the proposal of ferromagnetic dots working at room temperature is straightway acceptable.

20.5 Magnetic Quantum Cellular Automata (MQCA)

20.5.1 MQCA Versus QDCA

Cowburn and Welland [1] in 2000 used an oscillating magnetic field as a clock to demonstrate the operation of MQCA at room temperature. Submicrometer size ferromagnetic dots of circular shape were used. Circular nanomagnets have the shortcoming that they cannot benefit from the anisotropy dependent on shape of the magnet. This anisotropy is highly desirable because it makes the magnets intrinsically bistable. The bistability prevents the magnets from changing polarization states in response to thermal effects.

Bernstein et al. [2] in 2005 found that the magnetic interactions between nanomagnets were adequate to allow room temperature operation and that the coupling could be improved by tailoring their shapes. Imre et al. [3] in 2006 showed the working of a 3-input majority gate with nanomagnets. MQCA can be downscaled to the thermal limit $\sim 40\text{k}_{\text{B}}T$. A nanomagnet of size 5 nm^3 is required. At smaller sizes, the ambient temperature can cause erratic flips of spins of the nanomagnets. Nonetheless, the spin state is stable for ~ 1 ms. This time duration is sufficient for some information processes.

On the other side, as already mentioned in the chapter on QDCA, this logic has been demonstrated at cryogenic temperatures in subKelvin range. The enormous difference of operating temperatures between MQCA and QDCA arises from the large difference in switching energy for the two cases. To reiterate, the energy difference between the two polarization states in magnetic logic is $\sim$several eV for submicron-size permalloy (80 % Ni, 20 % Fe) magnets. Magnetic dots of this size are easily patternable by modern photolithographic techniques. In QDCA, for micron-sized dots with 50 nm × 50 nm tunnel barriers, the energy difference between ground and excited states, i.e., between the logic 1 and logic 0 states is infinitesimally small ~ 0.1 meV [4]. The dot and cell size must be decreased tremendously to the molecular scale to raise the switching energy sufficiently to enable operation at the room temperature level. Semiconductor process engineers are striving to surmount this formidable hurdle.

20.5.2 MQCA and CMOS

The switching energy for a magnetic grain is twice the anisotropy per unit volume of the grain multiplied by the volume of the grain. If a cubic ferromagnetic dot of size 5 nm × 5 nm × 5 nm is taken, this energy is estimated to be 0.8 eV, which is about 32 times the thermal energy $k_{\text{B}}T$ at room temperature. In comparison, the switching energy of CMOS logic is 5×10^4 times the room-temperature thermal energy. Hence, in case of ferromagnetic dots, the switching energy is less than for CMOS gates by a factor of 1562.5, which is more than three orders of magnitude.

This is a vital benefit favoring the adoption of magnetic logic *vis-a-vis* CMOS logic. If CMOS is scaled down below 10 nm, the heat dissipation will be too large to make its usage meaningless.

The insensitivity of magnetic materials to radiation makes MQCA particularly appealing for applications in radiation-contaminated hostile environments, notably for space missions, satellites and military operations [2].

On the downside, the precession speeds of the domains in the ferromagnet under the dipolar influence of nearby nanomagnets are typically of the order of several nanoseconds. As a result, MQCA is much slower than CMOS logic. Interfacing of MQCA logic with CMOS circuitry is difficult because MQCA is spin-based and CMOS is charge-based. Information conversion between the two paradigms is complex.

20.6 Reconfigurable Array of Magnetic Automata (RAMA)

20.6.1 RAMA for Logic Gates

MQCA architecture can be easily organized from a thin-film array of ferromagnetic nanopillars [5, 6]. This array is fabricated by polymeric self-assembly method on a substrate. It is called RAMA (Fig. 20.1). The nanopillars are made at the point of intersection of two sets of conducting nanowires. One set of nanowires runs below the lower surface of the nanopillars. The other set of nanowires runs above the upper surface of the nanopillars. The two sets of nanowires are mutually orthogonal to each other. The nanopillars are made of a ferromagnetic material, e.g., $CoFe_2O_4$ embedded in a matrix of ferroelectric or multiferroic material, e.g., $BiFeO_3$.

The magnetizations of the nanopillars are randomly arranged, either up or down (Fig. 20.2). Application of a fairly small electric field can change the direction of these magnetizations from perpendicular to the surface of the array, i.e., out-of-plane to parallel to the surface or in-plane.

To understand how the MQCA is configured from the above thin film array, let us look at its structure. Similar to the QDCA, an MQCA consists of square cells with four nanopillars at its four corners (Fig. 20.3). Two pairs of nanopillars are located on the opposite ends of the two diagonals of the square cell. The magnetizations of one pair of nanopillars are pointing upwards while those of the other pair of nanopillars are directed downwards. Thus there are two states of polarization according as the magnetizations of nanopillars on one diagonal are pointing up, or those of nanopillars on the other diagonal are pointing up. The MQCA works like QDCA. Magnetizations in opposite directions are attracted and drawn nearer. Those in same direction are repelled farther away.

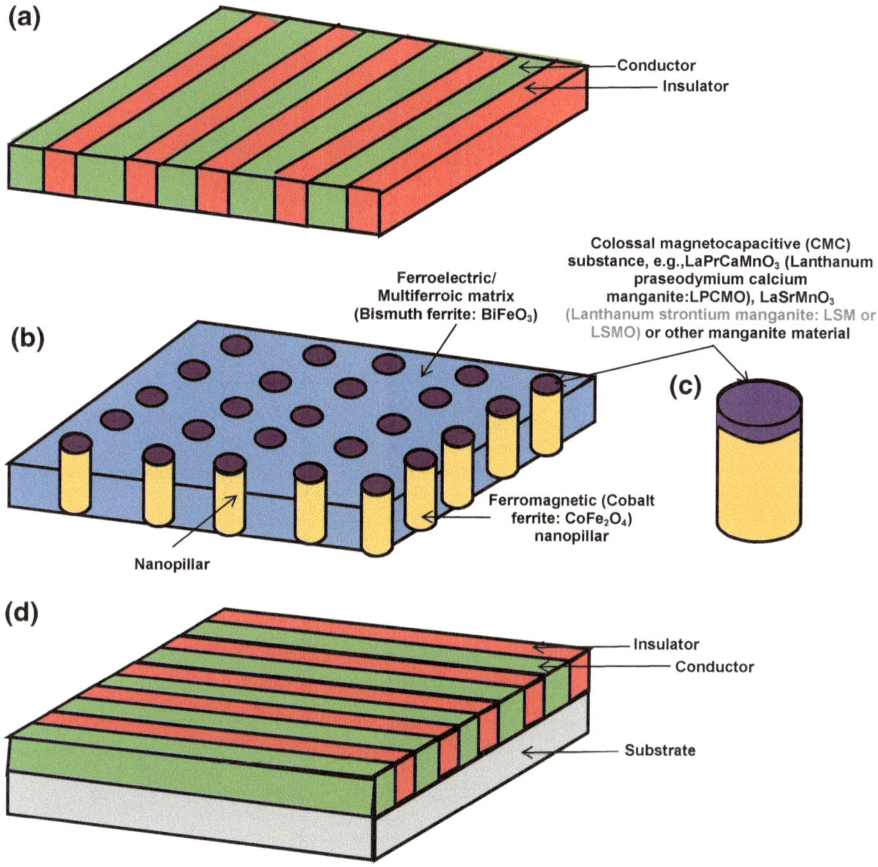

Fig. 20.1 Construction of reconfigurable array of magnetic automat (RAMA): **a** Top conductor and insulator arrangement, **b** nanopillars embedded in matrix, **c** CMC material in nanopillar and **d** bottom conductors and insulators supported on substrate

The reconfigurable array allows us to form any desired arrangement of nanopillars for MQCA. This is done by disabling the unwanted nanopillars in the array. At the interface between the nanopillars and the supporting matrix, electric fields are applied. These fields are applied column by column using the crossbars of nanowires. The interface is thereby strained. The energetically favored direction of spontaneous magnetization of the nanopillars is rotated from out-of-plane to in-plane condition. In this manner, all the unrequired nanopillars are disabled. Then the pattern to form a logic gate is superimposed on the array. The remnant polarization in the ferroelectric matrix preserves the magnetization direction in-plane unless a small electric field is applied in the reverse direction such as for creation of the pattern to form another logic gate.

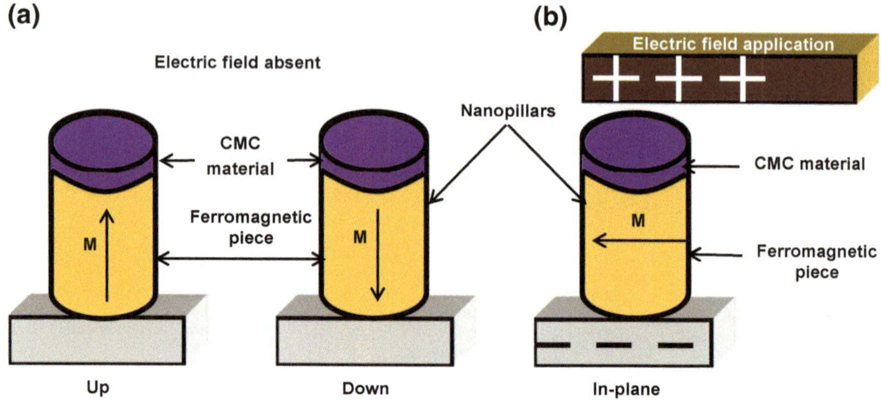

Fig. 20.2 Magnetization states of nanopillars: **a** in absence of electric of electric field, either polarized up or down; **b** in presence of an applied electric field, polarized in-plane

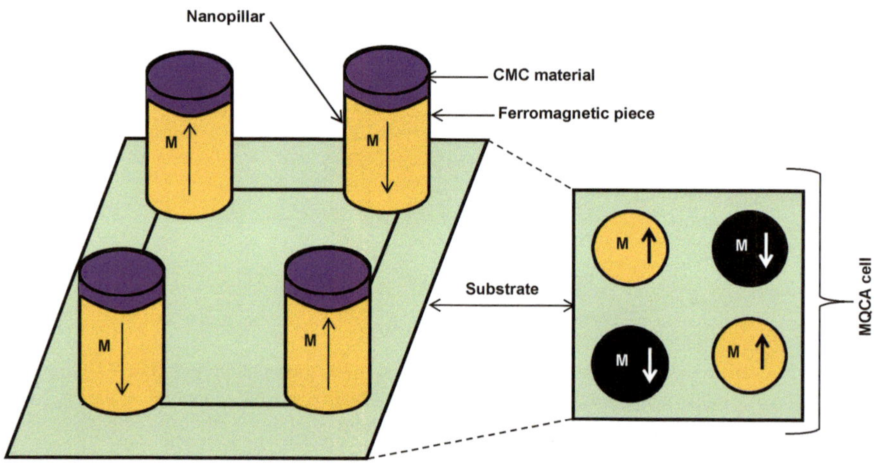

Fig. 20.3 Placement of four nanopillars inside a square cell with magnetizations in directions indicated by arrows to form a magnetic automata cell analogous to quantum dot cell

To begin the gate operations, an electric field is applied at the left topmost corner of the input cell. This electric field rotates the magnetization of the cell from out-of-plane to in-plane. As the electric field starts reversing, a small magnetic field is applied, either in upwards or downwards direction, and the nanopillars acquire their magnetization directions. To clock the array, synchronized electric fields are applied. No magnetic fields are used for clocking. By selectively enfeebling the magnetizations of the nanopillars, the electric fields bring down the barrier to rotation from out-of-plane to in-plane condition. Thus they gate the propagation of the spin flips.

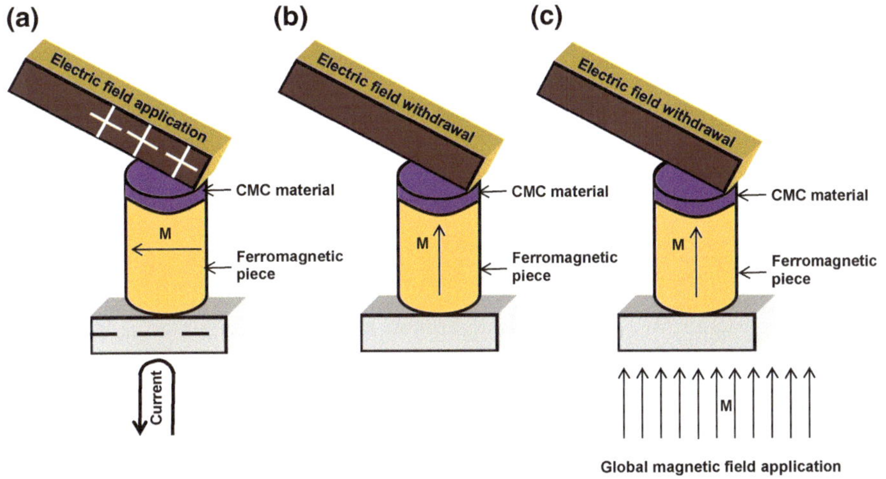

Fig. 20.4 Writing information into and reading it from nanopillars: **a** and **b** Write operations in which (**a**) an electric field is applied on the nanopillar to render the magnetization in-plane with a small bias current in the desired direction of writing, and **b** the electric field is withdrawn so that the input bit writing is completed. **c** Read operation in which the nanopillar is subjected to a global magnetic field and the information is recovered by measuring the super magnetocapacitor by addition or subtraction from ferromagnetism

The output signal is inductively coupled from the output cell to a pick up coil. It can also be obtained capacitively. Another method uses a magnetic tunnel junction (Sect. 11.3).

20.6.2 RAMA as a Memory Array

Each cell of the array can be addressed individually (Fig. 20.4). The cells retain their polarization states after the power is switched off. Deactivation of every alternate nanopillar will prevent mutual coupling between them. Then they can be easily written separately. If the array is to be used only as a memory, the inter-pillar spacing is increased. Then the nanopillars will couple weakly among themselves, and the complete array can be utilized for storing data. Thus a high-density array for non-volatile random access memory becomes available.

20.7 Discussion and Conclusions

A refreshing change from the routine charge-based nanoelectronics is provided by deviating from logic circuits utilizing switching of FETs or wirelessly coupled quantum dot cells. The nanomagnetic logic solves the problems of power dissipation of CMOS as well as low temperature requirement of QDCA with present technology. But all these merits are achieved at the sacrifice of operating speed. It appears while one paradigm gains in one aspect, it loses in another. So, an optimal combination of paradigms depending on application is the best choice.

Review Exercises

20.1 How is the performance of charge-based MOSFET nanoelectronics depending on flow of current limited by power dissipation as heat? How does quantum dot coupled automata model based on electrostatic coupling and requiring no current flow overcome this limitation? If it succeeds in doing so, what is preventing its widespread adoption?

20.2 Elaborate the statement, "Single-spin logic is highly energy-efficient." Put your arguments in favor of and against the proposal of single-spin logic.

20.3 What is a ferromagnetic material? Give some examples of such a material. What is the special property of this material?

20.4 A ferromagnetic dot may contain as high as 10^4 electron spins working cooperatively. Why is the flipping energy required to switch the dot from spin up to spin down state not 10^4 times the energy required to flip a single electron spin?

20.5 How is the proposal of room-temperature operation for nanomagnetic logic based on ferromagnetic dots justified from: (i) power dissipation viewpoint; (ii) thermal energy considerations?

20.6 Why is the anisotropy of shapes of nanomagnets desirable for MQCA logic?

20.7 Compare QDCA and MQCA in terms of the thermal environments required for their operations. Give reasons for your answer.

20.8 Compare MQCA with CMOS, pointing out their relative advantages and disadvantages.

20.9 How is MQCA obtained from the reconfigurable array of magnetic nanopillars? How is the gate operation initiated in MQCA? Explain the clocking of MQCA? How is the output signal obtained from MQCA?

20.10 Can RAMA be used as a memory array? If so, how can this be done?

20.11 Mention one important shortcoming of MQCA with respect to CMOS.

References

1. Cowburn RP, Welland ME (2000) Room temperature magnetic quantum cellular automata. Science 287(5457):1466–1468
2. Bernstein GH, Imre A, Metlushko V et al (2005) Magnetic QCA systems. Microelectron J 36:619–624

3. Imre A, Csaba G, Ji L et al (2006) Majority logic gate for magnetic quantum-dot cellular automata. Science 311:205–208
4. Orlov A, Imre A, Csaba G et al (2008) Magnetic quantum-dot cellular automata: Recent developments and prospects. J Nanoelectron Optoelectron 2:1–14
5. Wolf SA (2010) Reconfigurable array of magnetic automata (RAMA) and related methods thereof. WO 2010039871 A1, Publication date 8 April 2010 https://www.google.co.in/patents/WO2010039871A1?cl=en. Accessed on 11th May 2016
6. Kabir M, Stan MR, Wolf SA et al (2011) RAMA: a self-assembled multiferroic magnetic QCA for low power systems. GLSVLSI '11 Proceedings of the 21st edition of the Great lakes symposium on VLSI, ACM, New York, NY, pp 25–30

Chapter 21
Rapid Single Quantum Flux (RFSQ) Logic

Abstract The main circuit components of RFSQ logic are based on overdamped Josephson junctions. They produce, store, transfer, and reproduce voltage pulses. These pulses are of picosecond duration. They have quantized area, which is correlated with the transmission of a single quantum of magnetic flux across a Josephson junction. According to superconductivity theory, the magnetic flux threading a superconducting loop or a hole in a bulk superconductor cannot acquire continuous values. It only assumes integral values which are multiples of a quantum of magnetic flux = $h/(2q)$ where h stands for Plank's constant and q is the elementary electronic charge. Its value is 2.0678×10^{-15} Wb = 2.07 mV ps. The RFSQ elements represent an ultrafast digital electronics technology working at 100–700 GHz, and beyond. Although it is consuming very low power, a disadvantage of this technology is the need of liquid helium environment, which is expensive even for military applications.

21.1 Introduction

Rapid single flux quantum (RSFQ) logic is a superconductive technology working at cryogenic temperatures <123 K. In place of the customary transistor switches, the digital switches used in this technology are Josephson junctions (Fig. 21.1) comprising junctions between two metallic superconductors separated by a weak link which could be a thin insulating film or a constricted region. An example is a sandwich: superconductor–insulator (<3 nm thick)–superconductor.

21.2 Information Storage and Transference in RFSQ Logic

Information storage is done in the form of quanta of magnetic flux while it is transferred as voltage pulses of single quanta of flux, hence the term 'single flux quantum' logic. These voltage pulses have duration in picoseconds; hence known

V.K. Khanna, *Integrated Nanoelectronics*,
NanoScience and Technology, DOI 10.1007/978-81-322-3625-2_21

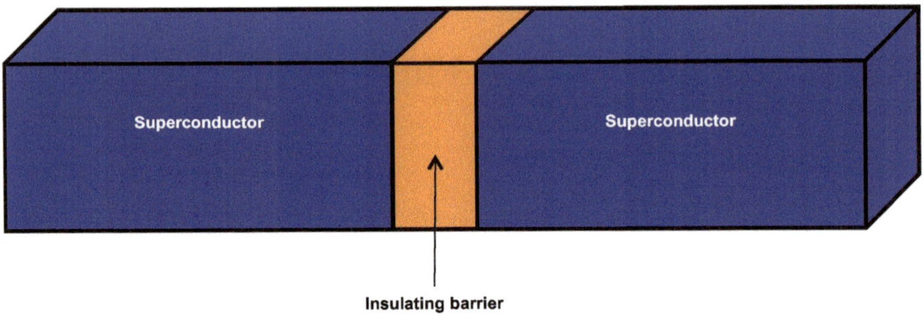

Fig. 21.1 A Josephson junction

as 'rapid single flux quantum' logic. The voltage pulse used in RFSQ logic is produced when the magnetic flux passing through a loop containing a Josephson junction undergoes a change by an amount equivalent to one quantum of magnetic flux. The change in magnetic flux occurs due to junction switching. The amplitude and duration of the RFSQ pulse are determined by the parameters of the Josephson junction. An event of switching of the junction is accompanied by the emission of a single quantum of magnetic flux. It tantamounts to logic 1. Absence of emission of this quantum, associated with nonswitching of Josephson junction, is equivalent to logic 0. Typically, the voltage pulse can have an amplitude of 2 mV or lower. It can be narrow as 1 ps and wider around 5–10 ps. The wires used for conveying signals are superconductive transmission lines producing negligibly small dispersion and attenuation of signals.

21.3 Components and Cells in RFSQ Logic

21.3.1 The Buffer Stage

In a Josephson junction, a super current flows across the junction by tunneling as long as the current produced by an applied DC voltage is less than a critical current I_C. Its magnitude is restricted within the range $\pm I_C$. No sooner than this critical current value is exceeded, an AC voltage V is produced across the junction with a frequency

$$\begin{aligned} f = (2q/h)V &= \frac{2 \times 1.6 \times 10^{-19}\,\text{Coulomb}}{6.626 \times 10^{-34}\,\text{J-s}} = 4.8295 \times 10^{14}\,\text{Hz/V} \\ &= 4.8295 \times 10^{14}\,\text{Hz}/(1000\,\text{mV}) = 4.83 \times 10^{11}\,\text{Hz/mV} \\ &= 483\,\text{GHz/mV} \end{aligned} \tag{21.1}$$

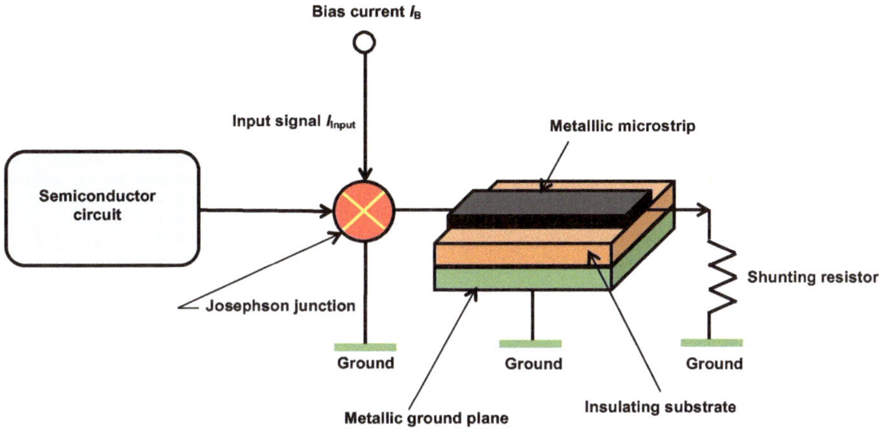

Fig. 21.2 The buffer stage used in RFSQ logic

In RFSQ logic, this oscillating current is overdamped [1], Fig. 21.2. The binary information is represented by pulses of very short duration ~ picoseconds, produced and processed by an elementary circuit, the buffer stage containing an overdamped Josephson junction fed by a current pulse from a semiconductor circuit [2]. The junction is overdamped by shunting it externally by a low-inductance metallic resistor. This obviously means that after passage of a pulse, the junction produces only one cycle of oscillation and is automatically reset to its original state, i.e., the phase changes by 360°.

21.3.2 *Josephson Transmission Line (JTL)*

Another important RSFQ circuit is the Josephson transmission line (JTL) shown in Fig. 21.3. It consists of several Josephson junctions, connected in parallel by low-inductance superconducting strips. The JTL is fed a biasing current I_B < the

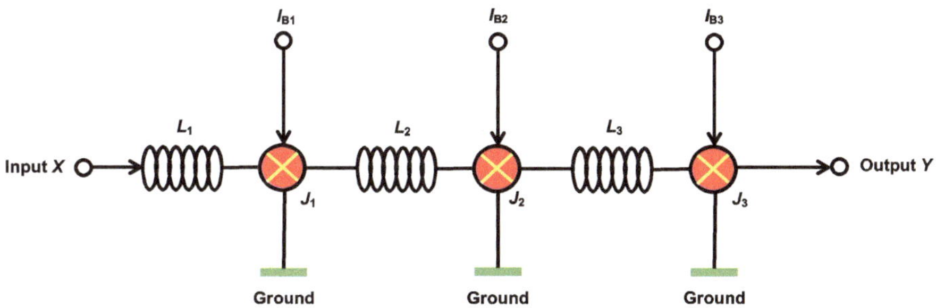

Fig. 21.3 The Josephson transmission line (JTL)

critical current I_C. The input signal first triggers the left Josephson junction, causing a 360° phase change. The voltage produced across junction J_1 activates junction J_2 and produces a similar 360° phase difference. Identical process takes place at J_3. The circuit amplifies RSFQ pulses producing current/power gain if the critical currents I_{C1}, I_{C2}, I_{C3} … of the junctions J_1, J_2, J_3, and the respective biasing currents I_{B1}, I_{B2}, I_{B3}, … increase in the direction of movement of the pulse, and the inductances L_1, L_2, L_3, … decrease proportionally. If the pulses are transmitted along microstrip lines, there is no amplification.

21.3.3 Pulse Splitter

The RSFQ pulse splitter (Fig. 21.4) is a generalization of the JTL circuit. It provides a reproduction of the input pulse at X at each of the two outputs Y and Z without any change in the amplitude of the voltage.

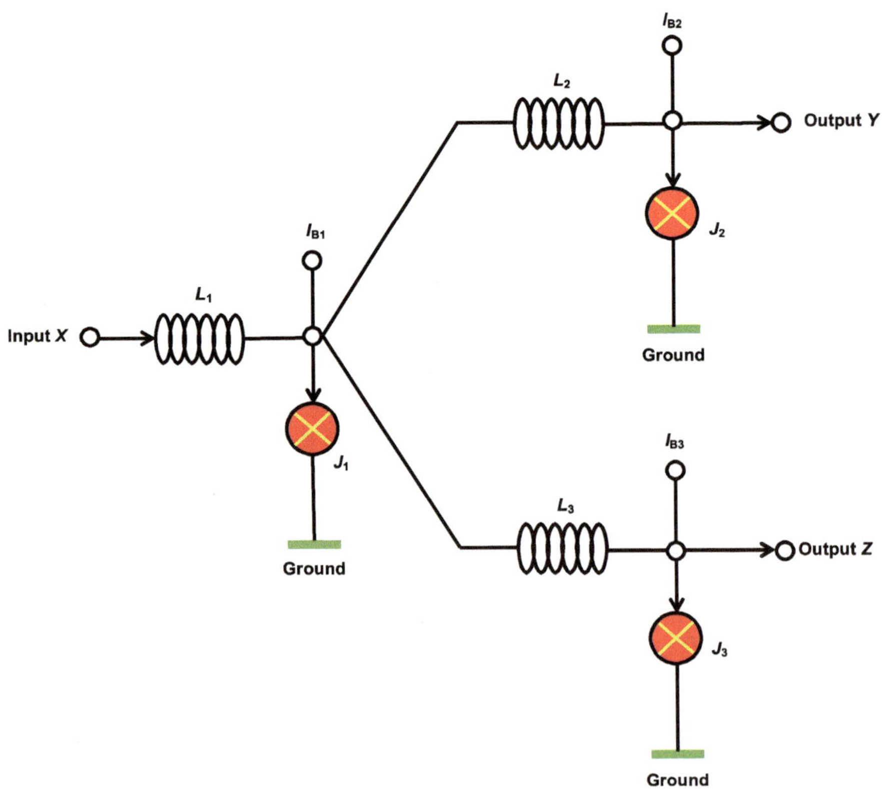

Fig. 21.4 The RFSQ pulse splitting circuit

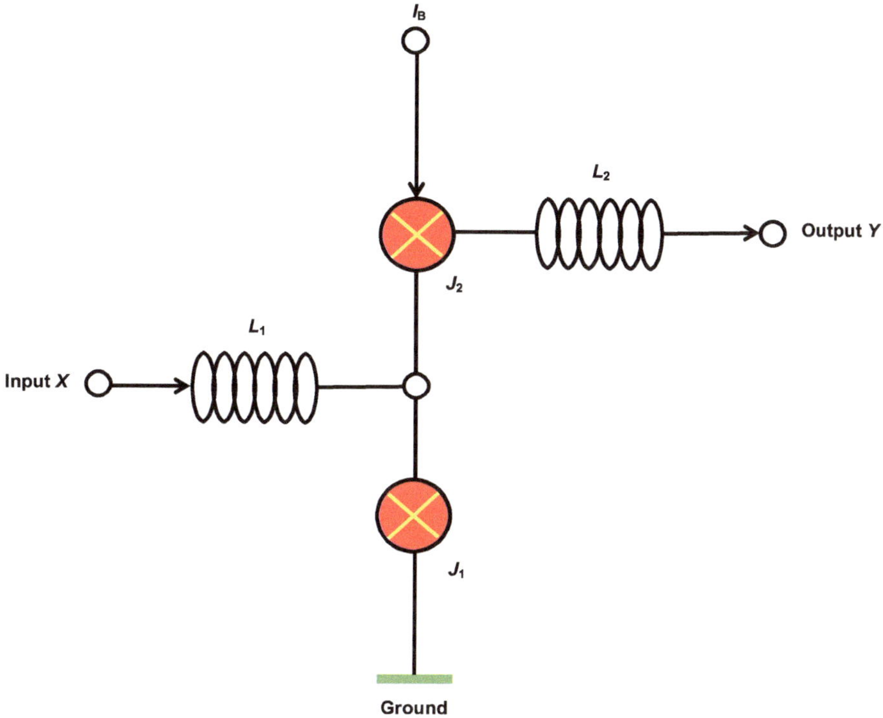

Fig. 21.5 The nonreciprocal RFSQ buffer stage

21.3.4 Non-reciprocal Buffer Stage

The above circuits being reciprocal cannot be employed for isolation. For making a non-reciprocal buffer stage (Fig. 21.5), the critical current of J_2 is taken slightly smaller than that of J_1, i.e., $I_{C2} < I_{C1}$. Now an input pulse at X acts only on J_1 producing a change of phase by 360° in J_1. It has no effect on J_2. The resultant RSFQ pulse is conveyed to output Y. On the opposite side, arrival of a pulse from output Y acts both on J_1 and J_2. Because $I_{C2} < I_{C1}$, the junction J_2 suffers the 360° phase change and voltage across J_1 remains near zero. Hence, the RSFQ pulse does not reach terminal X.

21.3.5 The Confluence Buffer

The confluence buffer (Fig. 21.6), a generalized form of aforesaid non-reciprocal buffer, allows the propagation of pulses from input terminals X and Y to output Z. The Josephson junctions J_3 and J_4 protect the inputs X and Y from any pulse coming from output Z.

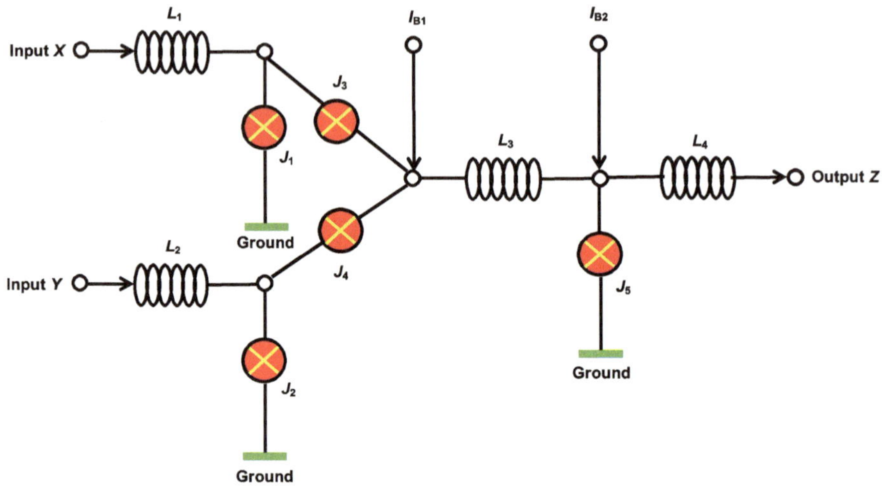

Fig. 21.6 The confluence buffer

21.3.6 The SQUID as an R-S Flip-Flop

A pervasive component in RSFQ logic is the DC superconducting quantum interference device (SQUID), i.e., the DC-SQUID consisting of two similar Josephson junctions (Fig. 21.7). The Stewart-McCumber parameter β of this SQUID is defined as

$$\beta = 2\pi L I_C / \Phi_0 \tag{21.2}$$

If β is chosen as 10 and $I_B \sim 0.8\ I_C$, this circuit exhibits two symmetrical stationary states (Fig. 21.8). These states differ in direction of the persistent current I_P flowing in the loop, and given by

$$I_P = \pm \Phi_0 / (2L) \tag{21.3}$$

One of these states pertains to the trapping of an extra single flux quantum in the superconducting loop of SQUID. Suppose that the current I_P is flowing in the anticlockwise direction, referring to binary 0 state. Then the current I_3 in J_3 becomes

$$I_3 = (I_B/2 + I_P) < I_C \tag{21.4}$$

At this instant, upon arrival of an RSFQ pulse at input S, junction J_3 undergoes a 360° phase jump whereas junction J_4 is unaffected because of the smaller DC current in it:

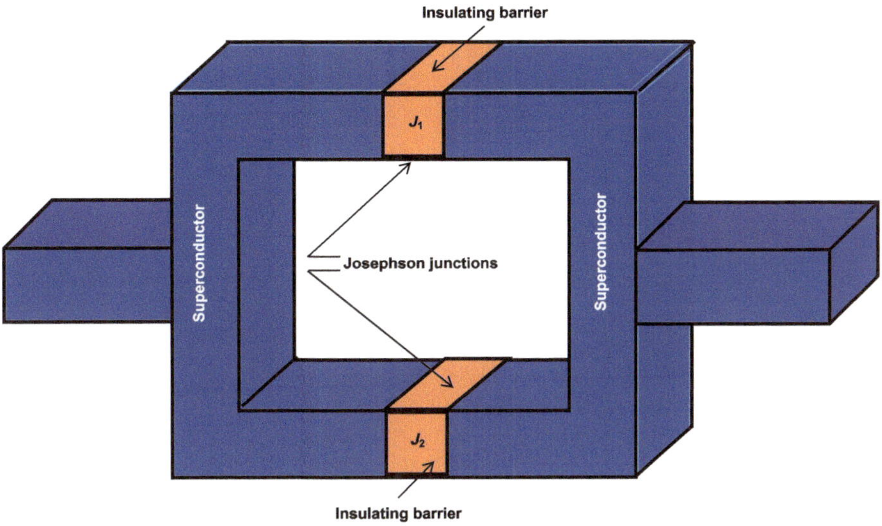

Fig. 21.7 A SQUID

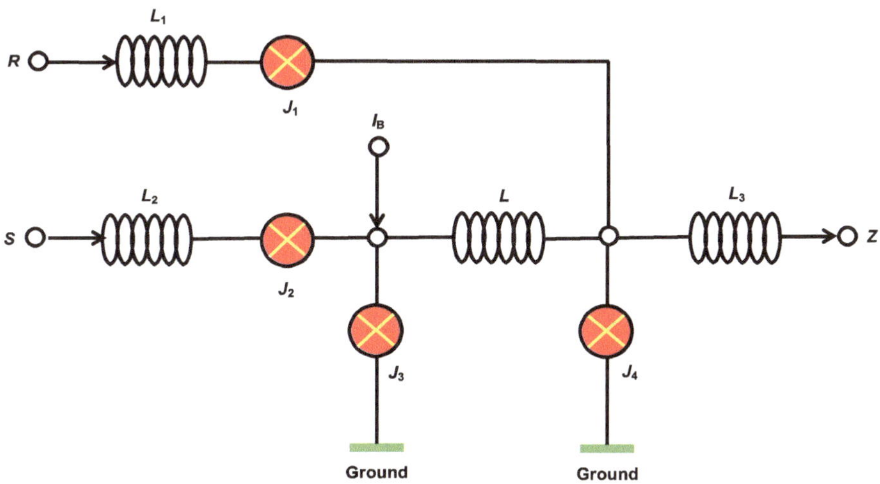

Fig. 21.8 The SQUID as an RFSQ R-S flip-flop

$$I_4 = I_{\mathrm{B}}/2 - I_{\mathrm{P}} \tag{21.5}$$

Consequently, the circuit changes its state to the condition in which the current I_{P} starts flowing in the clockwise direction, represented as binary 1 state. Thus a 0→1 switching has occurred. Now, the reverse 1→0 switching is initiated by an RSFQ pulse applied at the *R*-terminal. At the same time, an RSFQ pulse is produced across J_2. This RSFQ pulse acts as the output signal Z. The junctions J_1 and J_2 safeguard the RSFQ pulse sources against the back reaction of the SQUID in the event of an erroneous signal such as a signal at S-terminal during 1 state. Thus, the circuit operates like an R-S flip-flop for RSFQ pulses.

21.4 RFSQ Circuit and Convention

An RSFQ circuit consists of basic logic/memory cells (Fig. 21.9). These cells receive inputs from signal lines S_1, S_2, S_3, ... and a clock line T. The cells have usually two or more stable states. Every clock pulse marks a borderline separating two adjoining clock periods by setting the cell into the binary 0 condition. During the ensuing clock period, an RSFQ pulse may/may not arrive at the inputs. The consensual convention in RSFQ logic is that arrival of an RSFQ pulse on a terminal during the tenure of clock period represents binary 1 state whereas its nonarrival indicates binary 0. The convention does not make the temporal coincidence of the pulses mandatory. It only requires that the binary 1 pulse arrives within the clock period, irrespective of the time instant at which it reaches.

21.5 OR Gate

A two-input OR gate (Fig. 21.10) is formed by combining the confluence buffer circuit with the R-S flip-flop. When the RSFQ pulse is fed to any one of the two inputs X or Y, the buffer stage reproduces it and supplies it to the input of the flip-flop. If the pulse represents the first input pulse during the clock period, it causes a switching from binary 0 to binary 1 state. In case, the preceding pulse has already brought about the switching, the signal pulse creates a 360° jump in junction J_5. In the rare case of coincidence of several pulses in time, their overall

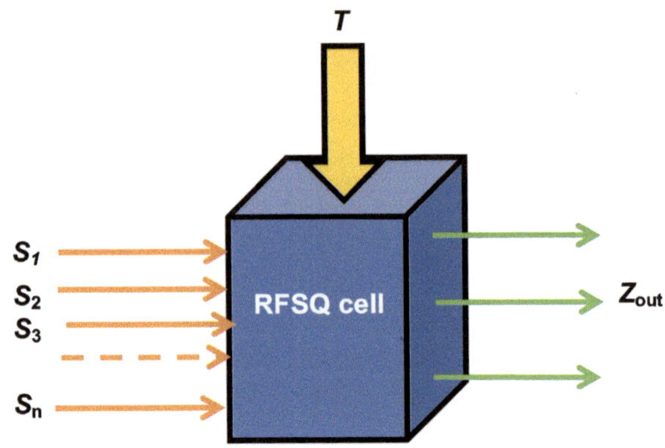

Fig. 21.9 An elementary cell of RFSQ circuit

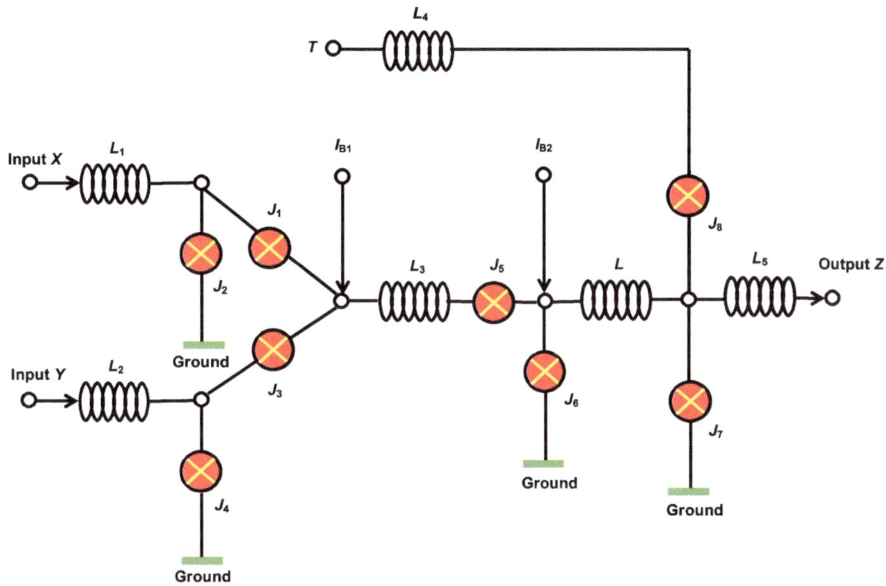

Fig. 21.10 An RFSQ OR gate

effect is the same as that of a single pulse. During the clock period, no RSFQ pulse makes appearance at the output terminal of the circuit. An output can appear only when at the minimum one RSFQ pulse has arrived during the clock period. This happens due to the resetting of the circuit by the clock pulse arriving at the point T, leading to cessation of the clock period.

21.6 NOT Gate

The NOT gate or inverter circuit (Fig. 21.11) contains one SQUID composed of the components (J_2, L_1, J_3) [3]. An extra Josephson junction J_1 is included. Suppose, in the beginning, the circuit is in the 0 binary state. In this condition, a large current is flowing through junction J_2 while the current across J_3 is very small. If there is no input signal X, the succeeding clock pulse activates the RFSQ pulse in J_1, not in J_3. Appearance of this pulse at the output implies that an input signal '0' provides output signal '1'. Arrival of the input signal X will cause switching of the SQUID into state '1'. In this state, the current in J_3 will be high. The following clock pulse will activate an RSFQ pulse across J_3, not J_1. Consequently, the circuit will be reset. It will not yield any output pulse implying that input signal '1' produces output signal '0'.

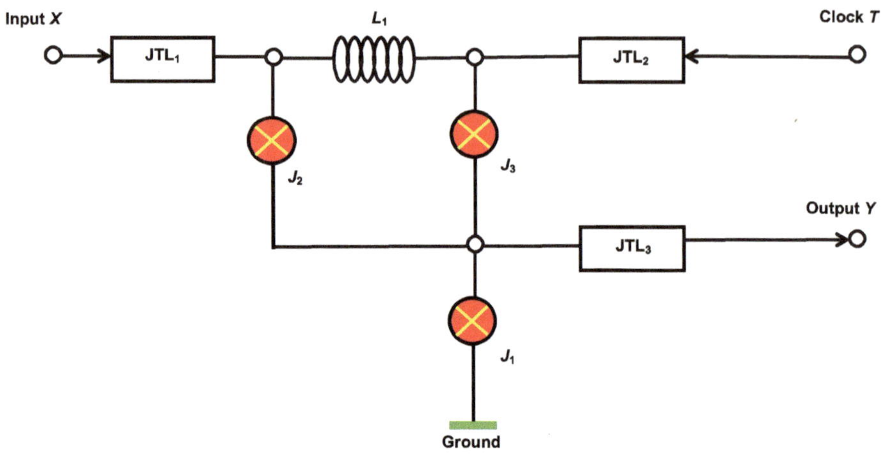

Fig. 21.11 RFSQ NOT gate

21.7 RFSQ IC Fabrication Techniques

The Josephson junctions used in RFSQ circuits are generally made from layers of niobium/aluminum/aluminum oxide/niobium [4]. The niobium and aluminum layers are deposited by sputtering. The aluminum is oxidized to aluminum oxide by exposure of aluminum layer to oxygen, thereby partially converting it into aluminum oxide. Generally, a 1 nm thick Al_2O_3 layer acts as a good insulating barrier. Finally, the Al_2O_3 film is coated with another sputtered niobium layer. For defining the Josephson junction pattern, photolithographic techniques are used, followed by reactive ion etching. Thus the RFSQ fabrication is essentially an "all thin film" technology. The critical current I_C of the Josephson junction is adjusted by controlling the thickness of the Al_2O_3 film.

21.8 Advantages and Applications of RFSQ Logic

The main advantage of RSFQ logic is its extremely low power consumption. It is lower by a factor of 10^5 times than CMOS circuits. In this comparison, the cooling costs are not taken into consideration. RFSQ circuits can function at frequencies of several tens to hundreds of GHz. They represent an ultrafast technology for digital systems working at sub-terahertz frequencies. Another advantage is that they can be used in combination with CMOS circuits. Furthermore, existing semiconductor manufacturing techniques can be adapted for large-scale production of RSFQ circuits. Over and above, these circuits are tolerant to inter-device variations and are self-clocking so that asynchronous designs can be utilized.

The RSFQ logic circuits have been used for digital signal processing applications. They have been applied in optical and other fast networks.

21.9 Disadvantages of RFSQ Logic

The main drawback of RFSQ logic is that these circuits are usually operated at liquid helium and liquid nitrogen temperatures. The cooling requirement wastes enormous power. To some extent, the problem can be assuaged by using high-temperature superconductors.

21.10 Discussion and Conclusions

Although RFSQ logic based on the manipulation and communication of quanta of magnetic flux, called flux shuttles or fluxons, can provide high speeds, the refrigeration costs required for the cryogenic bath have long thwarted its adoption. Nonetheless, commercial cryocoolers maintaining 4–5 K have made RFSQ circuits easily manageable and relatively cheaper. Further, being a thin film technology, high-temperature diffusion cycles, epitaxial and chemical vapor depositions are not necessary. The speed of RFSQ logic is derived not from aggressive downscaling but from physical phenomena because the flux quanta move at speeds approaching that of light. Hence, state-of-the-art microelectronic fabrication equipment can be used to make circuits surpassing the capabilities of silicon technology.

Review Exercises

21.1 What is a Josephson junction? How is it used as a switching element in RFSQ logic?

21.2 What is meant by the statement, "Magnetic flux is quantized"? Write the expression for a quantum of magnetic flux? State its value. Does this value vary with the superconductor?

21.3 What is a Josephson transmission line? Can it provide amplification of the signal?

21.4 Draw the circuit diagram of a non-reciprocal buffer stage and explain its operation. Hence, explain how a confluence buffer works?

21.5 What does the acronym "SQUID" stand for? Explain with the help of a diagram how this interferometer provides the function of an *R-S* flip-flop?

21.6 Explain with diagrams the operation of the following gates in RFSQ logic: (a) OR gate and (b) NOT gate.

21.7 Write two advantages of RFSQ logic.

21.8 Mention one shortfall of RFSQ logic.
21.9 Give two applications of RFSQ logic.
21.10 Explain the statement, “RFSQ IC fabrication is essentially a thin-film technology”.

References

1. Mukhanov OA, Polonsky SV, Sevenov VK (1991) New elements of the RFSQ logic family. IEEE Trans Magn 27(2):2435–2438
2. Likharev KK, Semenov VK (1991) RFSQ logic/memory family: a new Josephson junction technology for sub-tera-hertz-clock-frequency digital systems. IEEE Trans Appl Supercond 1 (1):3–28
3. Likharev KK (2016) Rapid single-flux-quantum logic. http://www.physics.sunysb.edu/Physics/RSFQ/Projects/WhatIs/rsfqre2m.html. Accessed 6 April 2016
4. Massachusetts Institute of Technology Lincoln Laboratory (2016) Forecasting superconductive electronics technology. The Next Wave 20(3). https://www.nsa.gov/research/tnw/tnw203/article2.shtml. Accessed 7 April 2016

Chapter 22
Molecular Nanoelectronics

Abstract Molecular electronics is a relatively young research area, which can be broadly defined as dealing with electronic devices in which molecular properties play a central role. The necessary criteria for considering a molecular system as a molecular device are decided in the context of the simplest conceivable molecular electronic device, the molecular switch. The main issues are placing the molecules in an immobile condition at pre-decided locations and connecting electrodes to them for current flow. The break junction method of nanogap electrode formation is described. The electrical properties of molecular contacts are treated in terms of HUMO and LUMO levels. Suitable materials serving as molecular wires and insulators in molecular devices are indicated. Appropriate molecules for forming N- and P-type regions are suggested. Two kinds of molecular switches are described, one triggered by electromagnetic radiation and the other via redox reaction. The pioneering theoretical and experimental work of Aviram and Ratner in the 1970s on molecular rectifying diode is elucidated with energy band diagrams. The fundamental demonstrations of the properties of electronic devices at the molecular scale make this field highly exciting and lucrative.

22.1 Introduction

Molecular electronics is the answer to the simple question: Can the intrinsic properties of a solitary molecule or tiny clusters of molecules be utilized to perform the tasks executed by contemporary electronic devices and circuits?

22.2 The Idea of Molecular Electronics

Molecular electronics, representing the zenith of miniaturization of electronics is a technology, which uses single molecules or small groups of molecules, to perform functions that are normally carried out by electronic devices such as diodes, bipolar

V.K. Khanna, *Integrated Nanoelectronics*,
NanoScience and Technology, DOI 10.1007/978-81-322-3625-2_22

or field effect transistors, etc. Therefore, it seeks to use individual molecules or small assemblies of molecules to fabricate components capable of emulating electronic functions. The expected outcome is replacement of existing silicon devices with molecular devices endowed with unforeseen competences beyond CMOS technology [1]. According to one school of thought, it is a technology utilizing only a single molecule. But this definition is very restrictive. Instead, a broader definition is more helpful. Molecular electronics is envisioned as a possible means of using nanosize objects like nanoparticles, nanotubes and nanowires with molecules to construct a variety of device structures and circuit architectures [2]. Herein, any device based on the utilization of molecular properties will be called a molecular electronic device.

22.3 Qualifying Characteristics of a Molecular Electronic Device and Related Hurdles

The pertinent question raised is: What properties a molecular device or component is supposed to possess in order to qualify as a molecular electronic device? This question can be answered in reference to the simplest device, a molecular switch [3]: (i) This switch should have two stable states, an on-state and an off-state. Thus it is a bistable arrangement. (ii) The on-state is characterized by the switch carrying out a useful task or enabling another device to do some work. In the off-state, this function must stop. (iii) The switch should never change its state from on- to off-condition of its own accord unless prompted to do so, i.e., it must remain at the position where it is set up. (iv) It must have a fixed physical location in space and should not move from this position. (v) It must be chemically stable. If it decomposes on raising the temperature, its functionality will be lost. (vi) It must be chemically inactive with respect to other similar molecules otherwise exchange of electrons with those molecules is likely to result in loss of the data represented by them.

Looking at the above set of conditions, it is evident that several disputes must be resolved. The concerns constitute some of the major impediments, which need to be satisfactorily addressed. Of course, solutions of many of these problems have been suggested from time to time. These will be treated in subsequent sections.

22.4 Placement/Positioning and Contacting of Molecules

Placing a molecule at a certain specified spot, i.e., a fixed *x*-*y* coordinates is a major issue confronting molecular electronics. Due to the extremely tiny size of a single molecule or even a small collection of molecules, providing macroscopic contacts to it is a hard nut to crack. Very complicated nanoscale structures and techniques must be contrived to accomplish this task. Attempts have been made at such positioning but the problems of localizing the molecules at chosen places have not been solved to a convenient level. Once the molecule has been correctly positioned, it may diffuse away from the selected site or wander away due to external forces. If that happens, the device will be untangled and will not function in the long run and its life will be limited. Therefore, it is required to constrain the molecule at the chosen site. Many groups have succeeded in trapping a single molecule between the claws of electrical contacts [4].

22.4.1 Top Junction Formation by Microscopic Technique

A molecule can be confined between a substrate and the apex of the tip of a microscope. The microscope used is either a scanning tunneling microscope (STM) or an atomic force microscope (AFM). The STM offers the advantage of viewing the molecule before it is contacted, revealing the structure and arrangement of the molecules. The substrate-molecule contact is well defined but the tip-molecule contact is ill-defined. This definition is improved either by embedding the molecule into a matrix of less conducting molecules or by attaching a gold nanoparticle to the molecule.

22.4.2 Nanogap Electrode Formation by Break Junction Method

A break junction consists of two metal wires separated by a small gap $\sim$ a few nm. A metallic wire is fixed on a flexural substrate inside a vacuum chamber maintained at liquid helium temperature (4.2 K) to allow condensation of impurities (Fig. 22.1). The low temperature guarantees thermal stability. The wire is pulled physically by bending the substrate until it breaks. The bending can be accurately varied using a piezoelement. The junction so formed by rupturing of a wire is called the break junction and the method of its formation is the mechanically controlled break junction (MCBJ) method.

The separation between the two broken wires is adjusted by measuring the electrical resistance of the break junction. Every value of separation is represented by a fixed value of resistance. As soon as the desired value of resistance is attained, the required separation has been achieved. Accuracy up to picometers is assured. The method gives freshly prepared metallic electrodes in ultraclean environment to which the desired molecule can be introduced, thus forming a molecular bridge. The molecules can be introduced into close proximity of the junction by passage through a capillary tube. Due to the vacuum conditions, the junction surfaces

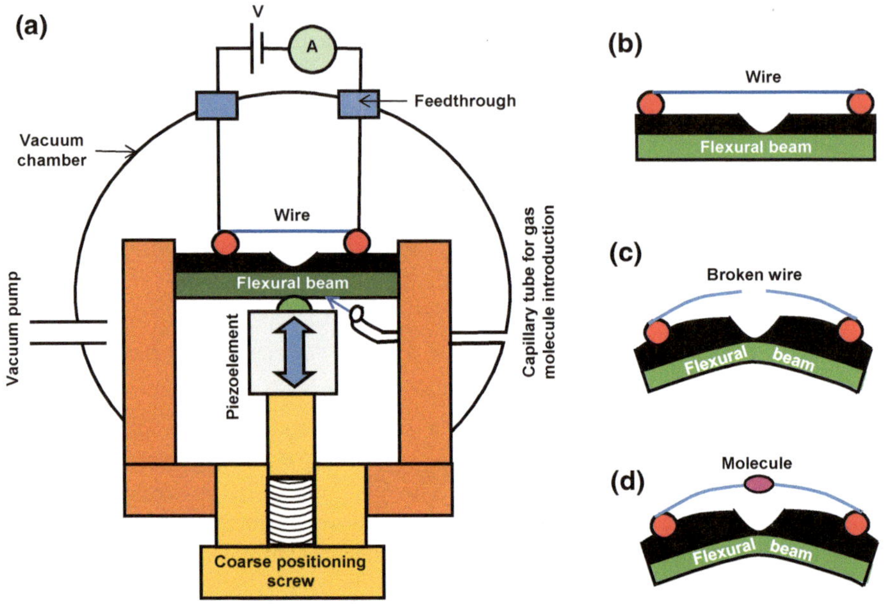

Fig. 22.1 Break junction method: **a** Setup for formation of the break junction. **b**, **c**, **d** Different stages in pulling of the wire: **b** Initial condition, **b** Wire breakage and **d** Molecular bridge construction

remain very clean for a long time. By squeezing the separation between electrodes, the junction can be reformed. In this way, the junction can be broken and reformed.

Alternatively, chemical methods are used to create the separation between metal wires. Serdio et al. [5] used electron beam lithography and self-terminating Au plating to produce electrodes at a distance of 3.0 ± 1.7 nm with 90% yield and good reproducibility.

22.5 Electrical Behavior of Contacts

A vital obstacle relates to capability of addressing an individual molecule electrically in order to interact with it at a microscopic level and also on the macroscopic scale. One must therefore look at the properties of 'molecular contacts'. The molecular orbitals are affected by the contacts. The important molecular orbitals for reactivity are the HOMO—highest occupied molecular orbital and LUMO—lowest unoccupied molecular orbital. These are called the frontier orbitals. The correspondence between the terminology of organic and inorganic semiconductors is as follows: HOMO (Organic) →Valence band (Inorganic) and LUMO (Organic) → Conduction band (Inorganic).

Hybridization takes place between the electronic states of the contact material and the orbitals of the molecule. As a result, the molecular orbitals are broadened and shifted. Usually, the Fermi level of the contacts is situated between the highest occupied molecular orbital (HOMO) and the lowest unoccupied molecular orbital (LUMO) of the molecule. At low values of applied voltages and when the gap intervening the HOMO and LUMO levels is adequate, the current flows from the contact to the molecule through quantum mechanical tunneling. The tunneling mechanism is coherent and nonresonant. When the charge transport occurs through LUMO, the current flow takes place by electron tunneling. When the charge is conveyed via HOMO, the mobile carriers are holes. Thus in the two tunneling cases, different charge carriers participate.

22.6 Conducting Molecular Wires for Interfacing

The interfacing of a molecular device with other such devices is equally important so that information can be exchanged amongst the different devices. Wiring up the molecule into a circuit can enable this action. A molecular wire is a molecular structure which can serve as a conduit for transfer of electrons. The wiring job can be performed by polyphenylene-based conductors. These are conjugated aromatic compounds based on benzene. They are composed of benzene rings. In these rings, one or two hydrogen atoms are eliminated. Conjugation means the partial overlap of p-orbitals on adjacent bonds with delocalization of electrons. Carbon nanotubes can also be used.

22.7 Insulators for Molecular Devices

Typical insulators are aliphatic organic molecules, e.g., $–CH_2–$, $–CH_2CH_2–$. They do not contain any benzene ring. Interconnections between carbon atoms are formed by straight chains, branched chains, or nonaromatic rings (alicyclic). Bonding is either by single bonds (alkanes), double bonds (alkenes), or triple bonds (alkynes). The elements most commonly bound to the carbon chain are hydrogen, oxygen, nitrogen, sulfur, and chlorine. In sigma-bonded organic molecule, the conducting channel outside the plane containing the nuclei is disrupted leading to discontinuity in electron density at the locations of the nuclei, and hence insulating behavior. A σ-bond is a covalent bond in which the molecular orbital is formed by end-to-end overlapping of atomic orbitals along the line connecting the two bonded atoms.

22.8 N- and P-Type Regions

It is possible to design molecules to be electron transporting or hole transporting. Conjugated molecules with an attached electron-donating moiety contain an excess of electrons. They are electron-rich or N-type. Examples of such electron-donating moieties are: $-NH_2$, $-OH$, $-CH_3$, $-CH_2CH_3$. Electron donors have a low first ionization potential. Conjugated molecules with electron-withdrawing moiety, e.g., $-NO_2$, $-CN$, $-CHO$, are electron-deficit or P-type. Electron acceptors have a high electron affinity.

22.9 Molecular Switch

22.9.1 Photochromic Switch

Photochromism is a reversible chemical transformation of a chemical entity between two forms induced by photoirradiation. These forms differ in their absorption spectra.

A photochromic switch consists of a dithienylethene ($C_{10}H_8S_2$) molecule [6]. If the molecule is exposed to visible light, the conjugation between its thienylene rings is broken. This is open state of the switch. When irradiated with UV light, the conjugation across the molecule is established again. This is closed state of the switch.

22.9.2 Redox Switch

A redox reaction is an oxidation-reduction reaction. The transfer of electrons between two chemical entities, named as the reductant or reducing agent, and the oxidant or oxidizing agent, is the main characteristic of this type of reaction. The reductant suffers a loss of electrons, whereas the oxidant gains electrons. Consequent upon this electron transference, the oxidation number of an atom, molecule or ion is altered.

van Dijk et al. reported an anthraquinone-based molecular redox-controlled switch [7]. It can be reversibly switched from cross-conjugated state to linear-conjugated state via two-electron reduction/oxidation reactions. The cross-conjugated state has high resistance; this is the off-state. The linear-conjugated state has low resistance; this is the on-state.

22.10 Molecular Rectifying Diode

The genesis of molecular electronics can be traced back to the work of Aviram and Ratner [8]. They pioneered a concrete idea of building an electronic device, the two-terminal diode, from a single molecule [9]. From computational analysis, they proposed a simple molecular device using a single organic molecule. This molecule consists of two π-conjugated segments: one with donor π moiety (D) and the other with an acceptor π moiety (A) (Fig. 22.2). A π-bond is a covalent bond in which the molecular orbital is formed by side-to-side overlapping of atomic orbitals. This overlapping is along a plane perpendicular to the line joining the nuclei of the atoms.

The two moieties are interconnected by a tunneling bridge. This bridge is through a σ-bond. A σ-bridge contains saturated bonds. This saturated covalent bridge acts a spacer preventing overlapping of the π-systems. It decouples the molecular orbitals of D and A. The σ-bonds provide an effective barrier for electron transfer between π-systems. It was shown that if contacts are established with the resulting D-σ-A molecule through metal electrodes, it will conduct current strongly in one direction only. It will thus serve as a molecular rectifier with a preferential direction of transference of charge. Here D = an organic donor = tetrathiafulvalene $(H_2C_2S_2C)_2$, an organosulfur compound, abbreviated as TTF; A = an organic acceptor = tetracyanoquinodimethane, $(NC)_2CC_6H_4C(CN)_2$, a cyanocarbon, in short form TCNQ; and the bridging was done by methylene ($-CH_2-$), an organic insulator.

The rectifying action of the molecular device can be understood from its orbital energy diagram [10]. But let us consider a more generalized situation of a monomolecular diode fabricated from a molecule in which donor and acceptor moieties are hooked up on the opposite sides of a σ-bridge (Fig. 22.3).

The orbital energy diagram of the structure without any bias is shown in Fig. 22.4. There are three potential barriers: (i) One central barrier for the insulating methylene bridge and (ii) Two side barriers. The left side barrier originates from the contact of molecule with left electrode and the right side barrier from the contact of molecule with right electrode. Due to the existence of these three potential barriers, the different parts of the structure are secluded and isolated from each other.

The energy levels in the electrodes are shown. Also distinctly marked on the diagram is the corresponding Fermi energy E_F. Consider both HOMO and LUMO levels. It is easy to appreciate the dependence of relative locations of energy levels on the left and right sides of the central barrier on the electron-donating and electron-withdrawing moieties. On the left side of the central barrier, electron-donating moiety is present. On the right side, the electron-withdrawing moiety can be seen. The consequence of the presence of two moieties is that energy levels on the left side are lifted up toward higher energies with respect to those on the right side. The reason behind this consequence is that the oxidation potential of the donor moiety is lower than that of acceptor moiety. It may be noted that the

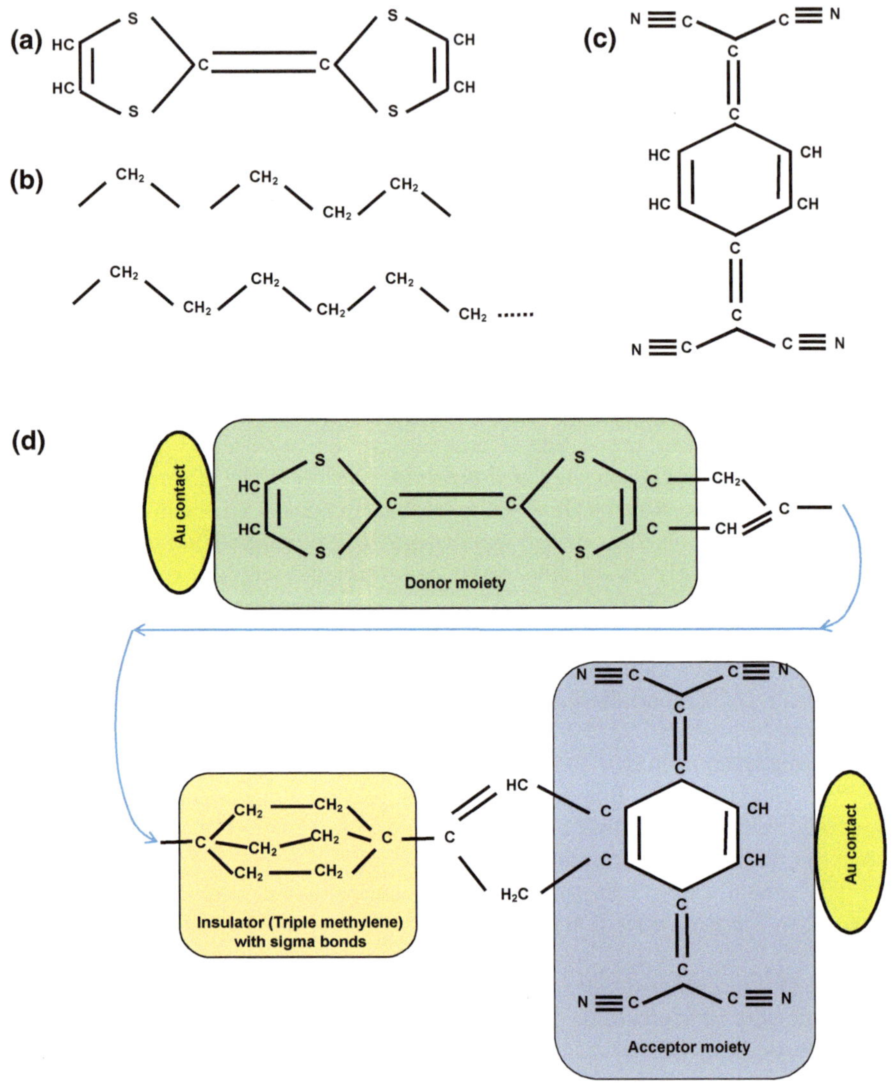

Fig. 22.2 Molecular diode: **a**, **b** and **c** Components, and **d** their assembly to form the structure: **a** Tetrathiafulvalene ($C_6H_4S_4$): Donor, **b** Methylene group chains of different lengths, **c** Tetracyanoquinodimethane ($C_{12}H_4N_4$): Acceptor, and **d** Donor-Sigma-Acceptor (D-σ-A) molecular diode

oxidation potential expresses the propensity of the functional group to lose electrons.

The energy difference between the LUMO levels on the two sides = ΔE_{Lumo}, constitutes the built-in potential across the molecule. This potential is the analog of the built-in potential of a P-N junction diode. An electronic current can flow from

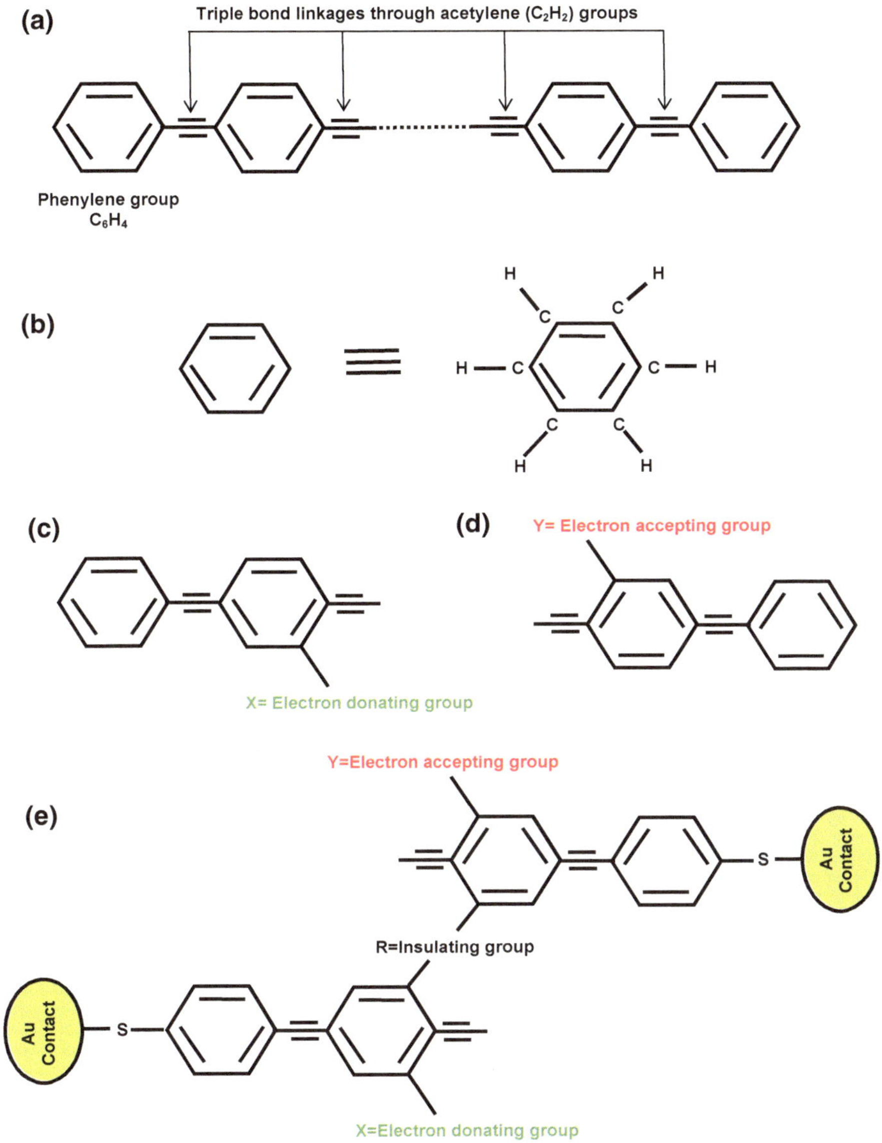

Fig. 22.3 **a** Polyphenylene wire, **b** the atomic arrangement in benzene represented by the ring symbol, **c** Section doped by donor moiety, **d** Section doped by acceptor moiety, **e** Mono-molecular diode. A monomolecular diode structure shown in (**e**) is produced from a Polyphenylene wire, as shown in (**a**) by doping with N-type or donor group *X*, as shown in (**c**), by P-type or acceptor group *Y*, as shown in (**d**), by separating by an insulating group *R*, and by applying gold contact electrodes through thiol (*S*) linkages, which are potential barriers to gold electrodes, helping to isolate the molecular wire

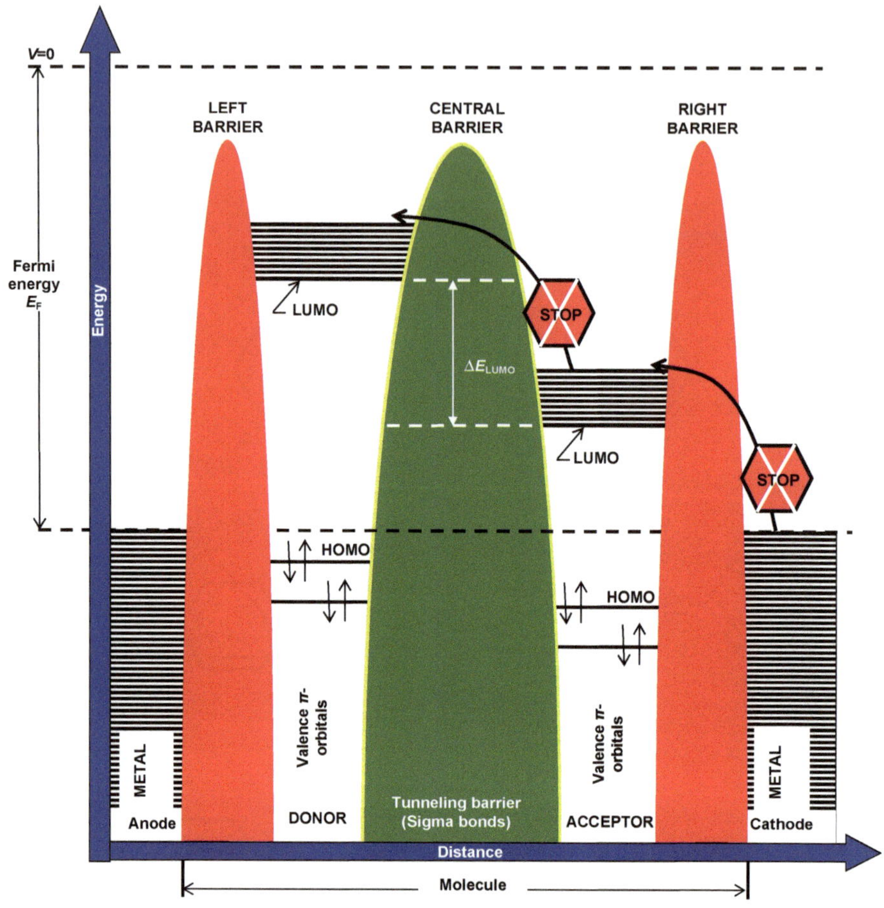

Fig. 22.4 Energy level diagram of a molecular rectifying diode under equilibrium at zero bias

the acceptor-doped layer to the donor-doped layer only under the condition that the potential barrier = ΔE_{Lumo} is overcome. This is the fundamental principle underlying the rectification property of the *D-σ-A* molecule.

Suppose the acceptor-doped side is at a lower potential V_F with respect to the donor-doped side (forward bias), Fig. 22.5. Then the energy levels of the acceptor side are elevated with respect to their initial positions under zero bias. Those of the donor side are lowered with respect to their original zero-bias positions. The central potential barrier is reduced from ΔE_{Lumo} to $(\Delta E_{\mathrm{Lumo}} - V_F)$. Consequently, an electronic current flows from the acceptor-doped side to the donor-doped side by tunneling, resulting in conventional current flow in the opposite direction. This current flow resembles the forward bias operation of a P-N junction diode.

Conversely, when the acceptor-doped side is at a higher potential V_R with respect to donor-doped side (reverse bias), Fig. 22.6, the energy levels of the

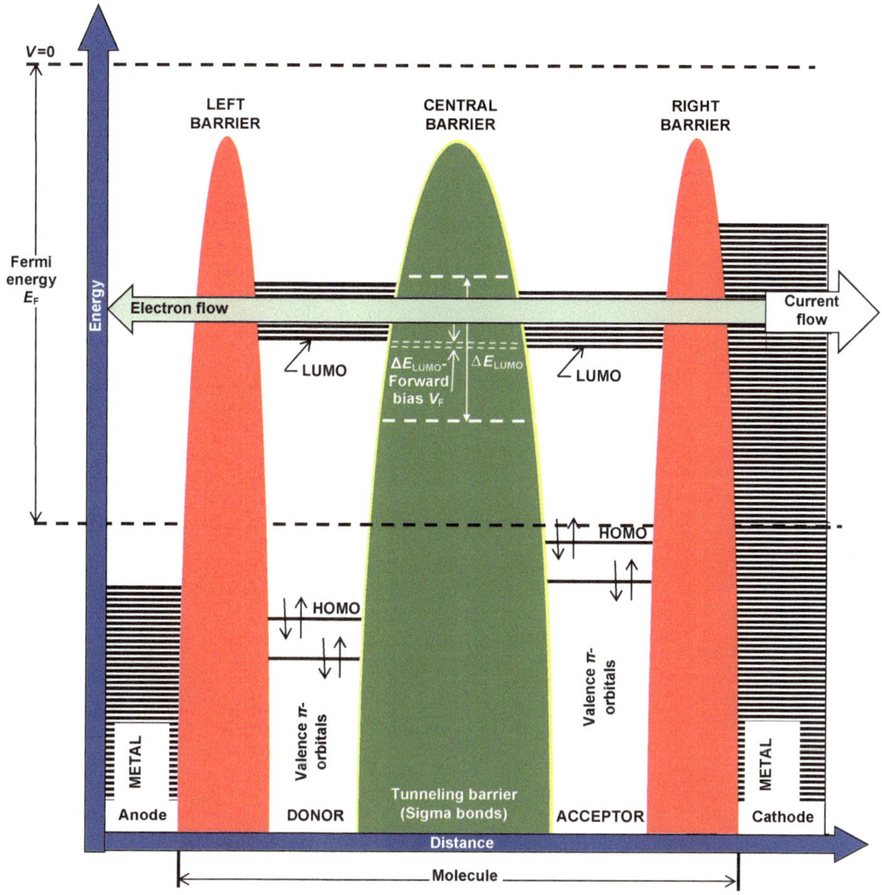

Fig. 22.5 Energy level diagram of a molecular rectifying diode under forward bias condition

acceptor side are lowered with respect to their starting zero-bias positions while those of the donor-doped side are raised with respect to it. Then the central potential barrier is increased from ΔE_{Lumo} to $(\Delta E_{\text{Lumo}} + V_R)$. An electronic current tries to flow from the donor-doped side to the acceptor-doped, which is at higher potential. But this current flow is impossible because the Fermi energy of the contact on donor-doped side is lower than the LUMO of this side. On increasing the applied reverse voltage, it may turn out that at a certain voltage, the Fermi energy of the contact on the donor-doped side comes in resonance with LUMO of the donor-doped side. Then the reverse current suddenly increases. This abrupt rise in reverse current is similar to the reverse-bias breakdown phenomenon in a P-N junction diode.

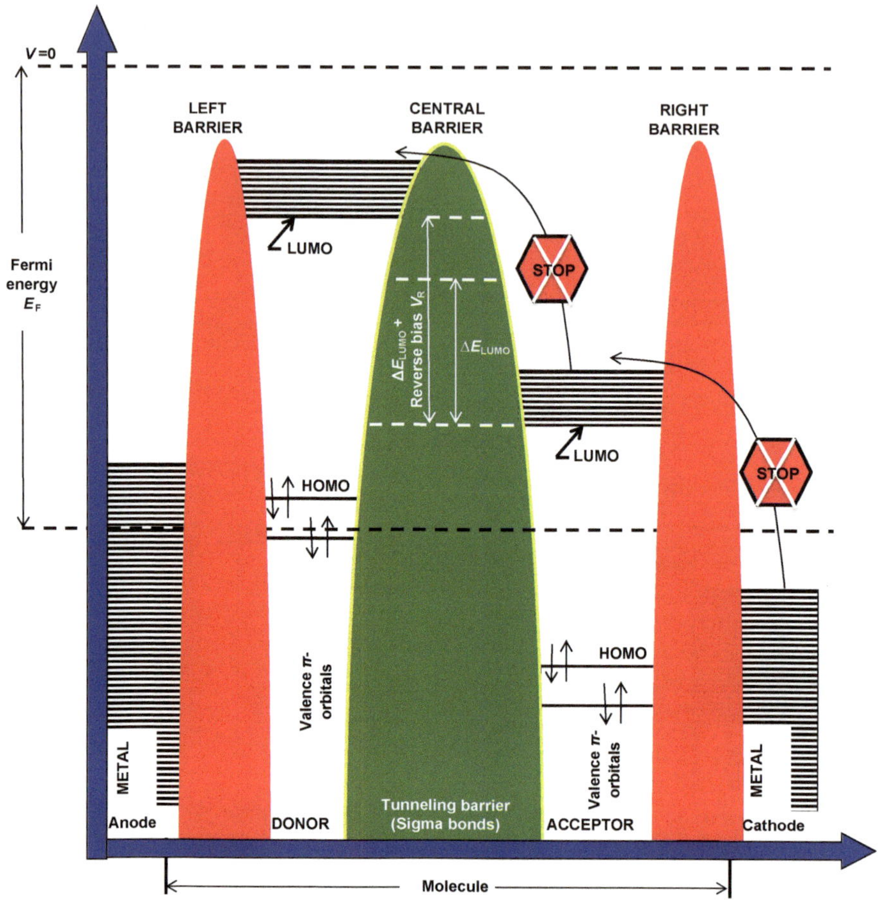

Fig. 22.6 Energy level diagram of a molecular rectifying diode under reverse bias condition

22.11 Discussion and Conclusions

The paramount issue in molecular electronics is measuring the conductance of a single molecule. For conductance measurement, the molecule must be trapped between contacts. Nanoscopic structures are used to connect the molecule to the outside world. A molecule can also be imprisoned between an STM tip and substrate. Break junction method is another technique to accomplish this task. Because of hybridization of orbitals of the molecule with the electronic states of the contacts, the charge transport occurs either via LUMO or HUMO. Polyphenylene-based conductors can satisfactorily perform the function of molecular wires while aliphatic organic molecules act as dielectrics. A photochromic switch is turned off when exposed to visible light. Exposure to UV radiation turns it on.

Anthraquinone-based molecular switch toggles between cross-conjugated and linear-conjugated states through redox mechanism. A molecular rectifier diode consists of two π-conjugate sections connected by a tunneling passage through a σ-bond. The understanding, fabrication, and control of electronic functions at single molecule level open up tremendous opportunities for aggressive downscaling of silicon-based nanoelectronics.

Review Exercises

22.1 Give a broad definition of molecular electronics. Explain with reference to a molecular switch, the essential properties that a molecular device must possess in order to qualify as a molecular electronic device.

22.2 Highlight the difficulties faced in the placement and contacting of molecules in molecular electronics. How is the top junction with a molecule formed using scanning probe techniques? Which type of SPM offers the opportunity to see the molecule before contacting?

22.3 What is a break junction? How is a mechanical break junction formed? How is the gap between the metallic wires established? What is the accuracy with which the distance between the electrodes can be varied? Suggest a means of forming a break junction without mechanical force.

22.4 What is the name of the molecular orbital in an organic semiconductor, which corresponds to the valence band of an inorganic semiconductor? Which molecular orbital of an organic semiconductor represents the conduction band of an inorganic semiconductor?

22.5 What is the condition under which the current flows from the contact to the molecule through tunneling mechanism? What charge carriers participate in the tunneling current flow between the contact material and the molecular orbitals through LUMO? What are the charge carriers taking part in the tunneling current flow through HUMO?

22.6 What is a molecular wire? Name an organic material, which can serve as a molecular wire.

22.7 What organic molecules are used as insulating materials in molecular electronics? Explain the origin of their insulating behavior.

22.8 (i) Give a few examples of electron-donating and electron-withdrawing moieties. (ii) "An electron-donating moiety should have a low ionization potential." Justify the statement. (iii) Explain why an electron-withdrawing moiety should have a high electron affinity.

22.9 What is photochromism? Explain the operation of a photochromic switch.

22.10 What is a redox reaction? Explain the operation of an anthraquinone-based redox switch.

22.11 What is the basic structure of a molecular device using a single organic molecule that was proposed by Aviram and Ratner from computational analysis? Elaborate the functions of the π-conjugating segments and the tunneling bridge made through σ-bond.

22.12 Explain with the help of energy band diagrams the rectifying behavior of a molecular device made from a non-centrosymmetric molecule with the donor and acceptor moieties fastened on the opposite sides of a σ-bridge.

References

1. Sun L, Diaz-Fernandez YA, Gschneidtner TA et al (2014) Single-molecule electronics: from chemical design to functional devices. Chem Soc Rev 43:7378–7411
2. Vuillaume D (2011) Molecular Nanoelectronics. Proc IEEE 98(12):2111–2123
3. Carroll RL, Gorman CB (2002) The genesis of molecular electronics. Angew Chem Int Ed 41:4378–4400
4. Sotthewes K, Geskin V, Heimbuch R et al (2014) Research update: molecular electronics: the single-molecule switch and transistor. APL Materials 2:010701–010701-11
5. Serdio VVM, Azuma Y, Takeshita S et al (2012) Robust nanogap electrodes by self-terminating electroless gold plating. Nanoscale 4:7161–7167
6. Kudernac T, Katsonis N, Browne WR et al (2009) Nano-electronic switches: light-induced switching of the conductance of molecular system. J Mater Chem 19:7168–7177
7. van Dijk EH, Myles DJT, van der Veen MH (2006) Synthesis and properties of an anthraquinone-based redox switch for molecular electronics. Org Lett 8(11):2333–2336
8. Aviram A, Ratner MA (1974) Molecular rectifiers. Chem Phys Lett 29(2):277–283
9. Sekulić D, Živanov M (2010) A model of the molecular rectifying diode type Aviram-Ratner. Electronics 14(2):23–26
10. Jagadesh Kumar M (2007) Molecular diodes and applications. Recent Pat Nanotechnol 1:51–57

Part V
Nanomanufacturing

Chapter 23
Top-Down Nanofabrication

Abstract Starting from a bulk material, the top-down fabrication process progresses to machine, modify, and shape it into the desired shape and size. In integrated circuit manufacturing, one takes a silicon wafer and carves patterns of specified dimensions by a series of lithographic steps through aligned masking levels, performs operations such as wet and dry chemical etching, ion implantation, diffusion, oxidation, metallisation and many others until the desired device/circuit has been obtained. The key to top-down nanofabrication has been the art of lithography which has been relentlessly improved to create patterns of smaller geometries with higher resolution. The illumination/irradiation source in lithography has been changed from an intense beam of deep UV photons to extreme UV photons, and focussed electrons. Due to its extremely short wavelength, the electron beam offers a very high diffraction-limited resolution but is a comparatively slow process. Another approach followed is to make patterns by mechanical pressure, e.g., by stamping and printing using designed templates. In block copolymer lithography, the directed self-assembly of block copolymers is synergistically integrated with common lithographic techniques for practical utilization by semiconductor industry. Scanning probe lithography can manipulate individual molecules but is a low throughput technique. The vast gamut of nanolithographic tools available to a semiconductor process engineer can be leveraged for fabrication of nanostructures of wide-ranging complexities.

23.1 Introduction

One type of sculpture involves making figures or designs in 2-D or 3-D forms by carving or chiseling a hard material such as wood, marble, or other stones. Thus, a big block of the material is given to the craftsman, and the shape is evolved from this block by cutting tools. This method of making smaller structures by starting from big pieces of raw material by using tools or implements for cutting or shaping falls under top-down approach. On the contrary, think of a mason erecting a house wall by laying down bricks, applying cement for gluing the next brick, and

V.K. Khanna, *Integrated Nanoelectronics*,
NanoScience and Technology, DOI 10.1007/978-81-322-3625-2_23

repeating the process until the complete wall stands up. Such a method in which one begins with elementary pieces and constructs a bigger structure by joining together these pieces together comes under the bottom-up approach. Taking cue from these examples of everyday life, nanoelectronics engineers have devised two modalities for fabricating devices and circuits, and named them as top-down and bottom-up approaches.

Nanoscale lithography is a large collection of top-down nanofabrication tools for realizing structures having feature sizes <100 nm. Most of these tools had their genesis in the semiconductor integrated circuit industry. In this chapter, the main lithographic techniques are described indicating their capabilities and limitations.

23.2 Optical Lithography

Used since 1960s, this technique involves shining deep ultraviolet (DUV) light from a KrF or ArF excimer laser source (wavelength 248 or 193 nm) through an illuminator onto a photomask consisting of a quartz plate selectively coated with chrome or emulsion in a defined pattern; earlier equipment used 436 nm (G-line) or 365 nm (I-line). From the photomask, the illumination pattern falls on a coating of a photosensitive polymer called photoresist on a wafer. The mask is placed in contact with the resist-coated wafer or in proximity to it. Another way is to pr6oject the image of the mask on the wafer through a lens. Accordingly, there are three classes of mask aligners: contact, proximity or projection. After exposure, the photoresist is developed. In the developer, either the exposed regions (positive photoresist) or the unexposed regions (negative photoresist) are removed (Fig. 23.1).

23.2.1 Key Metrics

Two main standards by which the performance of a lithographic system can be assessed [1] are: (i) Resolution of image: It is a resolving or distinguishing ability. It is defined as the ability to resolve or distinguish between any two small, adjacent objects in the image. (ii) Depth of focus: It is a tolerance parameter. It expresses the range or tolerance through which the image plane (surface of the wafer) can be positioned with respect to the lens with no accompanying loss of the focus or sharpness of image. Both these metrics depend on the wavelength of light used and the numerical aperture (NA) of the lens. The numerical aperture is expressed by the simple formula

$$\mathrm{NA} = n \sin \alpha. \tag{23.1}$$

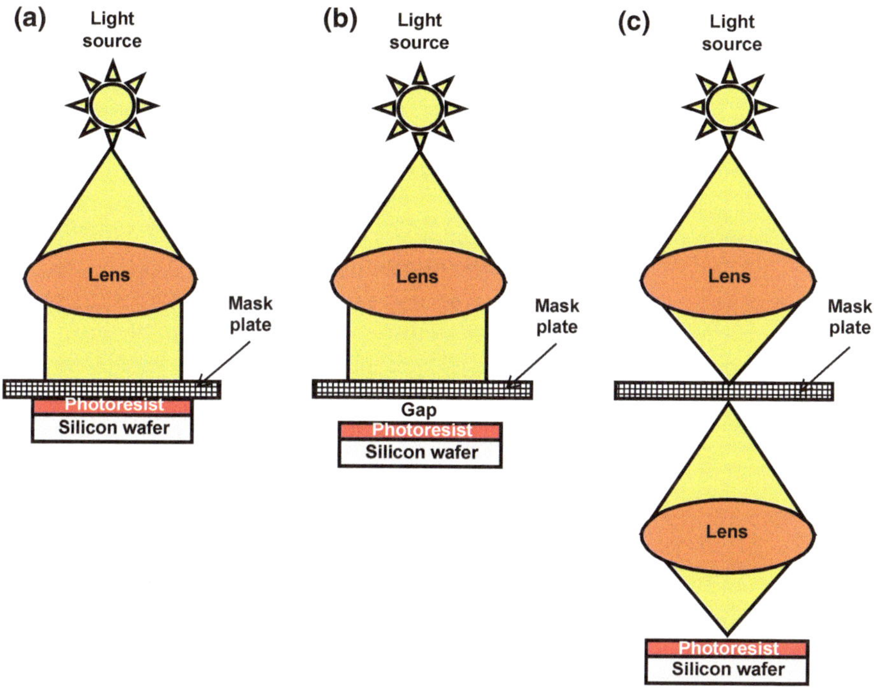

Fig. 23.1 Photolithography systems: **a** contact, **b** proximity and **c** projection

where n is the refractive index of the medium in which the lens is placed and α = one-half the angular aperture of the lens. The angular aperture, a measure of the light-gathering ability of a lens, is the angle subtended at the principal focus of the lens by the diameter of the entrance through which light is collected, representing the maximum angle of incidence formed by light with the normal to the surface of the wafer.

If λ is the wavelength of light used, the resolution is given by

$$\theta = k_1\lambda/\mathrm{NA} \tag{23.2}$$

where k_1 is a constant measuring the difficulty experienced in obtaining a particular dimension with a given combination of optical systems and photoresist used. This technology-based coefficient has been utilized in enhancement of resolution techniques, e.g., using phase-shift masks, by off-axis illumination, or by applying optical proximity correction.

Limit on resolution is imposed by the diffraction of light. The problem starts when the critical dimensions in the patterns exposed reach nearer to the wavelength of light. Then the wave properties of light become preponderant. This is readily understood from a simple analogy: How can the fine details of an image be painted with a coarse brush? Equation (23.2) shows that the resolution θ is improved by lowering the wavelength of light and raising the numerical aperture of the optical system.

The depth of focus is written as

$$\delta = k_2\lambda/(\mathrm{NA})^2. \tag{23.3}$$

where k_2 is proportionality constant called the process latitude factor similar to k_1. In advanced systems, $k_1 = 0.4$, $k_2 = 0.7$. Equation (23.3) shows that in a processed wafer, the surface undulations must be less than the depth of focus. Therefore, planarity of wafer surface is essential to achieve high resolution. Using high-resolution deep ultraviolet wavelength of 193 nm and 0.93 NA systems, features <100 nm are produced. Feature sizes of 1 μm can be resolved with $\lambda = 400$ nm.

23.2.2 Immersion Lithography

This is an enhanced resolution technique in which the photoresist-coated wafer is exposed using purified water or a liquid medium having refractive index >1 in place of the air gap separating the projection lens and the wafer [2]. Thereby the photolithographic process shifts from a dry to a wet version. It improves the limit of resolution from previous 65 to 40–45 nm. From Eq. (23.2) for resolution, higher is the numerical aperture, more is the resolution. For air as medium, $n = 1$. Typically, $\mathrm{NA} \geq 0.9$. But with water as medium, $n = 1.44$. So, NA is increased by 1.44 times to 1.35. Thus, resolution is increased without changing the wavelength of the laser. Further improvements are possible using fluids of refractive index >1.8.

23.2.3 Extreme UV (EUV) Lithography

This lithography is aimed at sub-10 nm resolution, and is the most likely successor of 193 nm immersion lithography [3]. In the extreme UV band (120–10 nm) lying between visible light and X-ray, light of wavelength 13.5 nm is used, which is soft X-ray and more than an order of magnitude shorter than 193 nm. The EUV photon energy = 91.8 eV is 40.7 times stronger than yellow visible light. EUV systems use a laser-produced plasma (LPP) source. In the LPP source, a high-power laser strikes a microscopic droplet of molten tin inside a large vacuum chamber to create the plasma. A condensing mirror collects the EUV light exiting from the plasma. The path of the light beam is completely confined inside the vacuum chamber because of heavy absorption of EUV light by air or nitrogen. The wafer to be exposed is also kept in vacuum. EUV masks work by reflection and absorption as opposed to transmission mode used traditionally. On a low thermal expansion substrate, a multilayer coating is formed along with an absorption layer. Conventional lenses cannot be used with EUV light because they too absorb EUV. So, refractive optics is avoided. Instead, reflective optics is employed using extremely flat, specially coated multilayer mirrors with ≥ 50 stacks of 6.88 nm thick sputtered Si-Mo bilayers. However, the number of mirrors used must be small because 30% EUV is absorbed by them. Building mirrors for precise reflection is challenging. For high-volume commercial production with EUV systems, an output power >200 W is required whereas the present systems provide ~ 10 W. So, insufficient brightness of the light source is a major obstacle because the exposure time and thereby the manufacturing time is unjustifiably lengthened. Furthermore, the intensive absorption of EUV in solid materials necessitates the use of reflective masks with a large number of reflecting coatings. Quest for new resist materials poses another bottleneck.

23.3 Electron Beam (E-Beam) Lithography

23.3.1 The Equipment and Method

The cornerstone of nanofabrication, electron beam lithography (EBL) is an automated maskless technique, which can directly write arbitrarily shaped custom structures of dimensions <10 nm that are digitally sketched on a computer using user-friendly graphical interface. The input is a CAD (computer-aided design) drawing file, as in a laser writer.

Compared to optical lithography, it is several orders of magnitude slower, providing low throughput. It is also complicated and expensive. One major use of this equipment is in fabrication of photomasks. The equipment is specially suited to prototyping nanostructures, and for research and development activities.

The EBL equipment consists of an electron writer, step motors for proper positioning of the stage, pico/nano ampere meter for reading the writing current, and controlling software. A finely focused electron beam performs 2-D scanning of a surface coated with a thin film of electron-sensitive polymer resist, point-by-point to write a complex pattern on it [4]. The electron sources are based on thermionic emission using lanthanum hexaboride (LaB_6) or field emission using W/ZrO_2. Typical energy for thermionic emission is ~ 100 keV. Writing currents are $\sim$nA–pA. Spot size $\sim$ a few nm (Fig. 23.2).

Polymethyl methacrylate (PMMA) is a common electron-resist or e-resist. The resist is chemically modified upon electron irradiation. The solubility of the exposed regions is either increased or decreased. Hence, it can be selectively removed in a solvent. This process is called developing. By etching, the pattern is engraved in the substrate material.

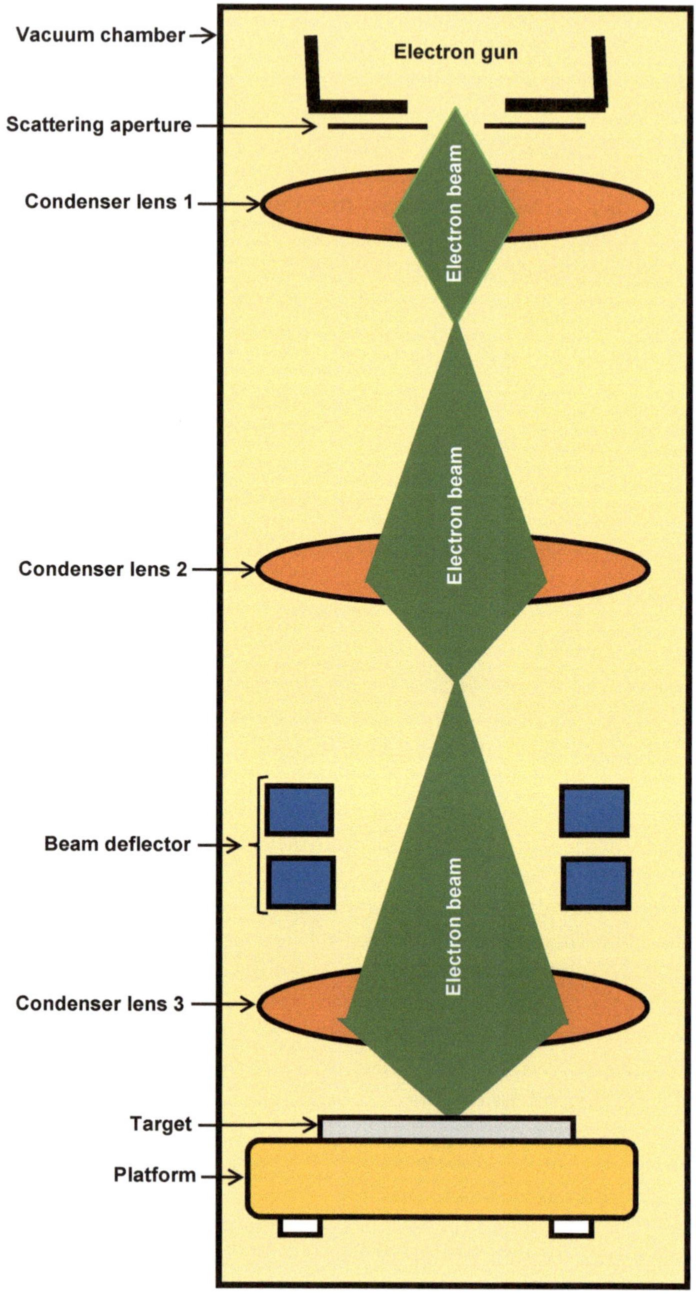

Fig. 23.2 Electron beam lithography

To expose large substrates to the narrow focal field of the electron beam, it is necessary to use high brightness electron sources together with high-resolution mechanical stages. A critical factor influencing the pattern is the electron optics. Equally important is the control of energy and dose of the electron beam. Do not forget the role choice of resist, substrate, and developer. Optimization of developing time and temperature too must be done [5].

23.3.2 Proximity Effect

Resolution of e-beam lithography is restricted by the proximity effect. This effect arises because the backscattered and secondary electrons in the substrate from neighboring features of a pattern return to the beam at a large distance from it, thereby increasing the effective beam diameter. Forward scattering of primary electrons in the thick resist in lateral directions from defined beam also contributes to this effect. Then the beam intensity distribution is represented by the summation of two Gaussian distribution functions. The electron spillover is corrected by modulating the dose. In EBL systems, which do not have facility of dose modulation, the size of a dense pattern is decreased to counteract the superfluous dose received by it. The problem is also solved in the inverse mode by computing the exposure function for the desired dose by proximity effect correction software.

23.3.3 Substrate Charging

On an insulating substrate such as a quartz plate, charging of the substrate occurs by the accumulation of electrons because the negative charges cannot easily flow to ground. To some extent, the charging is mitigated by the release of secondary electrons because these electrons leave the surface positively charged. The substrate charging affects the precision of focusing of the electron beam. It may displace or distort the structures. The issue can be avoided either by sputtering or thermally evaporating a thin conducting metal film or spin coating a conducting polymer over the surface of the resist.

23.3.4 Electron Projection Lithography (EPL)

This is a modified form of e-beam lithography aimed at providing faster throughput by taking advantage of the speed of the optical lithography and the resolution of e-beam lithography [6]. It follows optical lithography by using a physical mask to pattern a relatively large-sized electron beam. Quartz cannot be used as a masking material because electrons can penetrate only through a very short distance in quartz. The electrons differ from the massless photons. The energy of the electrons heats up the opaque regions of the mask causing thermal distortion. Therefore, a thin membrane mask or a stencil mask suffices. In the EPL system called Scattering with Angular Limitation in Projection Electron Beam Lithography (SCALPEL) system, a scattering mask is used consisting of a 100 nm thick nitride film overlaid with a 50 nm thick tungsten pattern. The masking principle is that a low atomic number material scatters the electron beam weakly at smaller angles whereas a high atomic number material scatters it strongly at larger angles. A plate containing an aperture is stationed at the back focal plane of the lens. This aperture plate stops the electrons that are scattered by large angles by high atomic number material. But it allows the electrons which are scattered through small angles by the low atomic number material. The electrons passing through the aperture form a high-definition image on the wafer or substrate. Since the energy of incident electrons is partly absorbed by the mask and partly by the aperture plate, the mask is not deteriorated by the heat generated (Fig. 23.3).

The field size of the projected electron beam is restricted by the distortion effects to 1 mm^2. To image large-size wafers, the wafer is subdivided into small fields which are exposed in sequence. The separate small fields are then stitched together accurately. Thus, a step-and-scan scheme is adopted.

In another strategy known as PREVAIL (Projection Reduction Exposure with Variable-Axis Immersion Lenses), fast electron beam scanning is combined with moderate speed mechanical scanning using variable-axis lenses. This lens electronically shifts the electron optical axis along a pre-decided curvature while at the same time deflecting the electron beam to exactly follow the curvilinear variable axis, whereby the beam is confined to the axis. Off-axis aberrations are removed. Effectively, e-beam projection is blended with e-beam scanning. A field size of 5 mm at 80 nm resolution was obtained by stitching 20 subfields of 0.25 mm 0.25 mm size [7].

23.4 Soft Lithography

The photolithographic methods described in preceding sections are highly capital-intensive requiring enormous installation, maintenance, and operational costs. Moreover, they are targeted for hard and flat semiconductor or glass surfaces. On several occasions, in biotechnology or flexible electronics, it is required to form sub-100 nm features on rough, uneven, or non-planar surfaces over large exposure areas. Cleanroom facilities may not be accessible to many chemists, biochemists and biologists. The time and cost requirements for photomasks used in photolithography are serious impediments to rapid prototyping of devices [8]. Therefore, aforementioned techniques cannot be adapted to such applications. Consequently, the process engineers resorted to the simplest, oldest, low-cost non-photolithographic methods dealing with soft materials, which are collectively included under the generic term, "soft lithography" [9]. These methods use

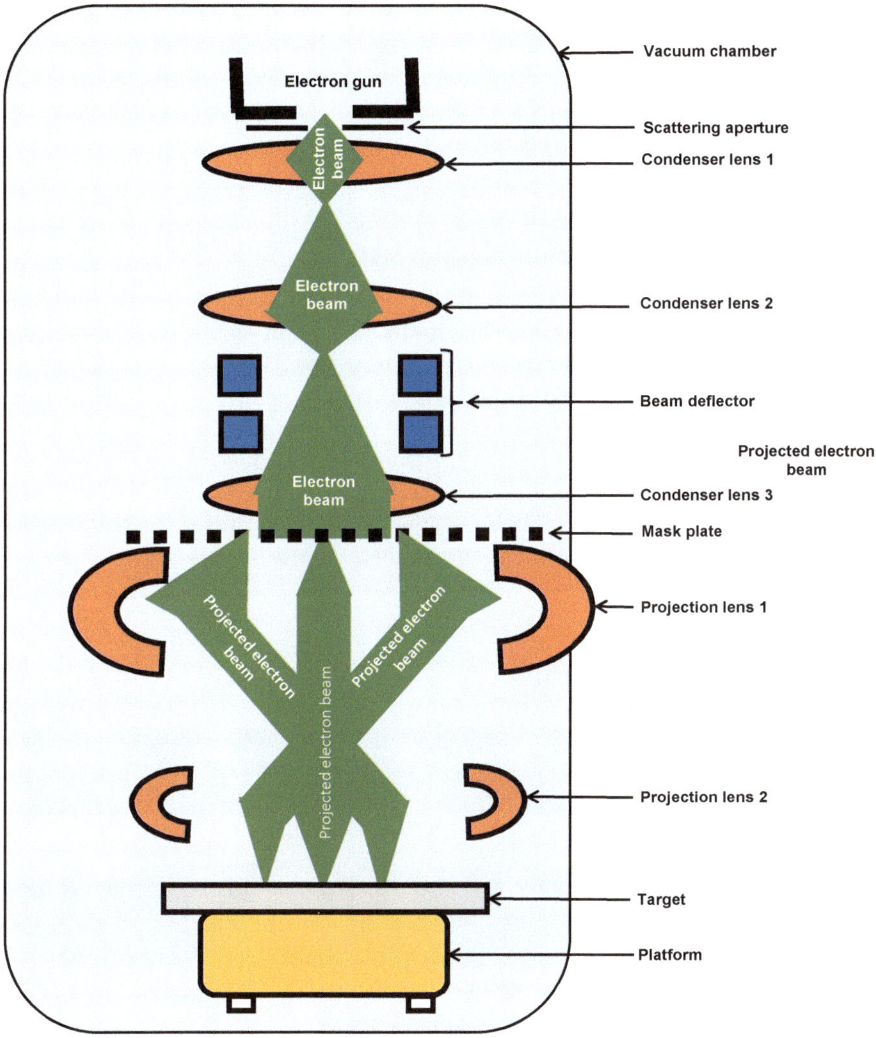

Fig. 23.3 Electron beam projection lithography

elastomeric stamps for printing, molding and embossing. They are capable of defining nanostructures in two and three dimensions on planar, curved, or bendable substrates.

Three common soft lithography techniques are: (i) Replica molding: It transfers a pattern from a hard mold to a substrate through liquid solidification in contact with the mold. PDMS (Polydimethylsiloxane) is poured into a master mold and allowed to cure. A negative of the master mold is thereby formed. PDMS is chosen due to its low surface free energy whereby it can conform to very minute features on the substrate. (ii) Microcontact printing (μCP): A stamp is rolled over a substrate in a manner similar to the printing process. It resembles the use of common stamp to transfer a pattern from the ink pad to the paper. (iii) Nanotransfer printing (nTP): Used to transfer a metal pattern by coating an elastomeric or hard mold with the metal and bringing it in contact with the substrate covered with an adhesion layer (Fig. 23.4).

23.5 Nanoimprint Lithography (NIL)

Optical lithography uses a photon beam to alter the properties of photoresist. Its resolution is limited by diffraction phenomenon. Electron beam lithography employs an electron beam to modify the properties of an e-resist. It performance is affected by scattering of the electron beam. In nanoimprint lithography, the resist is not exposed to any photon or electron irradiation [10]. Instead, it is subjected to mechanical deformation. Hence, nanoimprint lithography is not prone to diffraction or scattering affects. It is an economical method to delineate sub-10 nm polymeric structures over large areas at a high speed [11].

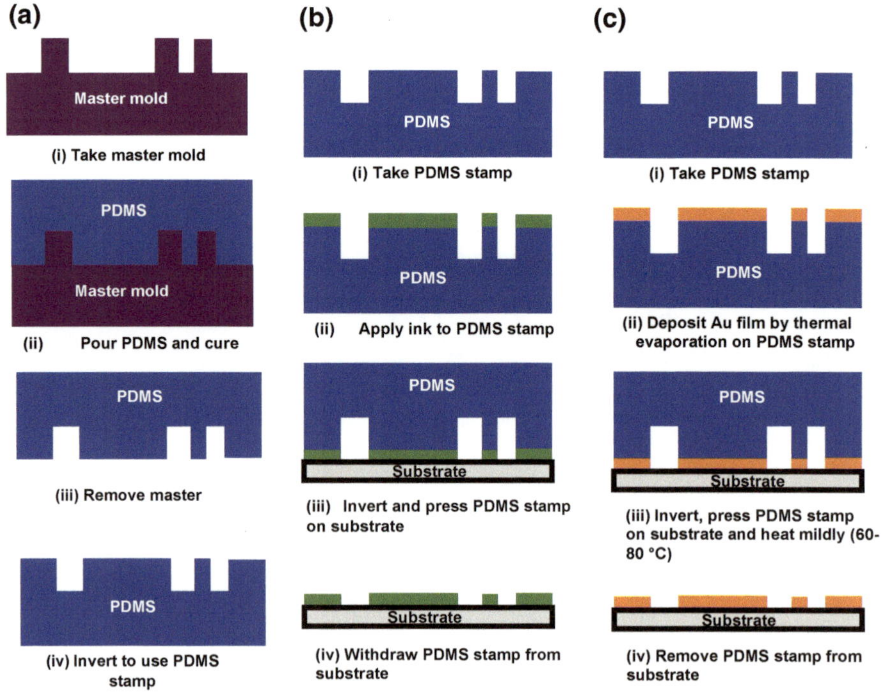

Fig. 23.4 Soft lithography: **a** Replica molding, **b** microcontact printing and **c** nanotransfer printing

There are two fundamental variants of NIL: (i) Thermal-NIL: Also called hot embossing, it uses a thermo-plastic resist. The resist is spin coated on the substrate. The mold containing the nanopattern is pressed against the resist-coated substrate. On heating the two together above the glass transition temperature, the pattern in the mold is transferred to the resist polymer. After cooling, the mold and the substrate are separated. By reactive ion etching, the substrate is etched according to the pattern in the resist. (ii) UV-NIL: in place of thermo-plastic resists, a UV-curable resist is used. Also, the mold is made of a transparent material, e.g., quartz. As before, the mold and the resist-coated substrate are pressed together. Then UV beam is shined on the two held together to cure the resist, which is cross-linked and hardened. Remaining process is same as for thermal-NIL (Figs. 23.5 and 23.6).

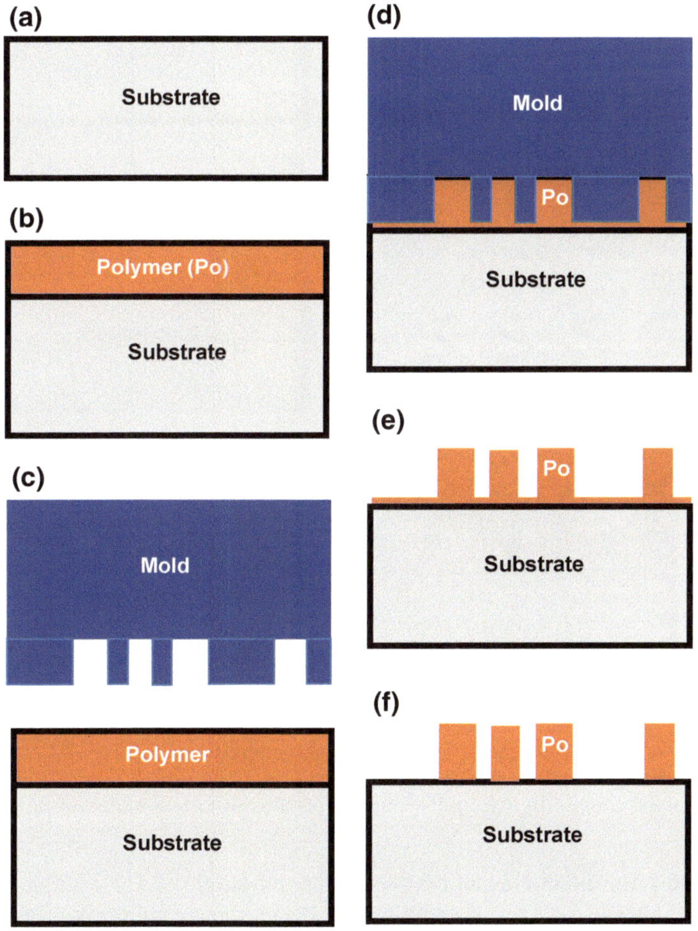

Fig. 23.5 Thermal nanoimprinting (hot embossing); imprint temperature ~150 °C and imprint pressure ~20–40 bars. **a** Take substrate. **b** Spin coat substrate with polymer. **c** Bring mold near polymer-coated substrate. **d** Heat and press mold against polymer-coated substrate. **e** Cool and demold (remove mold away from polymer-coated substrate). **f** Etch residual polymer

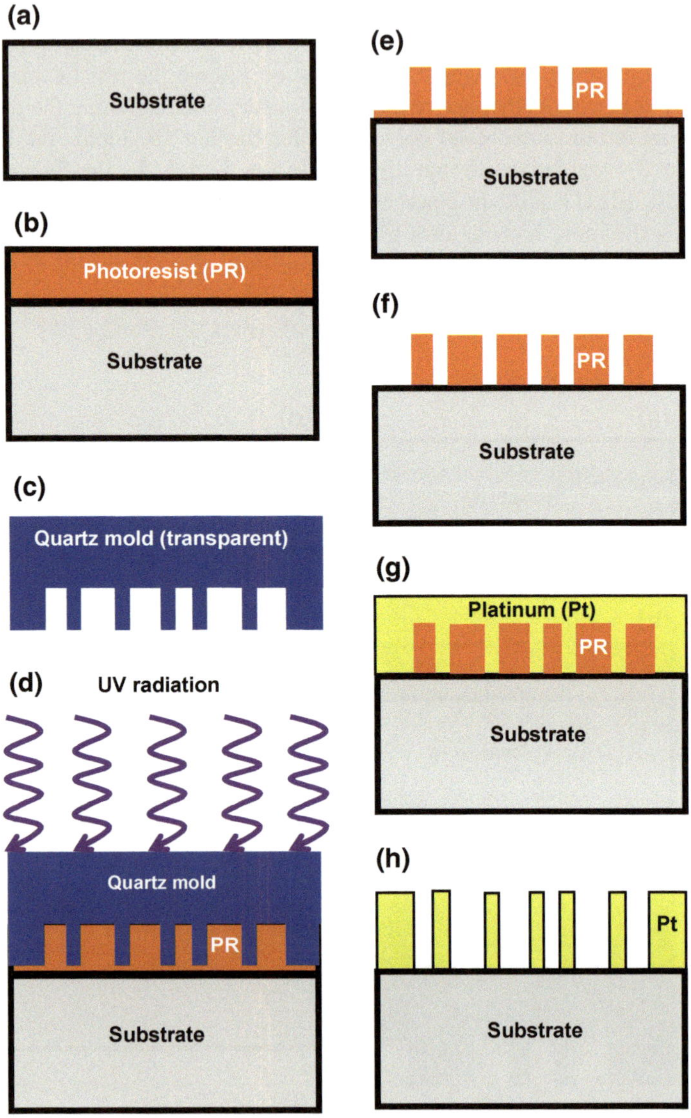

Fig. 23.6 Platinum lift-off process by UV nanoimprint lithography. **a** Take substrate. **b** Spin coat substrate with resist. **c** Take quartz mold. **d** Gently press and expose to UV. **e** Remove mold. **f** Residual resist removal by RIE. **g** Platinum evaporation. **h** Lift off with acetone

Thus in NIL, direct contact between a thermo-plastic or UV-curable resist and a template, either a mold or a stamp containing the nanopattern, is used to create a thickness contrast in the resist. Since NIL depends on the change in shape of a viscoelastic material squeezed between the mold and the substrate, the properties of the interface between the mold and the substrate critically impact the quality of the pattern.

23.6 Block Copolymer (BCP) Lithography

A molecule which binds with other similar molecules to form a polymer is called a monomer. If all the molecules of a polymer are identical, it is said to be a homopolymer. A polymer synthesized from two or more different monomers is known as a copolymer. A polymer in which the monomers are randomly placed is a random copolymer. A polymer in which all monomers of one kind are placed together in one single block, those of a second kind in another single block and similarly for other monomers, is a block copolymer. The self-assembling tendency of block copolymers to form ordered structures on the nanoscale enables their utilization for directed self-assembly (DSA) in combination with conventional lithography [12]. Surface patterns are formed by conventional lithography. These patterns direct the positions and orientations of block copolymer self-assembled nanodomains. As a consequence, laterally ordered, periodic arrays are generated. These arrays consist of spheres, cylinders and other shapes. Typical feature sizes are ~3–50 nm. The arrays serve as lithographic masks. They can be used to form geometries comprising parallel lines or circular dots.

In block copolymer lithography, the pattern transference is based upon selectivity of etching between the blocks [13]. It is commonly done with reactive ion etching. Depending on the feed gas plasmas used during reactive ion etching, each segment of the polymer has a different rate of degradation. Upon exposure of polystyrene-block-polybutadiene (PS-*b*-PB) block copolymer film to ozone plasma, the unsaturated bonds along the polybutadiene backbone break down. However, polystyrene is cross-linked. When the exposed film is developed, the cross-linked polystyrene regions are left behind. These are utilized as a sacrificial template. This mask is used for another pattern transfer through reactive ion etching.

The directed self-assembly initiated by the conventional lithography patterns falls into one of the two types: (i) Epitaxial self-assembly: When the period of the surface chemical pattern is in synchrony with the equilibrium period of the block copolymer self-assembled structure, an ordered nanopattern is spontaneously woven. The surface chemical pattern may be treated as a structure-directing surface chemical pattern. (ii) Graphoepitaxial self-assembly: It is controlled by the topographical features of the initial pattern. The initial pattern is essentially a structure-directing topographic pre-pattern. Selective wetting of a component of the block copolymer at the sidewalls of trenches constrains the lateral ordering of the block copolymer nanodynamics along the trenches. In this way, the pattern density is enhanced by subdivision of the initial pattern.

In epitaxial self-assembly, the 1:1 proportionality relationship of pattern period between the equilibrium block copolymer pattern and the surface pattern yields the same final pattern density as the surface pattern. It is possible to multiply the pattern density by processes such as thermal flow, lift-off or pattern trimming. In graphoepitaxy, disposable topographic pre-patterns can be made with organic photoresists. Further, epitaxial and graphoepitaxial self-assemblies can be combined synergistically with conventional lithography.

23.7 Scanning Probe Lithography (SPL)

This is a maskless method using the nanocharacterization tools called scanning probe microscopes (NSOM, STM, AFM) for nanolithography [14]. Obviously, the process is very slow. Based on the method of patterning nanoscale features on a surface, SPL is classified as: (i) Mechanical SPL: The probe tip ploughs a soft film. (ii) Thermal SPL: A heated tip melts the film. (iii) Thermo-mechanical SPL: It involves application of a mechanical force by a heated tip. (iv) Thermo-chemical SPL: The tip induces a thermally activated chemical reaction. (v) Dip-Pen SPL: The AFM probe pen is immersed in a solution containing the molecules. This solution is called an ink. On contacting a surface, the pen deposits the ink by diffusion through a water meniscus. This water meniscus is formed between the pen and the substrate under ambient conditions. (vi) Thermal Dip-pen: It extends the dip-pen technique to solid inks which are deposited as liquids on heating the probe. (vii) Oxidation SPL: It is local oxidation nanolithography. It works on an oxidation reaction confined in space. (viii) Bias-induced SPL: It applies the intense electric field produced at the apex of the probe to initiate a chemical reaction, as in (vii) above.

23.8 Discussion and Conclusions

Over the years, conventional optical photolithography has borne the major burden of semiconductor manufacturing, and has established itself as a rugged and reliable tool, whether it be in the form of deep UV lithography, immersion lithography and now extreme UV lithography. In the deep UV lithography, the 193 nm lithography has matured well and may lead to the extreme UV era, which is targeting sub-10 nm resolution patterning of integrated circuits. Electron beam lithography is derived from scanning electron microscopy. In electron beam lithography, the light diffraction limit is a non-issue. Resolution depends on the size of the electron beam and reaches up to 5 nm. It is suitable for defining nanostructures <20 nm. But the throughput is one or more orders of magnitude lower than optical lithography. Combined with its high cost and sophistication involved, it is being increasingly utilized in research environments or low-volume direct writing tasks. Soft and nanoimprint lithography techniques can delineate feature sizes <10 nm but are unsuitable for adoption by large-scale commercial industries. Moreover, they depend on other lithographic techniques for template generation. Integration of block copolymer self-assembly with traditional lithographic techniques has provided a high throughput nanolithographic tool at low cost. Positional order can be induced in the nanodomains, both by heteroepitaxy and graphoepitaxy. However, it is arduous to make self-assembled patterns recurring with variable periodicity. Scanning probe lithographic techniques are serial methods. Hence, they are inconveniently slow for bulk production. Parallelization is necessary for upscaling.

They have been used for realization of molecular electronics. They can support bionanofabrication entailing the fabrication of periodic nanostructures containing biomolecules.

Review Exercises

23.1. Distinguish between top-down and bottom-up approaches to nanofabrication.

23.2. What is the source of light used for optical lithography? What are the typical wavelengths used? What wavelengths were used in older equipment? What is the photomask plate made of? What is the coating material on the mask plate? What is the photoresist polymer spin coated on the silicon wafer called? Name the three kinds of mask aligners commonly used?

23.3. Define the following terms for a photolithographic system: (i) Resolution of image, (ii) Depth of focus.

23.4. What is meant by numerical aperture of a lens? How is it related to its angular aperture?

23.5. If θ denotes the resolution of an image, λ represents the wavelength of light used in the optical system and *NA* stands for the numerical aperture of the lens, write the formula connecting θ with λ and *NA*. What is effect of decreasing the wavelength on the resolution? How does an increase in numerical aperture affect the image resolution?

23.6. Why is it essential to use high-planarity wafers to achieve good resolution? What is the resolution obtained for a wavelength of 193 nm and numerical aperture 0.93?

23.7. What is immersion lithography? How is resolution of the image increased in immersion lithography without changing the wavelength of the laser source?

23.8. What is EUV band of wavelengths? What value of wavelength in the EUV band is used for lithography? How does the energy of a EUV photon compare with that of a yellow light photon?

23.9. What is the source of light used in EUV system? Do EUV masks work in the transmission mode? What are the main problems faced with present EUV systems?

23.10. Does an electron beam lithography system require masks? What is the input in this system? Which is faster lithographic technique: e-beam or optical?

23.11. What are the main components of an e-beam lithography system? Name a commonly used e-beam resist material.

23.12. Discuss proximity effect in the context of e-beam lithography? What is its basic cause? How is correction for proximity effect applied?

23.13. Explain the effect of substrate charging in e-beam lithography. How does it affect the image? Propose the required solutions to overcome this problem.

23.14. How does electron projection lithography combine the benefits of optical and e-beam lithographic techniques? Describe the operating principles of SCALPEL and PREVAIL systems for EPL.

23.15. What is the necessity of soft lithographic techniques? Name three such techniques. How do these techniques differ in their way of transferring a pattern on a substrate?

23.16. Name a lithographic technique based on mechanical deformation of resist. What are the two common versions of this technique? How do they differ?

23.17. What is a block copolymer? What property of block copolymers is utilized for nanolithography? How is a pattern transferred to a substrate in block copolymer lithography?

23.18. Name three scanning probe microscopy tools used for scanning probe lithography. Can SPL be used for modifying a surface chemically?

References

1. Rothschild M, Bloomstein TM, Fedynyshyn TH et al (2003) Recent trends in optical lithography. Lincoln Laboratory J 14(2):221–236
2. French RH, Tran HV (2009) Immersion lithography: photomask and wafer-level materials. Annu Rev Mater Res 39:93–126
3. Silverman PJ (2005) Extreme ultraviolet lithography: overview and development status. J Micro/Nanolithogr MEMS MOEMS 4(1):011006. doi:10.1117/1.1862647
4. Pease RFW (1981) Electron beam lithography. Contemporary Phys 22(3):265–290
5. Mohammad MA, Muhammad M, Dew SK et al (2012) 2:Fundamentals of electron beam exposure and development. In: Stepanova M, Dew S (eds) Nanofabrication: techniques and principles © Springer-Verlag/Wien, pp 11–41
6. Tseng AA, Chen K, Chen CD et al (2003) Electron beam lithography in nanoscale fabrication: recent development. IEEE Trans Electron Packag Manuf 26(2):141–149
7. Bindra A (2000) E-Beam projection prevails in nanometer lithography. Electronic Design. http://electronicdesign.com/archive/e-beam-projection-prevails-nanometer-lithography. Accessed 26 Sept 2015
8. Qin D, Xia Y, Whitesides GM (2010) Soft lithography for micro- and nanoscale patterning. Nat Protoc 5(3):491–502
9. Rogers JA. Nuzzo RG (2005) Recent progress in soft lithography. Materials Today February 50–56
10. Guo LJ (2007) Nanoimprint lithography: methods and material requirements. Adv Mater 19:495–513
11. Lan H, Ding Y (2010) 23: Nanoimprint lithography, lithography. In: Wang M (ed) InTech, http://www.intechopen.com/books/lithography/nanoimprintlithography. Accessed 26 Sept 2015
12. Jeong S-J, Kim JY, Kim BH et al (2013) Directed self-assembly of block copolymers for next generation nanolithography. Mater Today 16(12):468–476
13. Nunns A, Gwyther J, Manners I (2013) Inorganic block copolymer lithography. Polymer 54:1269–1284
14. Krämer S, Fuierer RR, Gorman CB et al (2003) Scanning probe lithography using self-assembled monolayers. Chem Rev 103(11):4367–4418

Chapter 24
Bottom-up Nanofabrication

Abstract A bottom-up approach to nanofabrication can be looked upon as a synthesis approach mimicking biological processes in which individual atoms are piled up one at a time on the substrate to form molecules. These molecules arrange themselves on their own into the desired form to yield the required nanostructures. The driving mechanisms for this molecular arrangement are the physical and chemical forces operative at the nanoscale. These mechanisms have been perfected by Mother Nature over a period of several millennia. Of particular interest to nanoelectronics are techniques such as sol-gel synthesis, vapour deposition, atomic layer deposition, molecular self-assembly, DNA-assisted assembly and many others. Sol-gel technique offers a simple process to produce nanoparticles. Two forms of vapour-phase techniques are physical vapour deposition in which the active species is evaporated into the vapour phase and chemical vapour deposition in which, a precursor is used which decomposes into the required species via a chemical reaction. Based on successive, self-restricting reaction cycles, atomic layer deposition provides thickness adjustment at nanometer level along with composition control. Molecular self-assembly exploits the organizational capability of matter to form homogeneous monolayers. Physical and chemical vapour deposition constitute self-assembly from gaseous phase. Artificial DNA nanostructures are used to arrange functional nanomaterials into nanoelectronic circuits.

24.1 Introduction

For practical nanoelectronic applications, structurally and dimensionally well-defined 1-D, 2-D and 3-D nanostructures are required. These nanostructures ought to be of controlled composition, mechanical, electrical, magnetic, optical, thermal and chemical properties. Furthermore, they need to be synthesized beginning from atomic and molecular species of sub-nm and nm sizes, and assembling them in a controlled fashion. Over a long span of time, a variety of techniques have evolved for this kind of synthesis in physical, chemical, and biological sciences. These growth techniques can control the shape, size and crystalline structure of the

V.K. Khanna, *Integrated Nanoelectronics*,
NanoScience and Technology, DOI 10.1007/978-81-322-3625-2_24

building block entities and unite them to produce the required nanostructures. All these processes start with a nucleation stage followed by successive stages in which the particles increase in size from the initial molecular sizes. They result in particles with a broad size distribution with restricted variation in the shape of the particles. Nonetheless, nanostructures obtained from this bottom-up approach have found diversified applications in nanoelectronics. Moreover, the bottom-up approach also provides respite from the high-priced top-down approach requiring state-of-the art facilities for large-scale fabrication of nanoelectronic circuits. Manipulating atoms and molecules using a scanning probe microscope can also enable this type of nanofabrication but the slow speed makes it very exhausting. In this chapter, only bottom-up nanofabrication methods based on natural self-assembly will be highlighted.

24.2 Sol-Gel Process

A sol is a liquid containing a stable suspension of solid nanoparticles (diameter 1–100 nm), e.g., black inkjet ink is a sol of carbon black in water. A gel is a solid interconnected network of nanostructures dispersed in a volume of liquid, e.g., gelatin. Sol-gel is a versatile and cost-effective chemical synthesis technique in which solid nanoparticles suspended in a liquid cluster together forming a continuous three-dimensional network of nanostructures dispersed throughout the liquid. It is erroneous, though commonly done, to refer to gels made by this technique as 'sol-gels' because all gels are made by this technique.

The sol-gel process consists of the following seven steps [1]:

(i) **Mixing** Commonly used precursors in the sol-gel synthesis are metal alkoxides, $M(OR)_n$, (M-metal and R-alkyl radical). For illustration of sol-gel process, we take silica-based gels as an example. The precursors are tetramethoxysilane, Si $(OCH_3)_4$, and tetraethoxysilane, $Si(OC_2H_5)_4$. They are known as TMOS and TEOS, respectively. They are represented as silicon alkoxide precursor $Si(OR)_4$.

The ingredients dissolved in a liquid are mixed. The silicon alkoxide precursor $Si(OR)_4$ reacts with water. It undergoes hydrolysis and polycondensation. Hydrolysis involves dissociation by reacting with water. Polycondensation entails the formation of a polymer by linking together monomers and elimination of a low molecular weight substance like water. The reactions are as follows:

Hydrolysis

$$Si(OR)_4 + 4\ H_2O \rightarrow Si(OH)_4 + 4R{-}OH \tag{24.1}$$

which can be written in a general form as:

$$-\text{MOR} + \text{H}_2\text{O} \rightarrow \text{MOH} + \text{ROH} \tag{24.2}$$

The hydrolysis reaction produces the reactive groups, e.g., the silanol groups. These are the groups which will form polymer links during polycondensation.

Polycondensation It converts monomers into oligomers (molecular complexes consisting of a few monomer units) and, finally, polymers, e.g., connective silicon-oxygen-silicon (MOM) bridges are formed via two different routes:

(a) Water condensation:

$$(\text{OR})_3\text{-Si-OH} + \text{HO-Si-}(\text{OR})_3 \rightarrow \left[(\text{OR})_3\text{Si-O-Si}(\text{OR})_3\right] + \text{H-O-H} \tag{24.3}$$

$$\text{In general, } -\text{MOH} + \text{HOM}- \rightarrow -\text{MOM} - + \text{H}_2\text{O} \tag{24.4}$$

(b) Alcohol condensation

$$(\text{OR})_3\text{-Si-OR} + \text{HO-Si-}(\text{OR})_3 \rightarrow \left[(\text{OR})_3\text{Si-O-Si}(\text{OR})_3\right] + \text{R-OH} \tag{24.5}$$

$$\text{Or generally, } -\text{MOH} + \text{ROM}- \rightarrow -\text{MOM} - + \text{ROH} \tag{24.6}$$

(ii) **Casting** The low-viscosity liquid obtained is poured into a non-sticking cast.

(iii) **Gelation** As time elapses, the nanoparticles formed by the polycondensation of silicon alkoxide link together. After reaching a critical size, the nanoparticles stop growing. Then they start agglomerating with other nanoparticles. When the number of nanoparticles agglomerated becomes so large that the entire liquid is spanned by the network, the gel has formed. The viscosity of the solution thereby increases. A nearly solid phase is produced fitting the shape of the mold. In case of silica-based gels, 3-D entangled networks of siloxanes are formed.

(iv) **Aging** After gelation, as time goes by, the solid network continues its evolution. The aging involves three sub-processes, viz., *(a) Polymerization*: Any unreacted hydroxyl groups left participate in polymerization. The connectivity of the network in the gel increases. The increase in connectivity is accompanied by a small degree of shrinkage of the network. *(b) Syneresis:* The gel irreversibly contracts of its own accord, exuding some liquid. Compressive stresses produced pull the solid network into the liquid. *(c) Coarsening or ripening:* Dissolution and re-precipitation occurs caused by solubility differences between surfaces having different radii of curvature. The growing small particles act as nutrients for the bigger crystals. The material is further stiffened.

(v) **Drying** The gel is dried to remove any liquid phases present. The gel decreases in volume by an amount equal to the volume of liquid lost.

(vi) **Dehydration** This is done at temperatures >800 °C. Any M–OH groups left are eliminated so that the gel does not rehydrate. The gel is thus chemically stabilized. Highly stable porous material is obtained.

(vii) **Sintering** Densification of the network is done at temperatures >800 °C. The solid network moves by viscous force or diffusion. The pores in the network collapse. Porosity is minimized. The residual organic contaminants are driven away. The gel is solidified.

24.3 Vapor Deposition (VD)

Vapor deposition is a process, either physical or chemical, usually carried out under vacuum, for deposition of a high-quality, adherent, thin film of a substance, element or compound, on a surface.

24.3.1 Physical Vapor Deposition (PVD)

Physical vapor deposition is a set of processes, usually performed under high to ultra-high vacuum environment, to deposit a thin solid layer of a substance (metal, semiconductor, insulator, ceramic, polymer, etc.), ranging in thickness from nm to μm, by condensation of the vapors of the given material on a wafer/substrate. The PVD is a three-step process: (i) Vaporization of the material, (ii) Its transportation to the substrate, (iii) Condensation of the material on the substrate. Because the process is executed strictly under vacuum, the vapors can easily transport to the substrate.

Two common forms of the PVD process are: thermal evaporation and sputtering. For thermal evaporation, the energy is supplied through resistive heating, by electron or ion beam or by laser source (Fig. 24.1). In sputtering, the vapors are generated from a target of the material by bombardment with ions of an inert gas such as argon ions accelerated in an electrical field (Fig. 24.2). Thus, the energy required to vaporize the material is supplied as heat or through gaseous plasma.

There are many variants of PVD process: *(i) Co-deposition from multiple sources*: Multi-component nanomaterials can be deposited by e-beam assisted co-deposition. These materials are simultaneously or sequentially evaporated from crucibles bombarded with e-beams. Similar remarks apply to co-sputtering. *(ii) Reactive sputtering:* Here, a reactive gas such as oxygen, nitrogen or hydrocarbon is introduced into the

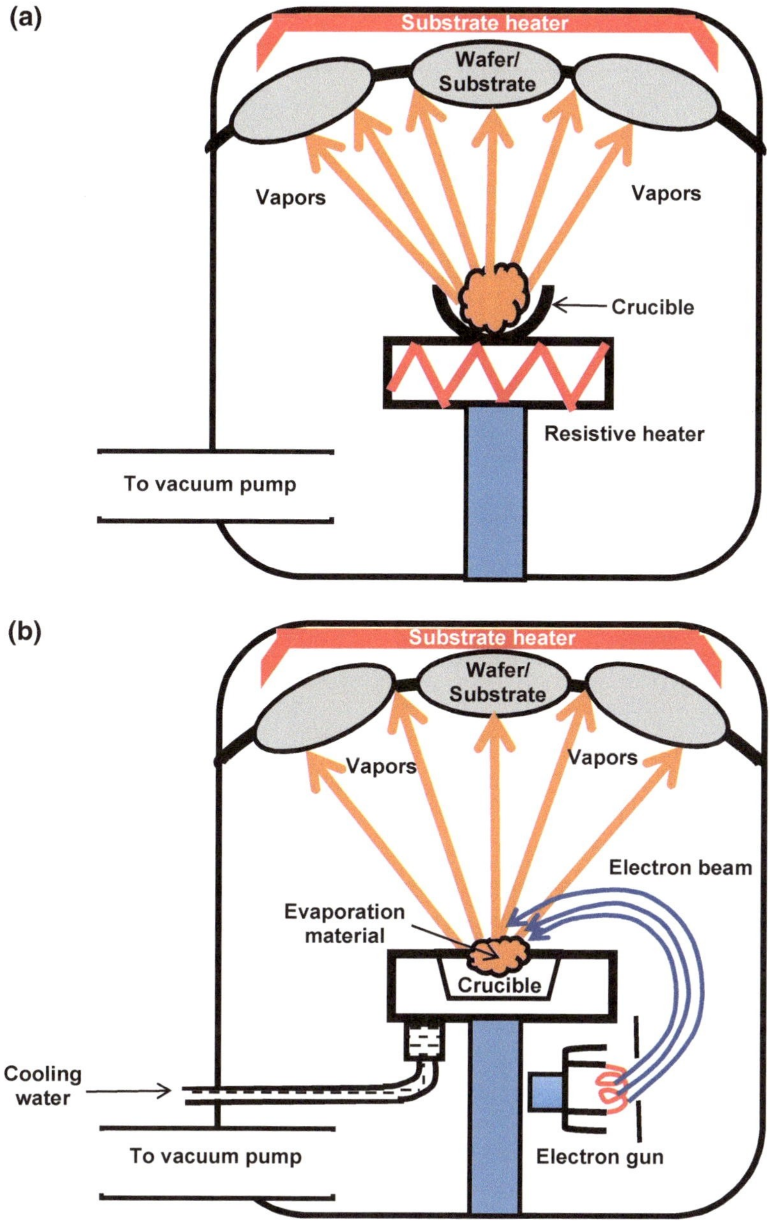

Fig. 24.1 Thermal evaporation with: **a** Resistive heating and **b** Electron beam heating

chamber, forming a compound with the vapors of sputtering target. *(iii) Substrate heating:* The properties of the film can be tailored by heating the substrate during deposition. *(iv) Post-deposition annealing:* The properties of the deposited film can be modified by heating in suitable gaseous ambient.

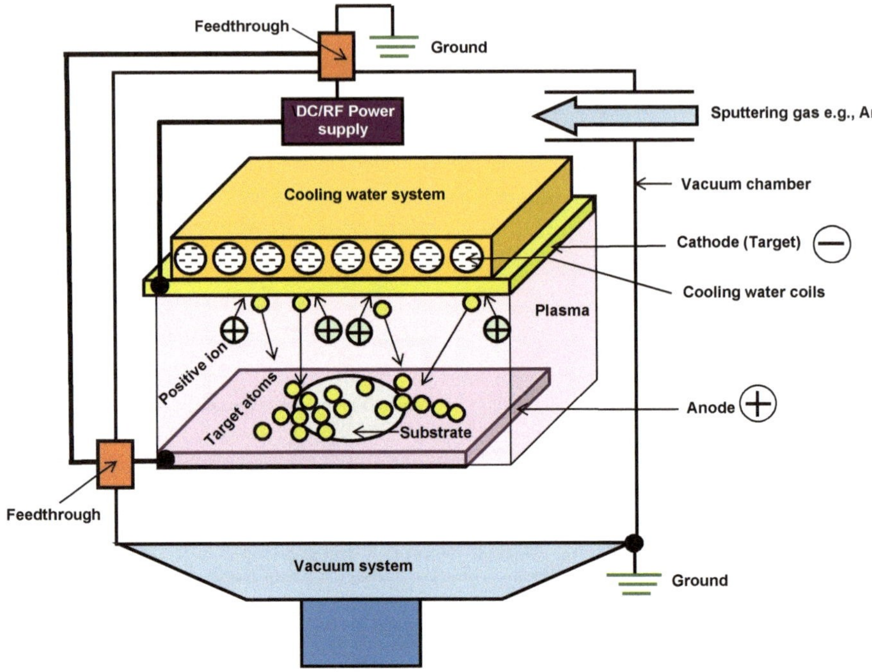

Fig. 24.2 Schematic of sputtering equipment

24.3.2 Chemical Vapor Deposition (CVD)

CVD is a material-processing technique which is carried out by: (i) transporting one or more gaseous precursors to the reaction chamber, (ii) activating the molecules of the gaseous reactants in the chamber thermally or otherwise by light or plasma discharge, and (iii) causing them to dissociate and react in vicinity of the substrates placed in the chamber, thereby (iv) depositing thin films formed by chemical reactions on the substrates [2]. Any unreacted precursor gases along with by-products of the reactions are exhausted from the chamber.

CVD reactors are of two types (Fig. 24.3): (i) Cold-wall reactor: RF induction or radiation lamps heat only the susceptor on which substrates are placed. Contamination from deposition on the reactor walls is reduced. This reactor is used for endothermic reactions. (ii) Hot wall reactor: The reactor is surrounded by a tube furnace. Deposition on the reactor walls contaminates the system. It is used for exothermic reactions.

Parameter variations of CVD process include: (i) Pressure range from sub-Torr to atmospheric and > atmospheric pressures; (ii) Temperature range from 200 to 1600 °C.

CVD process has been extensively applied to the synthesis of single- and multiwall carbon nanotubes, as well as graphene sheets.

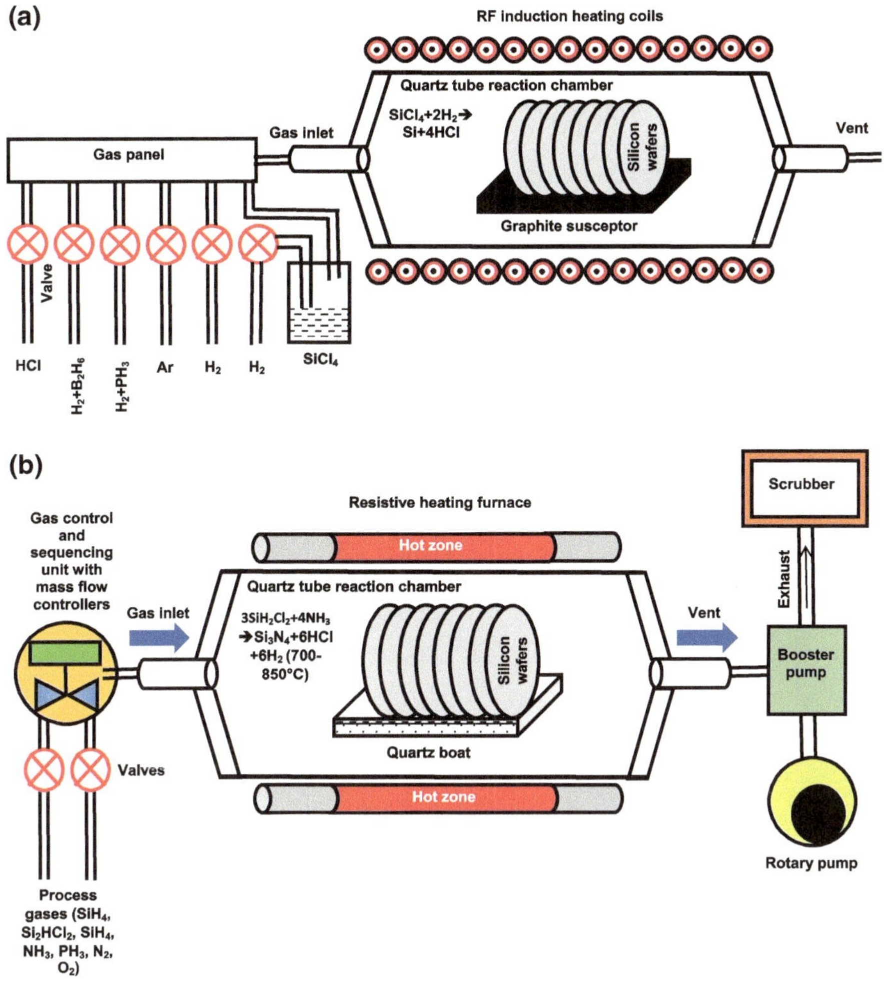

Fig. 24.3 Reactors for chemical vapor deposition: **a** Cold wall atmospheric pressure CVD and **b** Hot wall low pressure CVD

24.4 Atomic Layer Deposition (ALD)

24.4.1 ALD Process

ALD is a thin film technology. It is a chemical vapor deposition process based on a vapor phase technique comprising sequential, self-limiting reactions. It can provide control of thickness at nanometer scale. The films produced are free from any pinholes. The substrate can be of any shape or geometry; in porous, particulate or powder form. It may have trenches or cavities. Irrespective of surface topographical

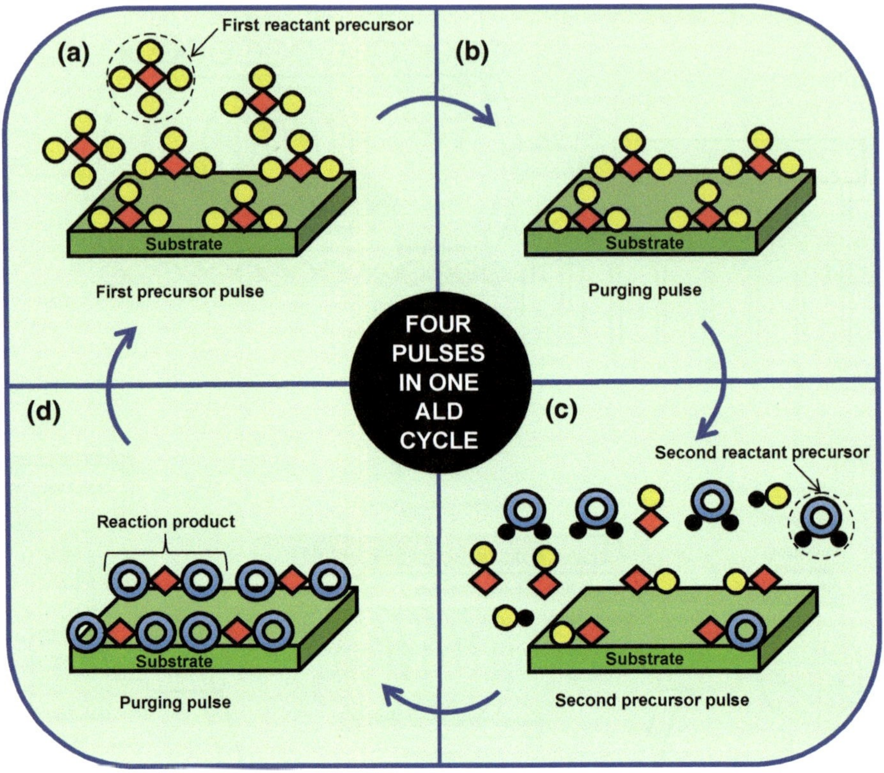

Fig. 24.4 Parts of a typical cycle of atomic layer deposition

features or aspect ratio, a conformal coating is always assured. Complex 3-D objects and large-area surfaces can be coated with repeatability. The process also offers scalability.

The obvious enquiry is: How does ALD restrict to precisely atomic layer-by-layer growth? In ALD, the CVD reaction is split up into two half reactions [3], Fig. 24.4. These half reactions are completed during alternating pulses in which separate gaseous precursors are fed into the reaction chamber. During the first pulse, one precursor is allowed into the chamber under vacuum <1 Torr. The reaction between the substrate and the precursor takes place for a pre-stipulated interval of time in which no more than a monolayer of the required material is formed. The chamber is purged with an inert gas such as nitrogen or argon to drive away any unreacted precursor left in the chamber as well as any by-products of the reaction. This completes one half of the reaction.

During the second pulse, the counter precursor, different from the first one, is introduced into the chamber. This precursor also reacts with the substrate to produce a single layer of the material. As in the previous case, purging follows the reaction. Such cycles of (one precursor + purging + second precursor + purging) are repeated until the desired film thickness is attained. Thus, ALD differs from conventional CVD in keeping the precursor materials separate during the reaction in contrast to CVD, which introduces multiple precursors together at the same time.

Typical process conditions for ALD are: (i) Pressure range: 0.1–10 Torr, or atmospheric pressure; (ii) Temperature range: 50–500 °C. The reaction temperatures for ALD are generally below 350 °C, varying with particular process. An 'ALD temperature window' defines the range of temperatures in which the growth is saturated. The process must be delimited within the prescribed ALD temperature window for achieving ALD-type growth at required growth rates.

24.4.2 Advantages of ALD

(i) Conformality of coatings: This is enabled by the self-limiting feature of the process whereby the surface reaction is restrained to one layer of the material. As the cycles are repeated, the reaction of the precursor with the complete surface is possible by its dispersal into deep trenches, permitting uniformity of growth. CVD may be uneven because of faster growth process. Physical vapor deposition (PVD) is prone to shadowing effects. (ii) Precise thickness control: For several ALD films, the growth rate is <10 nm/cycle. By varying the number of cycles, the thickness can be accurately monitored with a few nm tolerance. (iii) Composition control: This is done through ALD super cycles, which consist of a multiplicity of ALD processes, e.g., for obtaining a stoichiometric $SrTiO_3$ film, alternation of ALD super cycles for TiO_2 and $SrCO_3$ is carried out in 1:1 ratio. Post ALD, the $SrTiO_3$ film is annealed to get stoichiometry.

24.4.3 Disadvantages of ALD

The principal disadvantage of ALD is its slow deposition rate. ALD has an incredibly slow rate of film deposition ~100–300 nm/h, which makes it a prohibitively long procedure. Decisive factors of growth rate are the aspect ratio of the substrate and size of the reactor. For a larger aspect ratio substrate, the precursor pulse and purge times must be longer in order that the gases disperse into the deep crevices of the substrate. A larger volume reaction chamber too requires longer times for execution of pulse and purging operations. The growth period is shortened to typically 3600 nm/h by using spatial ALD. This ALD uses a spatially resolved head wherein the translation of the head around the substrate changes the precursor. The substrate is exposed to a given precursor for a time interval specified in the cycle. In another version, the substrate moves past the fixed precursor nozzles at a controlled speed, which varies the exposure time of substrate according to the cycle.

24.4.4 Applications of ALD

ALD has been extensively applied for growing a vast variety of materials, including metals, semiconductors, insulators, complex ternary materials and polymers [4]. To cite a few examples, mention may be made of: (i) Oxides: Al_2O_3, HfO_2, La_2O_3, Sc_2O_3, SiO_2, Ta_2O_5, TiO_2, ZnO, ZrO_2, etc. (ii) Nitrides: TiN, TaN, Si_3N_4, AlN, GaN, etc. (iii) Fluorides: MgF_2, AlF_3, CaF_2, LaF_3 etc. (iv) Sulfides: ZnS, SrS, etc. (v) Carbides: TaC, TiC, etc. (vi) Metals: W, Mo, Ir, Pd, Pt, Ru, etc. (vii) Polymers: PMDA–DAH, PMDA–ODA, etc. The list of materials is not exhaustive. ALD has been applied to growth of high-κ dielectrics in nanoelectronics. For silicon or germanium devices, Al_2O_3 and HfO_2 have been used. For III–V compound semiconductor devices, examples of high-κ dielectrics used are $TaSiO_x$, HfO_2, Al_2O_3 and HfAlO. High-κ HfO_2 coatings have been used in FINFETs for uniformity and conformality on the fin. ALD is promising for Bio NEMs, nanofluidics and NEMS applications.

24.4.5 Limitations of ALD

The main hindrance to ALD is the unavailability of reactants, which can facilitate the reaction pathway in a self-limiting fashion. In many cases, the choice is limited or does not exist at all. Sufficient volatility of the reactant is expected in order that it is in vapor phase at room/low temperature. Fast and irreversible surface reaction will enable quick saturation of growth. The substrate, the growing film or ALD reactor should not be dissolved, etched or damaged in any way by the reactant or any by-product liberated in the reaction. Cost factors cannot be ignored.

24.5 Molecular Self-Assembly

Molecular self-assembly is a spontaneous process of nanofabrication in which individual molecules combine together in consequence to specific local interactions among themselves to form more ordered structures. They do so without any prodding, guidance, management or any form of intervention from an external source. Self-assembly is a structural re-organization from a disordered molecular condition. The molecular binding is reversible. Non-covalent forces hold the molecules together. There are plenty of examples of self-assembly in nature. An interesting example is the lipid bilayer.

24.5.1 Lipid Bilayer Formation by Self-Assembly

The plasma membrane of the cells contains a lipid bilayer comprising two layers of phospholipids (molecules such as fats, waxes, sterols, glycerides, etc.), Fig. 24.5. Each phospholipid molecule is made up of two regions [5]: (i) A hydrophilic head region: It is a region which likes water. It contains a polar group, viz., the phosphate group, which can interact with water molecules forming hydrogen bonds with them. (ii) A hydrophobic tail region: It is the region, which does not like water or is afraid of water. It does not interact with water due to its non-polar nature. Thus, the phospholipid is said to be an amphipathic molecule, displaying partial hydrophilicity and partial hydrophobicity.

As we know, the liquid inside the cell is watery [6]. The liquid outside the cell too is watery. Due to the predominant prevalence of water, both in the interior or exterior of the cell, nature has evolved a technique in which the lipid molecules arrange themselves as a two-layer structure. The molecules align themselves naturally into bilayer formation. If the lipid molecules were placed in a single layer, the tails will be facing the water, which is an unstable situation. So, the lipid molecules are organized in two layers with the tails of the two layers on the inside and pointing towards each other. Hence, the tails avoid and hide away from water. The heads of the two layers are on the outside in opposite directions, facing the aqueous environment. So, the heads interact with water. For complete avoidance of water by the water-fearing tail, the lipid bilayer forms a closed sphere. The spherical vesicle

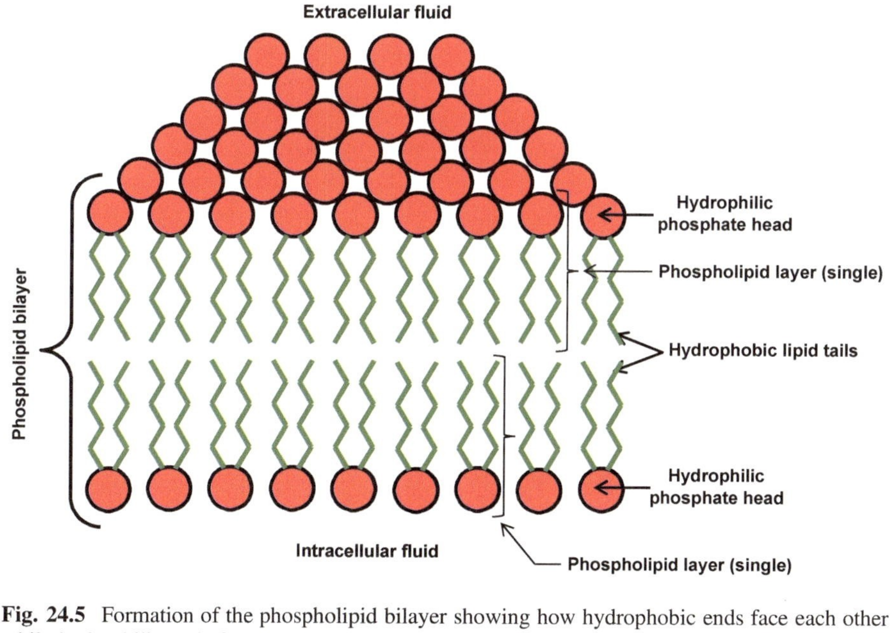

Fig. 24.5 Formation of the phospholipid bilayer showing how hydrophobic ends face each other while hydrophilic ends face the water environment

is called a liposome. Lipid bilayer is an eloquent example of self-assembly of molecules in nature to form a stable structure, which can survive with water on both sides. Nature is adept in self-assembling intricate structures from less complicated units.

24.5.2 Types of Molecular Self-Assembly

Molecular self-assembly is of two types: intermolecular (between different molecules) and intramolecular (between same types of molecules). The term 'self-assembly' is usually applied to the former class, while the latter is referred to as folding, e.g., protein folding. By folding, the molecule acquires its shape or conformation.

24.6 Driving Factors for Self-Assembly

24.6.1 Molecular Motion

Due to their thermal energy, the molecules are at all times in a state of incessant agitation and unremitting random motion. It is this thermal energy which is the main factor responsible for urging the molecules to come closer. It is also an organizational impetus. It motivates and forces the molecules to be arranged in the correct orientation for self-assembly.

24.6.2 Intermolecular Forces

These forces range from the weak van der Waals forces to the stronger hydrogen bonding and Coulomb forces. Hydrophobic interactions between molecules too play important roles. All these contributory forces may be themselves feeble but together they can add up to provide intense attractive forces binding the molecules together.

Thus the thermal motions bring the molecules near each other and intermolecular forces serve the function of the glue that ties together the molecules.

24.7 Approaches for Self-Assembly

Amongst the multitude of self-assembly approaches, two schemes which have received most attention are: (i) Electrostatic self-assembly, and (ii) Self-assembled monolayer (SAM) formation.

24.7.1 Electrostatic Self-Assembly

It is based on the alternate absorption of anionic and cationic groups on a suitable substrate [7]. A polyelectrolyte is a polymeric electrolyte behaving both as a polymer as well as an electrolyte. From the polymer viewpoint, it is a macromolecule with a high molecular weight. From the electrolyte standpoint, its repeating units contain electrolytic groups. These groups dissociate when dissolved in water, accompanied by the release of positive or negative ions. The ionic constituents of the polyelectrolyte make its aqueous solution conducting. Furthermore, their solution is often viscous, as one would expect for a polymeric solution.

The film is formed through the following sequential process: (i) A positively charged substrate is immersed into a solution of an anionic polyelectrolyte such as an anionic protein. By electrostatic attraction, a monolayer of the anionic polyelectrolyte is absorbed on the positively charged substrate. Over-absorption takes place so that the surface charge is reversed from positive to negative. (ii) The negatively charged substrate taken out from the anionic polyelectrolyte is rinsed thoroughly in water to wash away any unabsorbed or loosely held polyelectrolyte. (iii) After washing, the negatively charged substrate is dipped in a cationic polyelectrolyte, wherein a monolayer of the cationic polyelectrolyte is absorbed on its surface. Again by over-absorption, the surface charge polarity changes from negative to positive. (iv) The substrate is rinsed in water to remove the residual polyelectrolyte. (v) The steps from (i) to (iv) are repeated in the same order.

During every step, the charge on the surface undergoes reversal so that the oppositely charged polyelectrolyte can be absorbed in the subsequent step by electrostatic force. This alternation of the surface charge from positive to negative and vice versa enables a continuous increase in thickness till the desired thickness is attained. As the film is formed in a layer-by-layer growth process, this method is often called layer-by-layer (LbL) self assembly.

The LbL method can be carried out using beakers and tweezers to produce thin films of nanoscale precision on a wide variety of biomaterials. A proper choice of starting building block molecules is essential. In electronics, it is applied to the formation of enzyme structures on the active components of transistors to fabricate biosensors responsive to specific analytes.

24.7.2 Self-Assembled Monolayers (SAMs)

SAMs are highly ordered equilibrium nanostructures, typically 1–3 nm thick, comprising organic molecular assemblies [8]. They are formed on the surfaces of solids by the absorption of molecular constituents from an active surfactant, either in the solution or gaseous phase. A surfactant or surface-active agent is a substance which when added to a liquid lowers its surface tension, thereby facilitating its spreading. Unprompted and often epitaxial organization of the adsorbate on solid substrates yields semi-crystalline or crystalline SAM nanostructures. However, SAMs can also form in regular arrays on surfaces of liquids.

Extensive investigations have been carried out on the SAM formation by the adsorption of alkanethiols on metallic thin films (10–200 nm) such as gold, silver, platinum, palladium, copper, etc., deposited on silicon wafers, or substrates made of glass, mica, or plastic (Fig. 24.6). An alkanethiol is a compound containing an alkyl group connected to a sulfanyl or thiol group, –SH, e.g., methanethiol, CH_3 SH, an organosulphur compound. The common recipe for SAM formation involves immersing a clean or freshly prepared substrate in a dilute ethanolic solution of single/two or more different thiols (for mixed monolayers) at room temperature (25 °C). The period of immersion is ~12–18 h. In the beginning stages, the monolayer is formed at a very brisk pace, generally completing in a time span of seconds to minutes. This initially formed layer lacks order. It contains defects in its chains. With passage of time, ordering inside the layer improves. This process takes about 12 h to 2 days. SAM formation at temperatures >25 °C improves the kinetics of formation and minimizes defects. Use of contaminated thiol compounds gives a disordered, non-ideal SAM.

An alkanethiol molecule has three parts [9]: (i) A sulphur binding group,-SH; (ii) A spacer chain made of methylene groups, $(CH_2)_n$; and (iii) A functional head group. The sulphur atom in the thiol group and the carbon atoms in the chain provide the forces compelling the assembly. The terminal thiol group attaches the molecule to the metal surface. The head group plays the role of a platform. On this platform, any desired group can be affixed to create a surface displaying the

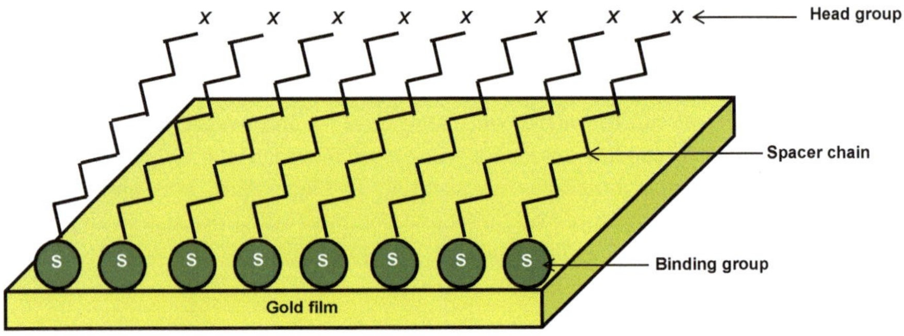

Fig. 24.6 Formation of self-assembled monolayer by alkanethiol molecules on a gold surface

required chemical properties. A methyl (–CH_3) head group makes the surface hydrophobic. A hydroxyl (–OH) or carboxyl (COOH) head group leads to a hydrophilic surface. An ethylene glycol head group [$C_2H_6O_2$] is used to produce a protein-resistant surface. By carefully tailoring the properties of the surface, a suitable surface can be custom designed for a specific application. SAMs therefore can provide tremendous opportunities in surface engineering.

Patterned mixed monolayers of dimensions 10–100 nm in the plane of the surface can be formed by using micro contact printing. Masters are made by photolithography or e-beam writing. Then stamps are fabricated by pouring an elastomer, e.g., PDMS into a master, curing and peeling. Each master can be used to make 50 stamps. Every stamp can be used several times. After inking the stamp with the ethanolic solution of concerned thiol, it is pressed against the gold surface. A gold-thiolate monolayer is formed at the contacting region in 10–20 s.

SAMs represent elementary forms of organic thin films. Studies on SAMs enhance the level of understanding on self-organization, surface wetting, interfacial phenomena, corrosion, etc. They inter-relate the electronic properties such as current–voltage characteristics, and optical properties, e.g., surface plasmon frequency of nanostructures with the external world. The commonly used head groups in molecular electronics are –CH_3, and –SH. Widespread utilization of SAMs has received emphasis, particularly due to their easy preparation technique without the need of any ultra-high vacuum or sophisticated equipment. SAMs can be formed on objects of all sizes such as thin films, nanowires and other nanostructures, imparting them necessary functionalities.

SAMs formed with thiols on gold have shown good stability for several days to weeks during interaction with complex liquid media. Under exposure to air, aqueous or ethanolic solutions, these SAMs are stable over several months. A high affinity exists between gold and thiols. Due to this affinity, any extraneous material on the gold surface is easily displaced by thiols. Further, gold does not show any unusual reactions with thiols, such as one which could lead to the production of a substitutional sulfide interface. In addition, gold has also been favored because it is non-toxic to cells, and gold thin films are the usual substrates for several spectroscopies and analytical tools such as surface plasmon resonance, quartz crystal microbalance and ellipsometry used in chemistry and biology.

24.8 DNA Nanoengineering

24.8.1 DNA Structure

DNA is the acronym for deoxyribonucleic acid. Deoxyribose ($C_5H_{10}O_4$) is a 5-carbon sugar present in DNA. It is a modified form of another sugar. The original sugar from which deoxyribose is obtained is called ribose ($C_5H_{10}O_5$).

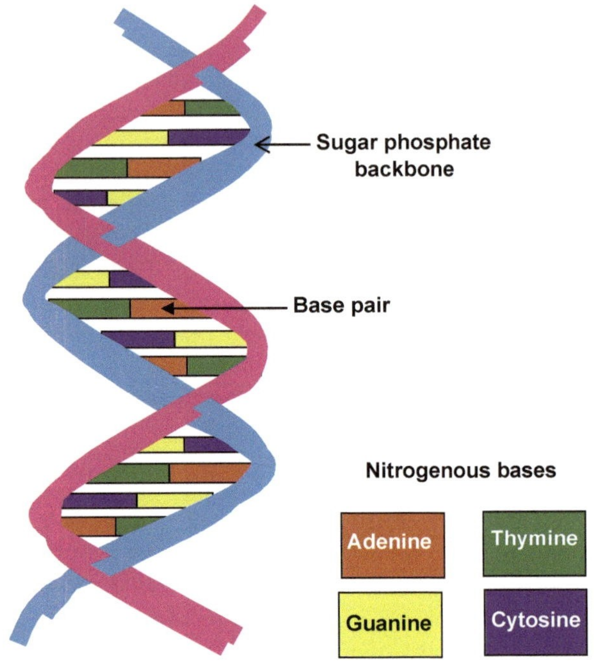

Fig. 24.7 DNA double helix structure

DNA molecule is a polymer of the class of nucleic acids. This class is known as polynucleotide (Fig. 24.7). The monomer in polynucleotide is nucleotide. Each nucleotide has three components: (i) A deoxyribose sugar. (ii) A base containing nitrogen. This base is attached to the deoxyribose sugar. (iii) A phosphate group (PO_4^{3-}). The nitrogen base in the nucleotide is one of the four types: (i) Adenine ($C_5H_5N_5$)-A, (ii) Guanine ($C_5H_5N_5O$)-G, (iii) Cytosine ($C_4H_5N_3O$)-C and (iv) Thymine ($C_5H_6N_2O_2$)-T. Hence, there are four types of nucleotides in DNA. In a DNA sequence, the order of nitrogen bases indicates the particular 'gene', much in the same way as the order of letters of alphabet in a group of letters represents a 'word'. The gene is a unit of heredity.

The nucleotides are covalently linked in two long strands or chains. This linkage is formed through a sequence of alternating sugar and phosphate molecules, which therefore form the DNA backbone. The two strands of DNA intertwine to form a spiral called the double helix. The double helix resembles a ladder. The rungs of this ladder are the base pairs. The sides of the ladder consist of sugar-phosphate backbone.

The intent here is not to look on DNA as a carrier of genetic information but as a constructional or building material. Copying from nature the use of DNA as a building block material in living organisms, the idea is to exploit it as a building block of nanostructures.

24.8.2 DNA Origami

Origami (ori-folding, kami-paper) is the art of creating decorative objects by folding paper. DNA origami is a DNA-manipulating technique which seeks to construct simple and intricate, two- or three-dimensional shapes by folding long DNA strand on the nanoscale, taking the help of multiple short DNA strands as fastening wires. Thus, the shape produced has two principal components: the long strand and the several short strands of DNA molecule. The short strands bind the long strand at different sites. The long strand is called the scaffold strand while the short strands are known as the staple strands. Recall that the scaffold is a temporary elevated platform. It is either supported from below or hung from above. Workers sit or stand on it to perform tasks at heights above the ground.

The question naturally arises: How are the strands bound to each other? For binding, the sticking property of complementary bases of DNA molecule is applied. The four bases of the DNA molecule have a natural tendency of binding in pairs: A→T, C→G. Thus, base T is the counterpart of base A, and base C that of base G. These relationships of bases are harnessed for self-assembly. Thus in a mixture of strands, each base searches for its counterparts. As soon as it is found, binding ensues. The base pairs: adenine–thymine, guanine–cytosine are called Watson –Crick base pairs. The linkage between the partners of each pair occurs through hydrogen bonds. There are 2 hydrogen bonds between A and T bases, and 3 hydrogen bonds between C and G bases. Such complementary base pairings impart to the DNA molecule its double-helix structure.

For DNA origami, the long scaffold strand may contain as many as 700 bases while the staple stands may contain 30–50 bases. To prepare a given shape by origami, computer-aided design software is used. The aspired shape may range in complexity from a simple 2-D sheet to an octahedron, and even greater complexity. Automated algorithms have been developed to trace the scaffold sequence around desired targets. The required 3-D shape is drawn on the computer. A mesh is formed from a series of polygons representing the figure. The polygons are connected together to obtain a 3-D picture [10]. The algorithm places the DNA strands along the edges of this mesh. This meshed drawing shows the scaffold and staple strands. The placement of each strand is examined for iteration, if any. Based on the finalized design, the list of DNA strands required is made.

After synthesizing the strands, they are mixed and subjected to standard warming and cooling procedures. The staple strands immediately rush to attach themselves to the correct places on the scaffold. There is a race among the strands to acquire their positions before all the places are filled up. This mixing does not require any guidance effort on the part of the experimenter. It takes place on its own, yielding the planned shape.

Thus the origami process entails folding a single DNA strand, the scaffold strand into a designed shape. The folding is accomplished through the shorter strands, the staple strands. The process is coaxed by complementary base pairing. The bases on the staple strands affix themselves to complementary bases on the scaffold strand.

24.9 Self Assembly of Nanocomponent Arrays on DNA Scaffolds

Dynamic random-access memory (DRAM) and programmable logic array (PLA) are two examples of systems that are laid out in regular two-dimensional (2D) arrays. Le et al. [11] reported the self-assembly of metallic nanoelectronic components into rows on DNA scaffolding. The nanocomponents were functionalized with DNA. They were attached to the scaffolding through DNA hybridization. In DNA hybridization, two complementary single-stranded DNAs are combined. By base pairing, a single double-stranded molecule is formed.

In the reported method, the DNA scaffolding was assembled from DNA strands. The assembly was done by slowly cooling a buffered solution from 90 °C to room temperature. This solution contained a stoichiometric mixture of the 21 DNA strands. The scaffolding was made of double-stranded DNA. But it had single-stranded hybridization sites on which complementary DNA from DNA-functionalized nanocomponents could attach. A suspension of this scaffolding was deposited on freshly cleaved mica. After its absorption on the surface for $\sim$1 min., the buffer salts were removed by rinsing in distilled water. The scaffolding was dried in nitrogen. Then DNA-Au conjugate in buffer solution was deposited on mica substrate. Hybridization was allowed for 5 min. Again, the surface was rinsed with water and dried.

24.10 Self-Assembled DNA Scaffolds for Nanoelectronic Circuit Boards

Kershner et al. [12] demonstrated the unification of self-assembly, a bottom-up technique with lithography, a top-down method. They could arrange DNA scaffolds on lithographically patterned SiO_2 surfaces used in present-day silicon device fabrication. The staple strands in these nanostructures can be modified with attachment groups such as avidin or single-stranded DNA hooks to serve as fixation sites for nanoelectronic components. Generally, 200 such directly accessible binding sites are possible at 6 nm resolution. On these sites, various nanodevices can be fixed besides carbon nanotubes, silicon nanowires, quantum dots or other nanoparticles. Precision of Watson-Crick base-pairing for assembly is limited only by the 0.34 nm separation between nucleotides.

The approach of Kershner et al. differs from that of previous workers in which DNA origami is synthesized in solution, and the haphazardly arranged scaffolds are irreconcilable with conventional microfabrication technology. They employed e-beam lithography to define the pattern and dry etching of silicon dioxide to produce DNA origami-shaped binding sites on SiO_2. These binding sites are etched patches of the same shape and size as the triangular origami structures. They serve as the templates on which the origami structures are to be fixed. The etched patches as well as the origami structures are negatively charged. High concentration of positively charged divalent magnesium ions are added to the saltwater solution containing the origami structures through 100 mM $MgCl_2$ (in Tris-acetate-EDTA [TAE] buffer). The incubation time ranges from minutes to hours. These positively charged ions bind to both the etched patches and the origami structures in the form: origami structure-magnesium ion–etched patch. A little adjustment allows the origami structures to align strictly with the outline of the etched patch [13].

The DNA scaffold with attached gluing groups can serve as a circuit board for assembly of integrated circuits used in computers. The technology can be used whenever it is necessary to place individual molecules in a well-defined pattern on a surface.

24.11 Discussion and Conclusions

Sol-gel nanofabrication is a versatile chemical synthesis process that can be applied for a variety of nanomaterials. It is very cost-effective. But it is presently difficult to upscale. Further advancements may provide scalability of this process. Molecular self-assembly is used to produce molecular nanopatterns. Extending it to multiple materials may lead to multifunctional nanosystems. Chemical vapor deposition is costly because of requirement of vacuum equipment, high deposition temperatures and hazardous/corrosive gases. DNA scaffolding is at an early stage and its compatibility with CMOS process needs to be investigated.

Presently, engineering nanostructures from bottom upwards is playing a leading role in nanoelectronics. However, bottom-up approach is not used alone but definitely in conjunction with top-down approach as a hybrid nanofabrication technology.

Review Exercises

24.1 Define the following terms: (i) Sol and (ii) Gel. Why is the term ‘sol-gel’ misleading?

24.2 What is meant by TMOS and TEOS? What is hydrolysis? Write the equation for the hydrolysis of silicon alkoxide precursor.

24.3 What is meant by polycondensation? Write the equation for water condensation of silicon alkoxide precursor. Also write the equation for alcohol condensation of this precursor.

24.4 After the polycondensation of silicon alkoxide, what process takes place inside the mixture during gelation?

24.5 Name and describe the three sub-processes that occur during aging of the silicon alkoxide mixture after gelation.

24.6 What is vapor deposition? Differentiate between physical vapor deposition and chemical vapor deposition.

24.7 What are the two common types of physical vapor deposition? In which type, the material is liberated by striking an electron beam on it? In which type, the material is released by bombardment of a target with inert gas ions? What is the difference between reactive sputtering and co-sputtering?

24.8 What are the main steps in chemical vapor deposition of a material? Describe the main features of hot wall and cold wall reactors.

24.9 Is "atomic layer deposition" a chemical vapor deposition process? For atomic layer deposition, is there any restriction on substrate shape, size or surface topography?

24.10 How does atomic layer deposition process ensure deposition of atomically thin layers of materials? How does it differ from traditional CVD?

24.11 Can ALD provide conformal coatings on irregular shaped substrates? How is thickness of deposited film controlled? How is its composition controlled?

24.12 (i) ALD is a fast process. True or false? (ii) What is spatial ALD?

24.13 Give a few examples of metallic and insulating films which can be grown by ALD. Can it be applied for depositing polymers?

24.14 Define molecular self-assembly and illustrate it by lipid bilayer formation. How does the phospholipid bilayer evolve into a spherical shape to avoid water by its water fearing tail?

24.15 How does molecular motion aid in molecular assembly? What is the role of intermolecular forces in molecular assembly?

24.16 What is a polyelectrolyte? How is it used in layer-by-layer self assembly of molecules? Give an example of application of this technique in sensors.

24.17 Define the following terms: (i) Surfactant and (ii) Self-assembled monolayer.

24.18 What is an alkanethiol? What are the three parts of an alkanethiol molecule? Write the recipe for using an alkanethiol for SAM formation?

24.19 How are patterned mixed monolayers formed by microcontact printing?

24.20 Why are studies on SAMs important? What has encouraged widespread utilization of SAMs? What are reasons of the increasing trends of SAMs formed by thiols on gold?

24.21 What is the full form of DNA? What are the four types of the nitrogen bases in the nucleotide of DNA? What does the order of the nitrogen bases in a DNA sequence indicate?

24.22 What is the meaning of 'Origami'? What is done in DNA origami? Distinguish between scaffold strand and staple strand in DNA origami.

24.23 How are the strands bound to each other in DNA origami? How is the double helix structure of DNA realized from complementary base pairings?

24.24 Starting from computer-aided design, describe the sequence of steps in the implementation of the DNA origami process?

24.25 How is the DNA scaffolding assembled from DNA strands? How are nanoelectronic components self-assembled into rows on DNA scaffolding?

24.26 How are self-assembled DNA scaffolds prepared for use as circuit boards for computer chips?

References

1. Overney R (2010) Nanothermodynamics and Nanoparticle Synthesis NME 498A/A 2010. http://courses.washington.edu/overney/NME498_Material/NME498_Lectures/Lecture4-Overney-NP-Synthesis.pdf. Accessed 30 Sept 2015
2. Jones AC, Hitchman ML (2009) Chapter 1: Overview of chemical vapour deposition. In: Jones AC, Hitchman ML (eds) Chemical vapour deposition: precursors, processes and applications, Royal Society of Chemistry. Springer, pp 1–36
3. Johnson RW, Hultqvist A, Bent SF (2014) A brief review of atomic layer deposition: From fundamentals to applications. Mater Today 17(5):236–246
4. Oxford Instruments: Atomic Layer Deposition © Copyright 2015. http://www.oxford-instruments.com/products/etching-deposition-and-growth/plasma-etch-deposition/atomic-layer-deposition. Accessed 27 Sept 2015
5. Lipid Bilayer: Definition, Structure & Function (Chapter 22/Lesson 11). http://study.com/academy/lesson/lipid-bilayer-definition-structure-function.html. Accessed 27 Sept 2015
6. Hardman J, Lipid Bilayer Copyright © 2015 Fastbleep Ltd. http://www.fastbleep.com/biology-notes/31/170/969. Accessed 27 Sept 2015
7. Ariga K, Hill JP (2008) Layer-by-Layer (LbL) assembly, A "gentle yet flexible" method toward functional biomaterials. Material Matters 3.3:57, 8 pp. http://www.sigmaaldrich.com/technical-documents/articles/material-matters/layer-by-layer-lbl.html. Accessed 28 Sept 2015
8. Love JC, Estroff LA, Kriebel JK et al (2005) Self-assembled monolayers of thiolates on metals as a form of nanotechnology. Chem Rev 105:1103–1169
9. Boeckl M, Graham D (2006) Self-assembled monolayers: advantages of pure alkanethiols. Mater Matters 1.2:3. http://www.sigmaaldrich.com/technical-documents/articles/material-matters/self-assembled-monolayers.html. Accessed 28 Sept 2015
10. Jex C (2015) Meet the nano-sized rabbit made of DNA. http://sciencenordic.com/meet-nano-sized-rabbit-made-dna. Accessed 29 Sept 2015
11. Le JD, Pinto Y, Seeman NC et al (2004) DNA-templated self-assembly of metallic nanocomponent arrays on a surface. Nano Lett 4(12):2343–2347
12. Kershner RJ, Bozano LD, Micheel CM et al (2009) Placement and orientation of individual DNA shapes on lithographically patterned surfaces. Nat. Nanotechnol. Aug 1–5. doi:10.1038/NNANO.2009.220
13. Self-assembled DNA Scaffolding Used To Build Tiny Circuit Boards. Copyright 2015 Science Daily. http://www.sciencedaily.com/releases/2009/08/090818130626.htm. Accessed 29 Sept 2015

Chapter 25
Nanocharacterization Techniques

Abstract During or after fabrication of nanoelectronic devices, it is necessary to qualitatively and quantitatively assess nanomaterial properties. A stringent monitoring of the process is necessary to assure compliance with design layout to achieve desired electrical performance. Besides, in several nanotechnology experiments, particularly those related to biosensors, biomolecules are used. During these experiments as well as in device fabrication, the nanotechnologist is required to perform measurements on both soft and hard samples and those in solid or fluid state. For these measurements, a vast gamut of simple/complex instrumentation is available. These equipments are used for surface topographical studies, grain and particle size determination, defect and elemental composition analysis. Under the microscopy head fall scanning probe microscopes, and scanning and transmission microscopes. X-ray-based analysis tools include energy dispersive X-ray analysis, X-ray diffraction, and X-ray photoelectron spectroscopy. Important optical spectroscopic techniques are infrared spectroscopy, ultraviolet and visible spectroscopy, and Raman spectroscopy. Size distribution of dispersions of nanoparticles in liquid media is studied by photon correlation spectroscopy. Stability of dispersions is predicted by zeta potential analysis. Noncontact vibratory motion measurements of objects are performed by laser Doppler vibrometry. An overview of these nanocharacterization techniques is presented herein. Operation, relative merits and demerits, limitations and applications of these techniques are described.

25.1 Introduction

Nanocharacterization is a broad-scope subject encompassing the multiple fields of nanoscience and engineering. A plethora of expensive, high-end analytical instruments is necessary to study material and surface properties, including high-resolution microscopes, diffractometers, spectrometers, and others. They are used to measure particle size, agglomeration state, morphology, surface topography, and elemental composition at the nanoscale. Generally, these equipments are housed and maintained as a central facility for access to academia and industry.

V.K. Khanna, *Integrated Nanoelectronics*,
NanoScience and Technology, DOI 10.1007/978-81-322-3625-2_25

25.2 Scanning Probe Microscopy (SPM)

Scanning probe microscopes refer to a group of microscopes. These microscopes are used to image fine details of a surface at the scale of molecules and atoms. They work by scanning a very small light source/sharp tip over the surface [1].

25.2.1 Near-Field Scanning Optical Microscopy (NSOM)

Due to diffraction of light, the resolution of an optical microscope is restricted to half the wavelength of light $\sim$250 nm. Practically, this limit is around 500 nm. NSOM can provide resolution up to <30 nm. In NSOM, a fiber-optic probe is used for scanning. Its surface is completely covered except for a nanometer-sized aperture $\ll$ wavelength of light. Light from a laser source is emitted from this aperture. An image with subwavelength resolution (size of the aperture) is generated if the distance between the aperture and the sample is < aperture diameter. There are three ways in which image data are collected: transmission, reflection, and fluorescence. Transmission mode is suitable for examination of transparent samples with low-to-medium light absorption, such as biological species. Reflection mode is suited for opaque samples.

25.2.2 Scanning Tunneling Microscopy (STM)

The STM was invented by scientists Gerd Binnig and Heinrich Rohrer in 1981. These scientists worked at IBM Zurich Laboratory, Switzerland. This achievement made the inventors the 1986 Nobel laureates in physics [2]. STM can be operated in vacuum, as well as in air, water, insulating gases or liquids. It can also work in ionic solutions. Its operation is not hampered at temperatures as low as 4 K to as high as 973 K. But it requires clean surfaces. Also, isolation from vibration is essential to ensure stability of operation. Resolution is typically 0.01 nm (vertical) and 0.1 nm (lateral).

The operation of STM is based on the quantum mechanical phenomenon called tunneling. In this microscope (Fig. 25.1), the tunneling current in the range of nano- to pico-amperes is measured between a metallic stylus (a hard, pointed tip), e.g., a tungsten or Pt–Ir alloy needle and a conducting specimen, as the moving stylus scans the surface of the specimen from a small distance $\sim$0.5–1.0 nm. The tunneling current is induced by a small bias voltage applied between the stylus and the surface. While, the stylus moves in *X*- and *Y*-directions, it is automatically lifted or lowered in the *Z*-direction by a piezoelectric drive controlled by a feedback mechanism to keep the tunneling current constant. Thus the controlling voltage of feedback electronics acquires the information about the surface topological features of the specimen. This information is used to construct 3D images of surface

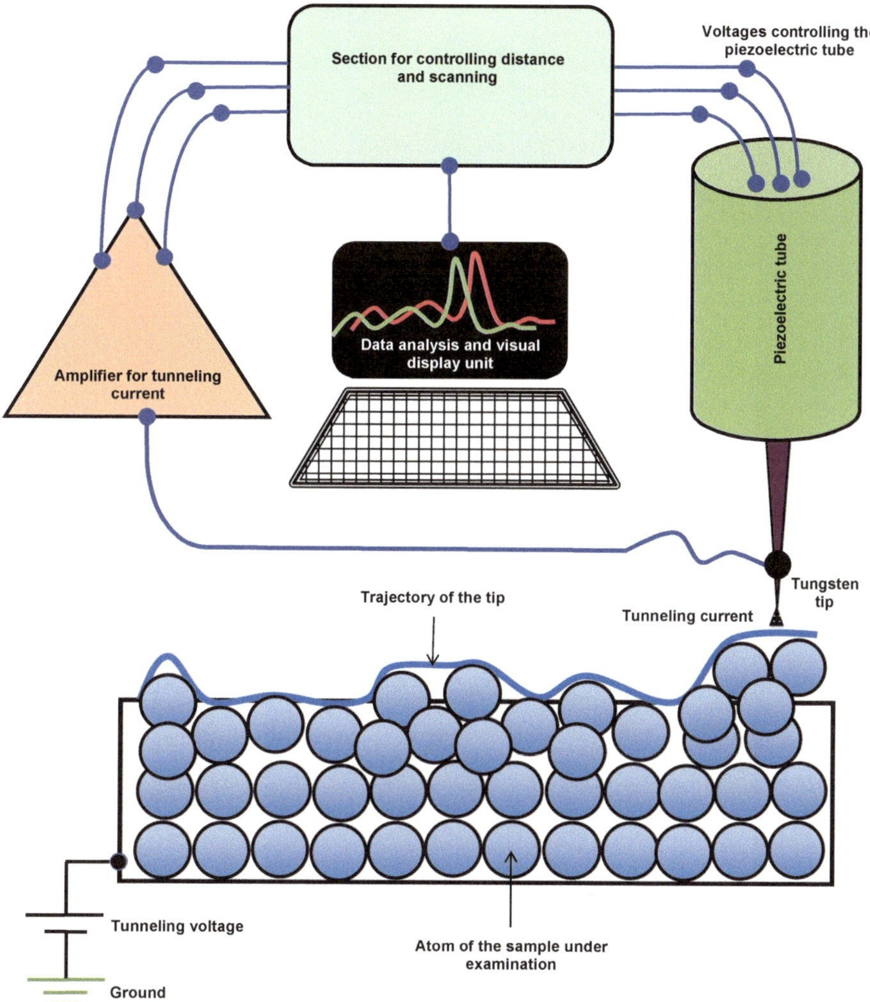

Fig. 25.1 Main parts of a scanning tunneling microscope

topography or contour maps, which are displayed on a computer screen as a grayscale image for visualization.

25.2.3 Atomic Force Microscopy (AFM)

The limitation of STM to image only metallic or semiconducting surfaces motivated the invention of AFM (Fig. 25.2). The AFM was invented in 1986 by Gerd Binnig

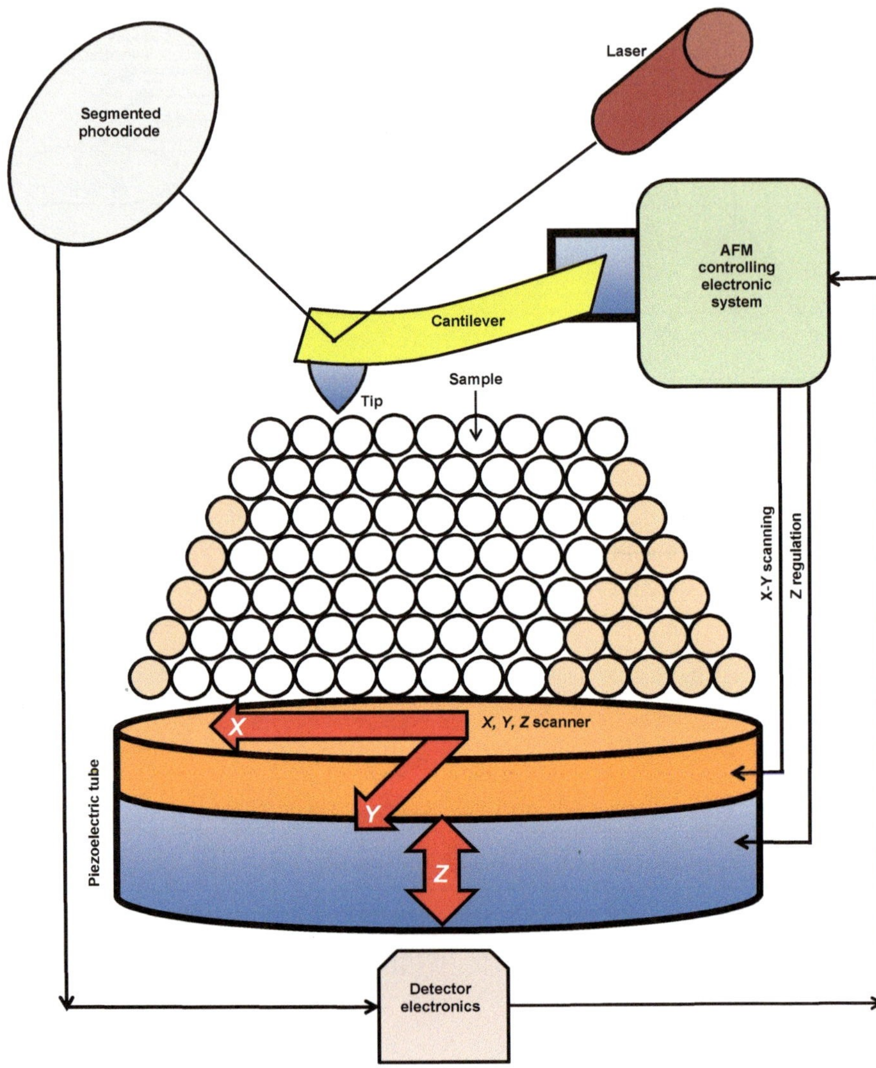

Fig. 25.2 Components of an atomic force microscope

and Calvin F. Quate of Stanford University, and Christoph Gerber of IBM San Jose Research Laboratory [3]. The AFM tip is a silicon or silicon nitride or CNT cantilever with 3-50 nm radius of curvature. Silicon tips are conical, less durable and have a hydrophilic surface to the sample. Silicon nitride tips are pyramidal in shape, comparatively durable and present a hydrophobic surface to the sample. CNT tips represent the ultimate in sharpness. Typical vertical resolution is 0.1 nm and lateral resolution is 3 nm. The AFM can operate in ultrahigh vacuum, in ambient atmosphere and in liquid media. High resonance frequency cantilevers are used for imaging in air. Low resonance frequency cantilevers are suitable for imaging in liquids. The AFM is called the "Eye of nanotechnology." But the image size obtained from single scan is very small ~150 μm × 150 μm. A scan generally covers 100 μm in *X*- and *Y*-directions and 4 μm in *Z*-direction.

There are three operational modes [4], Fig. 25.3: (i) DC mode (also termed Static mode or Contact mode or Constant force mode): The cantilever tip scans the surface of the sample in close contact with it. When the tip is in contact with the surface or at a very small distance from it, the force of interaction between them is strong and repulsive in nature. A flexible cantilever helps to reduce noise and drift. For avoiding collision between the tip and the sample and to prevent dragging of tip against the sample, the force is monitored and regulated by a piezoelectric positioning element controlled by a feedback amplifier. The piezoelectric element can move precisely along any of the three mutually orthogonal *X*, *Y*, and *Z* axes. The angular movement of the cantilever is measured by the deflection of a laser beam reflected off the cantilever surface into a position-sensitive photodetector consisting of an array of photodiodes. The measured deflection is compared with a preset value in a feedback amplifier. In case of a difference between the measured deflection and the preset value, this amplifier transmits the required voltage to the piezoelement. Accordingly, the piezoelement adjusts the position of the cantilever to maintain a constant deflection, either by moving upwards or downwards. The voltage values applied by the piezoelement provide indications of the heights of the different surface features of the sample. They are used to generate topographical maps of the scanned surface. Biological samples are very soft. They are easily destroyed if scanned in contact mode. (ii) AC mode or Noncontact mode or Attractive mode: The AFM tip flutters over the surface examined at a distance of 5–15 nm above it. At these distances, a feeble attractive force pulls the tip toward the surface. Surface topography maps are obtained from the Van der Waals forces acting between the hovering tip and the sample. As the interaction forces are very weak, the tip is made to undergo upward and downward vibrations at a frequency above its resonance frequency, with the signal provided by an oscillator circuit. So, AC methods can be applied to measure the amplitude, frequency or phase or the

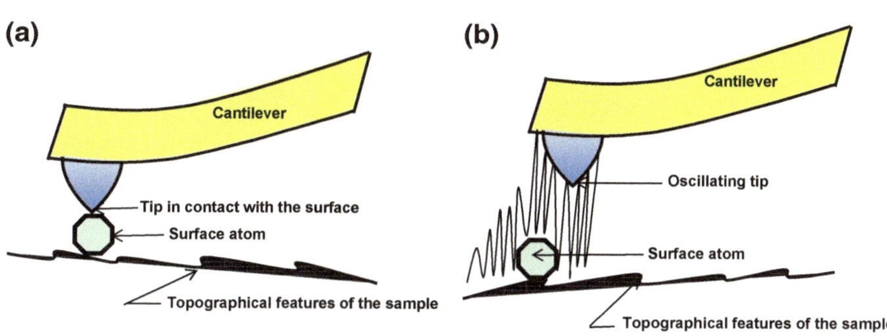

Fig. 25.3 Imaging modes of AFM: **a** Contact mode and **b** Noncontact or tapping mode

oscillating cantilever under the influence of force gradients. During scanning, the shift in the chosen parameter is set at a particular value and maintained at this level with the help of the feedback loop. (iii) Tapping or dynamic contact mode or intermittent contact mode: The vertically oscillating tip is alternately placed in contact with the surface to achieve high resolution and withdrawn from it to prevent any dragging effect. The force of interaction between the tip and the sample is strong and repulsive. By a feedback mechanism, the amplitude of oscillations is kept constant as the tip scans the surface. During scanning, when the tip encounters a cliff on the surface, the amplitude decreases. On coming across a valley, the amplitude increases. By recording the distance between the tip and the surface at every point, the image of the surface is constructed by the software.

Contact mode AFM can scratch the sample. It is prone to electrostatic and/or surface tension forces due to the gas layer adsorbed on the sample surface. Noncontact AFM gives less resolution. Any surface contamination can impact the oscillations. Tapping mode AFM provides high resolution. Chances for surface damage by friction are reduced.

25.3 Electron Microscopy

Electrical engineer Max Knoll and physicist Ernst Ruska, Berlin Technische Hochschule, Germany, invented the electron microscope in 1931. Instead of a beam of light used in an optical microscope, the electron microscope uses a beam of accelerated electrons to form an image of an object [5, 6]. In place of the customary glass lenses to guide the light beam, electrostatic and/or electromagnetic lenses are used to contrive the electron beam. The resolution of an electron microscope depends on the wavelength of electrons. Quicker the electrons propagate, the shorter the wavelength and therefore higher the resolution. The electron microscope provides several-fold larger magnification than an optical microscope. An optical microscope is restricted to magnifications $\sim$2000x. The resolution limit is $\sim$200 nm. In the transmission mode, an electron microscope can provide a resolution of 0.05 nm and magnification of 10^7x.

25.3.1 Transmission Electron Microscopy (TEM)

Suppose an electron beam accelerated at a voltage $\sim$40–400 keV irradiates a thin semitransparent specimen. After interaction with the specimen, the beam emerging out of it is attenuated according to the density and thickness of the specimen. Therefore, two-dimensional projection of the specimen can reveal its structure in the form of black-and-white images.

In a TEM system, the electron beam emitted from an electron gun fitted with a tungsten filament cathode is accelerated by the anode at typically +100 keV. It is

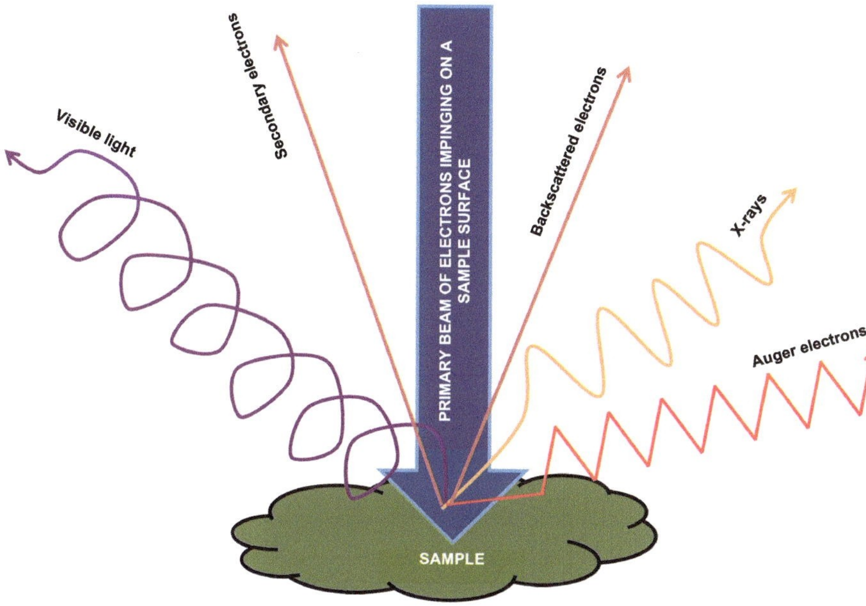

Fig. 25.4 Interaction of an electron beam with a specimen to produce different forms of particle/radiation beams

focused into a thin coherent beam by the lens assembly and made to strike the specimen. The transmitted beam is magnified and projected on a phosphor-coated fluorescent screen, a photographic film, or a CCD (charge-coupled device) camera.

25.3.2 Scanning Electron Microscopy (SEM)

Unlike the TEM which builds an image from transmitted electrons, SEM forms an image from high-energy backscattered and low-energy secondary electrons which are emitted from the surface of the specimen when a primary beam of electrons scans it in a raster pattern (Fig. 25.4). Unlike TEM, SEM uses a lower accelerating voltage ~1–5 kV. Resolution offered by SEM is an order of magnitude lower than that obtained from TEM. The main parts of a SEM are shown in Fig. 25.5.

25.3.3 Field Emission Scanning Electron Microscopy (FESEM)

An FESEM uses a more focused, narrower beam of electrons produced by a field emission cathode in the electron gun. This cathode consists of fine tungsten needles

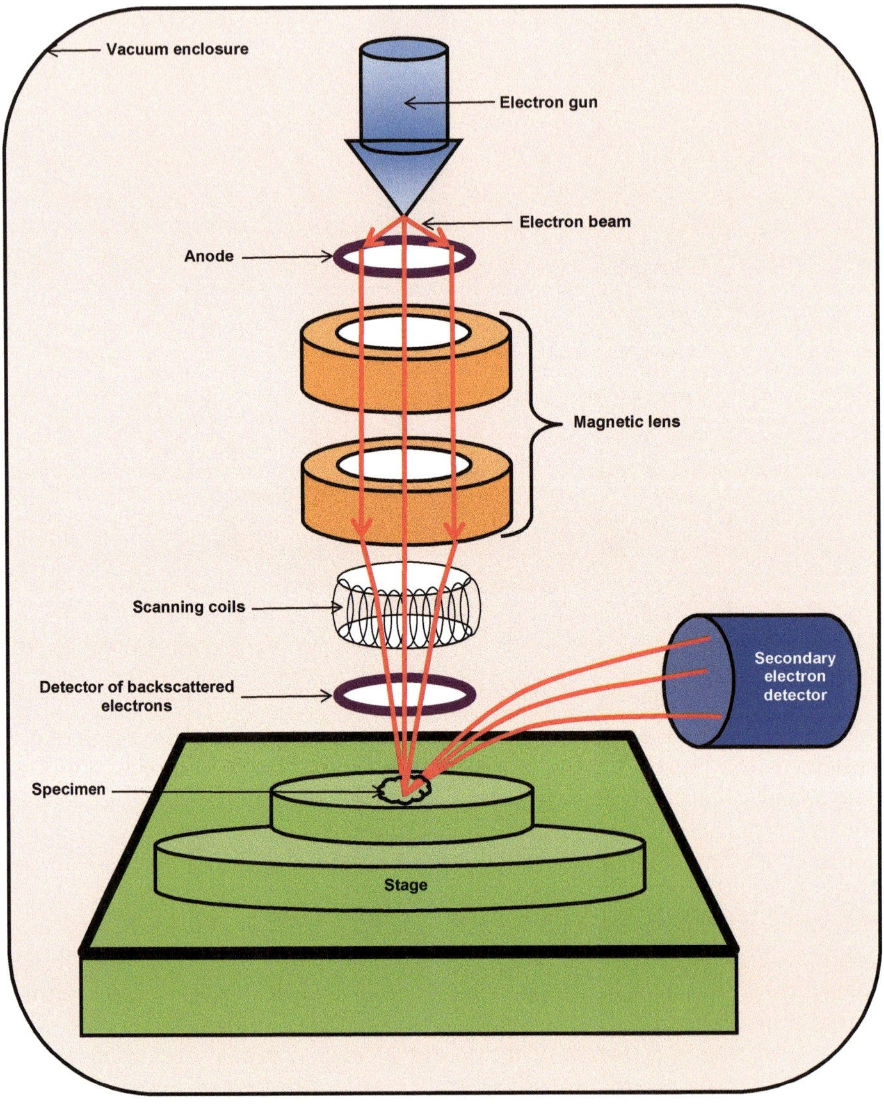

Fig. 25.5 Constructional features of a scanning electron microscope

coated with a low work function material such as zirconium oxide called a Schottky emitter. Low and high electron energies are generated at accelerating voltages from 0.5 to 30 kV. Magnifications up to 300,000x are achieved. Depth of field is virtually unlimited. Spatial resolution is improved up to 1.0 nm. Charging of the surface is reduced. Insulating materials need not be coated with conducting films. So, it is suitable for studying gate oxides of MOSFETs. Surface damage is minimal.

25.3.4 Focused Ion Beam Scanning Electron Microscopy (FIB-SEM)

Electrons are not the solitary charge carriers that can be manipulated by electrical and magnetic fields. Positively and negatively charged ions too can be controlled by these fields, and utilized advantageously to perform jobs demanding high accuracy. Constructionally, a focused ion beam system resembles an SEM. The only difference is that it uses a beam of ions in place of a beam of electrons. In an SEM, the electrons are used because they are low-mass ot light particles. Light particles are necessary because they should not inflict any damages on the sample examined. In an FIB system, the ion beam has a different objective. It is used to modify or mill the surface. This ion milling can be done with nanometer precision. An FIB-SEM is a dual-beam system combining an FIB system with a traditional SEM to simultaneously mill the surface and image the surface topology at nanoscopic scale. Complementary SEM imaging and FIB beam milling capabilities are thus combined in one machine.

25.3.5 Specimen Preparation for Electron Microscopy

As TEM/SEM operates in vacuum, one of the following methods is used to prepare the sample in a suitable form: (i) Fixation: Preservation of the sample to prevent degradation over time. (ii) Cryofixation: Preservation by freezing to liquid nitrogen temperature. (iii) Dehydration: Water removal. (iv) Embedding: Infiltration of tissue with a polymerizable resin, which is hardened for sectioning. (v) Sectioning: production of thin slices of the specimen. (vi) Staining: Contrast enhancement by using heavy metals such as lead and uranium to scatter electrons. (vii) Freeze fracturing and Freeze etching: Lipid membranes are cooled rapidly to liquid nitrogen temperature, fractured by breaking and etched by raising the temperature to −95 °C to show details. (viii) Sputtering of metal film: Deposition of thin Au film to prevent charging of an insulating surface when electron beam falls on it.

25.3.6 Electron Microscope Upkeep and Maintenance

Electron microscopes are expensive and sophisticated instruments which must be handled carefully by trained personnel. They require high-voltage supplies giving stable currents. Their vacuum systems require long continuous operation with cooling water circulating through the pumps. They must be shielded from magnetic fields. They must also be housed in vibration-free buildings. Sample preparations often involve lengthy, cumbersome processes.

25.4 X-Ray Techniques

25.4.1 Energy Dispersive X-Ray Analysis (EDX)

In an SEM/TEM, the scanning of the sample surface by an electron beam not only leads to the production of backscattered and secondary electrons. In addition, another type of radiation is produced. Consequent upon the secondary electrons leaving the sample surface, vacancies are created in the positions left by them in the atomic shells, which act as holes. These holes may sometimes be in the inner atomic shells. Consequently the sample atoms are in an unstable state. For stabilization, electrons from outer shells leap to the inner shells to fill the holes. During this transition, the difference of energy between the concerned energy levels is released in the form of X-rays. The X-rays emitted from a sample are characteristic in energy and frequency of the atomic number of the parent atom from which they are liberated. Therefore, they can disclose the identity of the parent atom [7].

Another possibility is that the superfluous energy is imparted to another electron, which is ejected from the surface. This electron is called an Auger electron and forms the basis of Auger electron spectroscopy (AES). AES is useful for recognition of low atomic number elements whereas X-ray emission helps for identification of high atomic number elements [8].

EDX systems are available in the form of attachments to SEM instruments to probe the elemental composition of samples.

25.4.2 X-Ray Powder Diffraction (XRD)

X-rays are electromagnetic waves of wavelength 0.1 nm. X-ray diffraction study is used for identification of crystalline form and its different polymorphs, distinguishing between crystalline and amorphous materials, and establishing the percentage of crystallinity of a material. It provides definitive structural information about the material including atomic arrangement, interatomic distances, bond angles, and imperfections. It does not give information about elemental analysis.

A beam of electrons emitted from a hot tungsten filament of a cathode ray tube bombards a Cu or Mo target. By this bombardment, electrons are ejected from 1 s orbitals of the target. Electrons fall from 2p or 3p levels to fill the vacancies created by the removal of 1 s electrons, accompanied by the emission of X-rays. The X-rays produced are filtered to obtain monochromatic radiation, collimated and directed toward the specimen. Interaction between the X-rays and the specimen leads to constructive interference for the values of scattering angle θ satisfying Bragg's law [9]:

$$n\lambda = 2d \sin \theta \tag{25.1}$$

where λ is the wavelength of X-rays, n is a positive integer, d is the distance between lattice planes and θ is the angle of scattering of X-rays. The diffracted rays from the specimen are detected for analysis.

By scanning the specimen through a range of angles, all the possible diffraction directions are obtained because the crystals in the powdered material are randomly oriented. The pattern recorded by measuring the intensity of scattered waves as a function of the scattering angle is referred to as the diffraction pattern. In the diffraction pattern, high intensity regions called diffraction peaks are clearly visible. These peaks represent the scattering angles which comply with Bragg's law. From the peaks of diffraction, the interplanar spacings d are derived. Knowledge of the d-spacing of the crystal gives a clue to its recognition because each crystal has a unique d-spacing. The diffraction peaks are related with d-spacings and Miller indices. The process of determining the dimensions of unit cell from the positions of diffraction peaks is called indexing. It can be done manually. As the manual process is time consuming, pattern-matching software is used for automation.

XRD is a powerful analytical tool for unambiguously identifying a mineral. Data interpretation is easy. But a reference file of inorganic compounds is necessary listing the d-spacings.

25.4.3 X-Ray Photoelectron Spectroscopy (XPS)

It was developed by Kai Siegbahn and team, University of Uppsala, Sweden in mid-1960s. Kai Siegbahn was awarded the Nobel prize for this work in 1981. In the beginning, it was called ESCA (Electron Spectroscopy for Chemical Analysis) emphasizing the chemical information provided by it. XPS spectra are recorded by irradiating the specimen with a focused beam of monoenergetic soft (1.5 kV) X-ray photons of energy $h\nu$ in ultrahigh vacuum $<10^{-9}$ mb, and measuring the kinetic energies and number of escaping photoelectrons. Let E_i be the initial energy of an atom and E_f its final energy after excitation by an X-ray photon. Then by energy conservation, the kinetic energy KE of the ejected photoelectron is given by

$$KE = h\nu - \left(E_f - E_i\right) \tag{25.2}$$

from which

$$h\nu - KE = E_f - E_i \tag{25.3}$$

i.e., the difference between energy of the incident X-ray photon and kinetic energy of the emitted electron = the binding energy of the orbital from which the electron is released = $E_f - E_i$. The energy of incident X-ray photon is known. The kinetic energy of the electron ejected is measured. From these data, the binding energy of the orbital is calculated. This binding energy has known values for various orbitals of particular elements. Therefore peaks of the XPS spectrum are correlated with specific atoms. The relative concentrations of the different elements are given by photoelectron intensities.

The information from XPS is related predominantly to the surface of the specimen up to a depth of 5 nm. Thus XPS is a surface-sensitive technique for determining the elemental composition of the top 0–10 nm of the surface. XPS does not detect hydrogen or lithium. It detects the elements from lithium to uranium. XPS is useful for both conducting and insulating materials. Detection limit is in parts per thousand and is extendable to parts per million range under special conditions.

25.5 Fourier Transform Infrared (FT-IR) Spectroscopy

Infrared radiation extends over the wavelength range 700 nm–1 mm (frequency 430 THz–300 GHz). IR spectroscopy is based on the change in dipole moment of the molecule by adsorption of IR radiation. IR spectrum of a specimen is a graph obtained by plotting absorbance of infrared radiation on the ordinate and its frequency or wavelength on the abscissa. Absorbance is the ratio of radiation entering the specimen to that leaving it. The absorption peaks represent the frequencies of resonance of IR with those of the vibrating bonds or groups. Each compound represents a distinguishing combination of atoms. Hence, each compound has a unique IR spectrum. The IR spectrum of a specimen represents its fingerprint. Older versions of IR spectrometer were of dispersive type. These spectrometers separated the individual frequencies of infrared energy by using a prism or grating. After passing through the specimen, the amounts of energy at different frequencies are measured to obtain the IR spectrum.

The main disadvantage of the dispersive IR spectrometer is the slow scanning because measurements for different frequencies are completed, one at a time. To alleviate this difficulty, FT-IR spectrometer was developed. This spectrometer uses an interferometer to collect the spectrum [10], Fig. 25.6. Its main parts are an infrared source, a beam splitter, and two mirrors (one stationary and one mobile). The incoming infra red radiation from a glowing black-body source passes through an aperture to strike a beam splitter, which subdivides it into two perpendicular beams. One beam is transmitted to the stationary mirror. The other beam is reflected to the mobile mirror, moving to and fro at a uniform speed. The two beams are reflected back from the respective mirrors. The returned beams are recombined at the beam splitter. The path length traversed by the beam returning from the stationary mirror is fixed. But the path length of the beam rebounding from the mobile mirror is variable. Therefore, recombination of the two beams creates an interference pattern or interferogram containing bright bands for constructive interference and dark bands for destructive interference. This interferogram has the property that various data points corresponding to dissimilar positions of the mobile mirror carry information regarding the different frequencies received from the source. From the beam splitter, the interferogram is conveyed to the specimen. On passage through the specimen, specific frequencies of energy, characteristic of the particular

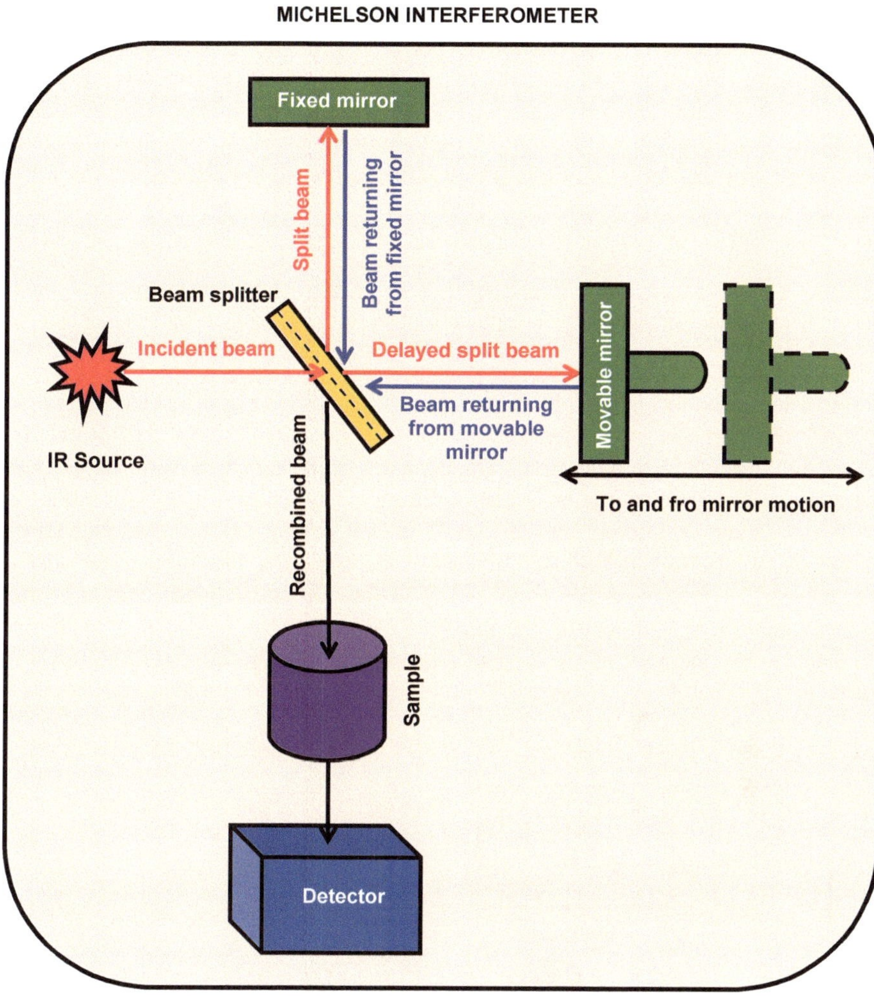

Fig. 25.6 Michelson interferometer configuration of an FTIR spectrometer

material, are absorbed and the remaining frequencies are transmitted to the detector. The transmitted energy received in the detector carries information about all the frequencies of the IR range. The detected signal is digitized and the individual frequencies in the interferogram are decoded by a mathematical technique called Fourier transformation. The Fourier transformation is implemented by a computer. To establish a comparative scale for absorption intensity, the above sequence of steps is also performed on a background or reference beam without loading the specimen. The ratio of the specimen beam to the background beam yields the % transmittance signal, which is converted into the absorbance signal by taking negative logarithm of the data points.

Besides the high throughput, the FT-IR spectrometer gives a higher signal-to-noise ratio than a dispersive spectrometer. Better signal quality is obtained because the FT-IR instrument uses less number of mirror surfaces than dispersive type. So, reflection losses are reduced. Consequently, the details of the spectrum are clearer and easily resolved. On the whole, the FT-IR instrument provides high spectral quality with good reproducibility at a fast speed. It is easy to use and maintain.

25.6 Ultraviolet and Visible (UV-Visible) Absorption Spectroscopy

Ultraviolet (200–400 nm) and visible (400–700 nm) wavelengths cause electronic transitions in organic molecules from lower to higher energy molecular orbital. This is dissimilar to the vibratory transitions caused by IR radiation. UV-visible spectroscopy is based on the principle that different molecules absorb ultraviolet and visible radiation at different frequencies [11]. Hence, absorption measurements of a beam of radiation at various frequencies will reveal the structural groups in the molecule. The absorption spectrum of UV-visible radiation for the carbonyl group in acetone will be identical to that in diethyl ketone.

The modus operandi of a UV-visible spectrometer (Fig. 25.7) is as follows: An ultraviolet or visible light beam is decomposed into its constituent frequencies by

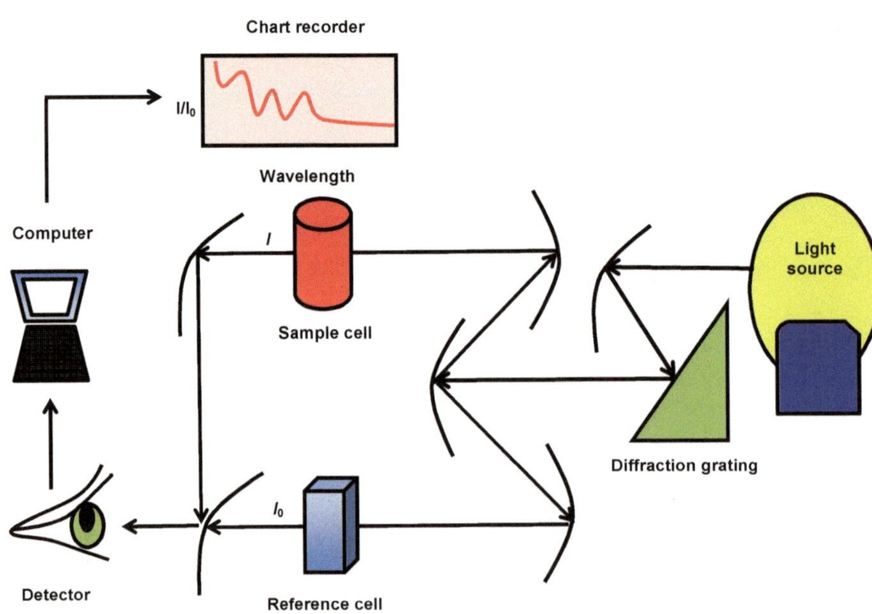

Fig. 25.7 UV-visible spectrometer

passing through a prism or grating. Beam of one single frequency is shone on a half-mirrored device, which splits it into two beams of equal intensity. One of these beams strikes a transparent solution of the unknown compound contained in a see-through vessel. This is the interrogating beam. The other beam is passed through the same solution in an identical container. This solution does not contain the unknown compound. The beam is called the reference beam. The outgoing beams from both containers are received by a detector which records their intensities. Let the intensity of the interrogating beam after emerging from the solution be I and that of the reference beam coming out from the solution be I_0. The ratio I/I_0 is calculated. Similar procedure is repeated for all the frequencies of the incident beam, and relative intensities are plotted against frequencies to sketch a spectrum for the sample.

25.7 Raman Spectroscopy

Raman effect is the change in frequency or wavelength of light when it is deflected by molecules [12–14]. Suppose light from a laser source with frequency ν_0 strikes the molecules of a substance. Since the laser beam is an oscillating electromagnetic wave, its interaction with the substance, deforms the molecules. As this deformation is periodic, the molecules undergo vibrations. Following three outcomes are likely (Fig. 25.8): (i) The scattered light has the same frequency ν_0 as the incident beam. This phenomenon is termed elastic Rayleigh scattering. (ii) The scattered light has a lower frequency ($\nu_0 - \nu_m$) than that of the incident light, where ν_m is the change in frequency. This phenomenon is known as Stokes Raman scattering. It is an inelastic scattering. (iii) The scattered light has a higher frequency ($\nu_0 + \nu_m$) than that of the incident light. This phenomenon is termed anti-Stokes scattering. It is also an inelastic scattering.

Only 10^{-5} fraction of the incident light undergoes inelastic Raman scattering with frequencies ($\nu_0 \pm \nu_m$). The remaining light produces elastic Rayleigh signal. The intensity of useful Raman signal is often dominated by the stray light from Rayleigh signal.

The Raman system (Fig. 25.9) consists of a laser excitation source in the visible, ultraviolet (UV) or near infrared (NIR) range, sample illumination and light collection optics, a filter or spectrometer for wavelength separation, light detecting device such as a photomultiplier tube, photodiode array or charge-coupled device, and associated electronics. Light beam falling on the sample undergoes scattering. The scattered light is collected by the lens and passed through the filter to obtain the Raman spectrum. The Raman spectrum is a graph of intensity of scattered light with respect to the frequency difference from incident light in units of wave numbers. The Raman spectrum is a fingerprint of the molecule. Raman spectroscopy is a simple analytical tool for molecular identification. It works on the change in polarizability of the molecule by absorption of radiation in contrast to IR spectroscopy where dipole moment is the changing parameter.

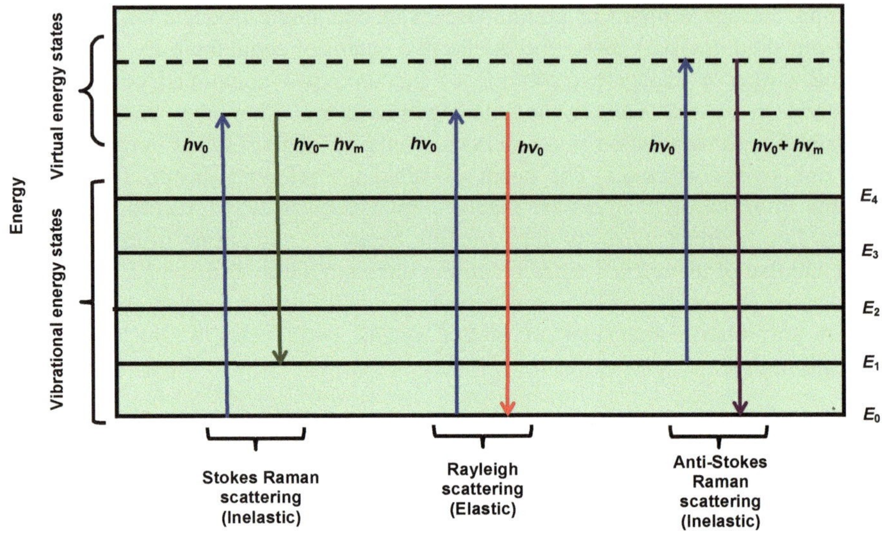

Fig. 25.8 Inelastic and elastic scattering of radiation by a substance

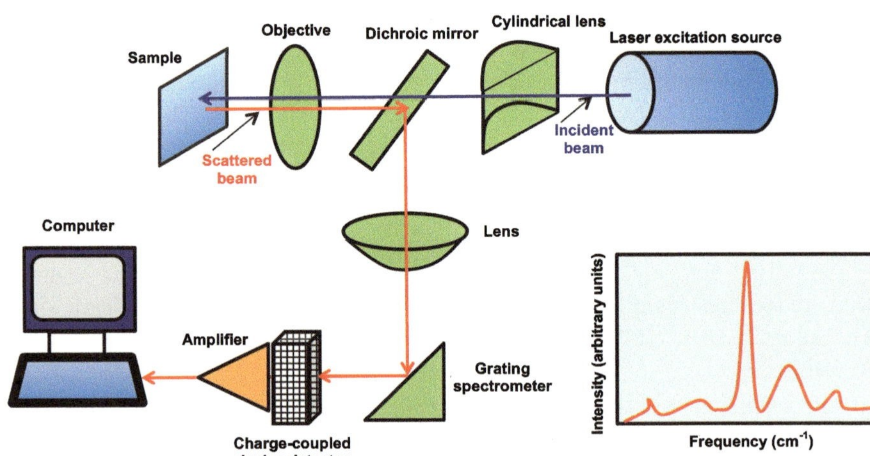

Fig. 25.9 Raman spectrometer instrumentation; inset shows a typical Raman spectrum

Raman spectroscopy offers several advantages over other techniques: (i) It is a noninvasive and nondestructive method. (ii) Samples can be solids, liquids, or gases. (iii) They can be in the form of particles, pellets, pastes, powders, gels, etc. (iv) Little or no sample preparation is necessary. (v) Insensitivity to aqueous absorption bands makes it possible to obtain spectra of species dissolved in water. (vi) Concentration of a species is obtainable from a fraction of 1 to 100% without dilution. (vii) Samples contained in glass, sapphire, and polymer containers can be analyzed in situ. (viii) Since spectra are recorded in time durations of seconds to minutes, Raman spectroscopy can provide real-time analysis. (ix) Raman spectroscopy is highly selective. Very similar molecules can be differentiated using Raman spectral libraries.

25.7.1 Resonance-Enhanced Raman Scattering Spectroscopy

The frequency or wavelength of the excitation source is chosen to be close to the frequency of an electronic absorption of the molecule. For such coincidence of frequencies, the scattering intensity is increased by a factor of 10^2–10^6. Much lower analyte concentrations can be detected than achieved with nonresonant Raman scattering. Also, the measurement time is reduced. Resonant Raman analysis is limited to samples that absorb light close to the frequencies of common excitation sources.

25.7.2 Surface-Enhanced Raman Scattering (SERS) Spectroscopy

SERS is based on the fact the Raman signal obtained from molecules adsorbed on certain metallic surfaces is 5–6 and even up to 14 orders of magnitude stronger than that from the same molecules in bulk form. Two mechanisms are responsible for SERS [15]: (i) Electromagnetic enhancement (EME) mechanism: This kind of enhancement takes place by interaction of the laser beam with the surface irregularities such as any nanostructures present or roughness profile of the surface. When the frequency of incident light equals that of the plasma frequency of the metal, the free electrons in the metal are excited into surface plasmon resonance. As a result, the adsorbed molecules experience an intense electromagnetic field, whereby their vibrations in the direction perpendicular to the surface are increased. (ii) Chemical enhancement (CE) mechanism: It takes place when a charge-transfer complex is formed between the surface and the analyte molecule. Enhancement occurs through resonance because the frequencies of electronic transitions in many charge-transfer complexes are in the visible range.

The major contribution to the enhancement of Raman signal in SERS comes from EME with CE increasing the signal by one or two orders of magnitude only. The substrate selection is crucial to the degree of enhancement. A popular SERS substrate consists of a silicon wafer coated with silver or gold nanopillars. SERS allows detection of trace molecules due to its high sensitivity.

A difficulty with SERS is spectrum interpretation because a SERS spectrum differs from normal Raman spectrum. Due to the dramatic signal enhancement, very frail and inconspicuous Raman bands in spontaneous spectrum can become prominent in SERS, and additional peaks can originate from contaminant molecules.

25.7.3 Confocal/Micro Raman Spectroscopy

It is Raman spectroscopy in which the probing laser beam is passed through a confocal microscope objective. The scattered light travels through a limiting aperture. This aperture allows a small specific region of the sample of microscopic dimensions to be viewed with high spatial resolution consistently up to 200 nm. This helps to produce a Raman spectrum characteristic of that region. Micron size samples too can be analyzed with good contrast. The higher resolution is possible due to blockage of out-of-focus light improving the visibility of very small regions. In a normal microscope, these regions are blurred by the out-of-focus glare. Thus confocal Raman spectroscopy extends the benefits of a confocal microscope to Raman microscopy, producing both optical and spectral images simultaneously. This enables a far better performance than when a conventional optical microscope is used. Micro-Raman spectroscopy is a powerful analytical technique.

25.8 Photon Correlation Spectroscopy

Also called dynamic light scattering (DLS) or quasi-elastic light scattering, it is a light scattering technique used to determine particle size down to 1 nm for particles suspended in the liquid [16]. These particles are in a state of incessant, chaotic motion called Brownian motion due to collisions with surrounding liquid molecules. This random motion at temperature T in Kelvin scale is described by the celebrated Stokes–Einstein equation

$$r_h = k_B T/(6\pi\eta D_t), \tag{25.4}$$

where r_h denotes the hydrodynamic radius of the particle, and D_t its translation diffusion coefficient, η is the dynamic viscosity of the liquid medium, and k_B stands for the Boltzmann constant.

To carry out the measurement, the dispersion of the particles in the liquid is illuminated with a monochromatic beam of light from a laser source. When light strikes the particles executing Brownian motion, the wavelength of light changes by Doppler effect. The change in wavelength depends on the size of the particle. A fast photon detector records the fluctuations of the intensity of scattered light at a known scattering angle. The intensity fluctuations are described by the intensity autocorrelation function. From the analysis of autocorrelation function, the diffusion coefficient of the moving particles is determined. Then the mean particle size within a given range is estimated. Elaborate instruments provide particle size distribution also. The experiment is almost fully automated and is completed in a short duration.

25.9 Zeta Potential Analysis by Laser Doppler Electrophoresis

Zeta potential is also called electrokinetic potential. Zeta potential of a solid particle immersed in a liquid is defined as the electrostatic potential on the particle at the shear or slipping plane (which is close to the surface of the particle) with respect to the bulk of the liquid. To elaborate, existence of a net charge on a particle is associated with attraction of ions of opposite charges from surrounding liquid [17]. Region over which the net charge exerts its influence is called the electrical double layer. It consists of an inner layer of tightly bound ions called the Stern layer, and an outer layer of loosely bound ions known as the diffuse layer. The particles move in the solution under gravity or an applied electric field. The ions too move along with the particle. At a given distance from the particle, there is a borderline. Beyond the borderline, the ions no longer move with the particle. This borderline is called the shear or slipping plane. It marks the surface of hydrodynamic shear or slipping. Its location is within the diffuse layer. Zeta potential is the potential at this shear or slipping plane.

Zeta potential analysis is an analytical tool to determine the surface charge of colloidal particles in a solution [18]. The magnitude of zeta potential, typically ranging between −100 and +100 mV, is an indicator of stability of the colloidal solution [13]. The higher is the Zeta potential, the greater is the repulsive force between charged particles. Therefore, greater is the opposition to van der Waals attractive forces. Consequently, less likely are the chances of the particles to aggregate. Hence, the solution is stable. Zeta potential values hold the key to understanding the dispersion and aggregation processes in colloidal solutions. Zeta potential depends on pH of the solution. The pH of the solution at which the zeta potential is zero is called the isoelectric point.

The zeta potential of a particle is measured by a technique known as laser Doppler electrophoresis. In this technique, a potential difference is applied across the opposite sides of a cell containing the suspension of particles. Under the influence of the applied voltage, the particles of specific charges start moving toward respective opposite polarity electrodes. This movement of charged particles in the applied electric field is called electrophoresis. The drift velocity of a charge carrier per unit field is called its mobility. For measurement of mobility, the cell containing the dispersion of particles is illuminated with a laser beam under the condition of applied field. If the particles were stationary, the scattered light will have the same frequency as incident light. But since the particles are moving, the scattered light is shifted in frequency with respect to the incident light due to Doppler effect. Since the frequency of light is very high ~1014 Hz, the frequency shift is determined by an interferometric technique. In this technique, a pair of coherent beams derived from a single source is used. One of these beams passes through the dispersion of particles. This beam undergoes scattering and is called the scattered beam. The other beam is circumnavigated around the cell through equal path length. This is the reference beam. Only during passageway through the particles, any path difference between the two beams is introduced. So, on combining the two beams together, a modulated beam is generated. This beam is produced by constructive and destructive interference. It has a much smaller frequency. So, it is easy to measure this frequency. The frequency equals the difference between the frequencies of the scattered and reference beams. The beat frequency gives the velocity of particles. Electrophoretic mobility μ_E of the particles is obtained from the known value of applied electric field. Then the zeta potential ζ is calculated using Henry's equation [17]:

$$\zeta = 3\mu_E\eta/\{2\varepsilon f(ka)\}, \tag{25.5}$$

where μ_E is the electrophoretic mobility of particles, η is the viscosity of the liquid medium, and ε is its relative permittivity. The function *f*(*ka*) is Henry's function. Either the value 1.0 or 1.5 is used. When zeta potential is measured in a nonpolar solvent, *f*(*ka*) is taken as 1.0. This is called Huckel approximation. For zeta potential measurement in aqueous electrolytic solution of moderate concentration, *f*(*ka*) is taken to be 1.5. This is known as Smoluchowski approximation.

Another electrophoretic technique uses laser Doppler velocimetry together with phase analysis light scattering (PALS). This technique provides accurate measurement of zeta potential for particles in the size range 6–6000 nm and suspensions of 200–2000 parts per million.

25.10 Laser Doppler Vibrometry (LDV)

Laser Doppler vibrometer (Fig. 25.10) is a noncontact optical instrument for measurement of parameters such as displacement and velocity of a vibrating object for NEMS applications. It works by measuring the Doppler frequency shift of a laser beam reflected from the vibrating object [19]. A helium–neon laser source having wavelength of 632.8 nm is generally used. For a wave of wavelength λ, the Doppler frequency shift Δf_D caused by an object moving with velocity v is given by

$$\Delta f_D = 2v\cos\theta/\lambda \tag{25.6}$$

where θ is the angle between the laser beam and direction of vibration of the object, i.e., the velocity vector. Applying this equation, the velocity is determined by

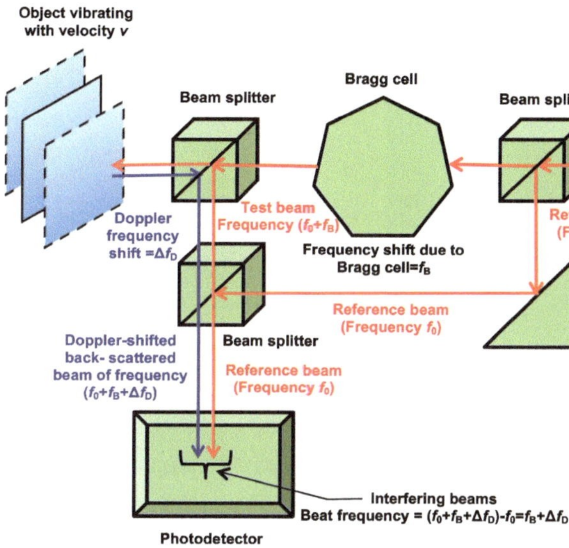

Fig. 25.10 Arrangement of components in an LDV

measuring Δf_D using interferometry. For interferometery, the laser light of frequency f_0 is divided by a beam splitter into two beams: the test beam and the reference beam. The test beam falls on the vibrating object after passing through an acousto-optic modulator (Bragg cell) which introduces a frequency shift f_B. The vibrating object causes a further frequency shift Δf_D by Doppler effect. Light is scattered by the moving object in all directions. A part of the scattered test beam is collected. It is reflected by the beamsplitter into the photodetector. In the photodetector, the test beam of frequency $f_0 + f_B + \Delta f_D$ is combined with the reference beam of frequency f_0, producing a beat frequency $(f_0 + f_B + \Delta f_D) - f_0 = (f_B + \Delta f_D)$. The frequency = $(f_B + \Delta f_D)$ is $\sim$40 MHz, which is much lower than the frequency of laser beam $\sim 10^{14}$ Hz. At the output of the photodetector, a frequency-modulated signal is received. The carrier frequency of this modulated signal is the frequency of the Bragg cell (f_B). The modulating frequency is the Doppler frequency shift (Δf_D). By demodulation of the signal, the velocity of the moving object is found as a function of time. Resolution in m/s (nanometer/second) in velocity and in pm (picometer) for displacement is achievable.

In the above measurement, the role of the Bragg cell (f_B) needs elaboration. When the vibrating object approaches the laser source, the frequency f_0 increases. In this situation, Δf_D is positive. When the vibrating object recedes away from the laser source, Δf_D is negative. Because the output signal of the detector is sinusoidal, it is not possible for the detector to discriminate whether the vibrating object is moving toward or away from the source. The Bragg cell is meant to provide directional sensitivity. It produces a carrier signal with frequency f_B. The frequency f_B of this carrier signal is modified by the vibrations of the object. The velocity of the object decides the magnitude of frequency deviation with respect to the frequency f_B as center frequency. This kind of interferometer is called a heterodyne optical interferometer [20].

25.11 Discussion and Conclusions

The familiar optical microscope available in most laboratories is hardly of any use to nanotechnologists. No work in nanoelectronics can progress without the help of precision characterization tools. Yet these tools are sometimes the costliest and most complicated of analytical instrumentation ever manufactured. Indeed, these tools are the eyes and ears of nanoelectronics engineer. No single tool can cater to the multifarious and multifaceted information required about the physical and chemical properties of nanostructures. So, the engineer must be familiar with a wide range of instruments. Before using an analytical instrument, it is essential to know its working, capabilities and the required sample preparation procedure. Not only are the operational precautions of the equipment important, it is equally imperative to be knowledgeable and trained about data acquisition, reading and interpretation.

Review Exercises

25.1 What is a scanning probe microscope? Name three types of SPM.

25.2 Resolution of an optical microscope is diffraction limited. How does a near-field scanning optical microscope overcome the diffraction limit to provide higher resolution? Describe its operation.

25.3 Can an STM be operated in air and water? Can it be operated at subzero °C temperatures? Can it work at 500 °C? Can it be used for imaging an insulating specimen?

25.4 How is the magnitude of tunneling current used to construct an image of surface topography of a given specimen by an STM?

25.5 Name three commonly used materials for fabricating an AFM cantilever. What are the typical lateral and vertical resolutions obtained from: (a) AFM and (b) STM? Which is superior resolution wise?

25.6 Does an electron microscope use glass lenses for beam focusing? How is the resolution of an electron microscope related to the electron velocity?

25.7 Compare a TEM with an SEM with regard to the following aspects: (i) method of image construction, (ii) acceleration voltages, and (iii) resolution achievable.

25.8 In what ways is an FESEM better than an ordinary SEM? Why?

25.9 What does an FIB system do? What is the advantage of building a combined FIB-SEM machine?

25.10 Can a TEM or SEM machine operate in air? Explain your answer with reason.
25.11 Mention 5 ways of sample preparation for electron microscopic examination.
25.12 Name any particle or radiation emitted inside a SEM machine due to sample irradiation with electron beam, apart from backscattered and secondary electrons. What is this emission used for? Explain the operation of the analytical instrument utilizing this phenomenon.
25.13 How does Auger electron spectroscopy differ from EDX analysis?
25.14 What properties of materials are obtained from XRD equipment? What type of information does this equipment fail to provide?
25.15 State Bragg's law. How are interplanar or *d*-spacings of a crystal determined by XRD analysis?
25.16 How is elemental composition of a material determined by an XPS machine? Can lithium be detected by XPS? Is XPS limited to conducting materials only?
25.17 How does a dispersive type IR spectrometer work? What is the main difficulty experienced with this spectrometer?
25.18 Describe the operation of an FT-IR spectrometer and explain how does it overcome the limitation of the dispersive type?
25.19 How does UV-visible spectroscopy differ from IR spectroscopy? How is the UV-visible spectrum of a sample recorded?
25.20 Distinguish between Rayleigh scattering, Stoke's Raman scattering and anti-Stoke's Raman scattering. What percentage of the incident light suffers Raman scattering?
25.21 Describe the setup used for recording a Rama spectrum of sample. List five advantages of Raman spectroscopy which have made it a popular analytical tool.
25.22 What is resonance-enhanced Raman spectroscopy? What is its advantage over ordinary Raman spectroscopy?
25.23 What is the typical order of magnitude improvement factor provided by SERS over ordinary Raman spectroscopy? What are the two mechanisms by which this improvement takes place? Which of the two is the dominant one?
25.24 What is a popular SERS substrate made of? What is the difficulty faced in interpreting a SERS spectrum?
25.25 How does confocal Raman spectroscopy differ from ordinary Raman spectroscopy? What makes it a powerful analytical instrument?
25.26 Write Stokes–Einstein equation and explain the symbols used. How is the size of particles in a colloidal dispersion of particles in a liquid determined? Describe the setup used for the experiment.
25.27 What is the zeta potential of a particle? What is its physical significance?
25.28 Write Henry's equation for zeta potential and explain the meaning of the symbols. Explain the laser Doppler electrophoresis for measuring zeta potential of a particle.

25.29 Write the equation for the Doppler frequency shift Δf_D in a wave of wavelength λ caused by an object moving with velocity v. Draw and explain the setup of laser Doppler vibrometry. What is the function of the Bragg cell?

References

1. Cross JW (2003) Scanning Probe Microscopy Copyright © John W. Cross, Revised: 13 June 2003. http://www.mobot.org/jwcross/spm/Accessed. Accessed 11 Jan 2016
2. Chen CJ (2008) Introduction to scanning tunneling microscopy. Oxford University Press, Copyright 2008 Oxford University Press, p 23
3. Binnig G, Quate CF, Gerber C (1986) Atomic-force microscope. Phys Rev Lett 56(9): 930–933
4. The Common AFM Modes. http://www.chemistry.uoguelph.ca/educmat/chm729/afm/details.htm#summary. Accessed on 11 Jan 2016
5. What is Electron Microscopy? https://www.jic.ac.uk/microscopy/intro_EM.html. Accessed 11 Jan 2016
6. Chaturvedi S, Dave PN (2012) Microscopy in nanotechnology, current microscopy contributions to advances in science and technology. In: Méndez-Vilas A (ed) © 2012 FORMATEX, pp 946–952
7. Hafner B, Energy dispersive spectroscopy on the SEM: a primer, characterization facility. University of Minnesota—Twin Cities. http://www.charfac.umn.edu/instruments/eds_on_sem_primer.pdf. Accessed 11 Jan 2016
8. Briggs D, Grant JT (eds) (2003) Surface analysis by Auger and X-ray photoelectron spectroscopy. IM Publications, Chichester, UK and Surface Spectra, Manchester, UK 900 pp
9. Dutrow BL, Clark CM, X-ray powder diffraction (XRD). http://serc.carleton.edu/research_education/geochemsheets/techniques/XRD.html. Accessed 12 Jan 2016
10. Introduction to fourier transform infrared spectrometry, Thermo Nicolet. http://mmrc.caltech.edu/FTIR/FTIRintro.pdf. Accessed 13 Jan 2016
11. Britton G (1995) UV/visible spectroscopy. Chem Inf 26(32):13–63
12. Raman CV (1928) A new radiation. Indian J Phys 2:387–398
13. Raman CV, Krishnan KS (1928) A new type of secondary radiation. Nature 121:501–502
14. Raman CV, Krishnan KS (1928) A new class of spectra due to secondary radiation, Part I. Indian J Phys 2:399–419
15. Campion A, Kambhampati P (1998) Surface-enhanced Raman scattering. Chem Soc Rev 27:241–250
16. Sartor M, Dynamic light scattering. University of California, San Diego. https://physics.ucsd.edu/neurophysics/courses/physics_173_273/dynamic_light_scattering_03.pdf. Accessed 14 Jan 2016
17. Zeta Potential Analysis of Nanoparticles, September 2012, V 1.1. https://cdn.shopify.com/s/files/1/0257/8237/files/nanoComposix_Guidelines_for_Zeta_Potential_Analysis_of_Nanoparticles.pdf. Accessed 14 Jan 2016
18. Zeta Potential-Electrophoresis. Escubed Ltd. http://www.escubed.co.uk/sites/default/files/zeta_potential_(an011)_elecrophoresis.pdf. Accessed 14 Jan 2016
19. Castellini P, Martarelli M, Tomasini EP (2006) Laser doppler vibrometry: development of advanced solutions answering to technology's needs. Mech Syst Signal Process 20:1265–1285
20. Topcu S, Chassagne L, Haddad D et al (2003) Heterodyne interferometric technique for displacement control at the nanometric scale. Rev Sci Instrum 74(11):4876–4880

Index

V.K. Khanna, *Integrated Nanoelectronics*,
NanoScience and Technology, DOI 10.1007/978-81-322-3625-2